Ultimate Limit State Analysis and Design of Plated Structures

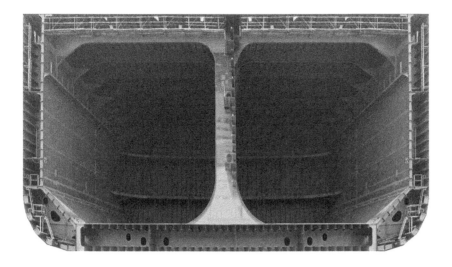

Ultimate Limit State Analysis and Design of Plated Structures

Second Edition

Jeom Kee Paik

Department of Naval Architecture and Ocean Engineering at Pusan National University, Korea
Department of Mechanical Engineering at University College London, UK

Edition History
Ultimate Limit State Design of Steel-Plated Structures, Edition One. January 2003. ISBN: 978-0-471-48632-9.
Jeom Kee Paik, Anil Kumar Thayamballi.

Registered Offices
John Wiley & Sons, Inc., 111 River Street, Hoboken, NJ 07030, USA
John Wiley & Sons Ltd, The Atrium, Southern Gate, Chichester, West Sussex, PO19 8SQ, UK

Editorial Office
The Atrium, Southern Gate, Chichester, West Sussex, PO19 8SQ, UK

For details of our global editorial offices, customer services, and more information about Wiley products visit us at www.wiley.com.

Wiley also publishes its books in a variety of electronic formats and by print-on-demand. Some content that appears in standard print versions of this book may not be available in other formats.

Library of Congress Cataloging-in-Publication Data

Names: Paik, Jeom Kee, author.
Title: Ultimate limit state analysis and design of plated structures / Jeom Kee Paik.
Description: 2nd edition. | Hoboken, NJ : John Wiley & Sons, 2018. | Includes index. |
Identifiers: LCCN 2017046042 (print) | LCCN 2017056354 (ebook) | ISBN
    9781119367765 (pdf) | ISBN 9781119367789 (epub) | ISBN 9781119367796 (cloth)
Subjects: LCSH: Building, Iron and steel. | Plates, Iron and steel.
Classification: LCC TA684 (ebook) | LCC TA684 .P24 2018 (print) | DDC
    624.1/821–dc23
LC record available at https://lccn.loc.gov/2017046042

Cover design: Wiley
Cover image: (Background) © fredmantel/Gettyimages; (Inset image) Courtesy of Jeom Kee Paik

Set in 10/12pt Warnock by SPi Global, Pondicherry, India

10 9 8 7 6 5 4 3 2 1

Contents

Preface

Plated structures are important in a variety of marine- and land-based applications, including ships, offshore platforms, box girder bridges, power/chemical plants, and box girder cranes. The basic strength members in plated structures include support members (such as stiffeners and plate girders), plates, stiffened panels, grillages, box columns, and box girders. During their lifetimes, the structures constructed with these members are subjected to various types of actions and action effects that are usually normal but sometimes extreme or even accidental.

In the past, criteria and procedures for designing plated structures were primarily based on allowable working stresses and simplified buckling checks for structural components. However, it is now well recognized that the limit state approach is a better basis for design because it is difficult to determine the real safety margin of any structure using linear elastic methods alone. It also readily follows that it is of crucial importance to determine the true limit state if one is to obtain consistent measures of safety that can then form a fairer basis for comparison of structures of different sizes, types, and characteristics. An ability to better assess the true margin of safety would also inevitably lead to improvements in related regulations and design requirements.

Today, the preliminary design of ships including naval and merchant vessels, offshore structures such as ship-shaped offshore installations, mobile offshore drilling units, fixed-type offshore platforms and tension leg platforms, and land-based structures such as bridges and box girder cranes tends to be based on limit state considerations, including the ultimate limit state.

To obtain a safe and economic structure, the limit state-based capacity and structural behavior under known loads must be assessed accurately. The structural designer can perform such a relatively refined structural safety assessment even at the preliminary design stage if simple expressions are available for accurate prediction of the limit state behavior. A designer may even desire to do this for not only the intact structure but also structures with premised damage to assess their damage tolerance and survivability.

Although most structural engineers in the industry are very skilled and well experienced in the practical structural design aspects based on the traditional criteria, they may need a better background in the concept of limit state design and related engineering tools and data. Hence, there is a need for a relevant engineering book on the subject that provides an exposition of basic knowledge and concepts. Many structural specialists in research institutes continue to develop more advanced methods for the limit state design of plated structures, but they sometimes lack the useful engineering data to validate them. Students in universities want to learn more about the fundamentals and practical procedures regarding the limit state analysis and design and thus need a book that provides useful insights into the related disciplines.

This book reviews and describes both the fundamentals and practical procedures for the ultimate limit state analysis and design of ductile steel-plated and aluminum-plated structures. Structural fracture mechanics and structural impact mechanics are also described. This book is an extensive update of my previous book *Ultimate Limit State Design of Steel-Plated Structures* (with Dr. A.K. Thayamballi), published in 2003. In contrast to the previous book, this update covers both steel- and aluminum-plated structures together with the latest advances and many newly added materials not included in the 2003 version. The book is basically designed as a textbook. The derivation of the basic mathematical expressions is presented together with a thorough discussion of the assumptions and the validity of the underlying expressions and solution methods.

I believe that the reader should be able to obtain insight into a wider spectrum of ultimate limit state analysis and design considerations in both an academic and a practical sense. In part, this book is an easily accessible analysis and design toolbox that facilitates learning by applying the concepts of the ultimate limit state for practice.

This book is primarily based on my own insights and developments obtained from more than 35 years of professional experience, as well as information and findings provided by numerous other researchers and limit state design practitioners. Wherever possible, I have tried my best to acknowledge the invaluable efforts of other investigators and practitioners, and, if I have failed anywhere in this regard, I did so inadvertently.

I gratefully acknowledge all those individuals who helped make this book possible. Most of all, Dr. A.K. Thayamballi, who was the coauthor of the previous book, provided valuable and comprehensive comments to improve this book. Finally, I take this opportunity to thank my wife Yun Hee Kim, my son Myung Hook Paik and my daughter Yun Jung Paik for their unfailing patience and support while this book was being written.

October 2017

Prof. Jeom Kee Paik, Dr. Eng., CEng, DHC (ULieg),
FRINA, LFSNAME
Pusan National University
and University College London

About the Author

Dr. Jeom Kee Paik is a professor and faculty member of both the Department of Naval Architecture and Ocean Engineering of Pusan National University (PNU) in Korea and the Department of Mechanical Engineering of University College London in the United Kingdom. He is an honorary professor both at University of Strathclyde in the United Kingdom and at Southern University of Science and Technology in China. He was a visiting professor at Technical University of Denmark, Virginia Polytechnic Institute and State University, USA, and University of Newcastle, Australia.

Prof. Paik founded two key institutions: the Korea Ship and Offshore Research Institute (KOSORI) (http://www.kosori.org) at PNU, which has been a Lloyd's Register Foundation Research Centre of Excellence (ICASS, International Centre for Advanced Safety Studies, http://www.icass.kr) since 2008, and the Forum for Safety of Fire and Explosion (http://www.safeforum.co.kr) under the Ministry of Interior and Safety of Korea. He also serves as president and chairman, respectively. He is founder and editor-in-chief of *Ships and Offshore Structures* (http://saos.edmgr. com), which is a peer-reviewed international journal published by Taylor & Francis, UK. He is cofounder and cochairman of the International Conference on Ships and Offshore Structures (http://www.iscos.info), which is an annual event associated with the *Ships and Offshore Structures* journal.

Prof. Paik received Bachelor of Engineering degree from Pusan National University, Korea and Master of Engineering and Doctor of Engineering degrees from Osaka University, Japan. Prof. Paik is a life fellow, fellows committee member, Marine Technology board member, and vice president of the US Society of Naval Architects and Marine Engineers (SNAME), and a fellow, council member, publications committee member, and Korean branch chairman of the UK Royal Institution of Naval Architects (RINA).

Prof. Paik's research interests include nonlinear structural mechanics, analysis, and design; advanced safety studies; limit state-based design; structural reliability; risk assessment and management; health condition assessment and management; fires, explosions, collisions, grounding, dropped objects, and impact engineering; corrosion assessment and management; structural longevity; inspection and maintenance; and decommissioning.

Prof. Paik has authored or coauthored more than 500 technical papers including over 270 peer-reviewed journal articles. He is the coauthor or coeditor of four books: *Ultimate Limit State Design of Steel-Plated Structures* (with A.K. Thayamballi), John Wiley & Sons, 2003; *Ship-Shaped Offshore Installations: Design, Building, and Operation* (with A.K. Thayamballi), Cambridge University Press, 2007; *Condition Assessment of Aged Structures* (with R.E. Melchers), CRC Press, 2009; and *Ship Structural Analysis and Design* (with O.F. Hughes), SNAME, 2013. He also obtained numerous patents based on his research studies over a wide range of topics in naval architecture and ocean engineering.

Among other recognitions, Prof. Paik received both the William Froude Medal of the RINA (2015) and the David W. Taylor Medal of the SNAME (2013), the two most prestigious medals in the global maritime community in recognition for his contributions to naval architecture and ocean engineering. He was conferred the Doctor Honoris Causa (Honorary Doctorate) by the University of Liege in Belgium (2012) in recognition for his contributions to international science, engineering, and technology. Prof. Paik was awarded the Republic of Korea Order of Science and Technology Merit (2014). He has received numerous (13) best paper awards and engineering prizes from the SNAME, the RINA, the UK Institution of Mechanical Engineers, the American Society of Mechanical Engineers, and the Society of Naval Architects of Korea. He was also awarded the Kyung-Ahm Prize (2013) from the Kyung-Ahm Education and Culture Foundation. As a very special honor for a living figure, the RINA created a prize named in honor of Prof. Paik, the *Jeom Kee Paik Prize*, which has been awarded each year since 2015 for the best paper on structures published by a researcher under 30 years of age; the prize is the first of its kind named for a non-Briton in the RINA's 156-year history.

Prof. Paik has served in numerous international engineering societies in various capacities. He served as editor-in-chief of UNESCO's Encyclopedia of Life Support System with EOLSS 6.177 Ships and Offshore Structures in 2006–2011. He served as chairman of the Korean Shipbuilding Advisory Committee of Registro Italiano Navale (Italian Classification Society) in 2013–2014 and the numerous technical committees of the International Ship and Offshore Structures Congress (ISSC) associated with Ship Collisions and Grounding (2000–2003), Condition Assessment of Aged Ships (2003–2006), and Ultimate Strength (2006–2012). He has presided numerous international conferences, including the International Conference on Thin-Walled Structures (ICTWS 2014, Busan, Korea) and the International Conference on Ocean, Offshore, and Arctic Engineering (OMAE 2016, Busan, Korea), and has cochaired the International Conferences on Ships and Offshore Structures. He will chair the upcoming International Symposium on Plasticity and Impact Mechanics (IMPLAST 2019, Busan, Korea). Currently, Prof. Paik heads the Korean Technical Committee of ClassNK (Japanese Classification Society), is a member of the Academic Advisory Council of Universiti Teknologi PETRONAS in Malaysia and of the ISSC Standing Committee, and is an editorial board member for more than 20 international journals.

How to Use This Book

Written to develop a textbook and handy source for the principles behind the ultimate limit state analysis and design of steel- and aluminum-plated structures, this book is designed to be well suited for university students approaching the related technologies. In terms of the more advanced and sophisticated analysis and design methodologies presented, this book should also meet the needs of structural analysts, designers, or researchers involved in the field of naval architecture and offshore, civil, architectural, aerospace, and mechanical engineering.

Hence, apart from its value as a ready reference and an aid to continuing education for established practitioners, this book can be used as a textbook for teaching courses on ultimate limit state analysis and design of plated structures at the university level, as it covers a wide enough range of topics that may be considered for more than one semester course.

A teaching course of 45 h for undergraduate students in structural mechanics or thin-walled structures may cover Chapter 1, "Principles of Limit State Design"; Chapter 2, "Buckling and Ultimate Strength of Plate–Stiffener Combinations: Beams, Columns, and Beam–Columns"; Chapter 3, "Elastic and Inelastic Buckling Strength of Plates Under Complex Circumstances"; Chapter 5, "Elastic and Inelastic Buckling Strength of Stiffened Panels and Grillages"; Chapter 7, "Buckling and Ultimate Strength of Plate Assemblies: Corrugated Panels, Plate Girders, Box Columns, and Box Girders"; and Chapter 8, "Ultimate Strength of Ship Hull Structures."

For postgraduate students who pass the teaching course for the undergraduate students noted previously, a more advanced course of 45 h may cover Chapter 1, "Principles of Limit State Design" (repeated); Chapter 2, "Buckling and Ultimate Strength of Plate–Stiffener Combinations: Beams, Columns, and Beam–Columns" (repeated); Chapter 4, "Large Deflection and Ultimate Strength Behavior of Plates"; and Chapter 6, "Large Deflection and Ultimate Strength Behavior of Stiffened Panels and Grillages."

In teaching courses, lecturers are advised to guide students to practice the derivations of important formulations described in each chapter together with practical problems for analysis and design of steel- and aluminum-plated structures. Students may submit homework reports to the lecturers, an exercise that would be helpful for students to better understand the fundamentals and practical applications.

Chapter 9, "Structural Fracture Mechanics," and Chapter 10, "Structural Impact Mechanics," should also be useful in association with fatigue limit state design and accidental limit state design, respectively. These two chapters are supplementary for the ultimate limit state analysis and design, as they describe the fundamentals and practices of fatigue and accidental limit states. Chapter 11, "The Incremental Galerkin Method"; Chapter 12, "The Nonlinear Finite Element Method"; and Chapter 13, "The Intelligent Supersize Finite Element Method," should be useful for postgraduate students, researchers, and practicing engineers given their more refined and sophisticated analyses of the ultimate strength behavior of plated structures.

The author has attempted to fulfill these many lofty aims in developing this book. He sincerely hopes his efforts prove successful, however modestly.

1

Principles of Limit State Design

1.1   Structural Design Philosophies

While in service, structures are likely to be subjected to various types of loads (or actions) and load effects (or action effects) due to operational and environmental conditions that are usually normal but are sometimes extreme or even accidental. The mission of the structural designer is to design a structure that can withstand the operational and environmental requirements designated throughout its expected lifetime.

The load effects or maximum load-carrying capacities or limit states of a structure are affected by a variety of factors that essentially involve a great deal of uncertainty, which include the following:

- Geometric factors associated with structural characteristics, buckling, large deformation, crushing, or folding
- Material factors associated with chemical composition, mechanical properties, yielding or plasticity, or fracture
- Fabrication related initial imperfections, such as initial distortion, welding induced residual stress, or softening
- Temperature factors, such as low temperatures associated with operation in cold waters or low-temperature cargo and high temperatures due to fire and explosions
- Dynamic or impact factors (e.g., strain rate sensitivity or inertia effect) associated with freak waves and impact pressure actions that arise from sloshing, slamming, or green water; overpressure actions that arise from explosion; and impact from collisions, grounding, or dropped objects
- Age related degradation factors, such as corrosion or fatigue cracking
- Accident induced damage factors, such as local denting, collision damage, grounding damage, fire damage, or explosion damage
- Human factors related to unusual operations (e.g., ship's operational speed compared with maximum permitted speed or acceleration, ship's heading, or loading or unloading conditions)

Uncertainties can comprise two groups: inherent uncertainties and modeling uncertainties. Inherent uncertainties are caused by natural variabilities in environmental actions and material properties, and modeling uncertainties arise from inaccuracy in engineering modeling associated with the evaluation and control of loads, load effects

Ultimate Limit State Analysis and Design of Plated Structures, Second Edition. Jeom Kee Paik.
© 2018 John Wiley & Sons Ltd. Published 2018 by John Wiley & Sons Ltd.

(e.g., stress, deformation), load-carrying capacities, or limit states and from variations in building and operational procedures. In design, a structure is thus required to have an adequate margin of safety against service requirements because of such inherent and modeling uncertainties.

A "demand" is analogous to load, and a "capacity" is analogous to the strength necessary to resist that load, both measured consistently (e.g., as stress, deformation, resistive or applied load or moment, or energy either lost or absorbed). In this regard, a performance function G of a structure can be given as follows:

$$G = C_d - D_d \tag{1.1a}$$

where C_d represents the "design" capacity and D_d represents the "design" demand. The terminology "design" implies that both demand and capacity are determined by accounting for the inherent and modeling uncertainties.

Because both C_d and D_d in Equation (1.1a) are a function of the basic variables, $X = (x_1, x_2, \ldots, x_i, \ldots, x_n)$, the performance function G can be rewritten as follows:

$$G = G(X) = G(x_1, x_2, \ldots, x_i, \ldots, x_n) \tag{1.1b}$$

When $G(X) > 0$, the structure is in the desired state. When $G(X) \leq 0$, the structure is in the undesired state. In industry practice, the performance function of a structure is sometimes defined in an opposite manner to Equation (1.1a) as follows:

$$G^* = D_d - C_d \tag{1.2}$$

where G^* is the performance function of a structure. In this case, the structure is in the desired state when $G^* < 0$, and it is in the undesired state when $G^* \geq 0$. Figure 1.1 illustrates the two performance functions associated with the desired and undesired states.

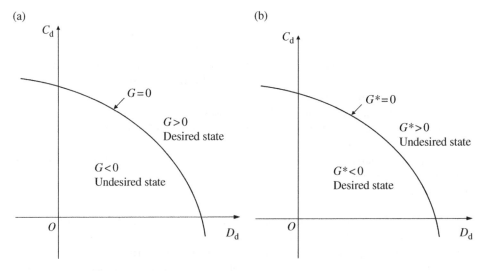

Figure 1.1 The performance functions associated with the desired and undesired states: (a) a performance function G, Equation (1.1a); (b) a performance function G^*, Equation (1.2).

1.1.1 Reliability-Based Design Format

The reliability-based design format usually involves the following tasks:

1) Definition of a target reliability
2) Identification of all unfavorable failure modes of the structure
3) Formulation of the limit state (performance) function for each failure mode identified in item (2)
4) Identification of the probabilistic characteristics (mean, variance, probability density distribution) of the random variables in the limit state function
5) Calculation of the reliability against the limit state with respect to each failure mode of the structure
6) Evaluation of the predicted reliability whether or not it is greater than the target reliability
7) Redesign of the structure otherwise
8) Evaluation of the reliability analysis results with respect to a parametric sensitivity consideration

Each of the basic variables in the reliability-based design format is dealt with in a probabilistic manner as a random parameter, where each random variable must be characterized by the corresponding probability density function that has a mean value and standard deviation. If the first-order approximation is adopted, the performance function $G(X)$ can be rewritten by the Taylor series expansion as follows:

$$G(X) \cong G(\mu_{x1}, \mu_{x2}, \ldots, \mu_{xi}, \ldots, \mu_{xn}) + \sum_{i=1}^{n} \left(\frac{\partial G}{\partial x_i}\right)_{\bar{x}} (x_i - \mu_{xi}) \tag{1.3}$$

where μ_{xi} is the mean value of the variable x_i, $\bar{x}$ is the mean value of the basic variables = $(\mu_{x1}, \mu_{x2}, \ldots, \mu_{xi}, \ldots, \mu_{xn})$, and $(\partial G/\partial x_i)_{\bar{x}}$ is the partial differentiation of $G(X)$ with respect to x_i at $x_i = \mu_{xi}$.

The mean value of the performance function $G(X)$ is then given by

$$\mu_G = G(\mu_{x1}, \mu_{x2}, \ldots, \mu_{xi}, \ldots, \mu_{xn}) \tag{1.4}$$

where μ_G represents the mean value of the performance function $G(X)$.

The standard deviation of the performance function $G(X)$ is calculated by

$$\sigma_G = \left[\sum_{i=1}^{n} \left(\frac{\partial G}{\partial x_i}\right)_{\bar{x}}^2 \sigma_{xi}^2 + 2 \sum_{i>j} \left(\frac{\partial G}{\partial x_i}\right)_{\bar{x}} \left(\frac{\partial G}{\partial x_j}\right)_{\bar{x}} \operatorname{covar}(x_i, x_j) \right]^{1/2} \tag{1.5a}$$

where σ_G is the standard deviation of $G(X)$, σ_{x_i} is the standard deviation of the variable x_i, $\operatorname{covar}(x_i, x_j) = E\left[(x_i - \mu_{xi})(x_j - \mu_{xj})\right]$ is the covariation of x_i and x_j, and $E[]$ is the mean value of [ ].

When the basic variables $X = (x_1, x_2, \ldots, x_i, \ldots, x_n)$ are independent of each other, $\operatorname{covar}(x_i, x_j) = 0$. In this case, Equation (1.5a) is simplified to

$$\sigma_G = \left[\sum_{i=1}^{n} \left(\frac{\partial G}{\partial x_i}\right)_{\bar{x}}^2 \sigma_{xi}^2 \right]^{1/2} \tag{1.5b}$$

If the so-called first-order second-moment method (Benjamin & Cornell 1970) is adopted, the reliability index for this case can be determined as follows:

$$\beta = \frac{\mu_G}{\sigma_G} \qquad (1.6)$$

where β represents the reliability index.

For a simpler case with a performance function $G(X)$ of two parameters, for example, capacity C and demand D, that are considered to be statistically independent, the reliability index β can be calculated as follows:

$$\mu_G = \mu_C - \mu_D \qquad (1.7a)$$

$$\sigma_G = \sqrt{(\sigma_C)^2 + (\sigma_D)^2} \qquad (1.7b)$$

$$\beta = \frac{\mu_C - \mu_D}{\sqrt{(\sigma_C)^2 + (\sigma_D)^2}} = \frac{\mu_C/\mu_D - 1}{\sqrt{(\mu_C/\mu_D)^2(\eta_C)^2 + (\eta_D)^2}} \qquad (1.7c)$$

where μ_C or μ_D are the mean values of C or D, σ_C or σ_D are the standard deviations of C or D, and η_C or η_D are the coefficients of variation (i.e., the standard deviation divided by the mean value) of C or D.

To achieve a successful design, the reliability index should be greater than a target reliability index:

$$\beta \geq \beta_T \qquad (1.8)$$

where β_T is the target reliability.

The target reliability or the required level of structural reliability may vary from one industry to another depending on various factors such as the type of failure, the seriousness of its consequence, or public and media sensitivity. Appropriate values of target reliability are not readily available and are usually determined by surveys or by examinations of the statistics on failures although the fundamental difference between a risk assessment and a reliability analysis needs to be acknowledged when interpreting such results. The methods to select the target safeties and reliabilities may be categorized into the following three groups (Paik & Frieze 2001):

- "Guesstimation": A "reasonable" value as recommended by a regulatory body or professionals on the basis of successful prior experience. This method may be employed for the new types of structure for which statistical database on failures does not exist.
- Calibration of design rules: The level of reliability is estimated by calibrating a new design rule to an existing successful one. This method is normally used for the revisions of existing design rules.
- Economic value analysis: The target reliability is selected to minimize total expected costs during the service life of the structure.

For elaborate descriptions in reliability analysis, interested readers may refer to Benjamin and Cornell (1970), Nowak and Collins (2000), Melchers (1999a), and Modarres et al. (2016), among others.

1.1.2 Partial Safety Factor-Based Design Format

In the partial safety factor-based design format, the design capacity or demand is defined by considering the corresponding partial safety factors that are associated with the inherent and modeling uncertainties. A characteristic or nominal value of capacity C_k or demand D_k is determined as the mean value of the corresponding random variable. A design capacity C_d or demand D_d is, however, defined to suit a specified percentage of the area below the probability curve for the corresponding random variable. For instance, a design strength or capacity C_d can be defined for a lower bound or 95% exceedance value, whereas a design load or demand D_d can be defined for an upper bound or a 5% exceedance value, as shown in Figure 1.2. In this regard, the design capacity or demand is defined as follows:

$$C_d = \frac{C_k}{\gamma_C} \tag{1.9a}$$

$$D_d = \gamma_D D_k \tag{1.9b}$$

where C_k is the characteristic (or nominal) value of capacity or μ_C in Equation (1.7a), D_k is the characteristic (or nominal) value of demand or μ_D in Equation (1.7a), γ_C is the partial safety factor associated with capacity, and γ_D is the partial safety factor associated with demand. Because the partial safety factors must be greater than 1.0, it is obvious that the characteristic value of capacity C_k is reduced and the characteristic value of demand D_k is amplified to determine their design values, C_d or D_d.

The measure of structural adequacy η can be determined as follows:

$$\eta = \frac{C_d}{D_d} = \frac{1}{\gamma_C \gamma_D} \frac{C_k}{D_k} \tag{1.10}$$

To achieve a successful design, the measure of structural adequacy η must be greater than 1.0 by a sufficient margin as follows:

$$\eta = \frac{C_d}{D_d} = \frac{1}{\gamma_C \gamma_D} \frac{C_k}{D_k} > 1 \tag{1.11}$$

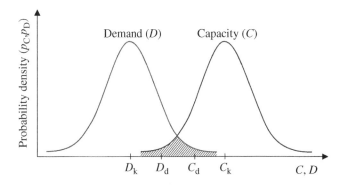

Figure 1.2 Probability density distributions of capacity and demand.

1.1.3 Failure Probability-Based Design Format

Whatever the level of uncertainty, every structure may have some probability of failure, which is the possibility of a load or demand exceeding its limit value or capacity. The probability of failure P_f for a particular type of failure in association with the performance function G, Equation (1.1), or G^*, Equation (1.2), is defined as follows:

$$\text{Probability of failure } P_f = \text{prob}(G \leq 0) = \text{prob}(G^* \geq 0) = \text{prob}(C_d \leq D_d) \quad (1.12a)$$

The safety of a structure is the converse, which is the probability that it will not fail, namely,

$$\text{Safety} = \text{prob}(G > 0) = \text{prob}(G^* < 0) = \text{prob}(C_d > D_d) = 1 - P_f \quad (1.12b)$$

The probability of failure can generally be calculated as follows:

$$P_f = \int_{G \leq 0} p_x(X)dx = \int_{G^* \geq 0} p_x^*(X)dx \quad (1.13)$$

where $p_x(X)$ and $p_x^*(X)$ are the joint probability density functions of the random variables, $X = (x_1, x_2, \ldots, x_i, \ldots, x_n)$, associated with demand and capacity, and $G(X)$ or $G^*(X)$ is the limit state (performance) function defined such that negative or positive values imply failure, respectively.

Since $G(X)$ or $G^*(X)$ is usually a complicated nonlinear function, it is not straightforward to perform the direct integration of Equation (1.13) associated with the joint probability density function, $p_x(X)$ or $p_x^*(X)$. Therefore, Equation (1.13) is often solved with approximate procedures, where the limit state (performance) function $G(X)$ or $G^*(X)$ is approximated at the design point by either a tangent hyperplane or hyperparabola, which simplifies the mathematics related to the calculation of failure probability. The first type of approximation with the tangent hyperplane is called the first-order reliability method (FORM), and the second type with the hyperparabola is called the second-order reliability method (SORM). Such methods facilitate the rapid calculation of the probability of failure by widely available standard software packages. In addition to the individual probability distributions of the random variables involved, the correlation between the "A" and "B" parameters can also be readily accounted for in such calculations.

Considering the probability density distributions of capacity and demand, as illustrated in Figure 1.2, the probability of a particular type of failure can be calculated as follows:

$$P_f = \int_0^\infty \left[\int_0^y p_C(x)dx \right] p_D(y)dy \quad (1.14)$$

where $p_C(x)$ is the probability density function of capacity associated with a variable x and $p_D(y)$ is the probability density function of demand associated with a variable y.

Although the mean value of capacity C_k is much greater than the mean value of demand D_k, there is still some possibility that the capacity is less than the demand. It is usually challenging to compute Equation (1.14), but it is interesting to note that the shaded area of the overlap in Figure 1.2 indicates an approximation of the probability of failure P_f. To achieve a successful design, the probability of failure should be minimized to a sufficiently low value.

1.1.4 Risk-Based Design Format

The risk-based design format usually involves the following five tasks: (i) hazard identification, (ii) risk calculation, (iii) establishment of a set of potential risk control options, (iv) cost–benefit analysis for the risk control options, and (v) decision making. In engineering community, risk is defined as a product of the frequency of the hazard and the level of consequence as follows:

$$R = F \times C \tag{1.15}$$

where R is the risk, F is the frequency of the hazard, and C is the level of consequence.

The frequency of the hazard represents the likelihood that the hazard will occur, and the level of consequence represents the impact or severity of consequence, indicating how bad the consequences would be if the hazard did occur in terms of casualties, property damage, and environmental pollution. The frequency of a hazard is usually measured by the number of occurrences per unit time (e.g., per year). The level of consequence is sometimes measured on a monetary basis (e.g., repair costs for accidental damage or insurance costs for pollution).

The characterization of the frequency and the consequences is required for risk assessment. Qualitative risk assessment techniques use simple methods that do not require numerical computations, but quantitative risk assessment requires more refined methods associated with numerical and experimental investigations. It is of course much more desirable to apply the quantitative risk assessment methods for more precise calculations of the risks in association with casualties, property damage, and environmental pollution.

According to Equation (1.15), it is obvious that one may need to reduce F or C or both to reduce risks. To achieve a successful design, fabrication, or operation, the risk should be minimized to an "as low as reasonably practicable (ALARP)" level. Undertaking activities to control risks is risk management, which involves risk control options. Cost–benefit analysis is undertaken to make a ranking between a set of potential risk control options, and a single or multiple options should be applied to best control the risks to meet the ALARP level. Risk assessment and management are recognized as the best tools for decision making in association with robust design, building, operation, or decommissioning of structures.

1.2 Allowable Stress Design Versus Limit State Design

Limit state design differs from the traditional allowable stress design. In the allowable stress design, the focus is on keeping the stresses from the design loads under a certain working stress level, which is usually based on successful similar experience. In industry practice, regulatory bodies or classification societies usually specify the value of the allowable stress as some fraction of the mechanical properties of materials (e.g., yield strength). The criterion of the allowable stress design is typically given by

$$\sigma < \sigma_a \tag{1.16}$$

where σ is the working stress and σ_a is the allowable stress.

In contrast to the allowable stress design, the limit state design is based on explicit consideration of the various conditions under which the structure may cease to fulfill its intended function. For these conditions, the applicable capacity or strength is estimated and used during design as a limit for such behavior.

For this purpose, a structure's load-carrying capacity is normally evaluated with simplified design formulations or more refined computations such as nonlinear elastic–plastic large-deformation finite element analyses with appropriate modeling related to geometric or material properties, initial imperfections, boundary conditions, load application, and finite element mesh sizes, as appropriate.

During the past several decades, the emphasis on structural design has moved from the allowable stress design to the limit state design because the latter approach makes possible a rigorously designed, yet economical, structure that directly takes into consideration the various relevant modes of failure.

A limit state is formally defined by the description of a condition for which a particular structural member or an entire structure would fail to perform the function designated beforehand. From the viewpoint of structural design, four types of limit states are relevant for structures:

- The serviceability limit state (SLS)
- The ultimate limit state (ULS)
- The fatigue limit state (FLS)
- The accidental limit state (ALS)

The SLS represents failure states for normal operations due to deterioration from routine functioning. SLS considerations in design may address the following:

- Local damage that reduces the structure's durability or affects the efficiency of structural elements
- Unacceptable deformations that affect the efficient use of structural elements or the functioning of equipment that relies on them
- Excessive vibration or noise that can cause discomfort to people or affect the proper functioning of equipment
- Deformations and deflections that may spoil the structure's aesthetic appearance

The ULS (also called ultimate strength) represents the collapse of the structure due to a loss of structural stiffness and strength. Such loss of capacity may be related to:

- A loss of equilibrium, of a part or of the entire structure, which is often considered as a rigid body (e.g., overturning or capsizing)
- Attainment of the maximum resistance of structural regions, members, or connections by gross yielding or fracture
- Instability, of a part or of the entire structure, from buckling and plastic collapse of plating, stiffened panels, and support members

The FLS represents the occurrence of fatigue cracking of structural details due to stress concentration and damage accumulation or crack growth under repeated loading.

The ALS represents excessive structural damage from accidents, such as collisions, grounding, explosion, and fire, that affect the safety of the structure, the environment, and personnel.

The partial safety factor-based criterion of the limit state design for a particular type of limit state is typically given from Equation (1.11) as follows:

$$C_d > D_d \quad \text{or} \quad \frac{C_k}{\gamma_C} > \gamma_D D_k \tag{1.17}$$

It is important to emphasize that in the limit state design, these various types of limit states may be designed against different safety levels, with the actual safety level to be attained for a particular type of limit state being an indirect and implicit function of its perceived consequences and the ease of recovery from that state to be incorporated in design. Within the context of Equation (1.17), useful guidelines for determination of the partial safety factors related to a structure's limit state design may be found in ECCS (1982), BS 5950 (1985), ENV 1993-1 (1992a, 1992b), ISO 2394 (1998), and NORSOK (2004), among others.

1.2.1 Serviceability Limit State Design

The structural design criteria used for the SLS design of structures are normally based on the limits of deflections or vibration for normal use. In reality, the excessive deformation of a structure may also be associated with excessive vibration or noise, and thus certain interrelationships may exist among the design criteria being defined and used separately for convenience.

The SLS criteria are normally defined by the operator of a structure or by established practice, with the primary aim being efficient and economical in-service performance without excessive routine maintenance or downtime. The acceptable limits necessarily depend on the type, mission, and arrangement of structures. Furthermore, in defining such limits, experts in other disciplines, such as machinery design, must also be consulted. As an example, the limiting values of vertical deflections for beams in structures as shown in Figure 1.3 are indicated in Table 1.1.

In Table 1.1, L is the span of the beam between supports. For cantilever beams, L may be taken as twice the projecting length of the cantilever. δ_{max} is the maximum deflection, which is given by $\delta_{max} = \delta_1 + \delta_2 - \delta_0$, where δ_0 is the pre-camber, δ_1 is the variation of the deflection of the beam due to permanent loads immediately after loading, and δ_2 is the variation of the deflection of the beam due to variable loading plus any subsequent variant deflections due to permanent loads.

For plate elements, criteria based on elastic buckling control are often used for SLS design, in some cases to prevent such an occurrence entirely and in other cases to allow

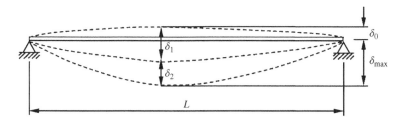

Figure 1.3 Nomenclature: lateral deflections of a beam.

Table 1.1 Serviceability limit values for vertical deflections of beams.

Condition	Limit for δ_{max}	Limit for δ_2
Deck beams	$L/200$	$L/300$
Deck beams that support plaster or other brittle finish or non-flexible partitions	$L/250$	$L/350$

elastic buckling to a known and controlled degree. Elastic plate buckling and its related effects, such as relatively large lateral deflections, must be prevented if such effects are likely to be detrimental. However, because a plate may have some reserve strength beyond elastic buckling until its ultimate strength is reached, allowing elastic buckling in a controlled manner can in some cases lead to a more economical structure. In Chapters 3 and 5 of this book, the use of such elastic buckling strength-based SLS design methods for plates and stiffened panels is described.

1.2.2 Ultimate Limit State Design

The structural design criteria to prevent the ULS are based on plastic collapse or ultimate strength. The simplified ULS design of many types of structures has tended to rely on estimates of the buckling strength of the components, usually from their elastic buckling strength adjusted by a simple plasticity correction, which is represented by point A in Figure 1.4. In such a design scheme based on the strength at point A, the structural designer does not use detailed information on the post-buckling behavior of the component members and their interactions. The true ultimate strength represented by point B in Figure 1.4 may be higher, although one can never be sure of this because the actual ultimate strength is not being directly evaluated.

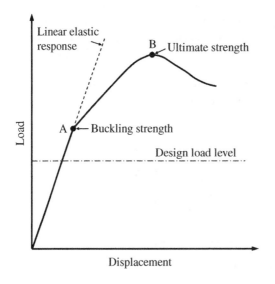

Figure 1.4 Structural design considerations based on the ultimate limit state.

In any event, as long as the strength level associated with point B remains unknown (as it is with traditional allowable stress design or linear elastic design methods), it is difficult to determine the real safety margin. Hence, more recently, the design of structures such as those of ships, offshore platforms, box girder bridges, and box girder cranes has tended to be based on the ultimate strength.

The safety margin of a structure can be evaluated by comparison of its ultimate strength with the extreme applied loads (or load effects, such as stress) as depicted in Figure 1.4. To obtain an economic yet safe structure, the ultimate strength and the design load must be assessed accurately. The structural designer may even desire to estimate the ultimate strength for not only the intact structure but also the structures with existing or in-service damage (e.g., corrosion wastage, fatigue cracking, or local denting damage) or even accident induced damage (e.g., due to collision, grounding, dropped object, fire, or explosion) to assess their damage tolerance and survivability.

The ULS design criterion can also be expressed by Equation (1.17). The characteristic measure of design capacity C_d in Equation (1.17) is in this case the ultimate strength, whereas D_d is the related load or demand measure. For ULS design, the partial safety factor γ_C is sometimes taken as $\gamma_C = 1.15$ for ships and offshore structures (NORSOK 2004).

It is important to note that any failure in a structure must ideally occur in a ductile manner rather than a brittle manner; the avoidance of brittle failure will lead to a structure that does not collapse suddenly, because ductility allows the structure to redistribute internal stresses and thus absorb greater amounts of energy before global failure. Adequate ductility in the design of a structure is facilitated by:

- Meeting the requisite material toughness requirements
- Avoiding failure initiation situations with a combination of high stress concentration and undetected weld defects in the structural details
- Designing structural details and connections to allow a certain amount of plastic deformation, that is, avoiding "hot spots"
- Arranging the members in such a manner that a sudden decrease in the structural capacity would not occur as a result of abrupt transitions or member failure

This book is primarily concerned with ULS design methods for structural members and systems composed of such ductile members, although other types of limit states are also described to some extent.

1.2.3 Fatigue Limit State Design

The FLS design is carried out to ensure that the structure has an adequate fatigue life. The predicted fatigue life can also be a basis for planning efficient inspection programs during the structure's operation. The design fatigue life for structural components is normally based on the structure service life required by the operator or by other responsible body such as a class society. For ship structures, the fatigue life is often considered to be 25 years or longer. The shorter the design fatigue life, or the greater the required reliability, the smaller the inspection intervals should be to assure an operation free from crack problems.

The FLS design and analysis should in principle be undertaken for every suspected location of fatigue cracking, which includes welded joints and local areas of stress concentration.

The structural design criteria for the FLS are usually based on the structure's cumulative fatigue damage under repeated fluctuation of loading, as measured by the Palmgren–Miner cumulative damage rule. A particular value of the Miner sum (e.g., unity) is taken to be synonymous with the formation or initiation of a crack. The structure is designed so that when it is analyzed for fatigue, a reduced target Miner sum results, implying that cracks will not form with a given degree of certainty.

The fatigue damage at a crack initiation site is affected by many factors, such as the stress ranges experienced during load cycles, the local stress concentration characteristics, and the number of stress range cycles. Two types of the FLS design approach are typically considered for structures:

- The S–N curve approach (S = fluctuating stress, N = associated number of cycles)
- The fracture mechanics approach

In the S–N curve approach, the Palmgren–Miner cumulative damage rule is applied together with the relevant S–N curve. This application normally follows three steps: (i) definition of the histogram of cyclic stress ranges, (ii) selection of the relevant S–N curve, and (iii) calculation of the cumulative fatigue damage.

One of the most important factors in fatigue design is the characteristic stress to be used both in defining the S–N curve (the capacity) and in the stress analysis (with the fluctuating local fatigue stresses being the demand on the structure). Four types of methods have been suggested on this basis:

- The nominal stress method
- The hot spot stress method
- The notch stress method
- The notch strain method

The nominal stress method uses the nominal stresses in the field far from the stress concentration area, together with S–N curves that must include implicitly the effects of both structural geometry and the weld. In the nominal stress method, therefore, the S–N curve should be selected for structural details depending on the detail type and weld geometry involved. Many S–N curves for various types of weld and geometry are generally needed and are available. When a limited number of standard S–N curves are used, any structural detail considered must be assigned to one of those categories, which requires a certain amount of judgment.

The hot spot stress method uses a well-defined hot spot stress in the stress concentration area to account for the effect of structural geometry alone, and the weld effect is incorporated into the S–N curve. This is currently a very popular approach, but certain practical difficulties must be conceded. The most basic of these pertains to the concept of hot spot stress itself, which is more appropriate for surface cracks than for imbedded cracks. Difficulties can also arise in the consistent definition of hot spot stresses across a range of weld and structural geometries and in the estimation of the hot spot structural stress needed for application of the technology in regions of stress concentration. For instance, attention should be paid to extrapolation of the stress to the weld toe for calculation of the stress concentration factor, and the need for appropriate selection of a relevant S–N curve from those for different weld types is still significant.

The notch stress method uses the stresses at the notch calculated by accounting for the effects of both structural geometry and the weld, whereas the S–N curve is developed to represent the fatigue properties of either the base material, the material in the heat-affected zone (HAZ), or the weld material, as appropriate. A significant advantage of the notch stress method is that it can address the specific weld toe geometry in the calculation of fatigue damage. A related difficulty is that the relevant parameters (e.g., the weld toe angle) in the case of the actual structure must be known with some confidence.

The notch strain method uses the strains at the notch when the low-cycle fatigue is predominant, because the working stresses in this case sometimes likely approach the material yield stress, and thus the stress-based approaches are less appropriate.

The fracture mechanics approach considers that one or more premised cracks of a small dimension exist in the structure and predicts the fatigue damage during the process of crack propagation, including any coalescence and breakthrough, and the subsequent fracture. In this approach to design, a major task is to preestablish the relevant crack growth equations or "laws." The crack growth rate is often expressed as a function of only the stress intensity factor range at the crack tip, on the assumption that the yielded area around the crack tip is relatively small. In reality, the crack propagation behavior is affected by many other parameters (e.g., mean stresses, load sequence, crack retardation, crack closure, crack growth threshold, and stress intensity range) in addition to the stress intensity factor range.

The structural fracture mechanics is dealt with in Chapter 9, and the *S–N* curve approach using nominal stresses is herein briefly described under the assumption of the linear cumulative damage rule, that is, the Palmgren–Miner rule. In the fatigue damage assessment of welded structural details, of primary concern are the ranges of the cyclic maximum and minimum stresses rather than the mean stresses, as shown in Figure 1.5, because of the usual presence of residual mean stresses near the yield magnitude. This tends to make the entire stress range damaging. The situation in non-welded cases is, of course, different, and, in such cases, the mean stresses can be important.

For practical FLS design using the nominal stress-based approach, the relevant *S–N* curves must be developed for various types of weld joints. To do this, fatigue tests are carried out for various types of specimens that are subjected to cyclic stress ranges of a uniform amplitude. As indicated in Figure 1.5, the maximum and minimum stresses

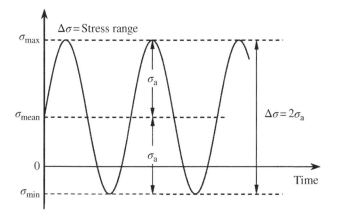

Figure 1.5 Cyclic stress range versus time.

are denoted by σ_{max} and σ_{min}, respectively. In such tests, the effect of the mean stress, $\sigma_{mean} = (\sigma_{max} + \sigma_{min})/2$, on fatigue damage can be quantified, which is necessary for non-welded cases. For convenience, the fatigue tests for specimens that incorporate non-welded geometries are usually carried out at either $\sigma_{min} = 0$ or $\sigma_{max} = -\sigma_{min}$ with a constant stress range, that is, $\Delta\sigma = \sigma_{max} - \sigma_{min} = 2\sigma_a$, where σ_a is the stress amplitude.

The number of stress cycles, N_I or N_F, with the former representing the crack initiation life, that is, until a crack initiates, and the latter representing the fracture life, such as until a small-scale test specimen is separated into two pieces, is obtained on the basis of the fatigue test results. With a series of such tests for a variety of stress ranges, $\Delta\sigma$, the S–N curves for the particular structural details may typically be plotted as shown in Figure 1.6. The curves for design are usually expressible by curve fitting the test results plotted on a log–log scale, namely,

$$\log N = \log a - 2s - m\log\Delta\sigma \tag{1.18a}$$

$$N(\Delta\sigma)^m = A \tag{1.18b}$$

where $\Delta\sigma$ is the stress range, N is the number of stress cycles with constant stress range, $\Delta\sigma$, until failure, m is the negative inverse slope of the S–N curve, $\log A = \log a - 2s$, a is the life intercept of the mean S–N curve, and s is the standard deviation of log N.

For the FLS design criterion based on the S–N curve approach, Equation (1.17) may be rewritten in the nondimensional form when the distribution of a long-term stress range is given by a relevant stress histogram in terms of a number of constant amplitude stress range blocks, $\Delta\sigma_i$, each with a number of stress fluctuations, n_i, as follows:

$$D = \sum_{i=1}^{B} \frac{n_i}{N_i} = \frac{1}{A}\sum_{i=1}^{B} n_i(\Delta\sigma_i)^m \le D_{cr} \tag{1.19}$$

where D is the accumulated fatigue damage, B is the number of stress blocks, n_i is the number of stress cycles in stress block i, N_i is the number of cycles until failure at the ith constant amplitude stress range block, $\Delta\sigma_i$, and D_{cr} is the target cumulative fatigue damage for design.

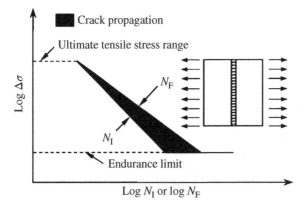

Figure 1.6 Typical S–N curves from constant amplitude tests.

To achieve greater fatigue durability in a structure, it is important to minimize stress concentrations, potential flaws (e.g., misalignment, poor materials), and structural degradation, including corrosion and fatigue effects. Fatigue design is interrelated with the maintenance regime to be used. In some cases, it may be more economical in design to allow the possibility of a certain level of fatigue damage, as long as the structure can continue to function after the fatigue symptoms are detected until repairs can be made. In other cases, fatigue damage may not be allowed to occur, if it is inconvenient to inspect the structure or interrupt production. The former approach may thus be applied as long as regular inspections and related maintenance are possible, whereas the latter concept is obviously more relevant if there are likely to be difficulties associated with inspections and thus a high likelihood of undetected fatigue damage.

Fatigue is sometimes classified into high-cycle fatigue and low-cycle fatigue. High-cycle fatigue indicates that a structure has a long fatigue life due to a small stress range, whereas low-cycle fatigue indicates that a structure has a short fatigue life due to a large stress range. The two are sometimes distinguished by the fatigue cycle of 10^4.

In Chapter 9, structural fracture mechanics and the ultimate strength of plate panels associated with fatigue cracking damage are described. For elaborate descriptions in fatigue damage analysis methods, interested readers may refer to Schijve (2009), Nussbaumer et al. (2011), and Lotsberg (2016), among others.

1.2.4 Accidental Limit State Design

The primary aim of the ALS design for structures may be characterized by the following three broad objectives:

- To avoid loss of life in the structure or the surrounding area
- To avoid pollution of the environment
- To minimize loss of property or financial exposure

In the ALS design, it is necessary to achieve a design in which the structure's main safety functions are not impaired during any accidental event or within a certain time after the accident. The structural design criteria for the ALS are based on limiting accidental consequences such as structural damage and environmental pollution.

Because the structural damage characteristics and behavior of damaged structures depend on the type of accidents, it is not straightforward to establish universally applicable structural design criteria for the ALS. Typically, for a given type of structure, the design of accidental scenarios and associated performance criteria must be decided on the basis of risk assessment.

In the case of ships or offshore platforms, possible accidental events that may need to be considered for the ALS include collisions, grounding, dropped objects, significant hydrodynamic impact (e.g., sloshing, slamming, or green water) that leads to buckling or structural damage, excessive loads from human error, berthing or dry docking, fires or internal gas explosions in oil tanks or machinery spaces, and underwater or atmospheric explosions. In land-based structures, the accidental scenarios may include fire, explosion, foundation movements, or related structural damage from earthquakes.

In selecting the design target ALS performance levels for such events, the approach is normally to tolerate a certain level of damage consistent with a greater aim such as survivability or minimized consequences; to do otherwise would result in an uneconomical structure.

The main safety functions of a structure that should not be compromised during any accident event or within a certain time after the accident include:

- Usability of escape ways
- Integrity of shelter areas and control spaces
- Global load-bearing capacity
- Integrity of the environment

Therefore, the ALS design criteria should be formulated so that the main safety functions mentioned previously will work successfully and the following points are considered to adequate levels:

- Energy dissipation related to structural crashworthiness
- Capacity of local strength members or structures
- Capacity of the global structure
- Allowable tensile strains to avoid tearing or rupture
- Endurance of fire protection

For the ALS design, the structure's integrity will typically be checked in two steps. In the first step, the structural performance will be assessed against design accident events, and post-accident effects such as damage to the environment are evaluated in the second step.

In the case of accidents to ships, for instance, the primary concern of the ALS design is to maintain the watertightness of the ship's compartments, the containment of dangerous or pollutant cargoes (e.g., chemicals, bulk oil, liquefied gas), and the integrity of the reactor compartments of nuclear-powered ships. To continue normal operations for the structure's mission, it is also important to maintain the integrity and residual strength of damaged structures at a certain level immediately after the accident occurs.

The different types of accident events normally require different methods to analyze the structure's resistance. For the ALS design criteria under predominantly impact-oriented loading, Equation (1.17) may typically be rewritten using energy dissipation-related criteria adopted with the view that the safety of the structure or the environment is not lost:

$$E_k \gamma_k < \frac{E_a}{\gamma_a} \tag{1.20}$$

where E_k is the kinetic energy lost during the accident, E_a is the available energy absorption capability until critical damage occurs, and γ_k and γ_a are partial safety factors related to kinetic energy loss and energy absorption capability, respectively.

The structure's dissipated energy during the accident may usually be calculated by integrating the area below the load–displacement curve of the structure under accidental loading, as shown in Figure 1.7. In Chapter 10, an elaborate description for the structural impact mechanics and the residual ultimate strength of plate panels with accident induced damage such as local denting is presented.

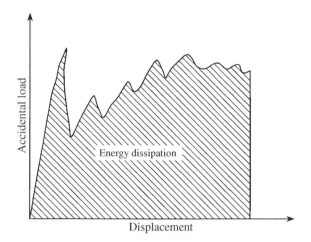

Figure 1.7 Energy absorption of the structure under accidental loading.

1.3 Mechanical Properties of Structural Materials

For materials of plated structures, steels or aluminum alloys are typically used. The specific gravity of aluminum alloys is about one-third that of steels, and thus aluminum alloys are primarily used in weight-critical structures. Aluminum alloys also have merits with their good resistance to corrosion by seawater and with an easier processing of extrusion, leading to the availability in a wide variety of section forms. However, the elastic modulus of aluminum alloys is only one-third that of steels, which is an apparent disadvantage of aluminum alloys.

In structural analysis and design, it is essential to define the material properties associated with the targeted structural systems. In industry practice, nominal values of material properties are often used in the analysis and design of a structure. When harsh environmental or operational conditions are of primary concern, however, the mechanical properties of the materials must be accurately quantified by considering the effects of such conditions. Because testing is only a method to quantify material properties, numerous test databases have been developed in the literature (e.g., Callister 1997); some are limited to specific conditions, and others are based on old materials that are no longer in use.

Modern material-manufacturing technologies have greatly advanced the material properties featured in old test databases, and today's structural systems are often exposed to the harsher environmental and operational conditions associated with their functional requirements. Thus, test databases for these volatile material properties should be continuously developed to meet such requirements (Paik et al. 2017).

1.3.1 Characterization of Material Properties

The mechanical properties of structural materials are characterized by testing predesignated specimens under monotonic tensile loading. Figure 1.8 shows an idealized

(a)

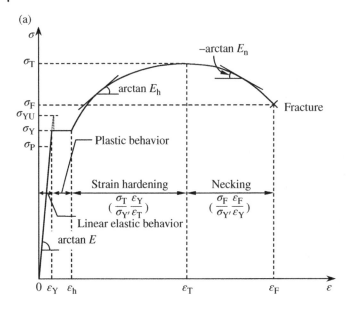

(b)

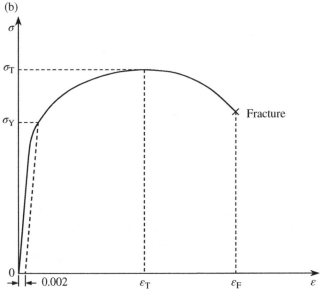

Figure 1.8 Schematic of engineering stress–engineering strain relationship for (a) ductile materials and (b) specially treated ductile materials.

engineering stress–engineering strain curve for structural metals. The material properties can be characterized using the following parameters:

- Young's modulus (or modulus of elasticity), E
- Poisson's ratio, ν
- Proportional limit, σ_P

- Upper yield point, σ_{YU}
- Lower yield point, σ_{YL} ($\approx \sigma_Y$)
- Yield strength, σ_Y
- Yield strain, ε_Y
- Strain-hardening strain, ε_h
- Strain-hardening tangent modulus, E_h
- Ultimate tensile strength, σ_T
- Ultimate tensile strain, ε_T
- Necking tangent modulus, E_n
- Necking stress at fracture (total breaking), σ_F
- Fracture (total breaking) strain, ε_F

1.3.1.1   Young's Modulus, *E*

The initial relationship between stress and strain is linear elastic, wherein the material recovers perfectly upon unloading. The slope of the linear portion of the stress–strain relationship in the elastic regime is defined as the modulus of elasticity, E (also called Young's modulus). Table 1.2 indicates typical values of Young's moduli for selected metals and metal alloys at room temperature. Young's modulus of aluminum alloys is about one-third that of steel.

1.3.1.2   Poisson's Ratio, *v*

Poisson's ratio is defined as the ratio of the transverse strain to the longitudinal strain of a material under tensile load in the elastic regime. Table 1.2 indicates typical values of Poisson's ratio for selected metals and metal alloys at room temperature.

1.3.1.3   Elastic Shear Modulus, *G*

The mechanical properties of materials under shear are usually defined using principles of structural mechanics rather than by testing. The elastic shear modulus is expressed by a function of Young's modulus, E, and Poisson's ratio, v, as follows:

$$G = \frac{E}{2(1+v)} \tag{1.21}$$

Table 1.2 Typical values of Young's moduli and Poisson's ratios for selected metals and metal alloys at room temperature.

Material	E (GPa)	v
Aluminum alloy	70	0.33
Copper	110	0.34
Steel	205.8	0.3
Titanium	104–116	0.34

1.3.1.4 Proportional Limit, σ_P

The maximum stress in the elastic regime, that is, immediately before initial yielding, is termed the proportional limit, σ_P.

1.3.1.5 Yield Strength, σ_Y, and Yield Strain, ε_Y

Strictly speaking, structural materials without special treatment (e.g., quenching, tempering) may have upper and lower yield points, as illustrated in Figure 1.8a. The lower yield point typically has an extended plateau in the stress–strain curve, which is approximated by the yield strength σ_Y and the corresponding yield strain, $\varepsilon_Y = \sigma_Y/E$.

The mechanical properties of structural materials vary with the amount of work and heat treatment applied during the rolling process. Typically, plates that receive more work have a higher yield strength than plates that do not. The yield strength of metals is usually increased by special treatment.

Figure 1.8b illustrates an idealized engineering stress–engineering strain curve of specially treated metals or metal alloys in which neither upper nor lower yield points appear until the ultimate tensile strength is reached. In this case, the yield strength is commonly defined as the stress at the intersection of the stress–strain curve and a straight line through an offset point strain, $(\sigma, \varepsilon) = (0, 0.002)$, that is, the proof stress at 0.2% strain, that is, with $\varepsilon = 0.002$, which is parallel to the linear portion of the stress–strain curve in the elastic regime.

It is important to realize that a material's yield strength is significantly affected by operational and environmental conditions, such as temperatures and loading speed (or strain rates), among others. For structural design purposes, regulatory bodies or classification societies identify the "minimum" requirements for the mechanical properties and the chemical composition of materials. For example, the International Association of Classification Societies (IACS) specify the minimum requirements of the yield strength, ultimate tensile strength, and fracture strain (elongation) of rolled or extruded aluminum alloys for marine applications, as indicated in Tables 1.3 and 1.4 (IACS 2014). Interested readers may also refer to Sielski (2007, 2008).

1.3.1.6 Strain-Hardening Tangent Modulus, E_h, and Strain-Hardening Strain, ε_h

Beyond the yield stress or strain, the metal flows plastically without appreciable changes in stress until the strain-hardening strain ε_h is reached. The slope of the stress–strain curve in the strain-hardening regime is defined as the strain-hardening tangent modulus E_h, which may not be constant, but rather dependent on different conditions.

Strain hardening may also be characterized as the ratio of the ultimate tensile stress σ_T to the yield stress σ_Y or as the ratio of the ultimate tensile stress ε_T to the yield strain ε_Y. The stress σ beyond the yield strength of the elastic–plastic material with strain hardening is often expressed at a certain level of plastic strain as follows:

$$\sigma = \sigma_Y + \frac{EE_h}{E - E_h}\varepsilon_p \tag{1.22}$$

where ε_p is the effective plastic strain.

1.3.1.7 Ultimate Tensile Strength, σ_T

When strain exceeds the strain-hardening strain, ε_h, the stress increases above the yield stress, σ_Y, because of strain hardening, and this behavior can continue until the ultimate

Table 1.3 Minimum requirements of the mechanical properties for rolled aluminum alloys (IACS 2014).

Grade	Temper	Thickness t (mm)	σ_Y (MPa)	σ_T (MPa)	ε_F (%) $t \leq 12.5$mm	$t > 12.5$mm
5083	O	$3 \leq t \leq 50$	125	275–350	16	14
	H111	$3 \leq t \leq 50$	125	275–350	16	14
	H112	$3 \leq t \leq 50$	125	275	12	10
	H116	$3 \leq t \leq 50$	215	305	10	10
	H321	$3 \leq t \leq 50$	215–295	305–385	12	10
5383	O	$3 \leq t \leq 50$	145	290	–	17
	H111	$3 \leq t \leq 50$	145	290	–	17
	H116	$3 \leq t \leq 50$	220	305	10	10
	H321	$3 \leq t \leq 50$	220	305	10	10
5059	O	$3 \leq t \leq 50$	160	330	24	24
	H111	$3 \leq t \leq 50$	160	330	24	24
	H116	$3 \leq t \leq 20$	270	370	10	10
		$20 < t \leq 50$	260	360	–	10
	H321	$3 \leq t \leq 20$	270	370	10	10
		$20 < t \leq 50$	260	360	–	10
5086	O	$3 \leq t \leq 50$	95	240–305	16	14
	H111	$3 \leq t \leq 50$	95	240–305	16	14
	H112	$3 \leq t \leq 12.5$	125	250	8	–
		$12.5 < t \leq 50$	105	240	–	9
	H116	$3 \leq t \leq 50$	195	275	10[1]	9
5754	O	$3 \leq t \leq 50$	80	190–240	18	17
	H111	$3 \leq t \leq 50$	80	190–240	18	17
5456	O	$3 \leq t \leq 6.3$	130–205	290–365	16	–
		$6.3 < t \leq 50$	125–205	285–360	16	14
	H116	$3 \leq t \leq 30$	230	315	10	10
		$30 < t \leq 40$	215	305	–	10
		$40 < t \leq 50$	200	285	–	10
	H321	$3 \leq t \leq 12.5$	230–315	315–405	12	–
		$12.5 < t \leq 40$	215–305	305–385	–	10
		$40 < t \leq 50$	200–295	285–370	–	10

Notes:
a) 8% for $t \leq 6.3$mm.
b) The mechanical properties for the O and H111 tempers are the same, but they are separated to encourage dual certification as these tempers represent different processing.

Table 1.4 Minimum requirements of the mechanical properties for extruded aluminum alloys (IACS 2014).

					ε_F (%)	
Grade	Temper	Thickness t (mm)	σ_Y (MPa)	σ_T (MPa)	$t \leq 12.5$mm	$t > 12.5$mm
5083	O	$3 \leq t \leq 50$	110	270–350	14	12
	H111	$3 \leq t \leq 50$	165	275	12	10
	H112	$3 \leq t \leq 50$	110	270	12	10
5383	O	$3 \leq t \leq 50$	145	290	17	17
	H111	$3 \leq t \leq 50$	145	290	17	17
	H112	$3 \leq t \leq 50$	190	310	–	13
5059	H112	$3 \leq t \leq 50$	200	330	–	10
5086	O	$3 \leq t \leq 50$	95	240–315	14	12
	H111	$3 \leq t \leq 50$	145	250	12	10
	H112	$3 \leq t \leq 50$	95	240	12	10
6005A	T5	$3 \leq t \leq 50$	215	260	9	8
	T6	$3 \leq t \leq 10$	215	260	8	6
		$10 < t \leq 50$	200	250	8	6
6061	T6	$3 \leq t \leq 50$	240	260	10	8
6082	T5	$3 \leq t \leq 50$	230	270	8	6
	T6	$3 \leq t \leq 5$	250	290	6	–
		$5 < t \leq 50$	260	310	10	–

tensile strength (also simply termed tensile strength), σ_T, is reached. The value of σ_T is obtained by the maximum axial tensile load divided by the original cross-sectional area of the test specimen. Tables 1.3 and 1.4 indicate the minimum requirements of the ultimate tensile strength for rolled or extruded aluminum alloys.

1.3.1.8 Necking Tangent Modulus, E_n

With further increase in strain, a large local reduction of the cross section occurs, which is termed necking or strain softening. The internal engineering stress decreases in the necking regime. The slope of the engineering stress–engineering strain curve in the necking regime is sometimes defined as the necking tangent modulus, E_n. Necking may also be characterized as the ratio of the fracture stress σ_F to the ultimate tensile stress σ_T or as the ratio of the fracture strain ε_F to the ultimate tensile strain ε_T.

1.3.1.9 Fracture Strain, ε_F, and Fracture Stress, σ_F

Fracture takes place when the strain reaches the fracture strain (elongation or total breaking strain), ε_F. The fracture stress σ_F is defined as the stress at fracture in the necking regime. Fracture strain is also significantly affected by operational and environmental conditions, such as temperatures and loading speed (or strain rates), among other factors.

Tables 1.3 and 1.4 indicate the minimum requirements of the fracture strain for rolled or extruded aluminum alloys.

1.3.2 Elastic–Perfectly Plastic Material Model

Figure 1.9 shows the illustrative effects of strain hardening on the elastic–plastic large-deflection behavior (i.e., average stress–average strain curve) of a steel rectangular plate under uniaxial compressive loads in the longitudinal direction, as obtained by the non-linear finite element analysis. The characteristics of the strain hardening are varied as shown in Figure 1.9a in the analysis. The plate is simply supported at all four edges, keeping them straight. It is evident that the strain-hardening effect can cause the plate ulti-mate strength to be greater than that obtained by neglecting it.

For the ULS assessment of structures made of ductile materials, an elastic–perfectly plastic material model, as shown in Figure 1.10, that is, one without strain hardening or necking, is often applied because strains are usually not significant. This material model may lead to a pessimistic estimation of the characteristic value of capacity. For the ALS assessment, however, the true stress–true strain relation with strain-hardening and necking effects should be considered because large plastic strains are usually involved.

1.3.3 Characterization of the Engineering Stress–Engineering Strain Relationship

When the details of the relationship between engineering stress σ versus engineering strain ε are unavailable, but such fundamental parameters as the elastic modulus E and the yield strength σ_Y are known, the relationship between engineering stress and engineering strain can often be approximated using the Ramberg–Osgood equation, which was originally proposed for aluminum alloys (Ramberg & Osgood 1943), as follows:

$$\varepsilon = \frac{\sigma}{E} + \left(\frac{\sigma}{B}\right)^n \tag{1.23}$$

where E is the elastic modulus at the origin of the stress versus strain curve, ε is the engi-neering strain, σ is the engineering stress, and B and n are constants to be determined by experiments.

Equation (1.23) is often simplified as follows (Mazzolani 1985):

$$\varepsilon = \frac{\sigma}{E} + 0.002\left(\frac{\sigma}{\sigma_{0.2}}\right)^n \tag{1.24a}$$

where $\sigma_{0.2}$ is the proof stress at 0.2% strain, that is, with $\varepsilon = 0.002$, which is usually taken as material yield stress σ_Y, that is, $\sigma_{0.2} = \sigma_Y$, as shown in Figure 1.11. Exponent n is given as a function of $\sigma_{0.2}$ and $\sigma_{0.1}$ as follows:

$$n = \frac{\ln 2}{\ln(\sigma_{0.2}/\sigma_{0.1})} \tag{1.24b}$$

where $\sigma_{0.1}$ is the proof stress at 0.1% strain, with $\varepsilon = 0.001$.

(a)

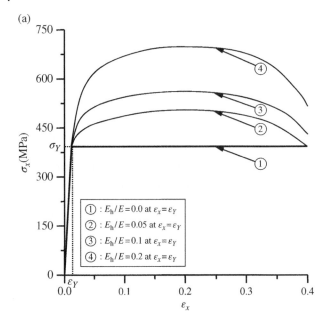

(b)

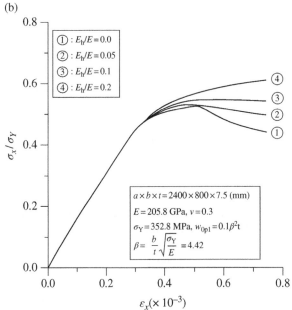

Figure 1.9 The effect of strain hardening on the ultimate strength of a steel plate under axial compression: (a) the engineering stress–engineering strain curves varying the strain-hardening characteristics; (b) a thin plate; (c) a thick plate (w_{0pl}, buckling mode initial deflection of the plate).

(c)

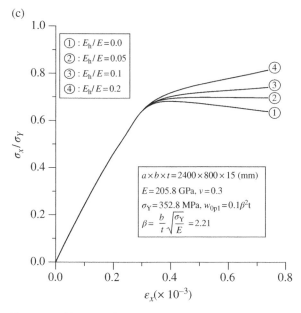

Figure 1.9 (Continued)

When the Ramberg–Osgood law is used, one practical difficulty is the determination of $\sigma_{0.1}$, in addition to E and $\sigma_{0.2}$ ($\approx \sigma_Y$). Without considering the strain-hardening effect, if the ratio $\sigma_{0.2}/\sigma_{0.1}$ approaches 1 (or $\sigma_{0.1} = \sigma_{0.2}$), the exponent becomes infinity, that is, $n = \infty$. This behavior corresponds to the elastic–perfectly plastic model of material, as illustrated in Figure 1.10, which can be expressed by

$$\varepsilon = \frac{\sigma}{E} + 0.002 \left(\frac{\sigma}{\sigma_{0.2}} \right)^{\infty} \tag{1.25}$$

For aluminum alloys, Steinhardt (1971) proposed an approximate method for determining exponent n without the value of $\sigma_{0.1}$ being known as follows:

$$0.1n = \sigma_{0.2} \left(\text{N/mm}^2 \right) \quad \text{or} \quad n = 10\sigma_{0.2} \tag{1.26}$$

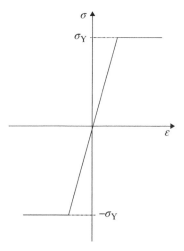

Figure 1.10 The elastic–perfectly plastic model of material.

1.3.4 Characterization of the True Stress–True Strain Relationship

For structural materials, the engineering stress–engineering strain relationship can be converted to the true stress–true strain relationship as follows:

$$\sigma_{\text{true}} = \sigma(1 + \varepsilon), \quad \varepsilon_{\text{true}} = \ln(1 + \varepsilon) \tag{1.27}$$

where σ_{true} is the true stress, $\varepsilon_{\text{true}}$ is the true strain, σ is the engineering stress, and ε is the engineering strain.

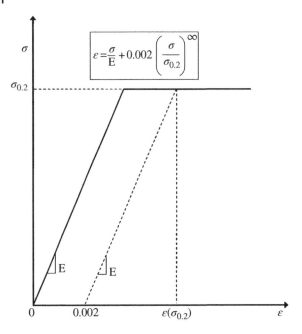

Figure 1.11 The Ramberg–Osgood law with the elastic–perfectly plastic model of material.

Figure 1.12 shows the engineering stress–engineering strain curve versus the true stress–true strain curve for mild steel and the aluminum alloy 5383-H116. It is recognized that Equation (1.27) tends to overestimate the strain-hardening and necking (strain-softening) effects. To resolve this issue, Paik (2007a, 2007b) suggested that Equation (1.27) be modified by the introduction of a knockdown factor that is a function of the engineering strain as follows:

$$\sigma_{\text{true}} = f(\varepsilon)\sigma(1+\varepsilon), \quad \varepsilon_{\text{true}} = \ln(1+\varepsilon) \tag{1.28a}$$

$$f(\varepsilon) = \begin{cases} \dfrac{C_1-1}{\ln(1+\varepsilon_T)}\ln(1+\varepsilon)+1 & \text{for } 0 < \varepsilon \leq \varepsilon_T \\[3mm] \dfrac{C_2-C_1}{\ln(1+\varepsilon_F)-\ln(1+\varepsilon_T)}\ln(1+\varepsilon)+C_1-\dfrac{(C_2-C_1)\ln(1+\varepsilon_T)}{\ln(1+\varepsilon_F)-\ln(1+\varepsilon_T)} & \text{for } \varepsilon_T < \varepsilon \leq \varepsilon_F \end{cases}$$

$$\tag{1.28b}$$

where $f(\varepsilon)$ is the knockdown factor as a function of the engineering strain, ε_F is the material's fracture strain (elongation), ε_T is the strain at the ultimate tensile stress, and C_1 and C_2 are the test constants affected by material type and plate thickness, among other factors.

Although the knockdown factor is governed by the characteristics of the material type and plate thickness, the test constants may be given as $C_1 = 0.9$ and $C_2 = 0.85$ for mild and high-tensile steel (Paik 2007a, 2007b). Figure 1.13 compares the original true stress–true strain curve versus the modified (knocked-down) true stress–true strain curve of mild steel and the aluminum alloy 5383-H116, where the constants $C_1 = 0.9$ and $C_2 = 0.85$ are applied for both mild steel and the aluminum alloy.

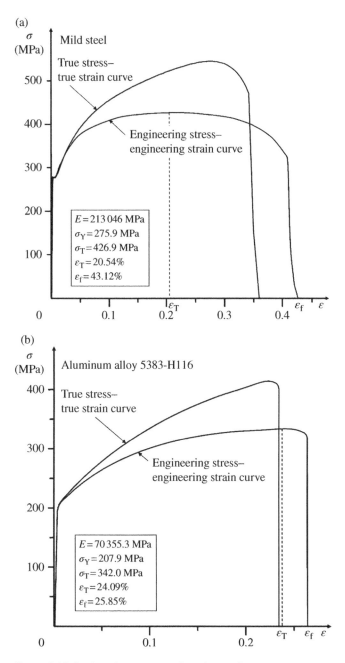

Figure 1.12 Engineering stress–engineering strain curve versus true stress–true strain curve for materials: (a) mild steel; (b) aluminum alloy 5383-H116.

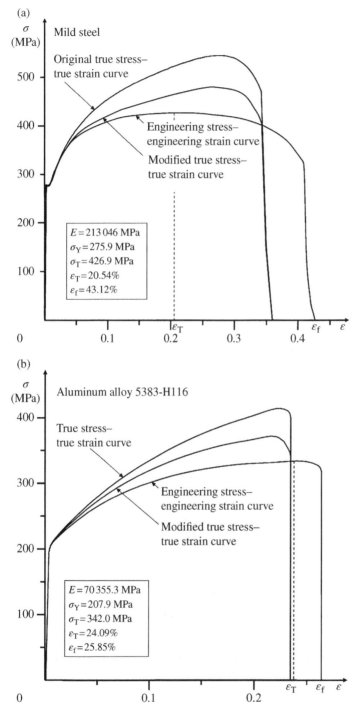

Figure 1.13 The original true stress–true strain curve versus the modified true stress–true strain curve for materials: (a) mild steel; (b) aluminum alloy 5383-H116.

1.3.5 Effect of Strain Rates

A material's mechanical properties are significantly affected by loading speed or strain rates $\dot{\varepsilon}$, which can be determined in an approximate fashion by assuming that the initial speed V_0 of the dynamic loads is linearly reduced to zero until the loading is finished, with average displacement δ, namely,

$$\dot{\varepsilon} = \frac{V_0}{2\delta} \tag{1.29}$$

In structural crashworthiness and/or impact response analysis, strain rate sensitivity plays an important role. Therefore, material modeling in terms of the dynamic yield strength and dynamic fracture strain must be considered. Figure 1.14 shows the engineering stress–engineering strain curves with varying strain rates obtained from experiments with mild steel (Grade A) and aluminum alloy 5083-O at room temperature, respectively (Paik et al. 2017).

As described in Section 10.3.2, the dynamic yield strength is often determined from the following Cowper–Symonds equation (Cowper & Symonds 1957):

$$\sigma_{Yd} = \left\{ 1 + \left(\frac{\dot{\varepsilon}}{C} \right)^{1/q} \right\} \sigma_Y \tag{1.30a}$$

where σ_Y is the static yield stress, σ_{Yd} is the dynamic yield stress, $\dot{\varepsilon}$ is the strain rate (1/s), and C and q are test constants, which may be taken as $C = 40.4/\text{s}$, $q = 5$ for mild steel, $C = 3200/\text{s}$, $q = 5$ for high-tensile steel, and $C = 6500/\text{s}$, $q = 4$ for aluminum alloys (Paik & Thayamballi 2007, Jones 2012, Paik et al. 2017).

The dynamic fracture strain is taken as the inverse of the Cowper–Symonds equation for the dynamic yield strength as follows:

$$\varepsilon_{Fd} = \left\{ 1 + \left(\frac{\dot{\varepsilon}}{C} \right)^{1/q} \right\}^{-1} \varepsilon_F \tag{1.30b}$$

where ε_F is the static fracture strain and ε_{Fd} is the dynamic fracture strain. It is noted that the test constants C and q for the dynamic fracture strain are different from those for the dynamic yield strength as described in Section 10.3.3.

Figures 1.15 and 1.16 show the effects of strain rates combined with cold temperatures on the yield strength or fracture strain obtained from experiments for mild steel, high-tensile steel, and aluminum alloy 5083-O, obtained from the experiments by Paik et al. (2017).

1.3.6 Effect of Elevated Temperatures

A material's mechanical properties are significantly decreased with elevated temperatures from operational and environmental conditions or accidents such as fires because the material's properties are associated with its thermal characteristics. Figure 1.17a shows the specific heat of steel, which varies with elevated temperature. The reduction factors of the proportional limit, Young's modulus, and yield strength for steel are indicated in Table 1.5 according to the ECCS Eurocode design manuals (Franssen & Real 2010).

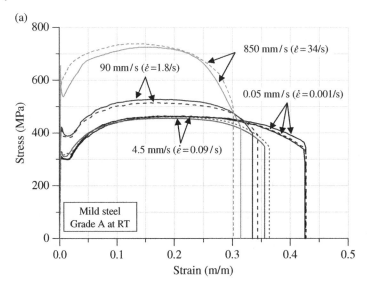

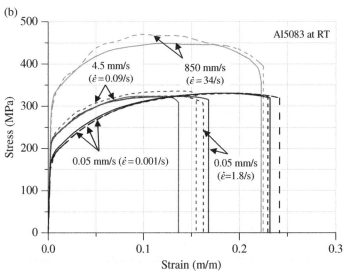

Figure 1.14 Engineering stress–engineering strain curves with different strain rates at room temperature (RT): (a) for mild steel (Grade A); (b) aluminum alloy 5083-O (Paik et al. 2017).

Figure 1.17b plots Table 1.5, showing that the mechanical properties of steel significantly decrease at temperatures above 400°C.

1.3.7 Effect of Cold Temperatures

The mechanical properties of materials are significantly affected by cold temperatures, which may be caused by operational conditions due to liquefied petroleum or natural gas

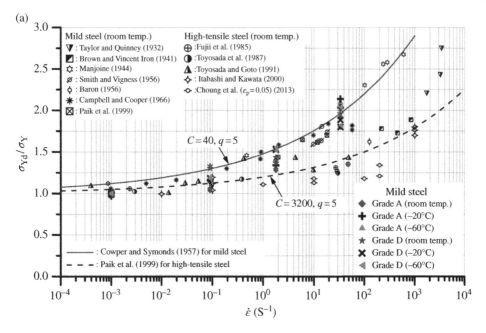

(a)

Mild steel (room temp.)
▼ : Taylor and Quinney (1932)
◪ : Brown and Vincent Iron (1941)
✿ : Manjoine (1944)
∅ : Smith and Vigness (1956)
◊ : Baron (1956)
✳ : Campbell and Cooper (1966)
⊠ : Paik et al. (1999)

High-tensile steel (room temp.)
⊕ :Fujii et al. (1985)
◐ :Toyosada et al. (1987)
▲ :Toyosada and Goto (1991)
◇ : Itabashi and Kawata (2000)
⬦ :Choung et al. ($\varepsilon_p = 0.05$) (2013)

$C = 40, q = 5$

$C = 3200, q = 5$

Mild steel
◆ Grade A (room temp.)
✚ Grade A (−20°C)
▲ Grade A (−60°C)
★ Grade D (room temp.)
✖ Grade D (−20°C)
◀ Grade D (−60°C)

——— : Cowper and Symonds (1957) for mild steel
- - - : Paik et al. (1999) for high-tensile steel

$\dot{\varepsilon}$ (S^{-1})

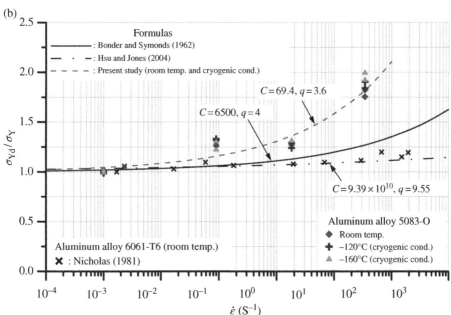

(b)

Formulas
——— : Bonder and Symonds (1962)
— · — : Hsu and Jones (2004)
- - - : Present study (room temp. and cryogenic cond.)

$C = 69.4, q = 3.6$

$C = 6500, q = 4$

$C = 9.39 \times 10^{10}, q = 9.55$

Aluminum alloy 5083-O
◆ Room temp.
✚ −120°C (cryogenic cond.)
▲ −160°C (cryogenic cond.)

Aluminum alloy 6061-T6 (room temp.)
✖ : Nicholas (1981)

$\dot{\varepsilon}$ (S^{-1})

Figure 1.15 Effect of strain rates and cold temperatures on yield strength of materials: (a) mild steel and high-tensile steel; (b) aluminum alloy 5083-O. (Cited references are from Paik et al. 2017.)

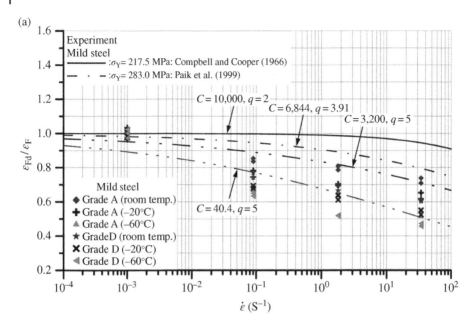

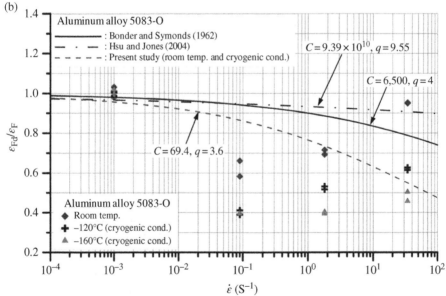

Figure 1.16 Effect of strain rates and cold temperatures on fracture strain of materials: (a) mild steel and high-tensile steel; (b) aluminum alloy 5083-O. (Cited references are from Paik et al. 2017.)

cargoes and by environmental conditions due to Arctic operations. Figures 1.18 and 1.19 show the combined effects of cold temperatures and strain rates on the yield strength or fracture strain of mild steel (Grade A) and aluminum alloy 5083-O, obtained from the experiments by Paik et al. (2017).

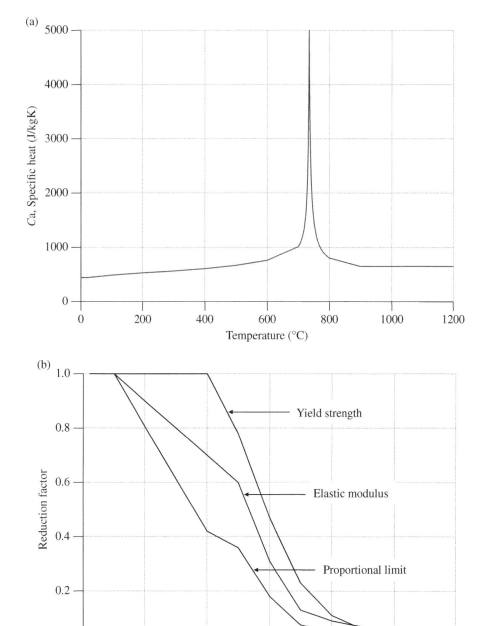

Figure 1.17 Effects of elevated temperature on properties of steel: (a) specific heat (ECCS 1982); (b) mechanical properties.

Table 1.5 Reduction factors of mechanical properties for carbon steels at elevated temperatures.

Steel temperature (°C)	Reduction factors at temperature relative to value of σ_Y, σ_P, or E at 20°C		
	σ_Y	σ_P	E
20	1.000	1.0000	1.0000
100	1.000	1.0000	1.0000
200	1.000	0.8070	0.9000
300	1.000	0.6130	0.8000
400	1.000	0.4200	0.7000
500	0.780	0.3600	0.6000
600	0.470	0.1800	0.3100
700	0.230	0.0750	0.1300
800	0.110	0.0505	0.0900
900	0.060	0.0375	0.0675
1000	0.040	0.0250	0.0450
1100	0.020	0.0125	0.0225
1200	0.000	0.0000	0.0000

Note: For intermediate values of the steel temperature, a linear interpolation may be used.

1.3.8 Yield Condition Under Multiple Stress Components

For a one-dimensional strength member under uniaxial tensile or compressive loading, the yield strength determined from a uniaxial tension test can be used to check the state of yielding, with the essential question to be answered being simply whether the axial stress reaches the yield strength.

A plate element that is the principal strength member of a steel- or aluminum-plated structure is likely to be subjected to a combination of biaxial tension/compression and shear stress, which can usually be considered to be in a plane stress state (as contrasted to a state of plane strain).

For an isotropic two-dimensional structural member for which the dimension in one direction is much smaller than those in the other two directions, and with three in-plane stress components (i.e., two normal stresses, σ_x, σ_y, and shear stress, τ_{xy}) or, equivalently, two principal stress components (i.e., σ_1, σ_2), three types of yield criteria are usually adopted as follows:

1) Maximum principal stress-based criterion: The material yields if the maximum absolute value of the two principal stresses reaches a critical value, namely,

$$\max(|\sigma_1|, |\sigma_2|) = \sigma_Y \tag{1.31a}$$

2) Maximum shear stress-based criterion (also called the Tresca criterion): The material yields if the maximum shear stress, τ_{max}, reaches a critical value, namely,

$$\tau_{max} = \left| \frac{\sigma_1 - \sigma_2}{2} \right| = \frac{\sigma_Y}{2} \tag{1.31b}$$

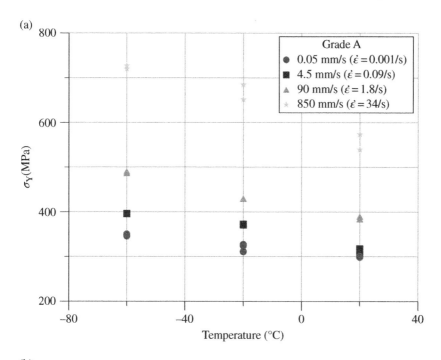

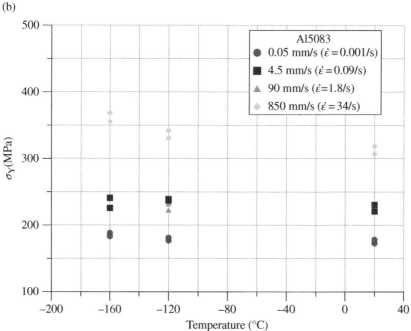

Figure 1.18 Effect of cold temperatures and strain rates on yield strength of materials: (a) mild steel (Grade A); (b) aluminum alloy 5083-O (Paik et al. 2017).

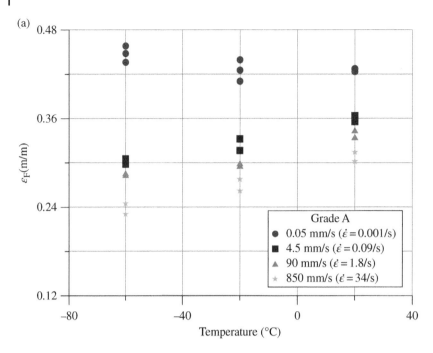

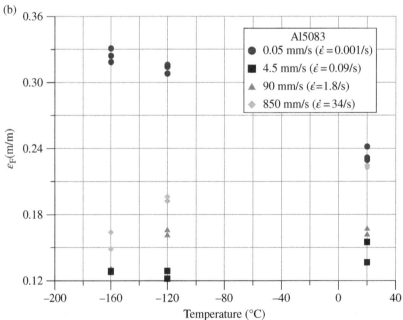

Figure 1.19 Effect of cold temperatures and strain rates on fracture strain of materials: (a) mild steel (Grade A); (b) aluminum alloy 5083-O (Paik et al. 2017).

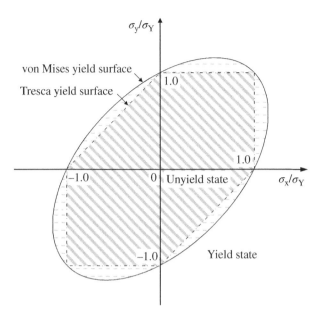

Figure 1.20 The von Mises and Tresca yield surfaces associated with two normal stress components.

3) Strain energy-based criterion (also called the Mises–Hencky or Huber–Hencky–Mises or von Mises criterion): The material yields if the strain energy due to geometric changes reaches a critical value, which corresponds to that at which the equivalent stress, σ_{eq}, reaches the yield strength, σ_Y, as determined from the uniaxial tension test as follows:

$$\sigma_{eq} = \sqrt{\sigma_x^2 - \sigma_x \sigma_y + \sigma_y^2 + 3\tau_{xy}^2} = \sigma_Y \tag{1.31c}$$

where σ_Y is the yield strength of material.

   It is recognized that the first yield condition, Equation (1.31a), is relevant for a brittle material and that the last two conditions, Equations (1.31b) and (1.31c), are more appropriate for a ductile material, although the von Mises condition, Equation (1.31c), is more popular for the analysis of plated structures. Figure 1.20 illustrates the von Mises and Tresca yield surfaces associated with two normal stress components, σ_x and σ_y. The shear yield stress, τ_Y, under pure shear can be determined by solving the von Mises condition, Equation (1.31c), with regard to τ_{xy} when $\sigma_x = \sigma_y = 0$, with the result as follows:

$$\tau_Y = \frac{\sigma_Y}{\sqrt{3}} \tag{1.31d}$$

1.3.9 The Bauschinger Effect: Cyclic Loading

During operation, structural members are likely to be subjected to load cyclic effects, as shown in Figure 1.21. If a material that has been plastically strained in tension is unloaded and then strained in compression, the stress–strain curve for the compression loading

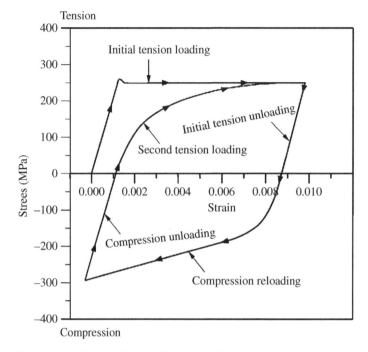

Figure 1.21 The Bauschinger effect in metals.

deviates from a linear relationship at stresses well below the yielding point of the virgin material, but it returns to the point of maximum stress and strain for the first tension loading cycle. The same effect is observed for the opposite loading cycle, that is, compression before tension. In this case, the modulus of elasticity is reduced, as shown by the shape of the stress–strain curve in Figure 1.21. This phenomenon is typically termed the Bauschinger effect (Brockenbrough & Johnston 1981). When stiffness is of primary concern, for example, in the evaluation of buckling or deflection, the Bauschinger effect may be of interest.

Within an acceptable level of accuracy, however, the mechanical properties of a particular type of steel or aluminum alloy as determined by uniaxial tension testing are also approximately accepted as being valid for the same type of the material under uniaxial compression.

1.3.10   Limits of Cold Forming

Cold forming is an efficient technique to form structural shapes, for example, a curved plate. However, it is important to realize that excessive strain during cold forming can exhaust ductility and cause cracking. Hence the strain in cold forming the structural shapes must be limited, not only to prevent cracking but also to prevent buckling collapse of structural elements subject to compressive loads. The cold-forming-induced strain is usually controlled by requiring the ratio of the bending radius to the plate thickness to be large, in the range of 5–10.

1.3.11 Lamellar Tearing

In most cases of plated structures, the behavior in the length and breadth of the plates related to load effects is of primary concern. The behavior in the wall thickness direction is normally not of interest. In heavy, welded structures, particularly in joints or connections with thick plates and heavy structural shapes, however, crack-type separation or delamination can take place in the wall thickness direction beneath the surface of plates or at weld toes. This failure is typically caused by large through-thickness strain, which is sometimes associated with weld metal shrinkage in highly restrained joints. This phenomenon is termed lamellar tearing. Careful selection of weld details, filler metal, and welding procedure and the use of steels with controlled through-thickness properties (e.g., the so-called Z grade steels) can be effective to control this failure mode.

1.4 Strength Member Types for Plated Structures

The geometric configuration of a steel- or aluminum-plated structure is determined primarily on the basis of the function of the particular structure. Figure 1.22 shows a basic part of a typical plated structure. A major difference between plated and framed structures is that the principal strength members of the former type of structure are plate panels together with support members, whereas those of the latter typically consist of truss or beam members for which the dimension in the axial direction is usually much greater than those in the other two directions.

Typical examples of plated structures are ships, ship-shaped offshore platforms, box girder bridges, and box girder cranes. Basic types of structural members that usually make up plated structures are as follows:

- Plate panels: Plating, stiffened panel, corrugated panel
- Small support members: Stiffener, beam, column, beam–column
- Strong main support members: Plate girder, frame, floor, bulkhead, box girder

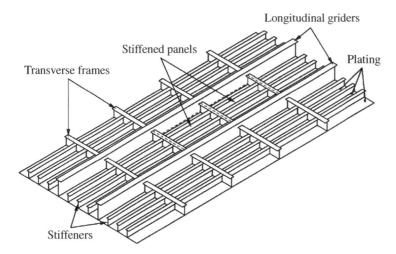

Figure 1.22 Typical plated structure.

To improve the stiffness and strength of plate panels, increasing the stiffener dimensions is usually more efficient than simply increasing the plate thickness, and thus the plate panel is usually reinforced by beam members (stiffeners) in the longitudinal or transverse direction. Figure 1.23a shows typical beam members used to stiffen the plating. A self-stiffened plate, such as the corrugated panel shown in Figure 1.23b, may also be used in some cases.

When the stiffened panels are likely to be subjected to lateral loads or out-of-plane bending or just require lateral support, they are supported by stronger beam members. Figure 1.23c shows typical strong main support members used to build plated structures. For ships and offshore structures, plate girders composed of deep webs and wide flanges are typically used for main support members. The deep web of a plate girder is often stiffened vertically and/or horizontally. Box-type support members that consist of plate panels are used for construction of land-based steel bridges or cranes. Diaphragms or transverse floors or transverse bulkheads are arranged at relevant spaces in the box girder.

Although plating primarily sustains in-plane loads, support members resist out-of-plane (lateral) loads and bending. A plate panel between stiffeners is called "plating," and plating with stiffeners is termed a "stiffened panel." A cross-stiffened panel is termed a "grillage," which in concept is essentially a set of intersecting beam members. When a one-dimensional strength member is predominantly subjected to axial compression, it is called a "column," whereas it is termed a "beam" when subjected to lateral loads or

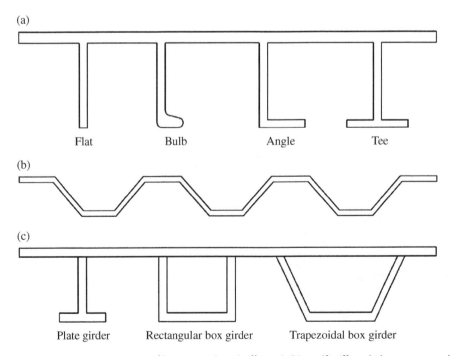

(a)

Flat   Bulb   Angle   Tee

(b)

(c)

Plate girder   Rectangular box girder   Trapezoidal box girder

Figure 1.23 (a) Various types of beam members (stiffeners); (b) a self-stiffened plate-corrugated panel; (c) various types of strong main support members.

bending. A one-dimensional strength member under combined axial compression and bending is called a "beam–column." When the strength member is subjected to combined bending and axial tension, it is called a "tension-beam."

Strong main support members are normally called "(longitudinal) girders" when they are located in the primary loading direction (i.e., the longitudinal direction in a box girder or a ship hull girder), whereas they are sometimes called "(transverse) frames" or main support members when they are located in a direction orthogonal to the primary load direction (i.e., in the transverse direction in a box girder or a ship hull girder).

For strength analysis of plated structures, stiffeners or some support members together with their associated plating are often modeled as beams, columns, or beam–columns, as described in Chapter 2.

1.5 Types of Loads

The terminology related to the classification of applied loads for ships and offshore structures is similar to that used for land-based structures. The types of loads to which plated structures or strength members are likely to be subjected may be categorized into the following four groups:

- Dead loads
- Operational or service (live) loads
- Environmental loads
- Accidental loads

Dead loads (also called permanent loads) are time-independent, gravity-dominated service loads. Examples of dead loads are the weight of structures or permanent items that remain in place throughout the life of the structure. Dead loads are typically static and can usually be determined accurately even if the weight of some of the items may in some cases be unknown until the structural design has been completed.

Operational or service loads are typically live loads by nature with gravity and/or thermal loads that vary in magnitude and location during the normal operation of the structure. Operational loads can be quasistatic, dynamic, or even impulsive in loading speed. Examples of operational loads are the weight of people, furniture, movable equipment, wheel loads from vehicles or cargoes, and stored consumable goods. In marine structures, pressure loads due to water and cargoes and thermal loads due to cargoes (e.g., liquefied petroleum gas, liquefied natural gas) are also examples of operational loads. In the design of land-based box girder bridges, highway vehicle loading is usually separately classified under highway live loads. Although some live loads (e.g., persons and furniture) are practically permanent and static, others (e.g., box girder cranes and various types of machinery) are highly time dependent and dynamic. Because the magnitude, location, and density of live load items are generally unknown in a particular case, the determination of operational loads for design purposes is not straightforward. For this reason, regulatory bodies sometimes prescribe design service loads based on experience and proven practice.

Environmental loads are actions related to wind, current, waves, snow, and earthquake. Most environmental loads are time dependent and repeated in some fashion, that is, cyclic. Environmental loads can thus be quasistatic, dynamic, or even impulsive in loading speed. The determination of design environmental loads is often specified by regulatory bodies or classification society rules, typically using the concept of a mean return period. The design loads of snow or wind, for instance, may be specified based on a return period of 100 years or longer, indicating that extreme snowfall or wind velocity that is expected to occur once in 100 years is used in the design.

Accidental loads are actions that arise from accidents such as collision, grounding, fire, explosion, or dropped objects. Accidental loads typically have a dynamic or impact effect on structural behavior with large strains. Guidelines to predict and account for accidental loads are more meager because of the unknown nature of accidents. However, it is important to treat such loads in design, particularly when novel types of structures are involved, about which experience may be lacking. This often happens in the offshore field, where several new types of structures have been introduced in recent decades. Experimental databases in a full-scale prototype or at least large-scale models are highly required to characterize and quantify the nonlinear mechanics of structures exposed to accidental conditions, as scaling laws to convert small-scale model test results to the actual full-scale structure are not always available.

The maxima of the various types of loads mentioned previously are not always applied simultaneously, but more than one type of load normally may coexist and interact. Therefore, the structural design must account for the effects of phasing for definition of the combined loads. Usually, this involves the consideration of multiple load combinations for design, each representing a load at its extreme value together with the accompanying values of other loads. The guidelines for relevant combinations of loads to be considered in design are usually specified by regulatory bodies or classification societies for particular types of structures.

1.6   Basic Types of Structural Failure

This book is concerned with the fundamentals and practical procedures for the ULS analysis and design of steel- and aluminum-plated structures. One primary task in ULS design is to determine the level of imposed loads that cause the structural failure of individual members and the overall structure. Therefore, it is crucial to better understand what types of structural failure can primarily occur. The failure of plated structures made of ductile materials is normally related to one or both of the following nonlinear types of behavior:

- Geometric nonlinearity associated with buckling or large deflection
- Material nonlinearity due to yielding or plastic deformation

For structural members, many basic types of failure are considered, the more important of which include:

- Buckling or instability
- Plasticity in local regions
- Fatigue cracking related to cyclic loading

- Ductile or brittle fracture, given fatigue cracking or preexisting defects
- Excessive deformations

The basic failure types mentioned previously do not always occur simultaneously, but more than one phenomenon may in principle be involved until the structure reaches the ULS. For convenience, the basic types of structural failure noted previously are sometimes described and treated separately.

As the external loads increase, the most highly stressed region inside a structural member will yield first, resulting in local plastic deformation, which decreases the member stiffness. With a further increase in the load, local plastic deformation will increase and/or occur at several different regions. The stiffness of the member with large local plastic regions becomes quite low, and the displacements increase rapidly, eventually becoming so large that the member is considered to have failed.

Buckling or instability can occur in any structural member that is predominantly subjected to load sets that result in compressive effects in the structure. In buckling-related design, two types of buckling are considered, bifurcation and non-bifurcation. The former type is seen for an ideal perfect member without initial imperfections, and the latter typically occurs in an actual member with some initial imperfections. For instance, a straight elastic column has an alternative equilibrium position at a critical axial compressive load that causes a bent shape to suddenly occur at a certain value of the applied load. This threshold load, which separates into two different equilibrium conditions, is called a bifurcation load.

An initially deflected column or beam–column induces bending from the beginning of the loading contrary to the straight column, and the lateral deflection increases progressively. The member stiffness is reduced by considerable deflection and local yielding, and it eventually becomes zero at a peak load. The deflection of the member with very low or zero stiffness becomes so great that the member is considered to have collapsed. In this case, an obvious sudden buckling point does not appear until the member collapses; this type of failure is called non-bifurcation instability or limit-load buckling (Galambos 1988).

Due to repeated fluctuation of loading, fatigue cracking can initiate and propagate in the structure's stress concentration areas. Fracture is a type of structural failure caused by the rapid extension of cracks. Three types of fracture are relevant, brittle fracture, rupture, and ductile fracture. Brittle fracture normally takes place at a very small strain in materials with a low toughness or below a certain temperature, when the material's ultimate tensile strength diminishes sharply. For materials with a very high toughness, rupture occurs at a very large strain by necking of the member, typically at room temperature or higher. Ductile fracture is an intermediate fracture mode between brittle fracture and rupture. In steels or aluminum alloys, the tendency to fracture is related not only to the temperature but also to the rate at which loading is applied. The higher the loading rate, the greater the tendency toward brittle fracture.

1.7 Fabrication Related Initial Imperfections

Welded metal structures always have initial imperfections in the form of initial distortions, residual stresses, or softening in the weld fusion zone or HAZ. Because such fabrication related initial imperfections may affect the structural properties and load-carrying capacities of structures, they must be dealt with as parameters of influence in structural analysis and design.

1.7.1 Mechanism of Initial Imperfections

When local heating is input to structural material, the heated part will expand, but because of adjacent cold parts, it will be subjected to compressive stress and distortion. When the heated part cools, it will locally shrink rather than revert to its initial shape and will thus be subjected to tensile stress. The strength of welded aluminum alloys in the HAZ is reduced by softening phenomenon in that melting temperature is reduced by improving fluidity, while it is recognized that the material strength in the softened zone is recovered by natural aging over a period of time (Lancaster 2003).

Experimental studies to examine the mechanism of initial imperfections have been undertaken in the literature with direct measurements of welding induced initial imperfections: Masubuchi (1980), Smith et al. (1988), Ueda (1999), and Paik and Yi (2016) for initial distortions of steel-plated structures; Paik et al. (2006), Paik (2007c, 2008), and Paik et al. (2008, 2012) for initial distortions of aluminum-plated structures; Masubuchi (1980), Smith et al. (1988), Cheng et al. (1996), Ueda (1999), Kenno et al. (2010, 2017), and Paik and Yi (2016) for residual stresses of steel-plated structures; and Paik et al. (2006), Paik (2008), and Paik et al. (2008, 2012) for residual stresses and softening of aluminum-plated structures, among others.

Based on the insights available in the literature, it is recognized that various types of welding induced distortions are relevant, as shown in Figure 1.24. In practice for the evaluation of structural capacity, both angular change and longitudinal bending distortion are of greater concern, as shown in Figure 1.24e, whereas the shrinkage in the longitudinal or transverse directions may often be neglected. Also, residual stress distributions in welded structural members represent the tensile residual stresses that develop in the HAZ and the compressive residual stresses that must then also exist to achieve self-equilibrium in the plane of the structural member, as shown in Figure 1.25. In welded aluminum structures with extruded stiffeners, residual stresses are developed as those shown in Figure 1.26. The breadth of the softened zone in welded aluminum structures almost equals that of the HAZ, as shown in Figure 1.27.

Figure 1.28 shows experimental and finite element method investigations of welding induced initial distortions and residual stresses in full-scale welded steel-stiffened plate structure models, obtained by Paik and Yi (2016) using modern technologies of the fabrication and measurements. Figure 1.29 shows experimental investigations of welding induced initial distortions and residual stresses in full-scale fusion-welded aluminum-stiffened plate structure models, obtained by Paik (2008).

1.7.2 Initial Distortion Modeling

Figure 1.30 shows some typical initial deflection shapes of welded one-dimensional members and their possible idealizations. For practical design purposes, the initial deflection shape of a welded one-dimensional member may be idealized as the dotted

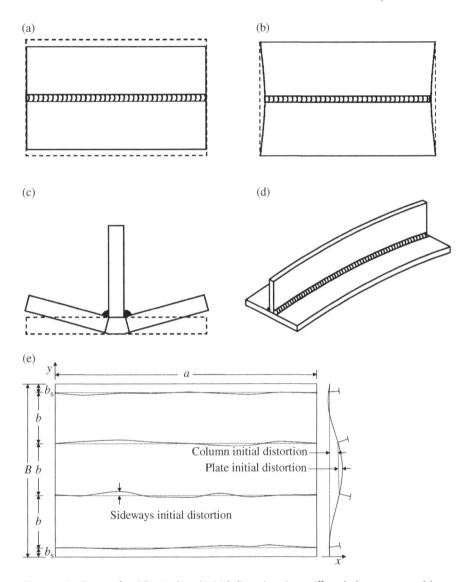

Figure 1.24 Types of welding induced initial distortions in a stiffened plate structure: (a) transverse shrinkage; (b) longitudinal shrinkage; (c) angular change; (d) longitudinal bending distortion; (e) three typical distortions.

line in Figure 1.30, which can be expressed approximately in mathematical form as follows:

$$w_0 = \delta_0 \sin\frac{\pi x}{L} \tag{1.32}$$

where w_0 is the initial deflection function; δ_0 is the initial deflection amplitude, which is often taken as $0.0015L$ for a practical strength calculation at an "average" level of imperfections; and L is the member length between supports.

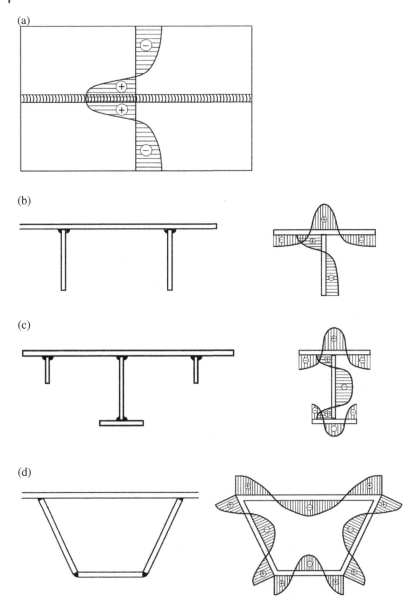

Figure 1.25 Distribution of the welding induced residual stresses in a stiffened plate structure: (a) butt-welded plate; (b) welded stiffened panel; (c) welded plate girder; (d) welded box section.

For welded stiffened plate structures, three types of initial distortions are relevant to welded metal-stiffened plate structures, as illustrated in Figure 1.31:

- The initial deflection w_{0pl} of the plating between the support members
- The column-type initial deflection w_{0c} of the support members
- The sideways initial deflection w_{0s} of the support members

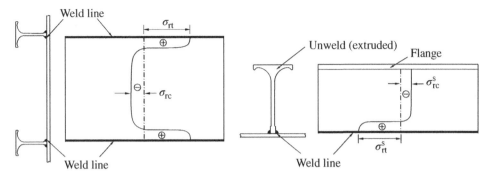

Figure 1.26 Distribution of the welding induced residual stresses in an aluminum plate welded at two edges and a stiffener web welded at one edge (+, tension, −, compression; left, plating, right, extruded stiffener web) (Paik et al. 2012, Hughes & Paik 2013).

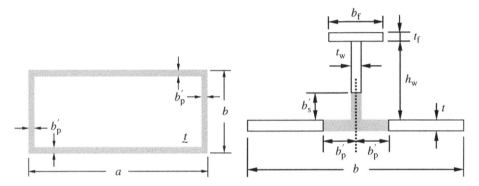

Figure 1.27 Breadths of the softening zones inside an aluminum plate welded at four edges and its counterpart in the extruded stiffener attachment to the plating (Hughes & Paik 2013).

The magnitude and shape of each type of initial distortion play important roles in buckling collapse behavior, and thus a better understanding of the actual imperfection configurations in the target structures is necessary (Paik et al. 2004). In fact, it is desirable to obtain precise information about the initial distortions of the target structure before structural modeling even begins. Considering the significant amount of uncertainty involved in fabrication related initial imperfections, existing measurements of the initial distortions in welded metal structures are often useful for the development of representative models.

1.7.2.1 Plate Initial Deflection

The shape of welding induced initial deflections for thin plates after support members are attached by welding is quite complex. The initial deflection of plating between stiffeners can be expressed as a Fourier series function as follows:

$$\frac{w_0}{w_{0\text{pl}}} = \sum_{i=1}^{M} \sum_{j=1}^{N} B_{0ij} \sin\frac{i\pi x}{a} \sin\frac{j\pi y}{b} \tag{1.33}$$

(a)

(b)

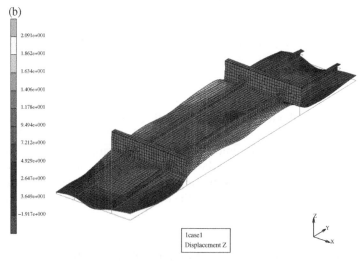

Figure 1.28 (a) Full-scale testing of initial imperfections in steel-stiffened plate structure models: (a) left: fabrication by auto flux-cored arc welding; right: three-dimensional scanning measurement of initial distortions; below: nondestructive measuring technique of welding induced residual stresses; (b) measurement of the welding induced initial distortions; (c) measurement and finite element method prediction of the welding induced residual stresses in the longitudinal stiffener direction; (d) measurement and finite element method prediction of the welding induced residual stresses in the transverse frame direction (Paik & Yi 2016).

(c)

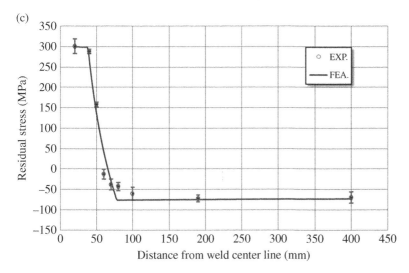

(d)

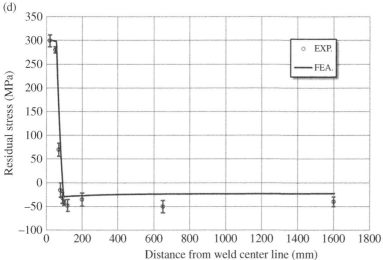

Figure 1.28 (Continued)

where a is the plate length and b is the plate breadth. B_{0ij} indicates the welding induced initial deflection amplitude normalized by the maximum initial deflection, w_{0pl}, which can be determined on the basis of the initial deflection measurements. The subscripts i and j denote the corresponding half-wave numbers in the x and y directions, respectively.

If measured databases for the initial deflection for plating are available, the initial deflection amplitudes of Equation (1.33) can be determined by expanding Equation (1.33) appropriately using a selected number of terms, M and N, depending on the complexity of the initial deflection shape.

For practical design purposes, further idealization may sometimes be necessary. The measurements of the initial deflection for plate elements in plated structures show that a

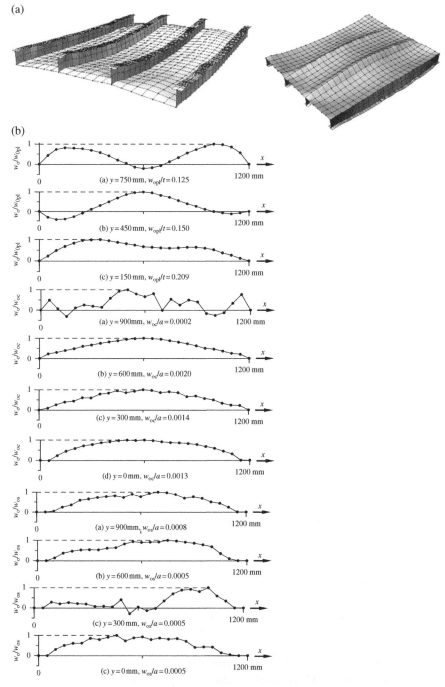

Figure 1.29 The welding induced initial distortions (amplified by 30 times) in a welded aluminum-stiffened plate structure: (a) shape of initial distortions (amplified by 30 times); (b) measurement of initial distortions (w_{0pl} = plate initial deflection; w_{0c} = column-type initial distortion of stiffener; w_{0s} = sideways initial distortion of stiffener); (c) measurement of residual stresses (Paik 2008).

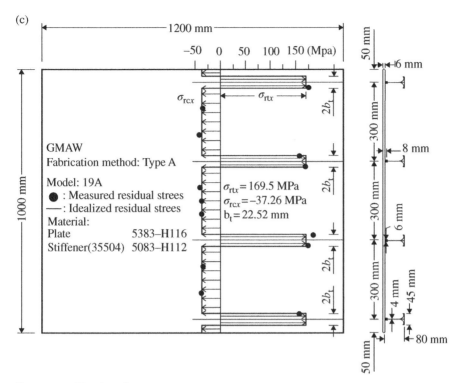

Figure 1.29 (Continued)

Figure 1.30 Idealization of initial deflection shapes for welded one-dimensional members.

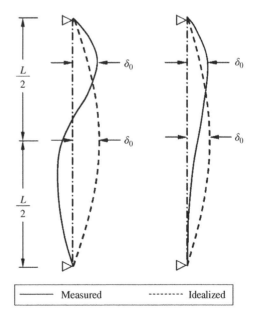

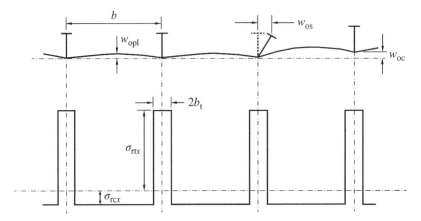

Figure 1.31 Three types of the welding induced initial distortions and residual stresses in a stiffened plate structure.

multiple-wave shape is predominant in the long direction, as shown in Figure 1.32, whereas one half wave is found in the short direction, as shown in Figure 1.31.

For a nearly square plate element, therefore, Equation (1.33) may be simplified by taking $M = N = 1$. For a long plate element with a multiple-wave shape in the x direction and one half wave in the y direction, Equation (1.33) becomes

$$\frac{w_0}{w_{0pl}} = \sum_{i=1}^{M} B_{0i}\sin\frac{i\pi x}{a}\sin\frac{\pi y}{b} \tag{1.34}$$

In practice, M in Equation (1.34) may be taken as an integer that corresponds to about three or more times the a/b ratio greater than 1 (Paik & Pedersen 1996). On this basis, B_{0i} in Equation (1.34) can be determined for the assumed M if the initial deflection measurements are available. The values of coefficients, B_{0i}, for the initial deflection shapes shown in Figure 1.32 are given in Table 1.6, by taking $M = 11$.

In current industry practice with regard to practical structural design and strength assessment, an average magnitude is assumed for these initial distortions, and their shape is assumed to be the buckling mode, because this shape usually has the most unfavorable consequences for the structure until and after the ULS is reached. The amplitude or maximum magnitude w_{0pl} of plate initial deflection w_0^P is often assumed to be the following:

$$w_0^P = w_{0pl}\sin\frac{m\pi x}{a}\sin\frac{\pi y}{b} \tag{1.35a}$$

$$w_{0pl} = C_1 b \tag{1.35b}$$

$$w_{0pl} = C_2\beta^2 t \tag{1.35c}$$

where w_0^P is the plate's initial deflection function, w_{0pl} is the maximum magnitude of plate initial deflection, b is the plate breadth along the short edge or the spacing between the

(a)

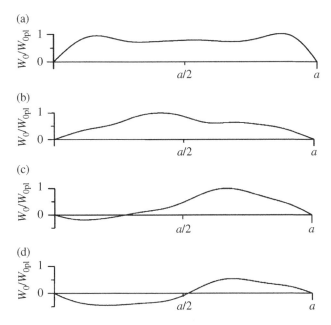

(b)

(c)

(d)

Figure 1.32 Some typical initial deflection patterns in welded plating between stiffeners in the long (plate length) direction (Paik & Pedersen 1996): initial deflection shape (a) 1; (b) 2; (c) 3 and (d) 4.

longitudinal stiffeners, t is the plate thickness, $\beta = (b/t)\sqrt{\sigma_Y/E}$ is the plate slenderness ratio, E is the material's elastic modulus, σ_Y is the yield strength, C_1 and C_2 are constants, and m is the buckling half-wave number of the plate.

It is interesting to note that the two alternative formulations, that is, Equations (1.35b) and (1.35c), have different usage backgrounds. Equation (1.35b), supported by some classification societies, states that w_{opl} is a function only of the plate breadth, whereas Smith et al. (1988) suggested that Equation (1.35c) gives a more precise representation of the plate characteristics because it is a function of the plate slenderness ratio.

In addition, the use of Equation (1.35b) may result in too small initial deflection for very thin plates and too large initial deflection for very thick plates. Equation (1.35c), in contrast, is suitable for both very thin and very thick plates. Nevertheless, the use of Equation (1.35b) remains more popular today in the construction of ships and offshore structures, as long as a moderate plate thickness is considered. This is partly because Equation (1.35b) is more suitable to specify construction tolerances regardless of the slenderness ratio-related characteristics of the plating.

The constants in Equations (1.35b) and (1.35c) may be determined on the basis of statistical analyses of the initial deflection measurements of the welded steel or aluminum plates. The following provides some additional guidance:

$C_1 = 0.005$ for an average level in steel plates

$$C_1 = \begin{cases} 0.0032 & \text{for a slight level} \\ 0.0127 & \text{for an average level} \\ 0.0290 & \text{for a severe level} \end{cases} \text{in aluminum plates (Paik 2007c)}$$

Table 1.6 Initial deflection amplitudes of Equation (1.35a) for various initial deflection shapes indicated in Figure 1.32.

Initial deflection shape	B_{01}	B_{02}	B_{03}	B_{04}	B_{05}	B_{06}	B_{07}	B_{08}	B_{09}	B_{010}	B_{011}
1	1.0	-0.0235	0.3837	-0.0259	0.2127	-0.0371	0.0478	-0.0201	0.0010	-0.0090	0.0005
2	0.8807	0.0643	0.0344	-0.1056	0.0183	0.0480	0.0150	-0.0101	0.0082	0.0001	-0.0103
3	0.5500	-0.4966	0.0021	0.0213	-0.0600	-0.0403	0.0228	-0.0089	-0.0010	-0.0057	-0.0007
4	0.0	-0.4966	0.0021	0.0213	-0.0600	-0.0403	0.0228	-0.0089	-0.0010	-0.0057	-0.0007

$$C_2 = \begin{cases} 0.025 & \text{for a slight level} \\ 0.1 & \text{for an average level} \\ 0.3 & \text{for a severe level} \end{cases} \text{ in steel plates (Smith et al. 1988)}$$

$$C_2 = \begin{cases} 0.018 & \text{for a slight level} \\ 0.096 & \text{for an average level} \\ 0.252 & \text{for a severe level} \end{cases} \text{ in aluminum plates (Paik et al. 2006)}$$

To determine the shape of the buckling mode initial distortions, eigenvalue computations are required. Based on these eigenvalue computations, the buckling modes of the stiffened plate structures can then be decomposed into the three types of initial distortions mentioned previously. Each type of initial distortion should be amplified up to the maximum target value, and the three resulting patterns should then be superimposed to provide a complete picture of the initial distortions. It is worthwhile to discuss here the classical theory of structural mechanics, which gives the buckling half-wave number of a simply supported plate element under longitudinal compression alone. This number is predicted as the minimum integer that satisfies the following condition, as described in Chapter 3 or 4:

$$\frac{a}{b} \le \sqrt{m(m+1)} \tag{1.36a}$$

where m is the number of buckling half waves of the plate in the longitudinal direction, whereas the number in the transverse direction is assumed to be unity.

The plate buckling half-wave number can then be determined under any combination of longitudinal compression σ_x and transverse compression σ_y, again as a minimum integer, but satisfying the following condition as described in Chapter 3 or 4:

$$\frac{(m^2/a^2 + 1/b^2)^2}{m^2/a^2 + c/b^2} \le \frac{\left[(m+1)^2/a^2 + 1/b^2\right]^2}{(m+1)^2/a^2 + c/b^2} \tag{1.36b}$$

where $c = \sigma_y/\sigma_x$ is the loading ratio. When $c = 0$, that is, under longitudinal compression alone, Equation (1.36b) simplifies to Equation (1.36a).

Classification societies or other regulatory bodies specify construction tolerances for strength members as related to the maximum initial deflection with the intention that the initial distortions in the fabricated structure must be less than the corresponding specified values. Some examples of the limit for the maximum plate initial deflection are as follows:

- NORSOK (2004):

$$\frac{w_{0pl}}{b} \le 0.01$$

- Japanese shipbuilding quality standards (JSQS 1985):

$$w_{0pl} \le 7 \,\text{mm for bottom plate}$$

$$w_{0pl} \le 6 \,\text{mm for deck plate}$$

- Steel box girder bridge quality standards (ECCS 1982):

$$w_{0pl} \le \min.\left(\frac{t}{6} + 2, \frac{t}{3}\right), \quad t \text{ in mm}$$

Related to this, it is of interest to note that quite often, specifications of quality to be achieved are developed (and used) without specific reference to the loads and load effects at a particular location. In that case, the corresponding specifications suggest what can be generally achieved in an economical way rather than what should be achieved in the context of a particular situation.

1.7.2.2 Column-Type Initial Deflection of a Stiffener

The column-type initial distortion of stiffeners is assumed as follows:

$$w_0^c = w_{0c} \sin \frac{\pi x}{a} \tag{1.37a}$$

$$w_{0c} = C_3 a \tag{1.37b}$$

where w_0^c is the column-type initial distortion of the support members, a is the length of the small stiffeners between two adjacent strong support members, and C_3 is a constant. The constant in Equation (1.37b) may be taken as follows:

$C_3 = 0.0015$ for an average level in steel plates

$$C_3 = \begin{cases} 0.00016 & \text{for a slight level} \\ 0.0018 & \text{for an average level} \\ 0.0056 & \text{for a severe level} \end{cases} \text{ in aluminum plates (Paik et al. 2006)}$$

1.7.2.3 Sideways Initial Distortion of a Stiffener

The sideways initial distortion of stiffeners is assumed as follows:

$$w_0^s = w_{0s} \frac{z}{h_w} \sin \frac{\pi x}{a} \tag{1.38a}$$

$$w_{0s} = C_4 a \tag{1.38b}$$

where w_0^s is the sideways initial distortion of the support member, z is the coordinate in the direction of stiffener web height, h_w is the stiffener web height, a is the length of the small stiffeners between two adjacent strong support members, and C_4 is a constant. The constant in Equation (1.38b) may be taken as follows:

$C_4 = 0.0015$ for an average level in steel plates

$$C_4 = \begin{cases} 0.00019 & \text{for a slight level} \\ 0.001 & \text{for an average level} \\ 0.0024 & \text{for a severe level} \end{cases} \text{ in aluminum plates (Paik et al. 2006)}$$

1.7.3 Welding Residual Stress Modeling

For practical design purposes, the welding residual stress distributions of a plate element between support members for which welding has been carried out along all four edges may be idealized to be composed of tensile and compressive stress blocks, such as those shown in Figure 1.33. Among them, Figure 1.33c is a typical idealization of the welding residual stress distribution in a plate element.

(a)　　　　　　　　　　　　　　　　(b)

(c)　　　　　　　　　　　　　　　　(d)

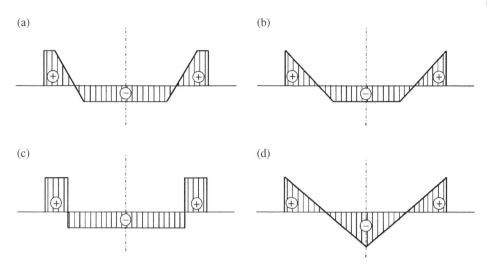

Figure 1.33 Idealized distributions of the welding induced residual stresses in a stiffened plate structure.

Welding residual stresses develop in both longitudinal and transverse directions if the support members are attached by welding in these two directions as shown in Figure 1.34. The breadth of the HAZ is denoted by b_t in the y direction or a_t in the x direction in which the residual stress in the HAZ is approximately equal to the tensile yield stress because the molten metal can expand freely, as a liquid, whereas after welding it quickly reverts to a solid and the shrinkage that occurs during cooling involves "plastic flow."

Along the welding line, tensile residual stresses usually develop with magnitude σ_{rtx} in the x direction and σ_{rty} in the y direction, with the welding being normally performed in

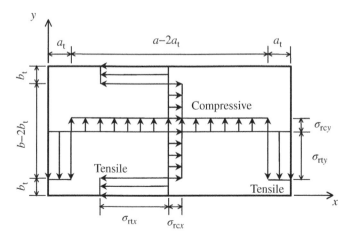

Figure 1.34 Typical idealization of the welding induced residual stress distribution inside the metal plate element in the x and y directions.

both x and y directions. To obtain equilibrium, corresponding compressive residual stresses with magnitude σ_{rcx} in the x direction and σ_{rcy} in the y direction develop in the middle part of the plate element.

As the tensile residual stress blocks are equivalent to the HAZ, their breadths can be estimated from the equilibrium between the tensile and compressive residual stresses as follows:

$$2b_t = \frac{\sigma_{rcx}}{\sigma_{rcx} - \sigma_{rtx}}b, \quad 2a_t = \frac{\sigma_{rcy}}{\sigma_{rcy} - \sigma_{rty}}a \tag{1.39}$$

where b_t and a_t are the breadths of the tensile residual stress block, σ_{rcx} and σ_{rcy} are the compressive residual stresses in the x or y directions, and σ_{rtx} and σ_{rty} are the tensile residual stresses in the x or y directions.

One can then define the residual stress distributions in the x and y directions as follows:

$$\sigma_{rx} = \begin{cases} \sigma_{rtx} & \text{for } 0 \le y < b_t \\ \sigma_{rcx} & \text{for } b_t \le y < b - b_t \\ \sigma_{rtx} & \text{for } b - b_t \le y \le b \end{cases} \tag{1.40a}$$

$$\sigma_{ry} = \begin{cases} \sigma_{rty} & \text{for } 0 \le x < a_t \\ \sigma_{rcy} & \text{for } a_t \le x < a - a_t \\ \sigma_{rty} & \text{for } a - a_t \le x \le a \end{cases} \tag{1.40b}$$

Smith et al. (1988) suggested the following formula to define the compressive residual stress σ_{rcx} in the x direction of a steel plate:

$$\sigma_{rcx} = \begin{cases} -0.05\sigma_Y & \text{for a slight level} \\ -0.15\sigma_Y & \text{for an average level} \\ -0.3\sigma_Y & \text{for a severe level} \end{cases} \tag{1.41a}$$

The counterpart of the compressive residual stress σ_{rcy} in the y direction may be assumed to be as follows:

$$\sigma_{rcy} = k\frac{b}{a}\sigma_{rcx} \tag{1.41b}$$

where k is a correction factor, which may take a value smaller than 1.0. When the residual stress is considered in the x direction alone, $k = 0$.

Paik and Yi (2016) suggested an advanced method to predict the welding induced residual stresses in a steel plate element. Based on the experimental and numerical investigations for steel plates with $a/b \ge 1$, they proposed the empirical formulations to predict the breadths of the HAZ as functions of the plate's slenderness ratio and the weld bead length (leg length) as follows:

$$b_t = c_1 L_w + c_2 \tag{1.42a}$$
$$a_t = d_1 L_w + d_2 \tag{1.42b}$$

where
$$c_1 = -0.4562\beta_x^2 + 4.1994\beta_x + 2.6354, \ c_2 = 1.1352\beta_x^2 - 4.3185\beta_x - 11.1750,$$
$$d_1 = -0.0399\beta_y^2 + 2.0087\beta_y + 8.7880, \ d_2 = 0.1042\beta_y^2 - 4.8575\beta_y - 17.7950,$$

$\beta_x = (b/t)\sqrt{\sigma_Y/E}$, $\beta_y = (a/t)\sqrt{\sigma_Y/E}$, and L_w is the weld bead length (mm), which is usually in the range 4–8 mm (4–5 mm for design requirement and 6 mm in average) for relatively thin plates in shipbuilding industry practice with one pass welding.

Once the breadths of the HAZ are determined from Equation (1.42), depending on the welding conditions and the plate's slenderness ratio, the welding induced compressive residual stresses can then be predicted from Equation (1.39) as follows:

$$\sigma_{rcx} = \frac{2b_t}{2b_t - b}\sigma_{rtx} \tag{1.43a}$$

$$\sigma_{rcy} = \frac{2a_t}{2a_t - a}\sigma_{rty} \tag{1.43b}$$

where $\sigma_{rtx} = \sigma_{rty} = \sigma_Y$ can be taken for steel.

1.7.4 Modeling of Softening Phenomenon

As previously noted, it is recognized that the strength of aluminum alloys in the softened zone may be recovered by natural aging over a period of time (Lancaster 2003). However, the ultimate strength of welded aluminum alloy-plated structures may be reduced by softening phenomenon as far as the material strength is not recovered.

The breadths of the softened zones approximately equal those of the HAZ in a welded aluminum structure. Paik et al. (2006) proposed the breadths of the softened zones with the nomenclature defined in Figure 1.27 as follows:

$$b'_p = b'_s = \begin{cases} 11.3 \text{ mm} & \text{for a slight level} \\ 23.1 \text{ mm} & \text{for an average level} \\ 29.9 \text{ mm} & \text{for a severe level} \end{cases} \tag{1.44}$$

The yield strength in the HAZ may be obtained as follows, depending on the type of aluminum alloy, following Paik et al. (2006):

a) Yield stress of the HAZ material for aluminum alloy 5083-H116

$$\frac{\sigma_{YHAZ}}{\sigma_Y} = \begin{cases} 0.906 & \text{for a slight level} \\ 0.777 & \text{for an average level} \\ 0.437 & \text{for a severe level} \end{cases} \text{ with } \sigma_Y = 215 \text{ N/mm}^2 \tag{1.45a}$$

b) Yield stress of the HAZ material for aluminum alloy 5383-H116

$$\frac{\sigma_{YHAZ}}{\sigma_Y} = \begin{cases} 0.820 & \text{for a slight level} \\ 0.774 & \text{for an average level} \\ 0.640 & \text{for a severe level} \end{cases} \text{ with } \sigma_Y = 220 \text{ N/mm}^2 \tag{1.45b}$$

c) Yield stress of the HAZ material for aluminum alloy 5383-H112

$$\frac{\sigma_{YHAZ}}{\sigma_Y} = 0.891 \text{ for an average level with } \sigma_Y = 190 \text{ N/mm}^2 \tag{1.45c}$$

d) Yield stress of the HAZ material for aluminum alloy 6082-T6

$$\frac{\sigma_{YHAZ}}{\sigma_Y} = 0.703 \quad \text{for an average level} \quad \text{with} \quad \sigma_Y = 240 \, \text{N/mm}^2 \tag{1.45d}$$

where σ_{YHAZ} is the yield strength in the softened zone and σ_Y is the yield strength in the base material.

The compressive residual stresses at the plate part and the stiffener web can then be determined from Equation (1.43) using Equations (1.44) and (1.45). Empirical formulations are also given regardless of the type of aluminum alloys as follows (Paik et al. 2006):

$$\sigma_{rcx} = \begin{cases} -0.110\sigma_{Yp} & \text{for a slight level} \\ -0.161\sigma_{Yp} & \text{for an average level} \\ -0.216\sigma_{Yp} & \text{for a severe level} \end{cases} \text{in the plate part} \tag{1.46a}$$

$$\sigma_{rcx} = \begin{cases} -0.078\sigma_{Ys} & \text{for a slight level} \\ -0.137\sigma_{Ys} & \text{for an average level} \\ -0.195\sigma_{Ys} & \text{for a severe level} \end{cases} \text{in the stiffener web} \tag{1.46b}$$

where σ_{Yp} and σ_{Ys} are the yield strengths of the plate part and the stiffener web, respectively.

1.8 Age Related Structural Degradation

In aging structures, defects related to corrosion and fatigue cracks are significant, especially in a marine environment (Paik & Thayamballi 2007, Rizzo et al. 2007, Paik & Melchers 2008). In a number of damage cases for aging marine and land-based structures that have been reported, it is possible that corrosion damage and fatigue cracks may have existed in the primary and other strength members. In any event, fatigue and corrosion are the two most important factors that affect structural performance over time.

It is therefore important for the structural designer and operator to have a complete understanding of the location and extent of structural damage formed during the structure's operation and how it can affect the structural capacity. One reason is that this knowledge is necessary to facilitate repair decisions, and another could be to support a structural life extension decision later in the structure's life. The structural capacity associated with Equation (1.17) needs to be determined by dealing with the age related degradation as a parameter of influence.

1.8.1 Corrosion Damage

Due to corrosion damage, the structural capacity can be decreased and/or leakages can take place in oil/watertight boundaries, with the latter possibly leading to undesirable pollution, cargo mixing, or gas accumulation in enclosed spaces. The corrosion process varies over time, and the amount of corrosion damage is normally defined by a corrosion rate with units of, say, millimeters per year, representing the depth of corrosion

(a)   (b)   (c)

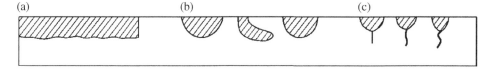

Figure 1.35 Typical types of corrosion damage: (a) general corrosion; (b) localized corrosion; (c) fatigue cracks from localized corrosion.

diminution per year. The corrosion rate itself can be a function of time in some cases, due to effects such as increased structural flexibility as the corrosion process proceeds.

Figure 1.35 shows some of the more typical types of corrosion-related damage that affect the strength of structures. "General" corrosion (also called "uniform" corrosion) uniformly reduces the thickness of structural members, as shown in Figure 1.35a, whereas localized corrosion (e.g., pitting or grooving) causes degradation in local regions, as shown in Figure 1.35b. Fatigue cracks may sometimes arise from localized corrosion, as shown in Figure 1.35c.

The corrosion damage to a structure is influenced by many factors, including the corrosion protection system and various operational parameters (Afanasieff 1975, Schumacher 1979, Melchers & Ahammed 1994, Paik & Thayamballi 2007). Generally used corrosion protection systems include coatings (paint) and anodes. The operational parameters include maintenance, repair, the use of heating coils, humidity conditions, water and sludge accumulation, microbial contamination, and the composition of inert gas. For ships and offshore structures, the percentage of time in ballast, the frequency of tank cleaning, and temperature profiles are also influential parameters. For the past several decades, several studies have been undertaken to understand the effects of many of these factors and their interactions.

To predict tolerance to likely corrosion damage, it is necessary to estimate the corrosion rates for various structural members grouped by type, location, and other parameters. To generalize this further, there are four aspects related to corrosion that one must ideally define for structural members:

- Where is corrosion likely to occur?
- When does it start?
- What is its extent?
- What are the likely corrosion rates as a function of time?

The first question would normally be answered using historical databases of some form, for example, the results of previous surveys. As to when corrosion starts, this again is information that should come from prior surveys for the particular structure. Lacking specific databases, assumptions for the time the corrosion will begin can of course be made, depending on the use of a protection system, the characteristics of the coatings, and the anode residence time.

For the residual strength and similar performance assessment of corroded structures, one must clarify how corrosion develops and proceeds in structural members, the spatial extents of member degradation, and the likely effects of such corrosion on structural performance measures such as strength and leakage characteristics. These considerations are complicated by the sheer number of factors that can potentially affect corrosion, including the type of protection, the type of cargo, temperature, and humidity. In addition, a probabilistic treatment is essential to account for the various uncertainties associated with corrosion.

The extent of corrosion presumably increases with time, but our ability to predict its spatial progress remains meager. The only real alternative is then to pessimistically assume an extent of corrosion than is actually likely, such as what one would do in the case of nominal design corrosion values. To put this in another way, one can assess the structural performance based on premised extents of corrosion when specific information on the extent of corrosion is lacking or unavailable.

Where coatings are present, the progress of corrosion would normally depend greatly on the degradation of such coatings. For this reason, most classification societies usually recommend maintenance of the corrosion protection system over time, and most owners carry out such maintenance, so the particular maintenance philosophy used also has a significant effect on structural reliability considering corrosion effects in the long term.

Figure 1.36 represents a plausible schematic of the corrosion process for a coated area in a structure. It is assumed in Figure 1.36 that there is no corrosion as long as the coating is effective and also during a short transition time after the coating breaks down. Therefore, the corrosion model accounts for three factors: (i) durability of coating (coating life), (ii) transition, and (iii) progress of corrosion.

The curve that shows the corrosion progression, indicated by a solid line in Figure 1.36, is a little convex, but it may in some cases be a concave curve in dynamically loaded structures, as indicated by the dotted line where flexing continually exposes additional fresh surface area to the effects of corrosion. However, one may take a linear approximation between them for practical assessment.

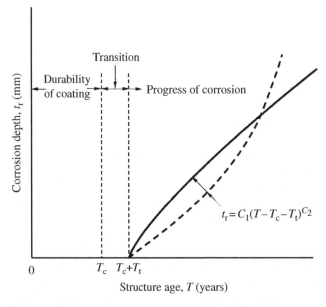

Figure 1.36 A schematic of the corrosion process for structures.

The life (or durability) of a coating essentially corresponds to the time when the corrosion begins after the contribution of a structure, or the application of a coating in a previously bare case, or the repair of a coating area to a good, intact standard. The life of a coating typically depends on the type of coating system used and the relevant maintenance, among other factors (Melchers & Jiang 2006). The coating life to a predefined state of breakdown is often assumed to follow a lognormal distribution, given by

$$f(T_c) = \frac{1}{\sqrt{2\pi}\sigma_c}\exp\left[-\frac{(\ln T_c - \mu_c)^2}{2\,\sigma_c^2}\right]$$

(1.47)

where μ_c is the mean value of $\ln T_c$ in years, σ_c is the standard deviation of $\ln T_c$, and T_c is the coating life in years.

The coating systems are sometimes classified by their target life. For example, IMO (1995) uses three groups (coating systems I, II, and III) for ships and offshore structures, for which the corresponding target durability is 5, 10, and 15 years, respectively. However, this particular classification is by no means universal. TSCF (2000) defines the requirements for 10-, 15-, and 25-year coating systems for ballast tanks in oil tankers. Generally, however, a 5-year coating life may be considered to represent an undesirable situation, whereas 10 years or longer would represent a relatively more desirable state of affairs. The selection of a target life to be achieved is primarily economical. Any given mean or median coating life is uncertain, and the coefficient of variation of the coating life is sometimes taken as $\sigma_c/\mu_c = 0.4$ for $\ln T_c$ (ClassNK 1995).

After the effectiveness of a coating is lost, some transition time, that is, the duration between the loss of coating effectiveness and the time the corrosion begins, is considered to exist before the corrosion "initiates" over a sufficiently large and easily measured area. The transition time is sometimes considered to be an exponentially distributed random variable. As an example, the mean value of the transition time for transverse bulkhead structures of bulk carriers was shown to be 3 years for deep-tank bulkheads, 2 years for watertight bulkheads, and 1.5 years for stool regions (Yamamoto & Ikegami 1998). When the transition time is assumed to be zero, that is, $T_t = 0$, it is implied that the corrosion will begin immediately after the effectiveness of the coating is lost.

As illustrated in Figure 1.36, the wear of plate thickness due to corrosion may be generally expressed as a function of the time after the corrosion starts (years), namely,

$$d_c = C_1 T_e^{C_2}$$

(1.48)

where d_c is the corrosion depth (or wear of plate thickness due to corrosion; mm); T_e is the exposure time after breakdown of the coating (years), which is taken as $T_e = T - T_c - T_t$; T is the age of the structure (years); T_c is the life of the coating (years); T_t is the duration of transition (years), which may be pessimistically taken as $T_t = 0$; and C_1 and C_2 are coefficients.

The coefficient C_2 in Equation (1.48) determines the trend of corrosion progress, whereas the coefficient C_1 is in part indicative of the annual corrosion rate that can be obtained by differentiating Equation (1.48) with respect to time. As may be surmised from Equation (1.48), the two coefficients closely interact, and they can be determined simultaneously based on the carefully collected statistical corrosion data of existing structures. However, this approach is not straightforward to apply in most cases, mainly because of differences in the database collection sites typically visited over the life of the structure.

That is, it is normally difficult to track corrosion at a particular site based on the typically available gauging data. This is part of the reason for the relatively large scatter of corrosion data in many studies.

An easier alternative is to determine the coefficient C_1 at a constant value of the coefficient C_2. This is mathematically a simpler model, but it does not negate any of the shortcomings due to the usual methods of data collection in surveys. It does, however, make possible the postulation of different modes of corrosion behavior over time depending on the value adopted for C_2 in an easy-to-understand way.

For corrosion of ships and offshore structures, studies have indicated that the coefficient C_2 can sometimes fall within the range of 0.3–1.5 (Yamamoto & Ikegami 1998, Melchers 1999b). This implies a behavior wherein the corrosion rates apparently decrease or stabilize over time. While such behavior is plausible for statically loaded structures, for dynamically loaded structures in which the corrosion scale is continually being lost and new material is being exposed to corrosion because of structural flexing, such values of C_2 may not always be appropriate or safe (Melchers & Paik 2009). For practical design purposes, $C_2 = 1$ is often adopted.

Figure 1.37 shows a schematic of the time-variant corrosion progress, which indicates that the probabilistic characteristics of the corrosion progress differ over time. Figure 1.38a shows evidence of this for time-variant corrosion progress in the ballast tank structures of bulk carriers (Paik & Kim 2012).

Paik and Kim (2012) derived a mathematical model to predict the time-variant corrosion wastage of the ballast tank structures of bulk carriers by accounting for the effects of the varying probabilistic characteristics with time, where the two-parameter Weibull

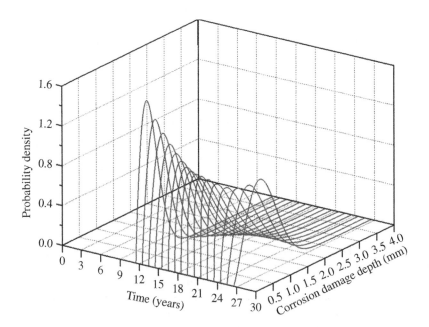

Figure 1.37 A schematic of the probabilistic characteristics of corrosion wastage progress over time (Paik & Kim 2012).

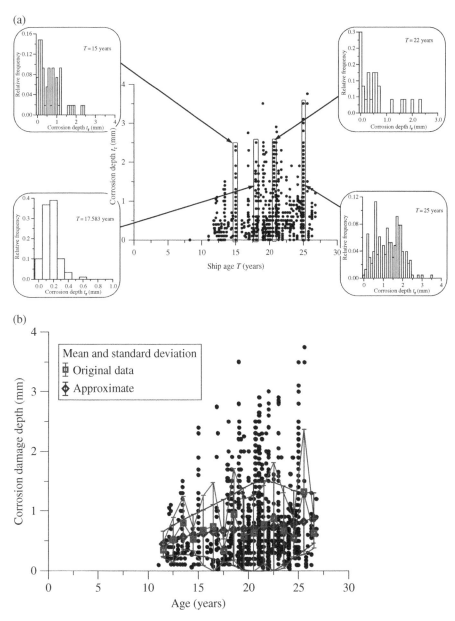

Figure 1.38 Probabilistic characteristics of corrosion wastage progress over time in ballast tank structures of bulk carriers: (a) measurement database; (b) comparison between measurement and prediction (Paik & Kim 2012).

function was realized by the goodness of fit tests to be the best suited to represent the corrosion wastage progress:

$$d_{\mathrm{c}} = \frac{\alpha}{\beta}\left(\frac{T_{\mathrm{e}}}{\beta}\right)^{\alpha-1}\exp\left[-\left(\frac{T_{\mathrm{e}}}{\beta}\right)^{\alpha}\right] \tag{1.49a}$$

where

$$\alpha = 0.0020T_e^3 - 0.0994T_e^2 + 1.5604T_e - 6.0025 \tag{1.49b}$$

$$\beta = 0.0004T_e^3 - 0.0248T_e^2 + 0.4793T_e - 2.3812 \tag{1.49c}$$

Figure 1.38b confirms the applicability of the approximate formula of Equation (1.49) by comparison with the original database of gathered corrosion measurements.

Hairil Mohd and Paik (2013) further applied this method to the time-variant corrosion damage prediction of subsea well tubes, where the two-parameter Weibull function was also found by the goodness of fit tests to be the best suited to represent the corrosion wastage progress. In this case, the coefficients α and β in Equation (1.49a) are now given as follows:

$$\alpha = -0.02287T_e^2 + 0.61835T_e - 0.94398 \tag{1.50a}$$

$$\beta = 0.001347T_e^2 + 0.004688T_e + 0.292059 \tag{1.50b}$$

Figure 1.39 confirms the validity of Equation (1.50) together with Equation (1.49a) for the time-variant corrosion wastage of subsea well tubes. Hairil Mohd et al. (2014) further

(a)

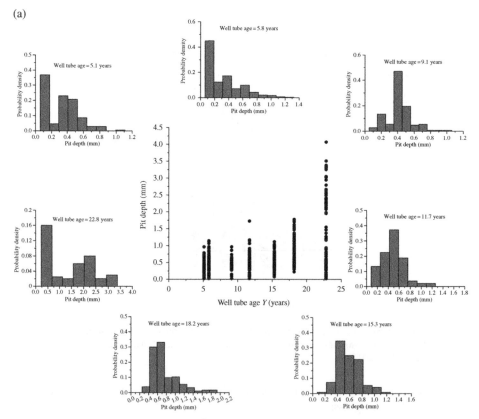

Figure 1.39 Probabilistic characteristics of corrosion wastage progress over time in subsea well tubes: (a) measurement database; (b) comparison between measurement and prediction in mean value; (c) comparison between measurement and prediction in probability (Hairil Mohd & Paik 2013).

(b)

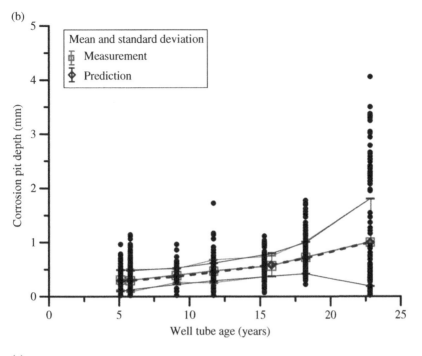

(c)

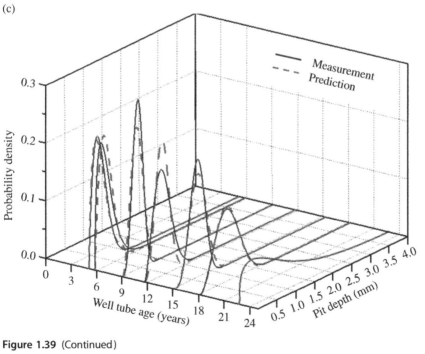

Figure 1.39 (Continued)

applied this approach to predict the time-variant corrosion wastage progress of subsea gas pipelines as shown in Figure 1.40, where the three-parameter Weibull function was found to be the best suited to represent the corrosion wastage progress as follows:

$$d_c = \frac{\alpha}{\beta}\left(\frac{T_e-\gamma}{\beta}\right)^{\alpha-1} \exp\left[-\left(\frac{T_e-\gamma}{\beta}\right)^{\alpha-1}\right] \tag{1.51a}$$

where

$$\alpha = 0.003337T_e^2 - 0.130420T_e + 2.4557 \tag{1.51b}$$

$$\beta = -0.000997T_e^2 + 0.013425T_e + 1.58201 \tag{1.51c}$$

$$\gamma = 0.0003455T_e^2 + 0.062137T_e - 0.365129 \tag{1.51d}$$

It is obvious that the characteristics of corrosion progress differ depending on the corrosion environment, which can differ at different locations of a structural member, even in the same structure. Paik et al. (2003a) divided the double-hulled oil tanker structures into a total of 34 structural member groups according to the different locations of the corrosion environment, as indicated in Figure 1.41 and Tables 1.7 and 1.8. Paik et al. (2003b) also divided the bulk carrier structures into a total of 23 structural member

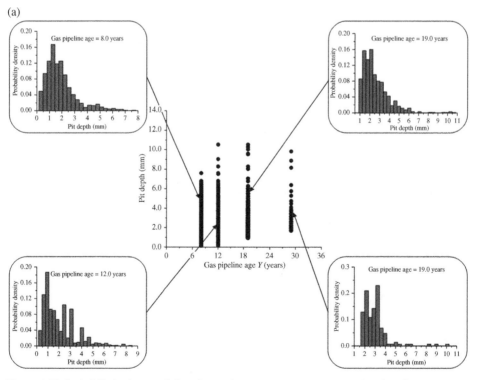

Figure 1.40 Probabilistic characteristics of corrosion wastage progress over time in subsea gas pipelines: (a) measurement database; (b) comparison between measurement and prediction in mean value; (c) comparison between measurement and prediction in probability (Hairil Mohd et al. 2014).

(b)

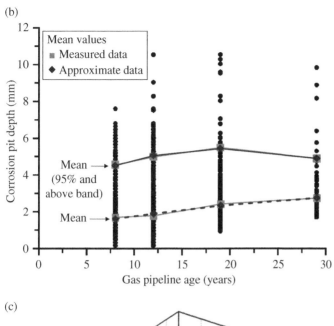

(c)

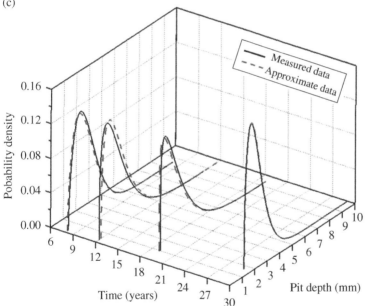

Figure 1.40 (Continued)

groups, as indicated in Figure 1.42 and Table 1.9. Each of the structural member groups has different corrosion characteristics.

1.8.2 Fatigue Cracks

Under repeated loading, fatigue cracks may be initiated in the structure's stress concentration areas. Initial defects or cracks may also form in the structure due to the

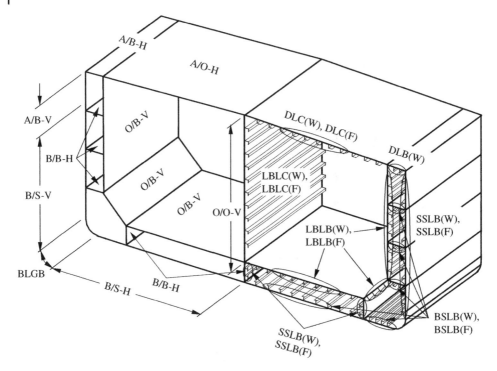

Figure 1.41 The 34 structural member groups (location and category groups) of double-hulled oil tanker structures (Paik et al. 2003a).

Table 1.7 Identification of 14 member location/category groups for the plating of tankers.

ID	Member type
B/S-H	Bottom shell plating (ballast tank)
A/B-H	Deck plating (ballast tank)
A/B-V	Side shell plating above draft line (ballast tank)
B/S-V	Side shell plating below draft line (ballast tank)
BLGB	Bilge plating (ballast tank)
O/B-V	Longitudinal bulkhead plating (ballast tank)
B/B-H	Stringer plating (ballast tank)
O/S-H	Bottom shell plating (cargo oil tank)
A/O-H	Deck plating (cargo oil tank)
A/O-V	Side shell plating above draft line (cargo oil tank)
O/S-V	Side shell plating below draft line (cargo oil tank)
BLGC	Bilge plating (cargo oil tank)
O/O-V	Longitudinal bulkhead plating (cargo oil tank)
O/O-H	Stringer plating (cargo oil tank)

Table 1.8 Identification of 20-member location/category groups for the stiffener webs and flanges of tankers.

ID (stiffener web)	Member types	ID (stiffener flange)	Member types
BSLBW	Bottom shell longitudinals in ballast tank—web	BSLBF	Bottom shell longitudinals in ballast tank—flange
SSLBW	Side shell longitudinals in ballast tank—web	SSLBF	Side shell longitudinals in ballast tank—flange
LBLBW	Longitudinal bulkhead longitudinals in ballast tank—web	LBLBF	Longitudinal bulkhead longitudinals in ballast tank—flange
BSLCW	Bottom shell longitudinals in cargo oil tank—web	BSLCF	Bottom shell longitudinals in cargo oil tank—flange
DLCW	Deck longitudinals in cargo oil tank—web	DLCF	Deck longitudinals in cargo oil tank—flange
SSLCW	Side shell longitudinals in cargo oil tank—web	SSLCF	Side shell longitudinals in cargo oil tank—flange
LBLCW	Longitudinal bulkhead longitudinals in cargo oil tank—web	LBLCF	Longitudinal bulkhead longitudinals in cargo oil tank—flange
BGLCW	Bottom girder longitudinals in cargo oil tank—web	BGLCF	Bottom girder longitudinals in cargo oil tank—flange
DGLCW	Deck girder longitudinals in cargo oil tank—web	DGLCF	Deck girder longitudinals in cargo oil tank—flange
DLBW	Deck longitudinals in ballast tank—web		
SSTLCW	Side stringer longitudinals in cargo oil tank—web		

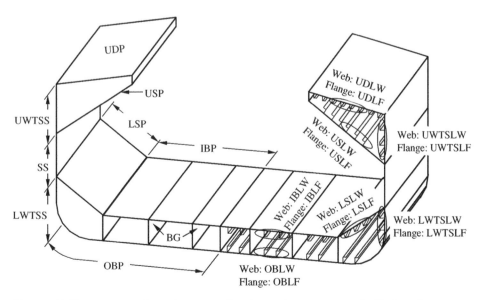

Figure 1.42 The 23 structural member groups (location and category groups) of bulk carrier structures (Paik et al. 2003b).

Table 1.9 Identification of the 23-member groups for a bulk carrier structure.

ID	Member type
OBP	Outer bottom plates
IBP	Inner bottom plates
LSP	Lower sloping plates
LWTSS	Lower wing tank side shells
SS	Side shells
UWTSS	Upper wing tank side shells
USP	Upper sloping plates
UDP	Upper deck plates
BG	Bilge girders
OBLW	Outer bottom longitudinals—web
OBLF	Outer bottom longitudinals—flange
IBLW	Inner bottom longitudinals—web
IBLF	Inner bottom longitudinals—flange
UWTSLW	Upper wing tank side longitudinals—web
UWTSLF	Upper wing tank side longitudinals—flange
USLW	Upper sloping longitudinals—web
USLF	Upper sloping longitudinals—flange
UDLW	Upper deck longitudinals—web
UDLF	Upper deck longitudinals—flange
LWTSLW	Lower wing tank side longitudinals—web
LWTSLF	Lower wing tank side longitudinals—flange
LSLW	Lower sloping longitudinals—web
LSLF	Lower sloping longitudinals—flange

fabrication procedures applied. In addition to their fatigue propagation under repeated cyclic loading, cracks, as they grow, may also propagate under monotonically increasing extreme loads, a circumstance that can eventually lead to the structure's catastrophic failure when given the possibility of rapid and uncontrolled crack extension without arrest or if the crack attains a length that results in significant degradation of the structural capacity.

It is obvious that fatigue cracking damage also varies with time. Figure 1.43 shows a schematic of fatigue-related cracking damage progress as a function of time (age) in structures (Paik & Thayamballi 2007). The fatigue damage progress can be separated into three stages: initiation (stage I), propagation (stage II), and failure (fracture) (stage III) (ISO 2394 1998). For assessment of residual strength in aging structures under extreme loads and under fluctuating loads, it is thus often necessary to account for an existing crack as a parameter of influence (Paik & Melchers 2008).

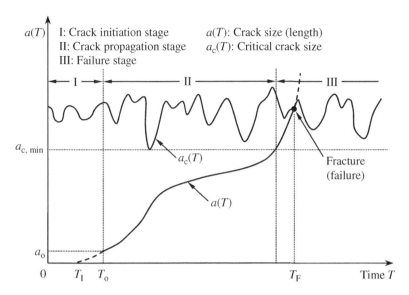

Figure 1.43 A schematic of crack initiation and growth for a structure with time.

1.9 Accident Induced Damage

The ultimate strength of a structure can be reduced as a result of accident induced damage. Potential accidents, such as collision, grounding, impact from dropped objects or mishandled cargo loading/unloading, fire, and explosions, can result in structural damage that reduces structural capacity (ultimate strength) or even leads to total loss of the structure.

Collision and grounding accidents typically result in crushing (folding), yielding, and tearing. Hydrodynamic impact can cause plastic deformation damage. Dropped objects can cause local dents and/or global permanent deformation. Fire or explosions can expose the structural material to elevated temperatures and explosions also being accompanied by blast. Exposure to a fire at high temperatures can cause not only structural damage but also metallurgical changes. For fire safety and the resistance of structures, refer to Lawson (1992), Nethercot (2001), and Franssen and Real (2010).

Ultimate strength of a structure with accident induced damage is often termed residual strength. The structural capacity associated with Equation (1.17) needs to be determined by dealing with the accident induced damage as a parameter of influence.

References

Afanasieff, L. (1975). Corrosion mechanisms, corrosion defense and wastage. Chapter 16 in *Ship structural design concepts*, Edited by Harvey Evans, J., Cornell Maritime Press, Cambridge, MA.

Benjamin, J.R. & Cornell, C.A. (1970). *Probability, statistics, and decision for civil engineers.* McGraw-Hill, New York.

Bodner, S.R. & Symonds, P.S. (1962). Experimental and theoretical investigation of the plastic deformation of cantilever beams subjected to impulsive loading. *Journal of Applied Mechanics*, **29**:719–728.

Brockenbrough, R.L. & Johnston, B.G. (1981). *USS steel design manual.* United States Steel Corporation, Pittsburgh, PA.

BS 5950 (1985). *The structural use of steelwork in building. Part 8.* British Standards Institution, London.

Callister, W.D. (1997). *Materials science and engineering.* Fourth Edition, John Wiley & Sons, Inc., New York.

Campbell, J. & Cooper, R.H. (1966). Yield and flow of low-carbon steel at medium strain rates. Proceedings of the Conference on the Physical Basis of Yield and Fracture, Institute of Physics and Physical Society, London, 77–87.

Cheng, J.J.R., Elwi, A.E., Grodin, G.Y. & Kulak, G.L. (1996). Material testing and residual stress measurements in a stiffened steel plate. In *Strength and stability of stiffened plate components*, **SSC-399**, Ship Structure Committee, Washington, DC.

ClassNK (1995). *Guidance for corrosion protection system of hull structures for water ballast tanks and cargo oil tanks.* Second Revision, Nippon Kaiji Kyokai, Tokyo.

Cowper, G.R. & Symonds, P.S. (1957). *Strain-hardening and strain-rate effects in the impact loading of cantilever beams*, Technical Report, **23**, Division of Applied Mathematics, Brown University, Providence, RI.

ECCS (1982). *European recommendations for the fire safety of steel structures*, ECCS Technical Committee, **3**, European Convention for Constructional Steelwork (ECCS), Brussels.

ENV 1993-1 (1992a). *Eurocode 3: design of steel structures, part 1.1 general rules and rules for buildings.* British Standards Institution, London.

ENV 1993-1 (1992b). *Eurocode 3: design of steel structures, part 1.2 fire resistance.* British Standards Institution, London.

Franssen, J.M. & Real, P.V. (2010). *Fire design of steel structures*, ECCS Eurocode Design Manuals, Ernst & Sohn, Berlin.

Galambos, T.V. (1988). *Guide to stability design criteria for metal structures.* John Wiley & Sons, Inc., New York.

Hairil Mohd, M., Kim, D.K., Kim, D.W. & Paik, J.K. (2014). A time-variant corrosion wastage model for subsea gas pipelines. *Ships and Offshore Structures*, **9**(2):161–176.

Hairil Mohd, M. & Paik, J.K. (2013). Investigation of the corrosion progress characteristics of offshore subsea oil well tubes. *Corrosion Science*, **67**:130–141.

Hsu, S.S. & Jones, N. (2004). Dynamic axial crushing of aluminum alloy 6063-T6 circular tubes. *Latin American Journal of Solids and Structures*, **1**(3):277–296.

Hughes, O.F. & Paik, J.K. (2013). *Ship structural analysis and design.* The Society of Naval Architects and Marine Engineers, Alexandria, VA.

IACS (2014). *Aluminum alloys for hull construction and marine structures.* International Association of Classification Societies, London.

IMO (1995). Resolution A.798(l9). In *Guidelines for the selection, application and maintenance of corrosion prevention systems of dedicated seawater ballast tanks.* International Maritime Organization, London.

ISO 2394 (1998). *General principles on reliability for structures.* Second Edition, International Organization for Standardization, Geneva, Switzerland.

Jones, N. (2012). *Structural impact.* Second Edition, Cambridge University Press, Cambridge.

JSQS (1985). *Japanese shipbuilding quality standards*. The Society of Naval Architects of Japan, Tokyo.

Kenno, S.Y., Das, S., Kennedy, J., Rogge, R.B. & Gharghouri, M.A. (2010). Distribution of residual stresses in stiffened plates with one or two stiffeners. *Ships and Offshore Structures*, **5**(3):211–225.

Kenno, S.Y., Das, S., Rogge, R.B. & Gharghouri, M.A. (2017). Changes in residual stresses caused by an interruption in the weld process of ships and offshore structures. *Ships and Offshore Structures*, **12**(3):341–359.

Lancaster, J. (2003). *Handbook of structural welding: processes, materials and methods used in the welding of major structures, pipelines and process plant*. Abington Publishing, Cambridge.

Lawson, R.M. (1992). Fire resistance and protection of structural steelwork. Chapter 7.3 in *Constructional steel design: an international guide*, Edited by Dowling, P.J., Hrading, J.E. & Bjorhovde, R., Elsevier Applied Science, London.

Lotsberg, I. (2016). *Fatigue design of marine structures*. Cambridge University Press, Cambridge.

Masubuchi, K. (1980). *Analysis of welded structures*. Pergamon Press, Oxford.

Mazzolani, F.M. (1985). *Aluminum alloy structures*. Pitman Publishing Ltd., London.

Melchers, R.E. (1999a). *Structural reliability analysis and prediction*. John Wiley & Sons, Ltd, Chichester.

Melchers, R.E. (1999b). Corrosion uncertainty modeling for steel structures. *Journal of Constructional Steel Research*, **52**:3–19.

Melchers, R.E. & Ahammed, M. (1994). *Nonlinear modeling of corrosion of steel in marine environments*, Research Report, **106.09.1994**, Department of Civil, Surveying and Environmental Engineering, The University of Newcastle, Callaghan.

Melchers, R.E. & Jiang, X. (2006). Estimation of models for durability of epoxy coatings in water ballast tanks. *Ships and Offshore Structures*, **1**(1):61–70.

Melchers, R.E. & Paik, J.K. (2009). Effect of flexure on rusting of ship's steel plating. *Ships and Offshore Structures*, **5**(1):25–31.

Modarres, M., Kaminskiy, M.P. & Krivtsov, V. (2016). *Reliability engineering and risk analysis: a practical guide*. Third Edition, CRC Press, New York.

Nethercot, D.A. (2001). *Limit states design of structural steelwork*. Third Edition Based on Revised BS 5950: Part I, 2000 Amendment, Spon Press, London.

NORSOK (2004). *Design of steel structures*. N-004, Rev.2, Standards Norway, Lysaker.

Nowak, A.A. & Collins, K.R. (2000). *Reliability of structures*. McGraw-Hill, Boston.

Nussbaumer, A., Borges, L. & Davaine, L. (2011). Fatigue design of steel and composite structures. In *Eurocode 3: design of steel structures*. ECCS Eurocode Design Manuals, Ernst & Sohn, Berlin.

Paik, J.K. (2007a). Practical techniques for finite element modeling to simulate structural crashworthiness in ship collisions and grounding (Part I: Theory). *Ships and Offshore Structures*, **2**(1):69–80.

Paik, J.K. (2007b). Practical techniques for finite element modeling to simulate structural crashworthiness in ship collisions and grounding (Part II: Verification). *Ships and Offshore Structures*, **2**(1):81–85.

Paik, J.K. (2007c). Characteristics of welding induced initial deflections in welded aluminum plates. *Thin-Walled Structures*, **45**:493–501.

Paik, J.K. (2008). *Mechanical collapse testing on aluminum stiffened panels for marine applications*, **SSC-451**, Ship Structure Committee, Washington, DC.

Paik, J.K., Andrieu, C. & Cojeen, H.P. (2008). Mechanical collapse testing on aluminum stiffened plate structures for marine applications. *Marine Technology*, **45**(4):228–240.

Paik, J.K. & Chung, J.Y. (1999). A basic study on static and dynamic crushing behavior of a stiffened tube. *Transactions of the Korean Society of Automotive Engineers (KSAE)*, **7** (1):219–238.

Paik, J.K. & Frieze, P.A. (2001). Ship structural safety and reliability. *Progress in Structural Engineering and Materials*, **3**:198–210.

Paik, J.K. & Kim, D.K. (2012). Advanced method for the development of an empirical model to predict time-dependent corrosion wastage. *Corrosion Science*, **63**:51–58.

Paik, J.K., Kim, K.J., Lee, J.H., Jung, B.G. & Kim, S.J. (2017). Test database of the mechanical properties of mild, high-tensile and stainless steel and aluminum alloy associated with cold temperatures and strain rates. *Ships and Offshore Structures*, **12**(S1):S230–S256.

Paik, J.K., Kim, B.J., Sohn, J.M., Kim, S.H., Jeong, J.M. & Park, J.S. (2012). On buckling collapse of a fusion-welded aluminum stiffened plate structure: an experimental and numerical study. *Journal of Offshore Mechanics and Arctic Engineering*, **134**:021402.1–021402.8.

Paik, J.K., Lee, J.M., Hwang, J.S. & Park, Y.I. (2003a). A time-dependent corrosion wastage model for the structures of single- and double-hull tankers and FSOs and FPSOs. *Marine Technology*, **40**(3):201–217.

Paik, J.K. & Melchers, R.E. (2008). *Condition assessment of aged structures*. CRC Press, New York.

Paik, J.K. & Pedersen, P.T. (1996). A simplified method for predicting the ultimate compressive strength of ship panels. *International Shipbuilding Progress*, **43**:139–157.

Paik, J.K. & Thayamballi, A.K. (2007). *Ship-shaped offshore installations: design, building, and operation*. Cambridge University Press, Cambridge.

Paik, J.K., Thayamballi, A.K., Park, Y.I. & Hwang, J.S. (2003b). A time-dependent corrosion wastage model for bulk carrier structures. *International Journal of Maritime Engineering*, **145**(A2):61–87.

Paik, J.K., Thayamballi, A.K. & Lee, J.M. (2004). Effect of initial deflection shape on the ultimate strength behavior of welded steel plates under biaxial compressive loads. *Journal of Ship Research*, **48**(1):45–60.

Paik, J.K., Thayamballi, A.K., Ryu, J.Y., Jang, J.H., Seo, J.K., Park, S.W., Seo, S.K., Andrieu, C., Cojeen, H.P. & Kim, N.I. (2006). The statistics of weld induced initial imperfections in aluminum stiffened plate structures for marine applications. *International Journal of Maritime Engineering*, **148**(Part A1):19–63.

Paik, J.K. & Yi, M.S. (2016). *Experimental and numerical investigations of welding induced distortions and stresses in steel stiffened plate structures*. The Korea Ship and Offshore Research Institute, Pusan National University, Busan.

Ramberg, W. & Osgood, W.R. (1943). *Description of stress-strain curves by three parameters*, Technical Note, **902**, National Advisory Committee on Aeronautics (NACA), Kitty Hawk, NC.

Rizzo, C.M., Paik, J.K., Brennan, F., Carlsen, C.A., Daley, C., Garbatov, Y., Ivanov, L., Simonsen, B.C., Yamamoto, N. & Zhuang, H.Z. (2007). Current practices and recent advances in condition assessment of aged ships. *Ships and Offshore Structures*, **2** (3):261–271.

Schijve, J. (2009). *Fatigue of structures and materials*. Second Edition, Springer, Cham, Switzerland.

Schumacher, M. (1979). *Seawater corrosion handbook*. Noyes Data Corporation, Park Ridge, NJ.

Sielski, R.A. (2007). Review of structural design of aluminum ships and crafts. *Transactions of the Society of Naval Architects and Marine Engineers*, **115**:1–30.

Sielski, R.A. (2008). Research needs in aluminum structure. *Ships and Offshore Structures*, **3** (1):57–65.

Smith, C.S., Davidson, P.C., Chapman, J.C. & Dowling, P.J. (1988). Strength and stiffness of ships' plating under in-plane compression and tension. *Transactions of the Royal Institution of Naval Architects*, **130**:277–296.

Steinhardt, O. (1971). Aluminum constructions in civil engineering. *Aluminum*, **47**:131–139; 254–261.

TSCF (2000). Guidelines for ballast tank coating systems and edge preparation. Tanker Structure Cooperative Forum, Presented at the TSCF Shipbuilders Meeting in Tokyo, Japan, October.

Ueda, Y. (1999). *Computational welding mechanics (a volume of selected papers in the commemoration of the retirement from Osaka University)*. Joining and Welding Research Institute, Osaka University, Osaka, Japan, March.

Yamamoto, N. & Ikegami, K. (1998). A study on the degradation of coating and corrosion of ship's hull based on the probabilistic approach. *Journal of Offshore Mechanics and Arctic Engineering*, **120**:121–128.

2

Buckling and Ultimate Strength of Plate–Stiffener Combinations

Beams, Columns, and Beam–Columns

2.1 Structural Idealizations of Plate–Stiffener Assemblies

A plated structure is composed of plate panels and rolled or built-up support members, usually termed stiffeners, as shown in Figure 2.1. The overall failure of the structure is affected by and can be governed by the buckling and plastic collapse of these individual members. In the ultimate limit state (ULS) design, therefore, one primary task is to accurately calculate the buckling and plastic collapse strength of such structural members.

The structural elements that make up plated structures do not work separately, which results in a high degree of redundancy and complexity, in contrast to those of framed structures. To enable the behavior of such structures to be analyzed, simplifications or idealizations must be made with consideration of the accuracy needed and the degree of complexity of the analysis to be used. Generally, a more complex analysis produces a greater degree of accuracy. However, the amount of structural simplification normally depends on the situation surrounding the problem. For instance, for an initial estimate, the ability to quickly provide a reasonable answer, with considerably less information, is often more important than extreme accuracy, whereas a final check solution should, of course, be as accurate as the circumstances allow.

A plated structure may be idealized into an assembly of many simpler "mechanical structural element models" or "idealized elements" or "engineering models"; each type displays similar behavior under a given load application, and the assembly behaves in (nearly) the same way as the actual structure.

Typical examples of structural idealization to model a continuous stiffened panel shown in Figure 2.1 are as follows:

- Plate–stiffener combination model (also called beam–column model)
- Plate–stiffener separation model
- Orthotropic plate model
- Pure plate element (segment) model

One of the most typical approaches is plate–stiffener (beam) combination idealization, which models a continuous stiffened panel as an assembly of possibly asymmetric I-beams together with their attached plating (i.e., flanges), assuming that the flanges

Ultimate Limit State Analysis and Design of Plated Structures, Second Edition. Jeom Kee Paik.
© 2018 John Wiley & Sons Ltd. Published 2018 by John Wiley & Sons Ltd.

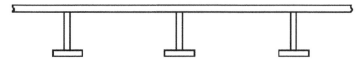

Figure 2.1 A continuous stiffened plate structure.

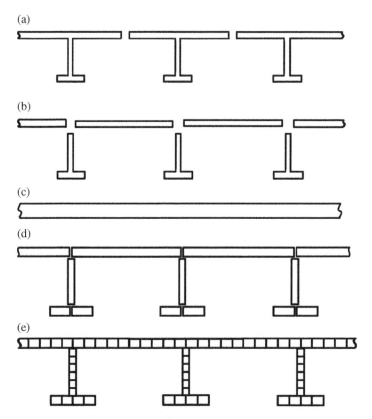

Figure 2.2 Four types of structural idealizations possible for a plated structure: (a) Plate-stiffener (beam) combinational model; (b) plate-stiffener (beam) separation model; (c) orthotropic plate model; (d) supersize finite element model (pure plate element model); (e) finite element model (pure plate element model).

support bending moments while the stiffener webs resist shear loads, as shown in Figure 2.2a. The torsional rigidity of the stiffened panel, the Poisson ratio effect, and the effect of the intersecting beams are all neglected in this modeling. The degree of accuracy for this approach may therefore become critical when the flexural rigidity of the stiffeners is lower than the plate stiffness, but the plate–stiffener combination model may be relevant when the support members (i.e., stiffeners) are of a medium or larger structural dimension so that they would behave as a beam–column together with their associated plating. The plate–stiffener combination approach may also be applied to model a cross-stiffened panel as a system of discrete intersecting beams (or grillage), with each beam composed of stiffeners and their associated effective plating. The primary concern of this chapter is to examine the behavior of the plate–stiffener combinations.

Alternatively, a mechanical idealization may be feasible by separating the support members from the plate panels at the plate–web junctions, as shown in Figure 2.2b. The so-called plate–stiffener (beam) separation model is more appropriate when the support members' structural dimensions are relatively large, so that the stiffener web and the plating between the stiffeners act as a plate panel. In this case, local buckling of the stiffener web and the attached plating signals a primary failure mode because the support members are sufficiently strong to remain straight well after plate failure. Buckling and collapse of the plating between stiffeners are described in Chapters 3 and 4, respectively.

In contrast, if the support members are relatively weak, they deflect together with the plate panels so that the behavior of the stiffened panel may in this case be idealized as that of an "orthotropic plate" by smearing the stiffeners into the plating, as shown in Figure 2.2c. The orthotropic plate approach is useful for computation of the panel ultimate strength based on the overall grillage collapse mode. In this approach, the orthotropic plate theory is used, which implies that the stiffeners are relatively numerous and small (i.e., they deflect together with the plating) and that they remain stable throughout the ranges of orthotropic plate behavior. The validity of representing the stiffened panel with an equivalent orthotropic plate normally depends on the number of stiffeners in each direction, their spacing, and whether their stiffness characteristics are identical. It is recognized that application of the orthotropic plate theory to cross-stiffened panels must be restricted to stiffened panels with more than three stiffeners in each direction (Smith 1966, Troitsky 1976, Mansour 1977). In addition, the stiffeners in each direction must be similar. The orthotropic plate approach is described in Chapters 5 and 6.

In the pure plate element (segment) model, each partitioned element such as plating between stiffeners, stiffener web, and one side of stiffener flange is idealized as one plate element or segment, as shown in Figure 2.2d. This idealization is rather similar to the finite element method modeling. The intelligent supersize finite element method (ISFEM) applies the pure plate element models, as described in Chapter 13. This modeling technique is also useful to automate the calculations of the cross-sectional properties for plate assemblies such as box girders or ship hull girders, where the stiffener flange may be modeled as one plate element (segment), as described in Chapters 7 and 8. The nonlinear finite element method idealizes the structures using finite element models with fine meshes as shown in Figure 2.2e. Finer mesh modeling results in more accurate solutions as described in Chapter 12.

It is therefore important to realize that different mechanical models for the same type of structure may be required to analyze the actual behavior under different structural dimensions or load applications. Clearly, in some cases it may be necessary to idealize a structure by combining the modeling methods mentioned previously. For instance, a longitudinally stiffened panel between strong transverse frames may be modeled with either an assembly of the plate–stiffener combinations or an orthotropic plate, whereas heavy transverse frames or plate girders may be idealized with the plate–stiffener separation model, where their webs are modeled as plate panels. Only plate elements can of course be used for modeling the entire structure as applied in the ISFEM described in Chapter 13.

In any event, the behavior of the idealized structure should be similar or nearly the same as that of the actual structure. This chapter describes the ultimate strength formulations of the plate–stiffener combination model under a variety of loading and end conditions. It is noted that the theories and methodologies described in this chapter can be commonly applied to both steel- and aluminum-plated structures.

2.2 Geometric Properties

In a continuous stiffened plate structure, a stiffener (support member) with attached plating is idealized by a plate–beam combination model whose span extends between two adjacent major support members in the other direction. The attached plating takes the "effective width" or "effective breadth" instead of the full width, as described in Section 2.6.

Figure 2.3 shows the geometric configurations of typical plate–stiffener combination sections together with their attached effective plating. For convenience, the x axis is in the longitudinal direction of the member, and the length (span) between supports is denoted by L. The full and effective widths of the attached plating are denoted by b and b_e, respectively.

2.3 Material Properties

Although the material of the stiffener's web and flange is usually the same, it sometimes differs from that of the attached plating (e.g., higher tensile steel used for the web and flange and mild steel used for the plating). For general purposes, the yield strengths

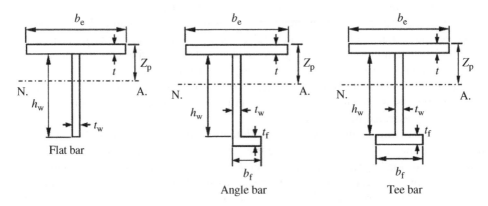

Figure 2.3 Typical types of plate–stiffener combination models composed of a stiffener and its attached effective plating (N.A., neutral axis).

Table 2.1 Properties of a plate–stiffener combination section with full or effective plating.

Property	Expression
Cross-sectional area	$A = A_p + A_w + A_f,\ A_e = A_{pe} + A_w + A_f$ where $A_p = bt,\ A_{pe} = b_e t,\ A_w = h_w t_w,\ A_f = b_f t_f$
Equivalent yield strength over the cross section	$\sigma_{Yeq} = \dfrac{A_p \sigma_{Yp} + A_w \sigma_{Yw} + A_f \sigma_{Yf}}{A}$
Distance from outer surface of attached plating to elastic horizontal neutral axis	$z_0 = \dfrac{0.5bt^2 + A_w(t + 0.5h_w) + A_f(t + h_w + 0.5t_f)}{A}$ $z_p = \dfrac{0.5b_e t^2 + A_w(t + 0.5h_w) + A_f(t + h_w + 0.5t_f)}{A_e}$
Moment of inertia	$I = \dfrac{bt^3}{12} + A_p\left(z_0 - \dfrac{t}{2}\right)^2 + \dfrac{h_w^3 t_w}{12} + A_w\left(z_0 - t - \dfrac{h_w}{2}\right)^2$ $+ \dfrac{b_f t_f^3}{12} + A_f\left(t + h_w + \dfrac{t_f}{2} - z_0\right)^2$ $I_e = \dfrac{b_e t^3}{12} + A_{pe}\left(z_p - \dfrac{t}{2}\right)^2 + \dfrac{h_w^3 t_w}{12} + A_w\left(z_p - t - \dfrac{h_w}{2}\right)^2$ $+ \dfrac{b_f t_f^3}{12} + A_f\left(t + h_w + \dfrac{t_f}{2} - z_p\right)^2$
Radius of gyration	$r = \sqrt{\dfrac{I}{A}},\quad r_e = \sqrt{\dfrac{I_e}{A}}$
Column slenderness ratio	$\lambda = \dfrac{L}{\pi r}\sqrt{\dfrac{\sigma_{Yeq}}{E}},\quad \lambda_e = \dfrac{L}{\pi r_e}\sqrt{\dfrac{\sigma_{Yeq}}{E}}$
Plate slenderness ratio	$\beta = \dfrac{b}{t}\sqrt{\dfrac{\sigma_{Yp}}{E}}$

Note: Subscript "e" represents the effective cross section.

of the stiffener web, flange, and attached plating are herein defined separately as σ_{Yw}, σ_{Yf}, and σ_{Yp}, respectively. The modulus of elasticity is E, and Poisson's ratio is ν. The shear modulus is $G = E/[2(1 + \nu)]$. Some important properties of the plate–stiffener combination sections with the attached full or effective plating are given in Table 2.1.

It is noted that the expressions of Table 2.1 are of course valid for flat bars with $b_f = t_f = 0$ and for symmetric I-sections. The equivalent yield strength over the cross section and the slenderness ratio of the attached plating between the stiffeners are calculated for the full section, that is, with the full width of the attached plating.

2.4 Modeling of End Conditions

The end conditions (also called boundary conditions) of the support members in plated structures are affected by the joining methods and rigidities of the support members in the other direction. In welded plated structures, the ends of the support members typically have a certain degree of restraint to rotation and/or translation that sometimes is not straightforward to model mathematically. For practical design purposes, however,

(a)   (b)   (c)   (d)   (e)

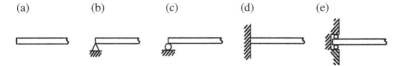

Figure 2.4 Typical idealized end conditions for plate–stiffener combination models: (a) free; (b) simply supported or pinned but translationally restrained; (c) simply supported or pinned; (d) fixed; (e) clamped.

the end condition of the plate–stiffener combination model is typically idealized by one or more of the five types, as shown in Figure 2.4, while Chapter 4 describes more refined treatments of the boundary conditions for plates surrounded by support members, which are partially rotation restrained.

At a free end, no restraints are present. The simply supported end represents a condition in which rotation freely takes place with zero bending moment, while the lateral deflection (translation) is fixed. At the fixed or clamped end, neither rotation nor lateral deflection is allowed. Depending on the possibility of axial movements, two different situations may be relevant, namely, Figure 2.4b or c for the simply supported end and Figure 2.4d or e for the fixed end. With axial restraints (this is termed fixed condition), membrane axial tension may develop as the member deflects, whereas a free axial movement can occur when axial restraints are not provided (this is termed clamped condition).

The same end condition may sometimes be applied at each end, but the possibility of different conditions must in general be considered in accordance with the dimensions and the joining methods of the support members. However, at least one end should accommodate the condition in which the translation is restrained to remove rigid-body motion.

Although both the upper and lower flanges of the I-beams in typical framed structures away from I-beam's ends can normally move freely, it is important to realize that the edges of the flanges (the attached plating) of the plate–stiffener combination models in continuous plated structures may be restricted from deforming sideways, as may be surmised from Figure 2.2a, because a symmetric condition is attained along the edges of the plate–stiffener combination, that is, the center line between two adjacent support members (stiffeners), even if the stiffener flanges are free to deflect vertically and rotate sideways. This essentially causes a different type of failure mode behavior for plate–stiffener combination models of plated structures than those for simple I-beams of framed structures.

2.5   Loads and Load Effects

The plate–stiffener combination model for plated structures is likely to be subjected to various types of loads, such as axial compression/tension, concentrated or distributed lateral load, and end moment, as shown in Figure 2.5.

Lateral loads distributed over the attached plating may typically be idealized as a lateral line load of $q = pb$ (i.e., multiplied by uniform lateral pressure p and the full breadth of the

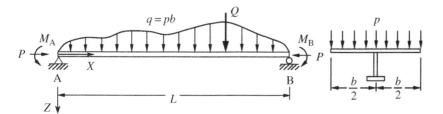

Figure 2.5 Typical load applications on a plate–stiffener combination model.

attached plating between the support members), assuming that the stiffener web resists all shear forces caused by the distributed lateral loads.

A one-dimensional structural member such as the plate–stiffener combination model is called a column under axial compression, whereas it is called a beam under lateral load or end moment after the occurrence of bending from the beginning of loading. The member is called a beam–column under combined axial compression and bending. Although axial tension normally stabilizes the behavior of beams, a beam under combined bending and axial tension is sometimes called a tension beam.

The load effects (e.g., stress, bending) of the plate–stiffener combination as an element of a complex structure are normally computed by global analysis with the linear elastic finite element method. They can also be analyzed with the classical theory of structural mechanics as described in many textbooks (e.g., Timoshenko & Goodier 1970, Chen & Atsuta 1976, 1977).

2.6 Effective Width Versus Effective Breadth of Attached Plating

The attached plating of the plate–stiffener combination model does not work separately from the adjacent members, and it is restricted from deforming sideways, whereas the stiffener flange may be free to deflect vertically and sideways. When a stiffened plate structure is idealized as an assembly of plate–stiffener combination models, therefore, one of the primary questions is to what degree and extent the attached plating reinforces the associated stiffener.

Related to this problem, two concepts, that is, effective "width" and effective "breadth," are relevant to characterize the ineffectiveness of the attached plating as a result of the nonuniform stress distribution (Paik 2008a). The effective width concept is used to model the effectiveness of plate elements that have buckled under predominantly axial compression or have inherently initial deflections after the occurrence of nonuniform stress distribution in the regime of post-buckling or large deflection. The effective breadth concept is primarily due to the action of lateral loads or out-of-plane bending. In a wide-flanged beam (i.e., plate–stiffener combination model) under out-of-plane bending, the classical beam theory provides a uniform distribution of longitudinal stress across the flange section. In reality, however, the nonuniform stress distribution can occur because the longitudinal stress caused by bending is transmitted to the flange in a nonuniform manner through shear at the junction between the flange and the web.

As a result, the distribution of the stress across the flange is not uniform, but greater at the edges (i.e., at the intersections between the plating and stiffener web) than at the middle, showing a stress lag with increasing distance from the web, as shown in Figure 2.6. The departure of the nonuniform stress from the uniform stress assumed by the classical beam theory is termed "shear lag," which essentially arises from the fact that the shear modulus of the material takes a finite value.

In summary, the reduced effectiveness of the attached plating or the flange breadth of the plate–stiffener combination model is called the "effective width" when it is due to compressive buckling, and it is called the "effective breadth" or "effective flange width" when it is due to shear lag arising from lateral loads or out-of-plane bending. In some situations, a nonuniform stress distribution and subsequent ineffectiveness of the attached plating in the plate–stiffener combination model are due to both compressive buckling and lateral pressure loads, such as those in the bottom plate panels of a ship in hogging condition.

The effective width and breadth concepts are useful for dealing with a plate that involves a nonuniform normal stress distribution as described in Chapter 4 because the plate with the effective width or breadth can be dealt with as a virtual plate with a uniform stress distribution that becomes a linear structural mechanics problem. For a plate involving a nonuniform shear stress distribution, the effective shear modulus concept (Paik 1995) can be applied, as described in Chapter 4.

The problem of the effective width, such as that for steel plating under in-plane compression, was initially raised by John (1877), a naval architect who investigated the strength of a ship that had broken into two pieces during heavy weather, presumably as a result of high stress induced by sagging moment. He pointed out that the light plating of the deck and topsides could not be considered fully effective under compression. To account for this effect in the calculation of the section modulus of the ship, he reduced the thickness of the plating and kept the stress (which could be calculated without considering buckling) unchanged.

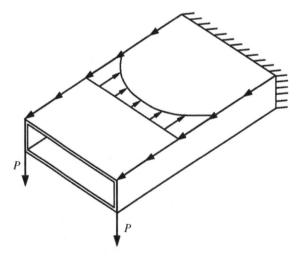

Figure 2.6 Nonuniform stress distribution induced by shear lag phenomenon.

A pioneer in the use of analytical approach for the plate effective width was Bortsch (1921), who used an approximate analytical effective width formula for practical problems related to bridge engineering. The modern era in the effective width concept was begun by von Karman (1924), who developed a general method to solve the problem theoretically and introduced for the first time the term "effective width." He calculated the stress distribution of two-dimensional problems using the stress function approach to evaluate the effective width. A remarkable advancement of the Karman method was achieved by Metzer (1929), who studied the effective flange width of simple beams and continuous beams.

In the 1930s, a large series of compression tests on steel plates was undertaken by Schuman and Back (1930), who noted that the buckled steel plate may behave as if only part of its width is effective in carrying loads. By applying the effective width concept, this phenomenon was investigated theoretically by von Karman et al. (1932), who obtained the first effective width expression of plating. Paik (2008a) reviewed some recent advances in the concepts of plate-effectiveness evaluation.

Figure 2.7 shows a typical nonuniform stress distribution of plating between stiffeners. The maximum membrane stress occurs at the intersection between the plate and the stiffener web, whereas the stresses inside the plate are comparatively smaller, which implies that the effectiveness of the plate in carrying the load through a uniform stress may be idealized as being confined to only a part near the plate–web junctions (plate edges). It is thus quite usual in such a situation that the total load would be carried by two strips of combined width, b_e, situated near the plate edges, carrying the maximum stress uniformly as a representative, instead of the actual stress distribution.

Regardless of the reason for the nonuniform stress distribution, the plate effectiveness may typically be characterized by a parameter b_e, which is the width (or breadth) over which the maximum membrane stress at the intersection of the flange and web is idealized to occur uniformly, with the total force thus carried being the same as that supplied by the (actual) nonuniform stress distribution across the flange (attached plating).

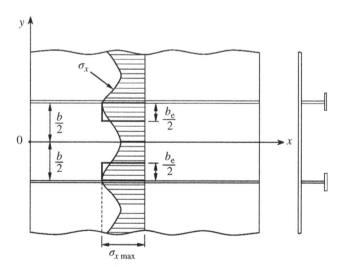

Figure 2.7 Effective width of the attached plating in a stiffened panel.

Following the coordinate system as shown in Figure 2.7, the effective width (or breadth), b_e, can be evaluated by

$$b_e = \frac{\int_{-b/2}^{b/2} \sigma_x dy}{\sigma_{xmax}} = b\frac{\sigma_{xav}}{\sigma_{xmax}} \tag{2.1}$$

where σ_x is the nonuniform normal stress, σ_{xav} is the average stress, and σ_{xmax} is the maximum normal stress at the plate–web junctions.

2.6.1 Shear Lag-Induced Ineffectiveness: Effective Breadth of the Attached Plating

An analytical formulation of the effective breadth for a plate–stiffener combination under predominantly shear lag or bending on a wide flange is now derived. For a review of various methods to derive analytical formulations of the shear lag-oriented effective breadth for wide-flanged beams in the plate–stiffener combination model, interested readers may refer to Troitsky (1976).

It is evident from Equation (2.1) that the nonuniform normal stress distribution must be known to calculate the effective breadth. To compute the stress distribution, the classical theory of elasticity (Timoshenko & Gere 1961) can be applied. For two-dimensional problems, the relationship between strains and displacements is given by

$$\varepsilon_x = \frac{\partial u}{\partial x}, \quad \varepsilon_y = \frac{\partial v}{\partial y}, \quad \gamma_{xy} = \frac{\partial u}{\partial y} + \frac{\partial v}{\partial x} \tag{2.2}$$

where ε_x and ε_y are the normal strains in the x or y direction, γ_{xy} is the shear strain, and u and v are the displacements in the x or y direction.

The relationship between the stresses and strains for two-dimensional problems is given by

$$\varepsilon_x = \frac{1}{E}(\sigma_x - v\sigma_y), \quad \varepsilon_y = \frac{1}{E}(\sigma_y - v\sigma_x), \quad \gamma_{xy} = \frac{2(1+v)}{E}\tau_{xy} \tag{2.3}$$

where σ_x and σ_y are the normal stresses in the x or y direction, τ_{xy} is the shear stress, and v is Poisson's ratio.

The stress distribution for two-dimensional problems can be obtained by solving the following compatibility equation:

$$\frac{\partial^4 F}{\partial x^4} + 2\frac{\partial^4 F}{\partial x^2 \partial y^2} + \frac{\partial^4 F}{\partial y^4} = 0 \tag{2.4}$$

where F is Airy's stress function, which satisfies the following condition:

$$\sigma_x = \frac{\partial^2 F}{\partial y^2}, \quad \sigma_y = \frac{\partial^2 F}{\partial x^2}, \quad \tau_{xy} = -\frac{\partial^2 F}{\partial x \partial y} \tag{2.5}$$

To calculate the nonuniform normal stress distribution of the attached plating in the x direction, it is assumed that the plate's lateral deflection is proportional to $\sin(2\pi x/\omega)$, where ω is the deflection wavelength, depending on the rigidities of the stiffener and the type of load application. For stiff transverse frames, one may approximately take $\omega = L$.

In this case, the axial displacement in the x direction, u, along the plate–web intersection, that is, at $y = \pm b/2$, may be calculated as follows (Yamamoto et al. 1986):

$$u = -u_0 \cos\frac{2\pi x}{\omega} \tag{2.6}$$

where u_0 is the amplitude of the axial displacement function.

The axial strain, ε_x, at $y = \pm b/2$ can then be calculated by substituting Equation (2.6) into Equation (2.2) as follows:

$$\varepsilon_x\big|_{y=\pm b/2} = \frac{\partial u}{\partial x}\bigg|_{y=\pm b/2} = \varepsilon_0 \sin\frac{2\pi x}{\omega} \tag{2.7}$$

where $\varepsilon_0 = u_0(2\pi/\omega)$.

To satisfy Equation (2.4), the stress function, F, may be expressed as follows:

$$F = f(y)\sin\frac{2\pi x}{\omega} \tag{2.8}$$

where

$$f(y) = C_1 \frac{2\pi y}{\omega}\sinh\frac{2\pi y}{\omega} + C_2 \cosh\frac{2\pi y}{\omega}$$

where C_1 and C_2 are constants to be determined by the boundary conditions.

To determine the two unknowns, C_1 and C_2, of Equation (2.8), two boundary conditions are applied. Although Equation (2.7) can be one boundary condition, the other one is provided so that the symmetric condition must be attained along the center line of the attached plating between two adjacent stiffeners, which is given by

$$\frac{\partial v}{\partial x}\bigg|_{y=0} = 0 \tag{2.9}$$

By substituting Equation (2.5) into Equation (2.3), the axial strain, ε_x, can be expressed in terms of Airy's stress function as follows:

$$\varepsilon_x = \frac{1}{E}\left(\frac{\partial^2 F}{\partial y^2} - v\frac{\partial^2 F}{\partial x^2}\right) \tag{2.10}$$

By substituting Equation (2.8) into Equation (2.10) and considering Equation (2.7), the first boundary condition can be written as follows:

$$\frac{d^2 f(y)}{dy^2} + v\left(\frac{2\pi}{\omega}\right)^2 f(y) = E\varepsilon_0 \quad \text{at } y = \pm\frac{b}{2} \tag{2.11}$$

The second boundary condition, Equation (2.9), can be rewritten using Equation (2.2) as follows:

$$\frac{\partial \gamma_{xy}}{\partial x} = \frac{\partial^2 u}{\partial x\partial y} + \frac{\partial^2 v}{\partial x^2} = \frac{\partial^2 u}{\partial x\partial y} = \frac{\partial \varepsilon_x}{\partial y} \quad \text{at } y = 0 \tag{2.12}$$

Substituting Equations (2.2), (2.3), (2.5), and (2.8) into Equation (2.12), the second boundary condition becomes the following third-order differential equation:

$$\frac{d^3 f(y)}{dy^3} - (2+v)\left(\frac{2\pi}{\omega}\right)^2 \frac{df(y)}{dy} = 0 \quad \text{at} \quad y = 0 \tag{2.13}$$

By substituting $f(y)$ in Equation (2.8) into Equations (2.11) and (2.13) and solving the set of the resulting two simultaneous equations with regard to C_1 and C_2, we get

$$C_1 = C_3 \sinh\frac{\pi b}{\omega},$$

$$C_2 = C_3\left[\left(\frac{1-v}{1+v}\right)\sinh\frac{\pi b}{\omega} - \frac{\pi b}{\omega}\cosh\frac{\pi b}{\omega}\right] \tag{2.14}$$

$$C_3 = E\varepsilon_0\left(\frac{\omega}{2\pi}\right)^2\left[\left(\frac{3-v}{2}\right)\sinh\frac{2\pi b}{\omega} - (1+v)\frac{\pi b}{\omega}\right]^{-1}$$

The normal stress, σ_x, can now be obtained by substituting Equations (2.8) and (2.14) into Equation (2.5) as follows:

$$\sigma_x = \left(\frac{2\pi}{\omega}\right)^2\left[C_1\frac{2\pi y}{\omega}\sinh\frac{2\pi y}{\omega} + (2C_1 + C_2)\cosh\frac{2\pi y}{\omega}\right]\sin\frac{2\pi x}{\omega} \tag{2.15}$$

By substituting Equations (2.14) and (2.15) into Equation (2.1), the effective breadth, b_e, can be calculated as follows:

$$b_e = \frac{4\omega\sinh^2(\pi b/\omega)}{\pi(1+v)[(3-v)\sinh(2\pi b/\omega) - 2(1+v)(\pi b/\omega)]} \tag{2.16}$$

The effective breadth normally varies along the span of the plate–stiffener combination model, but for practical design purposes, it may be taken to have the smallest value, which occurs at the location at which the maximum longitudinal stress develops. Because b_e must be smaller than b, Equation (2.16) may be approximated as

$$\frac{b_e}{b} = \begin{cases} 1.0 & \text{for } b/\omega \le 0.18 \\ 0.18L/b & \text{for } b/\omega > 0.18 \end{cases} \tag{2.17}$$

As previously noted, the wavelength, ω, in Equation (2.16) or (2.17) may approximately be taken as $\omega = L$ for the attached plating between two stiff transverse frames. Figure 2.8 shows the variation of the effective breadth (or effective flange width) from Equations (2.16) and (2.17) versus the ratio of the stiffener spacing to the span of the plate–stiffener combination model when ω is equal to L. It can be seen from Figure 2.8 that the normalized effective breadth significantly decreases as the breadth of the attached plating or the span length increases.

Equation (2.16) or (2.17) can be used to evaluate the effective breadth for the attached plating of a plate–stiffener combination model under predominantly out-of-plane bending.

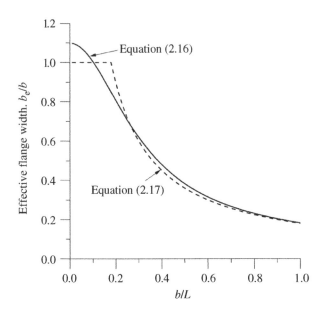

Figure 2.8 Variation of the effective breadth versus the ratio of stiffener spacing to the span of the plate–stiffener combination model when ω is equal to L.

2.6.2 Buckling-Induced Ineffectiveness: Effective Width of the Attached Plating

Strictly speaking, three different aspects of the effective width concept have been applied in the analysis of plate post-buckling behavior: the effective width for strength, the effective width for stiffness, and the reduced tangent-modulus width.

Immediately after buckling of a perfectly flat plate under axial compression, the maximum normal stress becomes greater than the average stress. It may be apparent in this case that the ratio of the effective width to the full width is the same as the ratio of the average stress to the maximum stress as defined in Equation (2.1). It has been shown that the maximum load-carrying capacity of a plate is near the load at which the maximum normal stress reaches the material yield stress. Because the effective width in terms of the maximum normal stress is useful for prediction of a plate's ultimate strength, it is termed the effective width for strength.

The tendency of increasing the average strain with the average stress is of course greater after buckling than before buckling. As long as the plate–web junction remains straight, the average value of the maximum normal stress along the plate edges may be obtained for a uniaxially compressed plate as follows:

$$\sigma_{x\max} = E\varepsilon_{xav} = E\frac{u}{L} \tag{2.18}$$

where ε_{xav} is the average axial strain of the attached plating, which may approximately be taken as the average value of the axial strain along the plate–web junctions, that is, $\varepsilon_{xav} = \varepsilon_x$ at $y = \pm b/2$, and u is the end displacement.

In this case, the effective width can also be evaluated from Equation (2.1) but replacing σ_{xmax} with the axial strain of Equation (2.18). An effective width for stiffness, that is, based on the average axial strain, may be used to characterize the overall stiffness of a buckled plate under predominantly axial compression.

The plate stiffness against axial compression is reduced immediately after buckling. Although this behavior may be characterized by the effective width for stiffness, it is sometimes of interest to know the magnitude of the tangent stiffness or the slope of the average stress–strain curve after buckling, which can be computed mathematically as $\partial \sigma_{xav}/\partial \varepsilon_{xav}$ in the post-buckling regime. The tangent stiffness after buckling is termed the "tangent effective width" or the "effective Young's modulus," E^*. Using this formulation, the ratio of the compressive stiffness after buckling to that before buckling is given by E^*/E. For a perfectly flat plate simply supported at four edges, it is known that $E^*/E \approx 0.5$ after buckling. As long as the unloaded edges remain straight so that some transverse stresses develop along the unloaded edges, it is recognized that E^*/E corresponds to $\partial \sigma_{xav}/\partial \varepsilon_{xav}$, whereas the former is always greater than the latter when the unloaded edges are free to move in plane with no stresses along them (Rhodes 1982). Chapter 4 describes this in detail.

Extensive reviews of the derivation of the effective width formulas for plates were made by Faulkner (1975) and Rhodes (1982). Ueda et al. (1986) derived the effective width formula for a plate under combined biaxial compression and edge shear accounting for the effects of initial deflections and welding induced residual stresses. Usami (1993) studied the effective width of plates buckled in compression and in-plane bending.

Although the concept of effective width is aimed at the evaluation of in-plane stiffness of plate elements buckled in compression, Paik (1995) suggested a new concept for the effective shear modulus to evaluate the effectiveness of plate elements buckled in shear stress. The effective shear modulus concept is useful to compute the post-buckling behavior of plates under predominant shear stresses.

One of the most typical effective width expressions for the compressive strength of long plates that are often used in industry is given in the following form:

$$\frac{b_e}{b} = \begin{cases} 1.0 & \text{for } \beta < 1 \\ C_1/\beta - C_2/\beta^2 & \text{for } \beta \geq 1 \end{cases} \tag{2.19a}$$

where C_1 and C_2 are constants that depend on the plate boundary conditions and β is the plate slenderness ratio for the full section, as defined in Table 2.1. Based on the analysis of available experimental data for steel plates with initial deflections at a moderate level but without residual stresses, Faulkner (1975) proposed $C_1 = 2.0$ and $C_2 = 1.0$ for plates simply supported at all (four) edges or $C_1 = 2.25$ and $C_2 = 1.25$ for plates clamped at all edges.

Although the original von Karman effective width expression of plates, that is, $b_e/b = \sqrt{(\sigma_{cr}/\sigma_Y)}$, where σ_{cr} is the critical stress and σ_Y is the yield stress, is considered to be reasonably accurate for relatively thin plates, it was found to be optimistic for relatively thick plates with initial imperfections. In this regard, Winter (1947) modified the von Karman equation as follows:

$$\frac{b_e}{b} = \sqrt{\frac{\sigma_{cr}}{\sigma_{max}}} \left(1 - 0.25 \sqrt{\frac{\sigma_{cr}}{\sigma_{max}}}\right) \tag{2.19b}$$

where σ_{max} is the maximum normal stress, which may be taken as $\sigma_{max} = \sigma_Y$.

Equation (2.19b) has been widely used to evaluate the post-buckling strength of cold-formed steel plates (AISI 1996, ENV 1993-1-1 1992). In some design codes, the term 0.25 in Equation (2.19b) is changed to 0.218 or 0.22.

Equation (2.19) may be used to evaluate the effective width of the attached plating of the plate–stiffener combination model under predominantly axial compressive loads. More refined expressions of the plate effective width (and the effective shear modulus) are described in Chapter 4.

2.6.3 Combined Shear Lag-Induced and Buckling-Induced Ineffectiveness

In reality, it is important to realize that the plating in plated structures is likely to be subjected to combined compressive loads and lateral pressure, resulting in both buckling and shear lag. In this case, evaluation of the effectiveness of the attached plating in the plate–stiffener combination model must account for both effects.

In this case, $\sigma_{x\max}$ of Equation (2.1) must be the maximum compressive stress, which is expressed as a function of the combined compressive loads, lateral pressure, and initial imperfections, as described in detail in Chapter 4.

2.7 Plastic Cross-Sectional Capacities

In the allowable working stress design method, "first yield" is often used as a design criterion, even though most metal structures can experience local yielding and still withstand some further increase of loading as the internal stresses are redistributed because of the ductility of material.

In contrast, the structural design criteria for the ULS design are based on maximum load-carrying capacity or ultimate strength based on the plastic theory. When the ultimate strength of a plate–stiffener combination model is being considered, plastic cross-sectional capacities are sometimes of interest when the effects of local buckling and strain hardening are not of primary concern.

In the subsequent sections, the cross-sectional capacities of the plate–stiffener combination model at either first or full (complete) yield are described under axial load, sectional shear, bending, or a combination of these.

2.7.1 Axial Capacity

The plastic capacity, P_P, for axial load is calculated by

$$P_P = \pm \left(A_p \sigma_{Yp} + A_w \sigma_{Yw} + A_f \sigma_{Yf} \right) \tag{2.20}$$

where Equation (2.20) is valid for axial tension and axial compression when local buckling does not take place, and a relevant sign convention is used (e.g., positive sign for tension or negative sign for compression).

2.7.2 Shear Capacity

It is practically considered that only the cross-sectional part parallel to the direction of a shearing force contributes to the shear structural resistance. When the vertical sectional shear force (with a positive sign for positive shear and a negative sign for negative shear) is considered, for instance, only the stiffener web cross-sectional area of a plate–stiffener combination model is included in the calculation of shear capacity, F_P, as follows:

$$F_P = \pm \left(A_w \tau_{Yw} \right) \tag{2.21}$$

where $\tau_{Yw} = \sigma_{Yw}/\sqrt{3}$ is the shear yield stress of web.

2.7.3 Bending Capacity

In a beam with uniform material properties, the first yield occurs at the outer fiber of the cross section at which the greatest bending develops in the span. With further loading, the cross section becomes entirely plastic. The plastic bending capacity of beams typically depends on their cross-sectional geometry and their material properties. The capacity formulas take a positive sign for positive bending and a negative sign for negative bending.

2.7.3.1 Rectangular Cross Section

Before the plastic bending capacity of the plate–stiffener combination model is calculated, a simpler case with rectangular cross section is considered. The neutral axis (N.A.) is in this case located at half the height of the web due to symmetry, as shown in Figure 2.9.

The first yield bending capacity, M_Y, can then be obtained by the first moment of axial stresses with regard to the neutral axis when either the upper or lower outer fiber yields as follows:

$$M_Y = \pm \int_{-h_w/2}^{h_w/2} \sigma_x t_w z dz = \pm Z_Y \sigma_{Yw} \tag{2.22}$$

where $Z_Y = t_w h_w^2/6$ is the first yield section modulus.

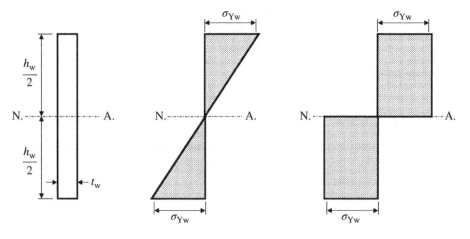

Figure 2.9 Stress distribution of a rectangular cross-sectional beam at first and full yielding.

The bending capacity at first yield can also be predicted by applying simple beam theory, leading to the following linear relationship between the bending moment and the bending stress:

$$M = \pm \frac{\sigma_x}{z} I \tag{2.23}$$

where M is the bending moment, σ_x is the bending stress, I is the moment of inertia, and z is the distance from the neutral axis.

In the case of the rectangular cross-sectional beam, because $I = t_w h_w^3/12$ and $\sigma_x = \sigma_{Yw}$ at the outer fiber, that is, $z = \pm h_w/2$, the first yield bending capacity obtained from Equation (2.23) corresponds to Equation (2.22). The position of the plastic neutral axis is determined so that the areas on two parts, that is, in the tension or compression side, are equal. For a symmetric rectangular cross-sectional beam, the plastic neutral axis is located at half the height of the web.

The full plastic bending capacity, M_P, is calculated by the first moment of axial stresses with regard to the plastic neutral axis when the cross section entirely yields, as shown in Figure 2.9, namely,

$$M_P = \pm \int_{-h_w/2}^{h_w/2} \sigma_x t_w z dz = \pm Z_P \sigma_{Yw} \tag{2.24}$$

where $Z_P = t_w h_w^2/4$ is the plastic section modulus.

2.7.3.2 Plate–Stiffener Combination Model Cross Section

The location of the elastic neutral axis from outer surface of the effective plating as shown in Figure 2.10 can be determined from Equation (7.44) or Table 2.1 as follows:

$$z_p = \frac{0.5 A_{pe} t + A_w (t + 0.5 h_w) + A_f \left(t + h_w + 0.5 t_f \right)}{A_{pe} + A_w + A_f} \tag{2.25}$$

The first yield bending capacity, M_Y, can be calculated from Equation (7.47) by applying classical simple-beam theory as follows:

$$M_Y = \pm \frac{I_e}{(z_p - 0.5t)} \sigma_{Yp} \quad \text{for first yield at attached plating} \tag{2.26a}$$

$$M_Y = \pm \frac{I_e}{\left(t + h_w + 0.5 t_f - z_p \right)} \sigma_{Yf} \quad \text{for first yield at stiffener flange} \tag{2.26b}$$

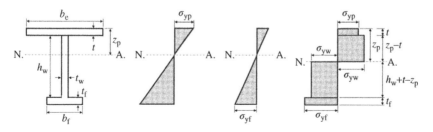

Figure 2.10 Stress distribution of the plate–stiffener combination section at full yield.

where I_e is defined in Table 2.1. It is noted that the segments have been assumed to remain elastic except the attached plating or stiffener flange as shown in Figure 2.10.

Equation (2.26) can be rewritten as follows:

$$M_Y = \pm Z_{Yp}\sigma_{Yp} \quad \text{for first yield at attached plating} \tag{2.27a}$$

$$M_Y = \pm Z_{Yf}\sigma_{Yf} \quad \text{for first yield at stiffener flange} \tag{2.27b}$$

where $Z_{Yp} = I_e/(z_p - 0.5t)$ is the first yield section modulus at the mid-thickness of the attached plating side and $Z_{Yf} = I_e/(h_w + t + 0.5t_f - z_p)$ is the first yield section modulus at the mid-thickness of the stiffener flange side.

The location of the plastic neutral axis from the outer surface of plating as shown in Figure 2.10 can be determined from Equation (7.50) as follows:

$$z_p = \frac{0.5A_{pe}\sigma_{Yp}t + A_w\sigma_{Yw}(t + 0.5h_w) + A_f\sigma_{Yf}(t + h_w + 0.5t_f)}{A_{pe}\sigma_{Yp} + A_w\sigma_{Yw} + A_f\sigma_{Yf}} \tag{2.28}$$

The full plastic bending capacity, M_P, can also be calculated from the simple-beam theory method from Equation (7.51) as follows:

$$
\begin{aligned}
M_P = {} & \pm \left[b_e t \sigma_{Yp}\left(z_p - \frac{t}{2}\right) + (z_p - t)t_w\sigma_{Yw}\frac{z_p - t}{2} \right.\\
& + (h_w + t - z_p)t_w\sigma_{Yw}\frac{h_w + t - z_p}{2} + b_f t_f \sigma_{Yf}\left(h_w + t - z_p + \frac{t_f}{2}\right) \Big]\\
= {} & \pm \left[b_e t \sigma_{Yp}\left(z_p - \frac{t}{2}\right) + \frac{(z_p - t)^2}{2}t_w\sigma_{Yw} + \frac{(h_w + t - z_p)^2}{2}t_w\sigma_{Yw} \right.\\
& + b_f t_f \sigma_{Yf}\left(h_w + t - z_p + \frac{t_f}{2}\right) \Big]
\end{aligned}
\tag{2.29}
$$

2.7.4 Capacity Under Combined Bending and Axial Load

When combined bending and axial loading is applied, the stress distribution at full yield of the cross section can be presumably considered to be as shown in Figure 2.11.

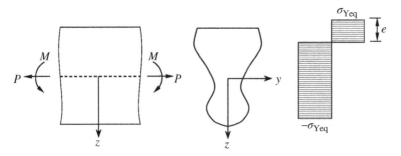

Figure 2.11 Stress distribution over an arbitrary cross section under combined bending and axial load.

The axial load, P, and bending moment, M, can then be calculated by integration of the stress distribution over the cross section as follows:

$$P = \int_A \sigma_x dA, \quad M = \int_A \sigma_x z dA \tag{2.30}$$

where $\int_A ()dA$ represents the integration over the cross-sectional area.

In Equation (2.30), P and M are expressed as a function of an unknown parameter, e, which is the distance from the outer fiber of the attached plating to the plastic neutral axis. Combining these two equations with regard to the unknown parameter, an interaction relationship between P and M can be obtained. It can be recognized that the plastic capacity of the sections under combined bending and axial loads is less than that for bending alone.

2.7.4.1 Rectangular Cross Section

Before the plastic capacity of the plate–stiffener combination model under combined M and P is calculated, a simpler case with a rectangular cross section is considered. The stress distribution over a rectangular cross section at full yield may be presumed as that shown in Figure 2.12. In this case, the stress distribution can be divided into two parts, one for pure bending stress and the other for pure axial stress.

Based on the presumed stress distribution, the reduced bending moment capacity, M, and the associated axial load, P, can be calculated by

$$P = t_w(h_w - 2e)\sigma_{Yw} \tag{2.31a}$$

$$M = M_P - M_{Pe} = M_P - Z_{Pe}\sigma_{Yw}$$

$$= \frac{t_w h_w^2}{4}\sigma_{Yw} - \frac{t_w(h_w - 2e)^2}{4}\sigma_{Yw} \tag{2.31b}$$

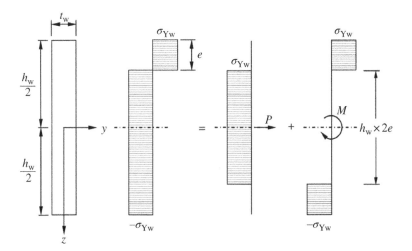

Figure 2.12 Stress distribution of a rectangular cross section under combined bending and axial load.

where $P_P = t_w h_w \sigma_{Yw}$, $M_P = (t_w h_w^2/4)\sigma_{Yeq}$, $M_{Pe} = Z_{Pe}\sigma_{Yw}$ is the plastic bending capacity for the cross section of e, and $Z_{Pe} = [t_w(h_w - 2e)^2]/4$ is the plastic section modulus for the cross section of e.

Combining Equations (2.31a) and (2.31b), the plastic capacity interaction equation of the plate–beam combination under M and P is given by

$$\left|\frac{M}{M_P}\right| + \left(\frac{P}{P_P}\right)^2 = 1 \tag{2.32}$$

Figure 2.13 shows the aforementioned interaction curve for rectangular cross sections under combined (positive) bending and (positive) axial load. As evident from Figure 2.13, the plastic bending capacity significantly decreases as the axial load increases.

2.7.4.2 Plate–Stiffener Combination Model Cross Section

The plastic stress distribution of a plate–stiffener combination model whose cross section is subject to combined bending and axial load may be described in terms of the position of the plastic neutral axis, e, which is the distance from the outer fiber of the attached plating to the plastic neutral axis, as those shown in Figure 2.14 (Ueda & Rashed 1984).

In contrast to the symmetric rectangular cross section, the expressions of the plastic capacity interaction relationships of the plate–stiffener combination model under M and P may differ in accordance with the direction of load application. Based on the presumed stress distribution for each state of load combination, the reduced bending moment, M, and the associated axial load, P, can be expressed as a function of the unknown, e. By omitting e between the two expressions, an interactive relationship for the plastic capacity is then derived.

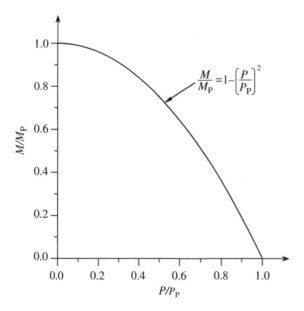

Figure 2.13 The interaction curve for a rectangular cross section under combined bending and axial load.

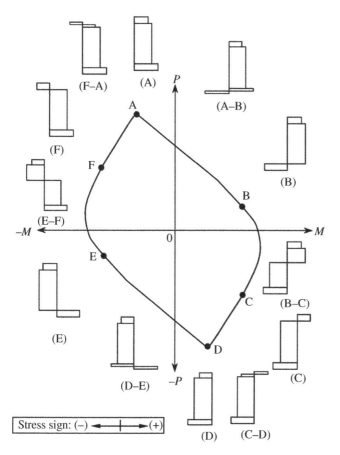

Figure 2.14 Presumed stress distributions over the plate–stiffener combination model cross section for varying states of combined bending and axial load.

In the interest of simplicity, Equation (2.32) is often used for the plastic capacity interaction formula of the plate–stiffener combination model under M and P, but using P_P and M_P for the plate–stiffener combination model.

2.7.5 Capacity Under Combined Bending, Axial Load, and Shearing Force

When combined bending, M, the axial load, P, and the shearing force, F, are applied, stress distributions similar to that for combined bending and axial load may be adopted, assuming that the shearing forces are sustained by the stiffener web alone. In this regard, a reduced yield strength, σ_{Yv}, for the stiffener web may be introduced by the Tresca yield criterion, Equation (1.31b), as follows (ENV 1993-1-1 1992):

$$\sigma_{Yv} = 2\left[\left(\frac{\sigma_{Yw}}{2}\right)^2 - \left(\frac{F}{A_w}\right)^2\right]^{0.5}$$

(2.33)

where F is the applied shearing force.

For combined bending, axial load, and shearing forces, therefore, the approximate reduced plastic bending capacity of the plate–stiffener combination model cross section can be obtained from Equation (2.32) with P_P and M_P for the plate–stiffener combination model, but replacing the stiffener web yield stress σ_{Yw} with σ_{Yv} of Equation (2.33).

2.8   Ultimate Strength of the Plate–Stiffener Combination Model Under Bending

A structure can collapse when it develops sufficient plastic hinges to form a plastic mechanism. The plastic collapse strength formulas of the plate–stiffener combination model under bending with many types of loading and end conditions can typically be derived by applying the rigid-plastic theory (Hodge 1959, Neal 1977), whereas the plate–stiffener combination model is dealt with as a beam.

Belenkiy and Raskin (2001) showed that the ultimate strength of the beams as determined by the rigid-plastic theory corresponds quite well to a "threshold" (ultimate) load obtained by nonlinear finite element analysis. This threshold load is defined to separate the linear elastic regime from the plastic regime. Figure 2.15 illustrates the threshold ultimate load concept for beams. In this figure, the load–deflection curve indicated by the solid line may be divided into three regimes: the linear elastic regime (0b), the transitional regime (bc) in which plastic deflection begins to grow, and the large-deflection regime (cd).

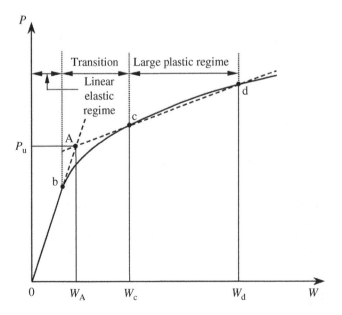

Figure 2.15 A schematic representation of the threshold (ultimate) load of beams.

The true load–deflection curve was approximated by a bilinear relation, that is, 0Ad, where the deflections w_c and w_d could be taken as $w_c = 0.005L$ and $w_d = 0.01L$, where L is the span of the beam. The threshold ultimate load is then defined as the load, P_u, at point A. Based on the comparisons of the ultimate loads of beams with various types of end conditions and load applications as determined by the rigid-plastic theory with the threshold loads obtained by nonlinear finite element analyses, Belenkiy and Raskin (2001) also developed some important insights: (i) the plastic deflection, w_A, that corresponds to the threshold loads is normally between $0.001L$ and $0.004L$, and (ii) the effects of strain hardening and membrane stress on the threshold loads are usually small.

For the use of the rigid-plastic theory to derive the ultimate strength formulations of beams, the basic assumptions noted in the succeeding text are typically assumed to be adequate:

- Strain-hardening effects can be ignored.
- The Tresca yield criterion is applicable.
- Small deformations are involved, and thus the membrane effects may be neglected.
- Local buckling does not take place.
- The localized plastic region does not expand into the longitudinal (axial) direction of beams, and thus the plastic hinge is considered to remain fixed at a particular cross section.
- The cross section of the beam remains in plane, that is, it does not distort in the axial direction.

In the following sections, the plastic strength formulas of beams are derived for various loading and end conditions, neglecting the effect of local buckling. In this case, the plastic strength formulas are expressed as functions of the beam's full plastic bending moment, M_P. The approximate ultimate strength of the beams taking into account the effects of local buckling may then be estimated from those plastic strength formulas but replacing M_P with the ultimate bending moment, M_u, which is determined by considering the local buckling effect at the cross section.

2.8.1 Cantilever Beams

The plastic collapse strength formulas for cantilever beams under various types of load applications as shown in Figure 2.16 are first derived. When a beam is subjected to point load Q at the free end as shown in Figure 2.16a, the bending moment along the beam span is given by

$$M = Q(L-x) \tag{2.34}$$

As the point load increases, the plastic region around the fixed end initiates and expands through the thickness. The beam then collapses via the formation of the plastic hinge mechanism if the cross section at the fixed end yields entirely, that is, when the bending moment at the fixed end reaches the plastic bending moment, M_P. Therefore, the plastic collapse load, Q_c, is in this case determined as follows:

$$Q_c = \frac{M_P}{L} \tag{2.35}$$

where M_P is defined in Section 2.7.3.

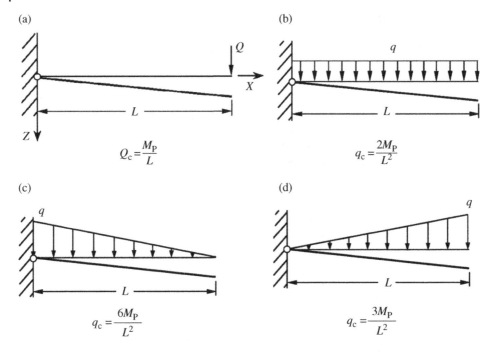

(a)

$$Q_c = \frac{M_P}{L}$$

(b)

$$q_c = \frac{2M_P}{L^2}$$

(c)

$$q_c = \frac{6M_P}{L^2}$$

(d)

$$q_c = \frac{3M_P}{L^2}$$

Figure 2.16 Plastic collapse loads of a cantilever beam under various types of load applications.

Using the same method as described previously, the plastic collapse loads of the cantilever beam under different load applications are determined as shown in Figure 2.16b–d.

2.8.2 Beams Simply Supported at Both Ends

The plastic collapse strength formulas for beams simply supported at both ends and subject to various types of load applications such as those shown in Figure 2.17 are now derived.

When a beam is subjected to uniformly distributed line loads, as shown in Figure 2.17a, it collapses if the cross section yields at any one location inside the span because both ends are already pinned. Due to the symmetric load application (similar to that shown in Figure 2.17a), maximum bending occurs at the mid-span, where the cross section yields first. The reaction forces at both ends and the bending moment distribution are in this case given by

$$R_A = R_B = \frac{qL}{2}, \quad M = R_A x - \frac{1}{2}qx^2 = \frac{1}{2}qx(L-x) \tag{2.36}$$

The maximum bending moment, M_{max}, at the mid-span, that is, $x = L/2$, is thus obtained as follows:

$$M_{max} = \frac{qL^2}{8} \tag{2.37}$$

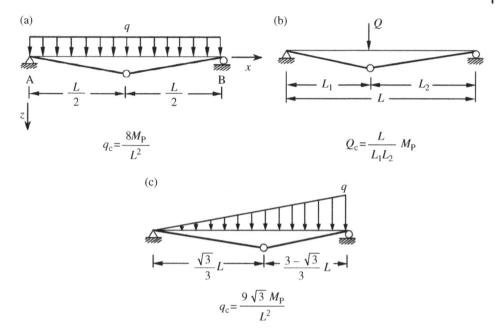

Figure 2.17 Plastic collapse loads of a beam under various types of load applications, simply supported at both ends.

By equating Equation (2.37) with the plastic bending capacity, M_P, the plastic collapse strength, q_c, is obtained by replacing q in Equation (2.37) with q_c as follows:

$$q_c = \frac{8M_P}{L^2} \tag{2.38}$$

The same procedure can also be applied to calculate the plastic collapse strengths of simply supported beams under different load applications, with results as shown in Figure 2.17b or c.

2.8.3 Beams Simply Supported at One End and Fixed at the Other End

The plastic collapse strength formulas for statically indeterminate beams, simply supported at one end and fixed at the other end, under various types of load applications, such as those shown in Figures 2.18 and 2.19, are now derived.

When a beam is subjected to uniformly distributed line loads as shown in Figure 2.18, the equilibrium condition gives the following equations for the reaction forces and end moment:

$$R_A = \frac{qL}{2} + \frac{M_A}{L}, \quad R_B = \frac{qL}{2} - \frac{M_A}{L} \tag{2.39}$$

where R_A and R_B are the reaction forces at points A and B and M_A is the redundant reaction (bending moment) at end A.

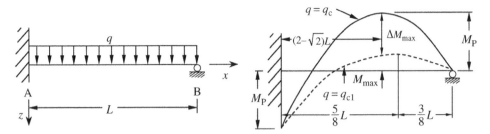

Figure 2.18 Moment distribution of a beam under uniform line load, simply supported at one end and fixed at the other end.

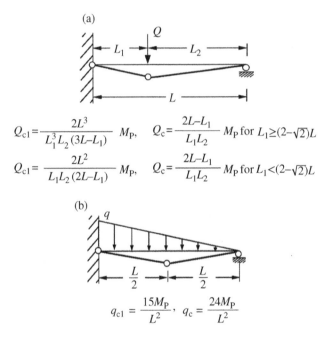

$$Q_{c1} = \frac{2L^3}{L_1^3 L_2 (3L-L_1)} M_P, \quad Q_c = \frac{2L-L_1}{L_1 L_2} M_P \text{ for } L_1 \geq (2-\sqrt{2})L$$

$$Q_{c1} = \frac{2L^2}{L_1 L_2 (2L-L_1)} M_P, \quad Q_c = \frac{2L-L_1}{L_1 L_2} M_P \text{ for } L_1 < (2-\sqrt{2})L$$

(b)

$$q_{c1} = \frac{15 M_P}{L^2}, \quad q_c = \frac{24 M_P}{L^2}$$

Figure 2.19 Plastic collapse loads of a beam under other types of load applications, simply supported at one end and fixed at the other end. (Note: Q_{c1} or q_{c1} indicates the critical load when the plastic hinge is only formed at the fixed end.)

The bending moment along the span can be expressed in terms of the redundant reaction, M_A, as follows:

$$M = -M_A + R_A x - \frac{q}{2}x^2 = -M_A\left(1 - \frac{x}{L}\right) + \frac{qL}{2}x - \frac{q}{2}x^2 \tag{2.40}$$

The bending strain energy, U, of a beam with the effective cross-sectional properties is given by

$$U = \frac{1}{2}\int_{\text{Vol}} \sigma_x \varepsilon_x d\text{Vol} = \frac{1}{2E}\int_{\text{Vol}} \sigma_x^2 d\text{Vol} = \frac{1}{2E}\int_{\text{Vol}} \left(\frac{M}{I_e}z\right)^2 d\text{Vol}$$

$$= \frac{1}{2E}\int_0^L \left(\frac{M}{I_e}\right)^2 \left(\int_{A_e} z^2 dA_e\right) dx = \frac{1}{2E}\int_0^L \frac{M^2}{I_e} dx \tag{2.41}$$

where $\int_{A_e} z^2 dA_e = I_e$, σ_x is the bending stress, ε_x is the bending strain, $\int_{Vol} (\ )d\text{Vol}$ indicates the volume integration over the beam, and $\int_{A_e} (\ )dA_e$ indicates the area integration over the effective cross section of the beam, where the subscript "e" represents that the attached plating of the plate–stiffener combination model (beam) has the effective breadth.

Applying the so-called Castigliano principle, the rotation at the fixed end, which may be calculated by differentiating the strain energy with regard to the associated bending moment, must be zero because of the fixed end condition, namely,

$$\theta_A = \frac{\partial U}{\partial M_A} = \int_0^L \frac{M}{EI_e} \frac{\partial M}{\partial M_A} dx = 0 \tag{2.42}$$

where θ_A is the rotation at end A.

By substituting Equation (2.40) into Equation (2.42), the redundant reaction moment, M_A, is determined by

$$M_A = \frac{qL^2}{8} \tag{2.43}$$

As the load increases, the maximum bending moment occurs at the fixed end, that is, $x = 0$. When the maximum bending moment at end A reaches the plastic bending capacity $|M_P|$, the fixed end becomes a plastic hinge. The critical lateral load in this state is defined by replacing q with q_{c1} as follows:

$$q_{c1} = \frac{8M_P}{L^2} \tag{2.44}$$

Even after the cross section at the fixed end has yielded, the beam may be able to sustain further loading because a plastic hinge mechanism has not yet formed. Until $q = q_{c1}$, the maximum bending moment inside the span, M_{max1}, occurs at $x = \frac{5}{8}L$. Because end A is now considered to be pinned, thus keeping the bending moment constant at $-M_P$, the additional bending moment, ΔM, due to the additional lateral load, that is, $q - q_{c1}$, is given as for a beam simply supported at both ends, namely,

$$\Delta M = \frac{1}{2}(q - q_{c1})(Lx - x^2) \tag{2.45}$$

The beam collapses if the total (accumulated) maximum bending moment inside the span reaches the plastic bending moment because a plastic hinge mechanism is then formed as follows:

$$M_{max} = M_{max1}^* + \Delta M_{max} \tag{2.46}$$

where M_{max} occurs at $x = (2 - \sqrt{2})L$, which does not correspond to the location of M_{max1}. M_{max1}^* is calculated by $M_{max1} = M$ using Equations (2.40) and (2.43) at $q = q_{c1}$ and $x = (2 - \sqrt{2})L$, and ΔM_{max} is obtained by $\Delta M_{max} = \Delta M$ using Equation (2.45) at $x = (2 - \sqrt{2})L$.

The beam eventually collapses if the accumulated maximum bending moment inside the span reaches the plastic bending moment, that is, $M_{max} = M_P$. In this state, the plastic collapse strength of the beam subject to evenly distributed line loads is given by

$$q_c = \frac{2(3 + 2\sqrt{2})M_P}{L^2} \tag{2.47}$$

For the other types of load applications, as shown in Figures 2.19a and b, the first critical and plastic collapse loads may be calculated using the same approach as before.

2.8.4 Beams Fixed at Both Ends

The plastic collapse load formulas for statically indeterminate beams fixed at both ends under various types of load applications, as shown in Figures 2.20 and 2.21, are now derived. In this case, the beams collapse if the cross sections at both ends and at any one location inside the span yield.

For a beam subjected to uniformly distributed line loads as shown in Figure 2.20, plastic hinges form simultaneously at both ends, where the maximum bending moments occur because of the symmetric loading and end conditions. Even after the formation of plastic hinges at both ends, the beam can withstand further loading until the cross section at the mid-span yields, leading to a plastic hinge mechanism.

In Figure 2.20, the bending moment along the span may be given by considering the symmetric load condition with regard to the mid-span as follows:

$$M = -M_A + \frac{qL}{2}x - \frac{q}{2}x^2 \tag{2.48}$$

where $M_A = M_B$ is the bending moment at beam ends.

Because the bending strain energy, U, of the beam with the effective cross section is calculated from Equation (2.41) and the rotation at fixed end A must be zero,

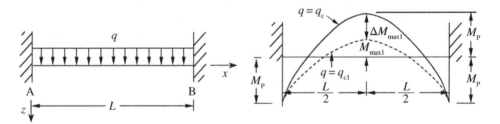

Figure 2.20 Moment distribution of a beam under uniform line load, fixed at both ends.

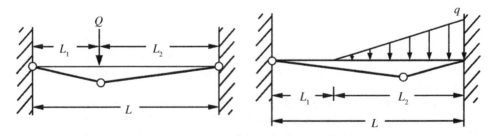

Figure 2.21 Plastic collapse loads of a beam under other types of load applications, fixed at both ends. (Note: Q_{c1} or q_{c1} indicates the critical load when the plastic hinge is only formed at both fixed ends.)

Equation (2.42) is satisfied. By solving Equation (2.42) together with Equation (2.48), M_A is determined by

$$M_A = \frac{qL^2}{12} \tag{2.49}$$

We now get a critical load, q_{c1}, when both ends just yield; that is, the end moment at $x = 0$ or L reaches the plastic bending moment, $-M_P$, namely,

$$q_{c1} = \frac{12M_P}{L^2} \tag{2.50}$$

The maximum bending moment, M_{max1}, which occurs at the mid-span, that is, $x = L/2$ until both ends just yield, is calculated from Equation (2.48) with Equation (2.49) because $q = q_{c1}$:

$$M_{max1} = \frac{q_{c1}L^2}{24} \tag{2.51}$$

Even after both ends have yielded, the beam may sustain further loading until the cross section at the mid-span yields. Although the end moment is kept constant at $-M_P$, the bending moment inside the span increases. Because the beam can now be considered to be simply supported at both ends, the additional bending moment, ΔM, inside the span due to further loading is given by neglecting the membrane stress effects, namely,

$$\Delta M = \frac{q - q_{c1}}{2}\left(Lx - x^2\right) \tag{2.52}$$

Because the maximum additional bending moment, ΔM_{max}, occurs at the mid-span, the total (accumulated) maximum bending moment, M_{max}, at the mid-span is obtained as follows:

$$M_{max} = M_{max1} + \Delta M_{max} = \frac{q_{c1}L^2}{24} + \frac{(q - q_{c1})L^2}{8} = \frac{qL^2}{8} - M_P \tag{2.53}$$

where $\Delta M_{max} = [(q - q_{c1})/8]L^2$.

Because a plastic hinge mechanism is formed when the cross section at the mid-span yields, with $M_{max} = M_P$, the plastic collapse load, q_c, of the beam is finally determined by

$$q_c = \frac{16M_P}{L^2} \tag{2.54}$$

Using a method similar to that used previously, the first critical or plastic collapse loads of the beams under other load applications, such as those shown in Figure 2.21a and b, can be calculated.

2.8.5 Beams Partially Rotation Restrained at Both Ends

When a beam is connected to adjoining structures, the rotation at the ends can be restrained to some degree. A beam partially rotation restrained at both ends is now considered. The beam is subjected to a lateral pressure distribution with a trapezoidal pattern that varies linearly between the two ends as shown in Figure 2.22a, which is given by

(a)

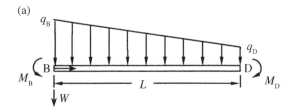

(b)

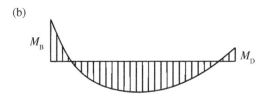

Figure 2.22 A beam elastically restrained at both ends and under lateral pressure: (a) loading; (b) elastic bending moment distribution.

$$q = -\frac{q_B - q_D}{L}x + q_B \tag{2.55}$$

where q_B and p_D are the lateral pressures at ends B and D, respectively.

The end moments arise from the constraints against rotation of the beam at the junctures of the beam and the adjoining structure, as indicated in Figure 2.22a. They thus depend on the torsional rigidity of the adjoining structures.

From the bending moment equilibrium condition, the constraints at the ends of the beam can be defined as follows:

$$\text{At end B}: \left(\frac{d^2w}{dx^2}\right)_{x=0} = \frac{C_B}{L}\left(\frac{dw}{dx}\right)_{x=0} \tag{2.56a}$$

$$\text{At end D}: \left(\frac{d^2w}{dx^2}\right)_{x=L} = \frac{C_D}{L}\left(\frac{dw}{dx}\right)_{x=L} \tag{2.56b}$$

where w is the lateral deflection of the beam and C_B and C_D are the respective constraint constants at the two ends of the beam. For simply supported or clamped ends, these two constants become either zero or infinity, respectively.

The elastic bending moment distribution of the beam is expressed by applying the simple beam theory as follows:

$$M = EI_e\frac{d^2w}{dx^2} = M_B - \frac{x}{L}(M_B - M_D) + \frac{q_B}{2}\left(x^2 - Lx\right) + \frac{q_B - q_D}{6}\left(Lx - \frac{x^3}{L}\right) \tag{2.57}$$

where I_e is the moment of inertia of the beam with effective section.

Figure 2.22b represents the beam's elastic bending moment distribution. It can be seen from the figure that three extreme values of the bending moments developed, at end B, at end D, and inside the span. By performing the double integration of Equation (2.57) and considering the end conditions, the lateral deflection, w, may be expressed by

$$
\begin{aligned}
w = & \frac{q_B + q_D}{24EI_e} \left(\frac{x^4}{2} - Lx^3 + \frac{L^3 x}{2} \right) + \frac{q_B - q_D}{24EI_e} \left(-\frac{x^5}{5L} + \frac{x^4}{2} - \frac{Lx^3}{3} + \frac{L^3 x}{30} \right) \\
& + \frac{M_B}{EI_e} \left(-\frac{x^3}{6L} + \frac{x^2}{2} - \frac{Lx}{3} \right) + \frac{M_D}{EI_e} \left(\frac{x^3}{6L} - \frac{Lx}{6} \right)
\end{aligned}
\tag{2.58}
$$

The end moments at ends B and D can be calculated as a function of the constraint coefficients by substituting Equation (2.58) into the equilibrium condition, Equation (2.56), as follows:

$$
M_B = \frac{C_B L^2}{120} \left[\frac{(q_B - q_D)(2 - C_D) + (q_B - q_D)(30 - 5C_D)}{12 + 4C_B - 4C_D - C_B C_D} \right]
\tag{2.59a}
$$

$$
M_D = \frac{C_D L^2}{120} \left[\frac{(q_B - q_D)(2 + C_B) - (q_B + q_D)(30 + 5C_B)}{12 + 4C_B - 4C_D - C_B C_D} \right]
\tag{2.59b}
$$

The extreme value of the bending moment inside the span occurs where the condition of $dM/dx = 0$ is satisfied. When both ends and any one point inside the span yield, a collapse hinge mechanism is formed. Depending on the end condition, the loading and other details related to the formation of the collapse mechanism may vary.

Under the ideal end conditions previously described in Section 2.8.2 or 2.8.3, the beam collapses if the following criteria are fulfilled:

1) A beam simply supported at both ends:

$$
\frac{q_B}{2} \left(x_p^2 - Lx_p \right) + \frac{q_B - q_D}{6} \left(Lx_p - \frac{x_p^3}{L} \right) = M_P
\tag{2.60}
$$

where

$$
x_p = \frac{L}{2} \qquad\qquad \text{for } q_D = q_B
$$

$$
x_p = \frac{L}{q_B - q_D} \left(q_B - \sqrt{\frac{q_B^2 + q_D^2 + q_D q_B}{3}} \right) \quad \text{for } q_B > q_D
$$

2) A beam fixed at end B and simply supported at end D:

$$
\frac{x_p \left[\left(L^2 - x_p^2 \right) q_D + \left(2L^2 - 3Lx_p + x_p^2 \right) q_B \right]}{6 \left(2L - x_p \right)} = M_P
\tag{2.61}
$$

where x_p is the distance from end B to the plastic hinge inside the span, which is taken so that q_B or q_D should be a minimum.

2.8.6 Lateral-Torsional Buckling

A beam under lateral load that is bent about its major axis can buckle sideways if its compression flange has insufficient stiffness in the lateral direction. At a critical load, the plate–stiffener combination model can become unstable because the compression flange may twist sideways. This phenomenon is sometimes termed lateral-torsional buckling (or tripping), which is generally considered to be one of the many types of behavior that may lead to the ULS of the plate–stiffener combination model. Section 5.8 describes lateral-torsional buckling of stiffeners with attached plating.

2.9 Ultimate Strength of the Plate–Stiffener Combination Model Under Axial Compression

The plate–stiffener combination model under axial compressive loads can be dealt with as a column. Unlike plate panels, which are described in Chapter 4, a column cannot be expected to have residual strength after the inception of buckling, and thus the buckling strength typically is considered to be synonymous with the ultimate strength for the column members.

In this section, the ultimate strength formulations for the plate–stiffener combination model under predominantly axial compressive loads are described.

2.9.1 Large-Deflection Behavior of Straight Columns

From the classical large-deflection column theory, the length, dL', of an infinitesimal element AB of a laterally deflected column, as shown in Figure 2.23, whose initial length was dx, can be calculated by (Shames & Dym 1993)

$$(dL')^2 = (dx)^2(1 + 2\varepsilon_x)$$ (2.62)

where $\varepsilon_x = du/dx + \frac{1}{2}(du/dx)^2 + \frac{1}{2}(dw/dx)^2$ is the axial strain of the straight column taking into account the large-deformation effects, u is the axial displacement, and w is the lateral deflection. (For a perfectly straight column, the added deflection equals the total deflection because no initial deflection exists.)

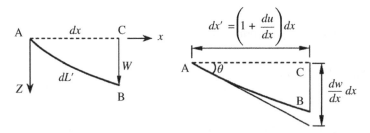

Figure 2.23 Large deflection of a laterally deflected column.

If the column neutral axis is supposed to be incompressible during its bending from the straight-line configuration, then $dL' = dx$. Therefore, Equation (2.62) becomes

$$\left(1 + \frac{du}{dx}\right)^2 + \left(\frac{dw}{dx}\right)^2 = 1 \tag{2.63}$$

From geometric consideration of Figure 2.23, the rotation of the segment AB can be calculated by

$$\sin\theta = \frac{(dw/dx)dx}{\overline{AB}} = \frac{(dw/dx)dx}{\sqrt{(1 + du/dx)^2 + (dw + dx)^2(dx)}} = \frac{dw}{dx} \tag{2.64}$$

because Equation (2.63) is satisfied as long as the column is incompressible.

After deformation, the curvature, $1/R$, of the column is given by considering $dL' = dx$, namely,

$$\frac{1}{R} \equiv \frac{d\theta}{dL'} = \frac{d\theta}{dx} = \frac{d^2w/dx^2}{\cos\theta} = \frac{d^2w/dx^2}{\sqrt{1 - (dw/dx)^2}} \tag{2.65}$$

because $\cos\theta\,(d\theta/dx) = d^2w/dx^2$ from Equation (2.64), $\cos\theta = \sqrt{(1 - \sin^2\theta)} = \sqrt{[1 - (dw/dx)^2]}$, and R is the radius.

When the column deflection is considered sufficiently small, Equation (2.65) is often simplified to

$$\frac{1}{R} \cong \frac{d^2w}{dx^2} \tag{2.66}$$

because $dw/dx \ll 1$ and hence $(dw/dx)^2 \approx 0$ in this case.

The total shortening of the entire column can be calculated by considering Equation (2.63) as follows:

$$u = \int_0^L \frac{du}{dx}dx = \int_0^L \left[\sqrt{1 - \left(\frac{dw}{dx}\right)^2} - 1\right]dx \cong -\frac{1}{2}\int_0^L \left(\frac{dw}{dx}\right)^2 dx \tag{2.67}$$

where it is important to realize that the aforementioned shortening expression accommodates only the effects of lateral deflection, whereas the column neutral axis is incompressible.

We now calculate the strain energy of the column due to bending. Using the Bernoulli–Euler hypothesis for the bending of beams as described in Section 7.5.1, the strain energy, U, can be obtained by

$$U = \frac{1}{2}\int_{\text{Vol}} \sigma_x \varepsilon_x d\text{Vol} = \frac{E}{2}\int_{\text{Vol}} \varepsilon_x^2 d\text{Vol} = \frac{E}{2}\int_{\text{Vol}} \left(\frac{z}{R}\right)^2 d\text{Vol} = \frac{EI_e}{2}\int_0^L \left(\frac{1}{R}\right)^2 dx \tag{2.68}$$

because $\varepsilon_x = z/R$, $\int_{A_e} z^2 dA_e = I_e$, and σ_x is the bending stress. The subscript e represents the effective section.

Substituting Equation (2.66) into Equation (2.68), the strain energy, U, may approximately be calculated by

$$U \cong \frac{EI_e}{2} \int_0^L \left(\frac{d^2 w}{dx^2} \right)^2 dx \tag{2.69}$$

In contrast, the external potential energy, W, can be obtained using Equation (2.67) as follows:

$$W = Pu = P \int_0^L \frac{du}{dx} dx \cong -\frac{P}{2} \int_0^L \left(\frac{dw}{dx} \right)^2 dx \tag{2.70}$$

The total potential energy, Π, can be given by a sum of the strain energy, U, and the external potential energy, W, as follows:

$$\Pi = U + W \tag{2.71}$$

which results in

$$\Pi = \frac{EI_e}{2} \int_0^L \left(\frac{1}{R} \right)^2 dx - P \int_0^L \frac{du}{dx} dx \cong \frac{EI_e}{2} \int_0^L \left(\frac{d^2 w}{dx^2} \right)^2 dx - \frac{P}{2} \int_0^L \left(\frac{dw}{dx} \right)^2 dx$$

for the straight column.

The large-deflection behavior of a column can then be analyzed by applying the principle of minimum potential energy to Equation (2.71). For instance, when the lateral deflection, w, is supposed to be a Fourier series function that satisfies the boundary condition of the column and includes several unknown constants, C_i, the function of w is substituted into Equation (2.71). The constants C_i are then determined by the principle of minimum potential energy because $\partial \Pi / \partial C_i = 0$; refer also to Equation (2.90).

2.9.2 Elastic Buckling of Straight Columns

To study the ultimate strength of columns, of primary interest is elastic buckling. To illustrate the column buckling phenomenon, a simply supported straight column is considered, as shown in Figure 2.24.

When the column deflection is considered sufficiently small, the following values for the column with effective section are obtained:

$$\text{Internal bending moment}: M = -EI_e \frac{1}{R} = -EI_e \frac{d^2 w}{dx^2} \tag{2.72a}$$

$$\text{External bending moment}: M = Pw \tag{2.72b}$$

because the curvature, $1/R$, is given by Equation (2.66).

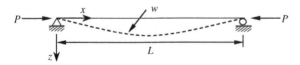

Figure 2.24 A straight column simply supported at both ends.

Considering the equilibrium condition for the bending moment, the following governing differential equation emerges:

$$-EI_e \frac{d^2 w}{dx^2} = Pw \quad \text{or} \quad \frac{d^2 w}{dx^2} + k^2 w = 0 \tag{2.73}$$

where $k = \sqrt{(P/EI_e)}$.

The general solution of Equation (2.73) reads

$$w = C_1 \sin kx + C_2 \cos kx \tag{2.74}$$

where C_1 and C_2 are constants to be determined in accordance with the end conditions.

Because both ends are simply supported, $w = 0$ at $x = 0$ and $x = L$. Substituting this end condition into Equation (2.74), the following two conditions emerge:

$$C_2 = 0, \quad C_1 \sin kL = 0 \tag{2.75}$$

Considering the first condition of Equation (2.75), the form of the solution becomes $w = C_1 \sin kx$. This means that the coefficient C_1 should not be zero; otherwise no deflection exists. Thus, the second condition of Equation (2.75) provides a nontrivial solution as follows:

$$\sin kL = 0 \quad \text{or} \quad kL = \pi, 2\pi, 3\pi, \ldots, n\pi, \ldots \tag{2.76}$$

The smallest value of the applied load, P, is given when $kL = \pi$. Thus, the so-called Euler buckling load, P_E, is calculated by

$$P_E = \frac{\pi^2 EI_e}{L^2} \tag{2.77}$$

After buckling, the deflection pattern of the column simply supported at both ends is therefore expressed as $w = \sin(\pi x / L)$.

2.9.3 Effect of End Conditions

The ends of a column are usually welded to other members and thus are restrained against rotation. The amount of this restraint typically varies with the properties of the different structural members involved.

For an infinitesimal part of the column with a certain type of end condition(s), as shown in Figure 2.25, the force equilibrium considerations lead to the basic fourth-order differential equation for flexural buckling of columns, when $dw/dx \approx 0$, as follows:

$$\frac{d^2}{dx^2} \left(EI_e \frac{d^2 w}{dx^2} \right) + P \frac{d^2 w}{dx^2} = 0 \tag{2.78}$$

When the cross section is uniform along the span, Equation (2.78) can be rewritten as follows:

$$EI_e \frac{d^4 w}{dx^4} + P \frac{d^2 w}{dx^2} = 0 \tag{2.79}$$

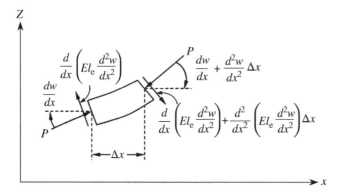

Figure 2.25 Force components acting on a column free-body element.

The general solution of Equation (2.79) is given by

$$w = C_1 \cos kx + C_2 \sin kx + C_3 x + C_4 \tag{2.80}$$

where k is defined in Equation (2.73).

Considering lateral deflection w, slope dw/dx, bending moment $EI_e d^2w/dx^2$, and shear force $EI_e d^3w/dx^3 + P\, dw/dx$, in the z direction, the end conditions of the column may be mathematically expressed by

$$\text{At the simply supported end}: w = 0, \quad \frac{d^2w}{dx^2} = 0 \tag{2.81a}$$

$$\text{At the fixed end}: w = 0, \quad \frac{dw}{dx} = 0 \tag{2.81b}$$

$$\text{At the free end}: \frac{d^2w}{dx} = 0, \quad \frac{d^3w}{dx^3} + k^2\frac{dw}{dx} = 0 \tag{2.81c}$$

Because two boundary conditions exist at each end, a total of four unknown constants in Equation (2.80) must be determined in any given case. The lateral deflection, w, then has nonzero solutions only if the determinant for a set of linear homogeneous equations with regard to C_1–C_4 vanishes. The buckling load can be calculated as the minimum value that satisfies the condition that the determinant becomes zero.

For instance, buckling of a column is now considered when one end is fixed and the other end is free, such as that shown in Figure 2.26. The end conditions are in this case given by

$$\text{At } x = 0 \text{ (free end)}: \frac{d^2w}{dx} = 0, \quad \frac{d^3w}{dx^3} + k^2\frac{dw}{dx} = 0 \tag{2.82a}$$

Figure 2.26 A cantilever column.

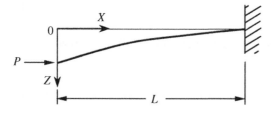

At $x = L$(fixed end) : $w = 0$, $\dfrac{dw}{dx} = 0$ (2.82b)

By substituting these boundary conditions into Equation (2.80), we get

$$C_1 = C_3 = 0, \quad C_2 \sin kL + C_4 = 0, \quad C_2 k \cos kL = 0 \tag{2.83a}$$

The second and third conditions of Equation (2.83a) become

$$C_4 = -C_2 \sin kL = C_2(-1)^n, \quad C_2 \neq 0, \quad k \neq 0$$

$$\cos kL = 0 \rightarrow kL = (2n-1)\frac{\pi}{2}, \quad n = 1, 2, 3, \dots \tag{2.83b}$$

The buckling load is then obtained from the last equation of Equation (2.83b) when the smallest load is obtained, with $n = 1$, as follows:

$$kL = \frac{\pi}{2} \rightarrow P_E = \frac{\pi^2 EI_e}{4L^2} = \frac{\pi^2 EI_e}{(2L)^2} \tag{2.83c}$$

After buckling, the lateral deflection pattern of the cantilever column is given by $w = 1 - \sin(\pi x/2L)$. As the rotational restraints of the adjacent members increase, the column ends may approach fixed conditions. In this case, the buckling wavelength between the points of inflection decreases. For instance, the buckling wavelength of the cantilever column, as shown in Figure 2.27b, may become 200% of the original length as evident from Equation (2.83c). For the column fixed at both ends as shown in Figure 2.27c, the buckling wavelength is 50% of the original length. When one end is fixed and the other is pinned, as shown in Figure 2.27d, the buckling wavelength becomes 70% of the original length.

It is apparent that the buckling wavelength decreases as the rotational restraints at the column ends increase. Also, the shorter the buckling wavelength, the larger the buckling load. For convenience, the term "effective length" (also called "buckling length") is typically used to account for the effects of the column end conditions so that the elastic buckling load of a column with various types of end conditions can be determined by the Euler formula, but replacing the original (or system) length, L, by the effective length, L_e, as follows:

$$P_E = \frac{\pi^2 EI_e}{L_e^2} = \frac{\pi^2 EI_e}{(\alpha L)^2} \quad \text{or} \quad \sigma_E = \frac{P_E}{A} = \frac{\pi^2 E}{(\alpha L/r_e)^2} \tag{2.84}$$

where α is a constant to account for the effect of the column end condition.

For various end conditions, the applicable theoretical values of the effective length, L_e, or constant α, are given as those of Figure 2.27.

2.9.4 Effect of Initial Imperfections

The actual columns in welded plated structures may have initial imperfections in the form of initial deflections (out of straightness) and residual stresses that affect the structural behavior and load-carrying capacity.

(a)
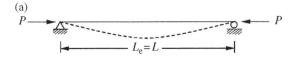

Figure 2.27 Effective length of a column varying the end conditions: (a) both ends pinned ($\alpha = 1$); (b) one end free, the other fixed ($\alpha = 2$); (c) both ends fixed ($\alpha = 0.5$); (d) one end pinned, the other fixed ($\alpha = 0.7$).

(b)
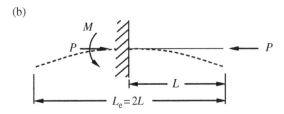

(c)
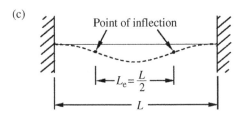

(d)
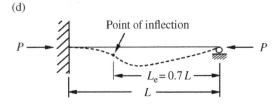

An initially deflected column simply supported at both ends as shown in Figure 2.28 is now considered. The geometric configuration of the initial deflection, w_0, which takes a half sinusoidal wave pattern, may approximately be defined as follows:

$$w_0 = \delta_0 \sin\frac{\pi x}{L} \tag{2.85a}$$

where δ_0 is the amplitude of the initial deflection.

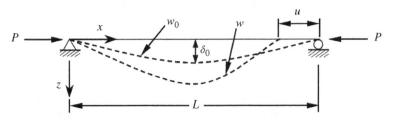

Figure 2.28 An initially deflected column simply supported at both ends.

The total (initial plus added) deflection, w, after buckling, may also take a similar shape to the initial deflection as follows:

$$w = \delta \sin\frac{\pi x}{L} \tag{2.85b}$$

where δ is the amplitude of the total deflection.

The bending moment equilibrium is in this case given by

$$EI_e \frac{d^2(w - w_0)}{dx^2} + Pw = 0 \tag{2.86}$$

where the first term on the left side represents internal bending moments due to the added deflection alone, because the initial deflection does not contribute to internal bending, whereas the second term is the external bending moment imposed by the total deflection.

The axial strain of a column with initial deflection may be given as similar to that of a straight column defined in Equation (2.62) by taking into account the large-deflection effects but by neglecting the large axial displacement effects as follows:

$$\varepsilon_x = \frac{\partial u}{\partial x} + \frac{1}{2}\left(\frac{\partial^2 w}{\partial x^2}\right)^2 - \frac{1}{2}\left(\frac{\partial^2 w_0}{\partial x^2}\right)^2 \tag{2.87}$$

where the first term on the right side of Equation (2.87) represents the small strain component and the second and third terms portray the large-deflection effects.

To determine the amplitude of total deflection in Equation (2.85), the strain energy-based approach is used. Considering that the strain energy is due to the contribution associated with added deflection, that is, $w - w_0$, the elastic strain energy, U, of Equation (2.69) for the initially deflected column can be rewritten as follows:

$$U = \frac{EI_e}{2}\int_0^L \left(\frac{\partial^2 w}{\partial x^2} - \frac{\partial^2 w_0}{\partial x^2}\right)^2 dx \tag{2.88a}$$

Substituting Equations (2.85a) and (2.85b) into Equation (2.88a) and performing the integration along the column, the strain energy is obtained as follows:

$$U = \frac{\pi^4 EI_e}{4L^3}(\delta - \delta_0)^2 \tag{2.88b}$$

because

$$\int_0^L \sin^2\frac{\pi x}{L}dx = \int_0^L \frac{1}{2}\left(1 - \cos\frac{2\pi x}{L}\right)dx = \frac{L}{2}.$$

On the other hand, the external potential energy, W, of the compressive load, P, related to the added deflection is calculated from Equation (2.70) together with Equations (2.85a) and (2.85b) by neglecting the small strain component:

$$W = Pu = -\frac{P}{2}\int_0^L \left[\left(\frac{dw}{dx}\right)^2 - \left(\frac{dw_0}{dx}\right)^2\right]dx = -\frac{P\pi^2}{4L}(\delta - \delta_0^2) \tag{2.89}$$

The total potential energy, Π, of the initially deflected column is obtained from Equation (2.71), but using the strain energy, U, of Equation (2.88) and the external potential energy, W, of Equation (2.89). Applying the principle of minimum potential energy, the amplitude of the total deflection can be found as follows:

$$\frac{\partial \Pi}{\partial \delta} = 0 = \frac{\pi^4 E I_e}{2L^3}(\delta - \delta_0) - \frac{P\pi^2}{2L}\delta \quad \text{or} \quad \delta = \frac{\delta_0}{1 - P/P_E} = \phi \delta_0 \tag{2.90}$$

where P_E is the Euler buckling load as defined in Equation (2.77) and $\phi = 1/(1 - P/P_E) =$ magnification factor.

Substituting Equation (2.90) into Equation (2.85b), the total deflection can now be obtained as follows:

$$w = \frac{\delta_0}{1 - P/P_E}\sin\frac{\pi x}{L} = \phi \delta_0 \sin\frac{\pi x}{L} \tag{2.91}$$

Figure 2.29 plots Equation (2.91), representing the applied compressive load versus the total deflection of the column, varying the magnitude of initial deflection. As evident from Figure 2.29, the deflection increases progressively from the very beginning of compressive loading when initial deflection exists and a bifurcation buckling point does not exist in this case. Also, the load-carrying capacity decreases as the magnitude of initial deflection increases.

The existence of any compressive residual stresses further reduces the column's buckling strength. For practical design purposes, the effect of compressive residual stress on the column buckling strength is sometimes included by deducting the same value of the compressive residual stress from the computed buckling strength.

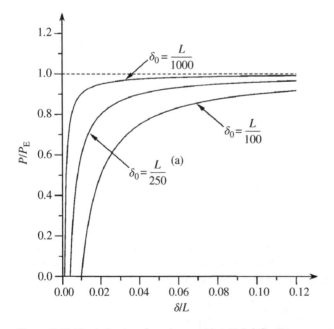

Figure 2.29 The behavior of a column with initial deflection.

2.9.5 Collapse Strength of Columns

The elastic buckling strength formulas derived thus far are valid as long as the material remains in the elastic regime. This may be true for a slender column without initial deflection. Because the Euler buckling stress must be less than the proportional limit of the material, σ_P, a limit of the column slenderness ratio for use of the Euler formula may be given from Equation (2.84) as follows:

$$\frac{L}{r_e} \geq \frac{\pi}{\alpha} \left(\frac{E}{\sigma_P}\right)^{1/2} \tag{2.92}$$

because $\sigma_E \leq \sigma_P$.

For instance, if the proportional limit of material is taken as $\sigma_P = 200$ MPa, $E = 210$ GPa and $\alpha = 1.0$, the column slenderness ratio must satisfy $L/r_e \geq \pi (210\,000/200)^{1/2} = 101.7$ so that the Euler formula result is valid to use. For a stocky or imperfect column, however, the elastic proportional limit is often exceeded, and a certain degree of plasticity takes place before the inception of buckling. As a result, the real buckling load in this case is less than the Euler buckling load.

Therefore, the Euler buckling formula is not directly available for a stocky or imperfect column, both of which are more common in actual structures. However, because the Euler formula provides very useful insight into the column buckling behavior, many researchers have attempted to use it to the extent possible, even for elastic–plastic buckling of a column, with corrections applied to some of the parameters involved. For instance, some classical theories such as the double-modulus theory or tangent-modulus theory that resemble the original Euler formulation have been suggested to deal with elastic–plastic effects on column buckling (Bleich 1952).

In reality, a stocky column with a high value of computed elastic buckling strength does not buckle in the elastic regime, but rather reaches the ultimate strength with a certain degree of plasticity. To account for this behavior, some approximate formulas based on the insights developed from experiments, such as the so-called Gordon–Rankine formulation, the Tetmajer formulation, and the Johnson–Ostenfeld formulation, are sometimes used, although the Johnson–Ostenfeld formulation is the most popular in today's industry practice.

For modern practical design purposes, the various available ultimate strength formulations for the plate–stiffener combination model under predominantly axial compressive loads are often based on one of the three common approaches, namely,

- The Johnson–Ostenfeld formulation method
- The Perry–Robertson formulation method
- A purely empirical formulation method.

The Johnson–Ostenfeld formulation method takes into account the effects of plasticity in the elastic buckling strength. The resulting "elastic–plastic" buckling strength is termed the "critical" buckling strength, which is approximately regarded as the ultimate strength.

The Perry–Robertson formulation method considers that the plate–stiffener combination model collapses when the maximum compressive stress at the extreme fiber reaches the yield strength of the material. Two possible collapse modes are considered for the Perry–Robertson formulation method of the plate–stiffener combination model, depending on the compressed side, namely, plate-induced failure (PIF) or stiffener-induced failure (SIF); the former is initiated by compression at the attached plating side, and the latter is initiated by compression at the stiffener flange side.

In a purely empirical formulation method, the ultimate strength formulations are developed by curve fitting based on mechanical collapse test results and/or numerical computations. These types of empirical formulas can be cast as simple closed-form expressions, which have certain advantages in getting first-cut estimates, whereas their use may be restricted to a specified range of dimensions or be subject to other limitations.

2.9.5.1 The Johnson–Ostenfeld Formulation Method

The critical buckling strength based on the Johnson–Ostenfeld formulation is given as follows:

$$
\sigma_{cr} = \begin{cases} \sigma_E & \text{for } \sigma_E \leq \alpha\sigma_F \\ \sigma_F[1 - \sigma_F/(4\sigma_E)] & \text{for } \sigma_E > \alpha\sigma_F \end{cases} \tag{2.93}
$$

where σ_E is the elastic buckling stress, σ_{cr} is the critical (or elastic–plastic) buckling stress, σ_F is the reference yield stress, $\sigma_F = \sigma_Y$ for compressive normal stress and $\sigma_F = \tau_Y = \sigma_Y/\sqrt{3}$ for shear stress, σ_Y is the material yield stress, and α is a constant depending on the material proportional limit, which is usually taken as $\alpha = 0.5$ or 0.6.

For a plate–stiffener combination model with different materials (e.g., mild steel for the attached plating and high-tensile steel for the stiffener), σ_Y may be taken as the equivalent yield stress, that is, $\sigma_Y = \sigma_{Yeq}$ as defined in Table 2.1. In using Equation (2.93), the sign of the compressive stress is taken as positive. Also, Equation (2.93) can be applied for a plate or stiffened panel as well as a column.

2.9.5.2 The Perry–Robertson Formulation Method

In the Perry–Robertson formulation, it is assumed that a column collapses if the maximum compressive stress at the extreme fiber of the column cross section reaches the yield stress. For an initially deflected column simply supported at both ends as shown in Figure 2.28, the maximum bending moment, M_{max}, can be calculated using the total deflection, δ, at the mid-span, as given by Equation (2.90), as follows:

$$
M_{max} = P\delta = \frac{P\delta_0}{1 - P/P_E} \tag{2.94}
$$

where it is assumed that local buckling or lateral-torsional buckling does not occur.

The maximum compressive stress at the outer fiber of the cross section can be obtained by the sum of axial stress and bending stress as follows:

$$
\sigma_{max} = \frac{P}{A} + \frac{M_{max}}{I_e}z_c = \frac{P}{A} + \frac{z_c}{I_e}\frac{P\delta_0}{1 - P/P_E} = \sigma + \frac{A\delta_0 z_c}{I_e}\frac{\sigma}{1 - \sigma/\sigma_E} \tag{2.95}
$$

where $\sigma = P/A$ and z_c is the distance from the elastic neutral axis to the outer fiber of the compressed side.

Following the Perry–Robertson formulation method, the ultimate strength of a column is determined from Equation (2.95) by replacing σ with σ_u when σ_{max} reaches the equivalent yield stress, σ_{Yeq}, namely,

$$\sigma_{\max} = \sigma_{\text{Yeq}} = \sigma_{\text{u}} \left(1 + \frac{\eta}{1 - \sigma_{\text{u}}/\sigma_{\text{E}}} \right) \tag{2.96}$$

where $\eta = A\delta_0 z_{\text{c}}/I_{\text{e}} = \delta_0 z_{\text{c}}/r_{\text{e}}^2$.

The real ultimate strength, σ_{u}, is taken as the minimum value of the two solutions obtained by solving Equation (2.96) with regard to σ_{u} as follows:

$$\frac{\sigma_{\text{u}}}{\sigma_{\text{Yeq}}} = \frac{1}{2}\left(1 + \frac{1+\eta}{\lambda_{\text{e}}^2} \right) - \left[\frac{1}{4}\left(1 + \frac{1+\eta}{\lambda_{\text{e}}^2} \right)^2 - \frac{1}{\lambda_{\text{e}}^2} \right]^{0.5} \tag{2.97}$$

where $\lambda_{\text{e}} = (L/\pi r_{\text{e}})\sqrt{(\sigma_{\text{Yeq}}/E)} = \sqrt{(\sigma_{\text{Yeq}}/\sigma_{\text{E}})}$.

For a straight column, that is, without initial deflection, the constant η becomes $\eta = 0$. Therefore, it is evident that Equation (2.97) is reduced to the Euler formula when $\lambda_{\text{e}} \geq 1$, namely,

$$\frac{\sigma_{\text{u}}}{\sigma_{\text{Yeq}}} = \frac{1}{\lambda_{\text{e}}^2} \tag{2.98}$$

The direction of the column deflection is governed by that of the initial deflection, as long as lateral loads are not applied. Because the nature of the initial deflection is somewhat uncertain, the failure mode of the plate–stiffener combination model may be either PIF or SIF. It is for this reason that the ultimate strength for the Perry–Robertson formulation method may be determined as the minimum value of the two strengths.

In a continuous stiffened plate structure, SIF is a trigger to the collapse of the entire panel. The original concept of the Perry–Robertson formulation method assumes that SIF occurs if the tip of the stiffener yields. This assumption may in some cases be too pessimistic in terms of the collapse strength predictions. Rather, plasticity may grow into the stiffener web as long as lateral-torsional buckling or stiffener web buckling does not take place, so that the stiffener may resist further loading even after the first yielding occurs at the extreme fiber of the stiffener. In this regard, only the PIF-based Perry–Robertson formula, that is, excluding the SIF, is sometimes adopted for prediction of the ultimate strength of the plate–stiffener combination model as representative of a continuous stiffened panel.

2.9.5.3 The Paik–Thayamballi Empirical Formulation Method for a Steel Plate–Stiffener Combination Model

Although a vast number of empirical formulations (sometimes called column curves) for the ultimate strength of a column used in framed structures have been developed (e.g., Chen & Atsuta 1976, 1977, ECCS 1978, among others), relevant empirical formulas that can be used to predict the ultimate strength of a plate–stiffener combination model associated with plated structures are also available (Lin 1985, Paik & Thayamballi 1997, Zhang & Khan 2009).

As an example, Paik and Thayamballi (1997) developed an empirical formula to predict the ultimate strength of a plate–stiffener combination model under axial compression as a function of the column and plate slenderness ratios. The Paik–Thayamballi empirical formulation method is derived by curve fitting of a mechanical collapse test database for the ultimate strength of steel-stiffened panels under axial compression and with initial imperfections (initial deflections and residual stresses) as follows:

$$\frac{\sigma_u}{\sigma_{Yeq}} = \frac{1}{\sqrt{0.995 + 0.936\lambda^2 + 0.170\beta^2 + 0.188\lambda^2\beta^2 - 0.067\lambda^4}} \leq \frac{1}{\lambda^2} \tag{2.99}$$

where λ and β are the column and plate slenderness ratios for the full section, respectively, as defined in Table 2.1. Because the ultimate strength of a column cannot be greater than the elastic column buckling strength, $\sigma_u/\sigma_{Yeq} = 1/\lambda^2$ should be taken if $\sigma_u/\sigma_{Yeq} \geq 1/\lambda^2$.

Equation (2.99) implicitly includes the possible effects of local buckling or lateral-torsional buckling and initial imperfections (initial deflection and welding residual stress). Also, both the column and plate slenderness ratios used in Equation (2.99) are calculated for the full section, that is, without evaluating the effective width of the attached plating. This may sometimes be of benefit when evaluation of the plate effective width is difficult.

2.9.5.4 The Paik Empirical Formulation Method for an Aluminum Plate–Stiffener Combination Model

Paik (2007, 2008b) derived empirical formulations to predict the ultimate compressive strength of an aluminum plate–stiffener combination model that are similar to Equation (2.99) but have two different expressions for extruded or built-up T-bar cross section and flat-bar cross section:

Aluminum extruded or built-up T-bars:

$$\frac{\sigma_u}{\sigma_{Yeq}} = \frac{1}{\sqrt{1.318 + 2.759\lambda^2 + 0.185\beta^2 - 0.177\lambda^2\beta^2 + 1.003\lambda^4}} \leq \frac{1}{\lambda^2} \tag{2.100a}$$

Aluminum flat bars:

$$\frac{\sigma_u}{\sigma_{Yeq}} = \text{Min.} \begin{cases} \dfrac{1}{\sqrt{2.500 - 0.588\lambda^2 + 0.084\beta^2 + 0.069\lambda^2\beta^2 + 1.217\lambda^4}} \leq \dfrac{1}{\lambda^2} \\ \\ \dfrac{1}{\sqrt{-16.297 + 18.776\lambda + 17.716\beta - 22.507\lambda\beta}} \end{cases} \tag{2.100b}$$

Figure 2.30a–d confirms the applicability of Equations (2.100a) and (2.100b) by comparison with experiments and nonlinear finite element method solutions together with Equation (2.99), in which experiments and nonlinear finite element solutions are all obtained for aluminum-stiffened plate structures (Paik 2008b, Paik et al. 2012). For flat-bar type of stiffeners, aluminum-stiffened plate structures reach the ultimate compressive strength by local web buckling when the column slenderness ratio is small, in contrast to steel-stiffened plate structures.

It is interesting to note that the Paik–Thayamballi formula of Equation (2.99), which was originally derived for a steel plate–stiffener combination model, may also be applied to an aluminum plate–stiffener model when the plate slenderness ratio of the attached plating is relatively large or the attached plating is thin. It is realized that the bias and coefficient of variation are 1.032 and 0.101, respectively, for Equation (2.100a) and 1.020 and 0.114, respectively, for Equation (2.100b). Figure 2.31 compares the

(a)

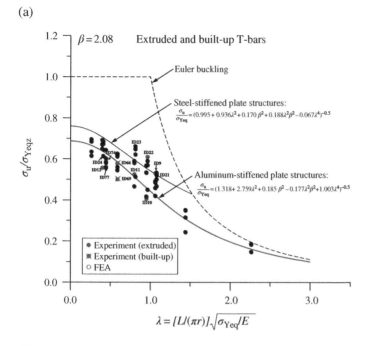

(b)

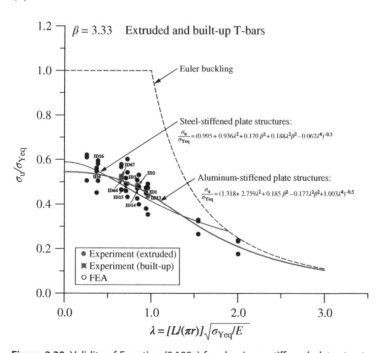

Figure 2.30 Validity of Equation (2.100a) for aluminum-stiffened plate structures: (a) extruded and built-up bars with β = 2.08; (b) extruded and built-up bars with β = 3.33; (c) flat bars with β = 2.08; (d) flat bars with β = 3.33 (Paik 2008b).

(c)

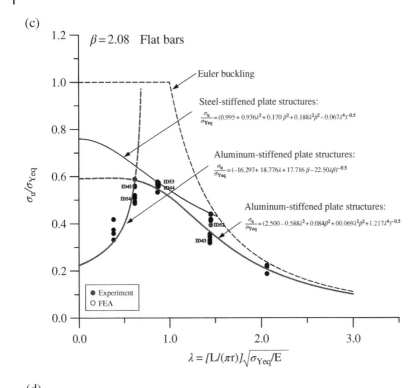

(d)

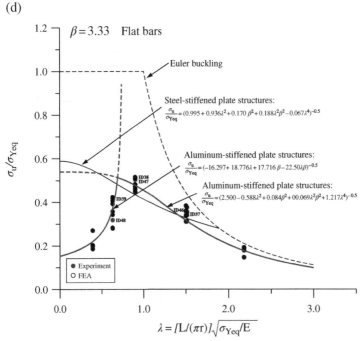

Figure 2.30 (Continued)

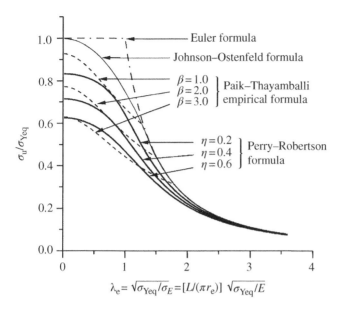

Figure 2.31 A comparison of the ultimate strength formulations for a steel plate–stiffener combination model under axial compression.

Johnson–Ostenfeld formulation method, the Perry–Robertson formulation method, and the Paik–Thayamballi empirical formulation method for the column ultimate strength for a steel plate–stiffener combination model varying the column slenderness ratios for selected initial eccentricity and plate slenderness ratios. For convenience in the present comparisons, it was assumed that $\lambda_e = \lambda$.

2.9.6 Local Web or Flange Buckling Under Axial Compression

In some cases, local buckling can take place in the web or flange of a stiffener in a stiffened panel. Once such stiffener web or flange buckling occurs, the stiffened panel may easily fall into the overall collapse mode because the stiffeners may no longer function as support members. As described in Chapters 5 and 6, local web buckling is a collapse mode of a stiffened panel, because the plating is left with essentially no stiffening once the stiffener web buckles in the elastic–plastic regime, and a global buckling mode may thus follow immediately.

Chapter 6 describes the ultimate strength calculation methods that account for local buckling of the stiffener web, where the local web or flange buckling strength formulations are presented in Chapter 5. It is noted that the plate–stiffener combination model usually does not account for local web or flange buckling of stiffeners, but the purely empirical formulation methods such as the Paik–Thayamballi formulation method have been established based on the experimental database and/or numerical computations that take into account the effects of local web or flange buckling of stiffeners.

2.9.7 Lateral-Torsional Buckling Under Axial Compression

In some cases, the stiffener web can twist sideways if the stiffener flange is not sufficiently strong to remain straight. This phenomenon is called lateral-torsional buckling (or tripping), which occurs suddenly and results in subsequent unloading of the support member. As described in Chapters 5 and 6, lateral-torsional buckling is a collapse mode of a stiffened panel, because the plating is left with essentially no stiffening once the stiffener twists sideways, and a global buckling mode may thus follow immediately.

Chapter 6 describes the ultimate strength calculation methods that account for lateral-torsional buckling of the stiffener, where the lateral-torsional buckling strength formulations of stiffeners are presented in Chapter 5. It is again noted that the plate–stiffener combination model usually does not account for lateral-torsional buckling of stiffeners, but the purely empirical formulation methods such as the Paik–Thayamballi formulation method has been established based on the experimental database and/or numerical computations that take into account the effects of lateral-torsional buckling of stiffeners.

2.10 Ultimate Strength of the Plate–Stiffener Combination Model Under Combined Axial Compression and Bending

In this section, the ultimate strength formulations for a plate–stiffener combination model under combined axial compression and bending are described by accounting for the effects of initial imperfections. In this case, the plate–stiffener combination model is dealt with as a beam–column.

2.10.1 The Modified Perry–Robertson Formulation Method

The original Perry–Robertson formulation method is available to calculate the collapse strength of the plate–stiffener combination model under axial compressive loads only, where it is assumed that local buckling such as local web or flange buckling and lateral-torsional buckling of stiffeners does not take place. It is considered that the column collapses if the maximum compressive stress at the outer fiber of the cross section reaches the yield stress.

For a plate–stiffener combination model under combined axial compression and bending as a beam–column, the original Perry–Robertson approach may also be applied to calculate the ultimate strength, but the maximum compressive stress at the outer fiber of the plate–stiffener combination model cross section is now a function of bending and axial compression where the bending arises from lateral pressure loads. This approach is called the modified Perry–Roberson formulation method.

For a plate–stiffener combination model (beam–column) under combined axial compression, P, and lateral line load, q, as defined in Figure 2.5, the internal bending moment along the span can be obtained by

$$M = M_q + Pw \tag{2.101}$$

where M_q is the bending moment due to the lateral line load q and w is the total lateral deflection.

Using Equations (2.72a) and (2.101), the bending equilibrium of the plate–stiffener combination model with the effective cross section reads

$$EI_e \frac{d^2 w}{dx^2} = -M = -M_q - Pw \quad \text{or} \quad \frac{d^2 w}{dx^2} + k^2 w = -\frac{M_q}{EI_e} \tag{2.102}$$

where k is defined in Equation (2.73).

The total lateral deflection and bending moment distribution of the plate–stiffener combination model can be obtained by solving Equation (2.102) under prescribed boundary and loading conditions. For instance, the total lateral deflection and bending moment of the plate–stiffener combination model simply supported at both ends under combined axial compressive load, P, and lateral line load, q, are obtained as follows:

$$w = \frac{q}{Pk^2}\left\{1 - \frac{\cos[k(L/2-x)]}{\cos(kL/2)}\right\} + \frac{q}{2P}x(L-x), \quad M = \frac{q}{k^2}\left\{1 - \frac{\cos[k(L/2-x)]}{\cos(kL/2)}\right\} \tag{2.103}$$

Because the maximum lateral deflection, w_{max}, or the maximum bending moment, M_{max}, occurs at the mid-span, that is, $x = L/2$, the following values are obtained:

$$w_{max} = C_1 w_{q\,max}, \quad M_{max} = C_2 M_{q\,max} \tag{2.104}$$

where

$$C_1 = \frac{384}{5k^4 L^4}\left[\sec\left(\frac{kL}{2}\right) - 1 - \frac{k^2 L^2}{8}\right],$$

$$C_2 = \frac{8}{k^2 L^2}[1 - \sec(kL/2)], \quad w_{q\,max} = \frac{5qL^4}{384EI_e}, \quad M_{q\,max} = \frac{qL^2}{8},$$

$w_{q\,max}$ and $M_{q\,max}$ are the maximum lateral deflection and the maximum bending moment, respectively, caused by lateral line load q alone. The coefficients C_1 and C_2 portray the magnification factors for the lateral deflection and bending moment, respectively. As is apparent, the magnification factors may differ depending on the load applications or end conditions.

When applying the Perry–Robertson formulation method, it is assumed that the plate–stiffener combination model collapses when the maximum compressive stress at the outer fiber of the cross section reaches the yield stress. Depending on the direction of lateral loading, the compressed side of the cross section can be automatically determined.

For practical design purposes, specifically when the direction of lateral loading is previously unknown, the maximum stress at the cross section may be taken as the larger value of the stresses at the two extreme fibers, namely,

$$\sigma_{max} = \frac{P}{A} + \frac{M_{max}}{I_e}z_{max} = \sigma_{Yeq} \tag{2.105}$$

where z_{max} is the larger value of z_p or $h_w + t + t_f - z_p$ and z_p is defined in Table 2.1.

The ultimate axial compressive stress, σ_u, is obtained as the solution of Equation (2.105) with regard to $\sigma = P/A$. An iterative process may be needed to solve Equation (2.105) with regard to the axial load because M_{max} is a nonlinear function of P.

To obtain a closed-form expression of the ultimate compressive strength for the plate–stiffener combination model, a simplification can herein be made. It is assumed that the maximum bending moment of the plate–stiffener combination model is the sum of the bending moment due to lateral loads plus that due to geometric eccentricity, which may include the lateral deflection caused by an external load as well as the initial deflection, namely,

$$M_{max} = M_{q\,max} + P\phi\left(w_{q\,max} + \delta_0\right) \tag{2.106}$$

where $M_{q\,max}$ is the maximum bending moment due to the lateral load alone, $w_{q\,max}$ is the maximum deflection (amplitude) due to the lateral load alone, δ_0 is the initial deflection, and ϕ is the magnification factor as defined in Equation (2.90).

To check the accuracy of Equation (2.106), an example is considered when a beam–column (plate–stiffener combination model) is subjected to uniform lateral line load, q, and axial compression $P = 0.5P_E$. We assume that initial deflection does not exist, that is, $\delta_0 = 0$, so that the exact solution of Equation (2.104), which is $M_{max} = 2.030\,M_{q\,max}$, can be compared directly with Equation (2.106). Because $w_{q\,max} = 5qL^4/(384EI_e)$ and $\phi = 1/(1 - P/P_E) = 2$, the maximum bending moment by Equation (2.106) results in

$$M_{max} = M_{q\,max} + \frac{5qL^4}{384EI_e}\frac{\pi^2 EI_e}{L^2} = M_{q\,max}\left(1 + \frac{5\pi^2}{48}\right) = 2.028 M_{q\,max} \tag{2.107}$$

It is evident that Equation (2.106) is sufficiently accurate in this case because the difference between Equations (2.104) and (2.106) is less than 0.1% in this example. Because the ultimate strength is reached when the maximum stress equals the yield stress, an equation similar to Equation (2.105) appears, namely,

$$\sigma_{max} = \frac{P}{A} + \frac{M_{q\,max}}{I_e}z_{max} + \frac{P}{1 - P/P_E}\left(w_{q\,max} + \delta_0\right)\frac{z_{max}}{I_e} = \sigma_{Yeq} \tag{2.108}$$

By introducing the following nondimensional parameters,

$$R = \frac{\sigma_u}{\sigma_{Yeq}}, \quad \lambda_e = \sqrt{\frac{\sigma_{Yeq}}{\sigma_E}} = \frac{L}{\pi r_e}\sqrt{\frac{\sigma_{Yeq}}{E}}, \quad \eta = \frac{Az_{max}}{I_e}\left(w_{q\,max} + \delta_0\right), \quad \mu = \frac{M_{q\,max}\,z_{max}}{\sigma_{Yeq}\,I_e}$$

Equation (2.108) may be expressed as a quadratic function of the axial compressive stress and lateral load, namely,

$$\eta R - (1 - R - \mu)\left(1 - \lambda_e^2 R\right) = 0 \tag{2.109}$$

Regarding the lateral load as a constant dead load, the ultimate compressive strength of the plate–stiffener combination model under combined axial compression and lateral load is obtained as the minimum value of the two solutions of Equation (2.109) with regard to R, namely,

$$R = \frac{1}{2}\left(1 - \mu + \frac{1+\eta}{\lambda_e^2}\right) - \left[\frac{1}{4}\left(1 - \mu + \frac{1+\eta}{\lambda_e^2}\right)^2 - \frac{1-\mu}{\lambda_e^2}\right]^{0.5} \tag{2.110}$$

Figure 2.32 shows the variation of R so obtained, versus the column slenderness ratio, for selected values of η and μ. To approximately account for the effects of welding

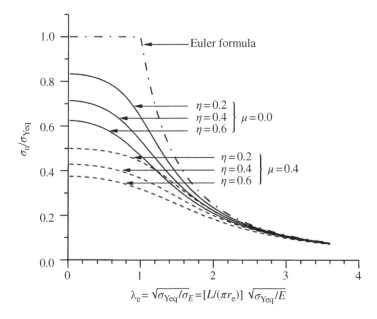

Figure 2.32 Variation of the ultimate compressive strength from the Perry–Robertson formula versus the column slenderness ratio for plate–stiffener combinations under combined axial compression and lateral load.

residual stresses, Equation (2.110) may be modified by multiplying a knockdown factor, α, as follows:

$$R = \alpha \left\{ \frac{1}{2}\left(1 - \mu + \frac{1+\eta}{\lambda_e^2} \right) - \left[\frac{1}{4}\left(1 - \mu + \frac{1+\eta}{\lambda_e^2} \right)^2 - \frac{1-\mu}{\lambda_e^2} \right]^{0.5} \right\} \tag{2.111}$$

where the knockdown factor α due to compressive residual stress, σ_{rsx}, may sometimes be taken as $\alpha = 1.03 - 0.08|\sigma_{rsx}/\sigma_{Yeq}| \leq 1.0$ for built-up sections.

2.10.2 Lateral-Torsional Buckling Under Combined Axial Compression and Bending

The lateral-torsional buckling (also called tripping) strength, σ^T, for a plate–stiffener combination model under combined axial compression and bending can be calculated as described in Chapter 5, where the bending arises from lateral pressure loads.

To illustrate the length effect of a plate–stiffener combination model under either axial compression alone or combined axial compression and bending, Figure 2.33 shows the variations of the elastic tripping strengths and the inelastic tripping strengths predicted with the Johnson–Ostenfeld formulation method. For comparison, the ordinary Euler column buckling strengths and the plasticity effect are also plotted.

It is observed from Figure 2.33 that the effects of lateral-torsional buckling are significant in a relatively stocky column. The lateral-torsional buckling strength of a stocky column is much lower than the ordinary Euler column buckling strength, which does

(a)

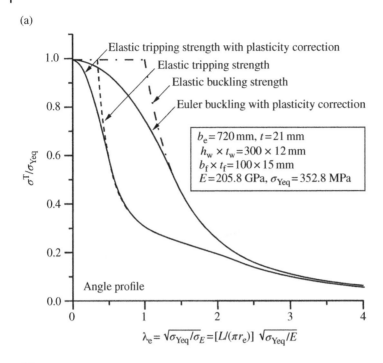

(b)

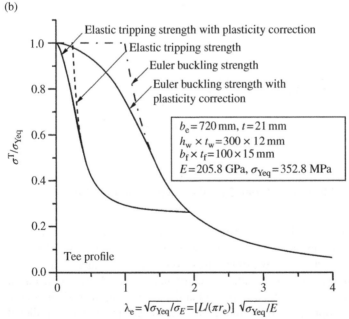

Figure 2.33 Variation of the lateral-torsional buckling strength versus the column slenderness ratio for a plate–stiffener combination model under axial compression alone: (a) angle section stiffener with attached effective plating; (b) Tee-section stiffener with attached effective plating.

(a)

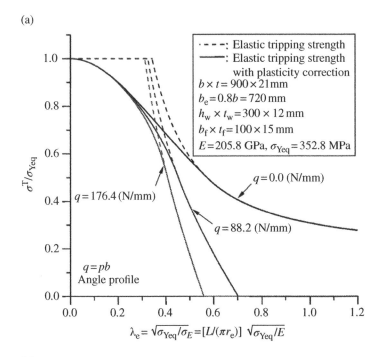

(b)

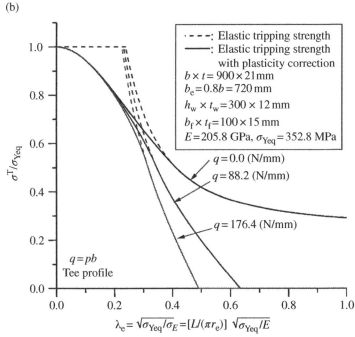

Figure 2.34 Variation of the tripping strength versus the column slenderness ratio for a plate–stiffener combination model under combined axial compression and bending: (a) angle section stiffener; (b) Tee-section stiffener.

not accommodate the lateral-torsional deformations of the stiffener. For a slender column, however, the lateral-torsional deformation effect is negligible, at least as far as these examples (with an identical stiffener flange breadth) are concerned.

It is realized that a one-sided (asymmetric) stiffener flange (e.g., angle section stiffener) can have more desirable performance than a symmetric stiffener flange (e.g., Tee-section stiffener) when the column slenderness ratio is low or when the column is short in length. On the contrary, a symmetric stiffener flange can provide more desirable performance than an asymmetric stiffener flange when the column slenderness ratio is high or when the column is long in length. As shown in Figure 2.33b, ordinary flexural buckling may be a more dominant failure mode than lateral-torsional buckling for the symmetric section stiffener when the column slenderness ratio is high.

Figure 2.34 shows the effects of the lateral load on the lateral-torsional buckling strength of a plate–stiffener combination model. It is observed from Figure 2.34 that the lateral loads can significantly reduce the lateral-torsional strength of the plate–stiffener combination model under axial compression.

References

AISI (1996). *Specification for the design of cold formed steel structural members.* American Iron and Steel Institute, New York.

Belenkiy, L. & Raskin, Y. (2001). Estimate of the ultimate load on structural members subjected to lateral loads. *Marine Technology*, **38**(3):169–176.

Bleich, F. (1952). *Buckling strength of metal structures.* McGraw-Hill, New York.

Bortsch, R. (1921). Die mitwirkende Plattenbreite. *Der Bauingenieur*, **23**:662–667 (in German).

Chen, W.F. & Atsuta, T. (1976). *Theory of beam-columns, Vol. 1, In-plane behavior and design.* McGraw-Hill, New York.

Chen, W.F. & Atsuta, T. (1977). *Theory of beam-columns, Vol. 2, Space behavior and design.* McGraw-Hill, New York.

ECCS (1978). *European recommendations for steel construction.* European Convention for Constructional Steelwork, Brussels.

ENV 1993-1-1 (1992). *Eurocode 3: design of steel structures, Part 1.1 general rules and rules for buildings.* British Standards Institution, London.

Faulkner, D. (1975). A review of effective plating for use in the analysis of stiffened plating in bending and compression. *Journal of Ship Research*, **19**(1):1–17.

Hodge, P.G. (1959). *Plastic analysis of structures.* McGraw-Hill, New York.

John, W. (1877). On the strains of iron ships. *RINA Transactions*, **18**:98–117.

Lin, Y.T. (1985). *Ship longitudinal strength modelling.* Ph.D. Dissertation, University of Glasgow, Scotland.

Mansour, A.E. (1977). *Gross panel strength under combined loading, SSC-270.* Ship Structure Committee, Washington, DC.

Metzer, W. (1929). *Die mittragende Breite.* Dissertation, der Technischen Hochschule zu Aache (in German).

Neal, B.C. (1977). *The plastic methods of structural analysis.* Third Edition, Chapman & Hall, London.

Paik, J.K. (1995). A new concept of the effective shear modulus for a plate buckled in shear. *Journal of Ship Research*, **39**(1):70–75.

Paik, J.K. (2007). Empirical formulations for predicting the ultimate compressive strength of welded aluminum stiffened panels. *Thin-Walled Structures*, **45**:171–184.

Paik, J.K. (2008a). Some recent advances in the concepts of plate-effectiveness evaluation. *Thin-Walled Structures*, **46**:1035–1046.

Paik, J.K. (2008b). *Mechanical collapse testing on aluminum stiffened panels for marine applications*, **SSC-451**. Ship Structure Committee, Washington, DC.

Paik, J.K., Kim, B.J., Sohn, J.M., Kim, S.H., Jeong, J.M. & Park, J.S. (2012). On buckling collapse of a fusion-welded aluminum stiffened plate structure: An experimental and numerical study. *Journal of Offshore Mechanics and Arctic Engineering*, **134**:021402.1–021402.8.

Paik, J.K. & Thayamballi, A.K. (1997). An empirical formulation for predicting the ultimate compressive strength of stiffened panels. *Proceedings of International Offshore and Polar Engineering Conference*, Honolulu, IV: 328–338.

Rhodes, J. (1982). Effective widths in plate buckling. Chapter 4 in *Developments in Thin-Walled Structures*, Edited by Rhodes, J. & Walker, A.C., Applied Science Publishers, London, 119–158.

Schuman, L. & Back, G. (1930). *Strength of rectangular flat plates under edge compression*, NACA Technical Report, **356**. National Advisory Committee for Aeronautics, Washington, DC.

Shames, I.H. & Dym, C.L. (1993). *Energy and finite element methods in structural mechanics*. McGraw-Hill, New York.

Smith, C.S. (1966). Elastic analysis of stiffened plating under lateral loading. *Transactions of the Royal Institution of Naval Architects*, **108**(2):113–131.

Timoshenko, S.P. & Gere, J.M. (1961). *Theory of elastic stability*. Second Edition, McGraw-Hill, New York.

Timoshenko, S.P. & Goodier, J.N. (1970). *Theory of elasticity*. Third Edition, McGraw-Hill, New York.

Troitsky, M.S. (1976). *Stiffened plates: bending, stability and vibrations*. Elsevier, Amsterdam.

Ueda, Y. & Rashed, S.M.H. (1984). The idealized structural unit method and its application to deep girder structures. *Computers & Structures*, **18**(2):277–293.

Ueda, Y., Rashed, S.M.H. & Paik, J.K. (1986). Effective width of rectangular plates subjected to combined loads. *Journal of the Society of Naval Architects of Japan*, **159**:269–281 (in Japanese).

Usami, T. (1993). Effective width of locally buckled plates in compression and bending. *ASCE Journal of Structural Engineering*, **119**(5):1358–1373.

von Karman, T. (1924). *Die mittragende Breite. Beitrage zur Technischen Mechanik und Technischen Physik, August Foppl Festschrift*. Julius Springer, Berlin, 114–127 (in German).

von Karman, T., Sechler, E. E. & Donnell, L.H. (1932). Strength of thin plates in compression. *Transactions of the American Society of Civil Engineers*, **54**(5):53–57.

Winter, G. (1947). *Strength of thin steel compression flanges*. Reprint, **32**, Engineering Experimental Station, Cornell University, Ithaca, NY.

Yamamoto, Y., Ohtsubo, H., Sumi, Y. & Fujino, M. (1986). *Ship structural mechanics*. Seisantou Publishing Company, Tokyo (in Japanese).

Zhang, S. & Khan, I. (2009). Buckling and ultimate capability of plates and stiffened panels in axial compression. *Marine Structures*, **22**(4):791–808.

3

Elastic and Inelastic Buckling Strength of Plates Under Complex Circumstances

3.1 Fundamentals of Plate Buckling

The behavior of a plated structure is considered at three levels: the bare plate element level, the stiffened panel level, and the entire plated structure level. This chapter describes buckling strength at the first level, that is, the plating between longitudinal stiffeners and transverse frames. As the predominantly compressive stress reaches a critical value, the plate buckles, resulting in a rapid increase in lateral deflection after a significant decrease in the in-plane stiffness.

The phenomenon of buckling is normally described by categorizing plasticity into three classes—elastic buckling, elastic–plastic buckling, and plastic buckling—with the latter two considered inelastic buckling. Elastic buckling occurs solely in the elastic regime. Elastic–plastic buckling occurs after a local region inside the plate undergoes plastic deformation. Plastic buckling indicates that buckling occurs in the regime of gross yielding, that is, after the plate has yielded over large regions. Thin plates normally show elastic buckling, whereas thick plates usually exhibit inelastic buckling.

rBuckling of a plate between stiffeners, which is a basic failure mode in a stiffened panel, is a good indication for the serviceability limit state (SLS) design. To understand the ultimate limit state (ULS)-based design procedure, basic knowledge on the buckling of plates is essential. The buckling behavior of plates normally depends on a variety of influential factors, including geometric or material properties, loading characteristics, boundary conditions, initial imperfections, and local damage (e.g., perforations).

This chapter describes classical and more advanced formulations of buckling strength for plates under simple and more complex circumstances. Generally, in this chapter, new or less well-known results are emphasized. Multiple load components are treated to the greatest possible extent. The effects of boundary restraints other than idealized simply supported or fixed conditions and the effects of lateral pressure, perforations, and residual stresses are also treated.

The coverage in this chapter is extensive, but as may be surmised, plate elements constitute the major portion of the structural weight of complex plated structures. By extension, it also follows that there are significant benefits to be gained by designing them in an optimal and appropriate manner. It is noted that the theories and methodologies described in this chapter can be commonly applied to steel and aluminum plates.

Ultimate Limit State Analysis and Design of Plated Structures, Second Edition. Jeom Kee Paik.
© 2018 John Wiley & Sons Ltd. Published 2018 by John Wiley & Sons Ltd.

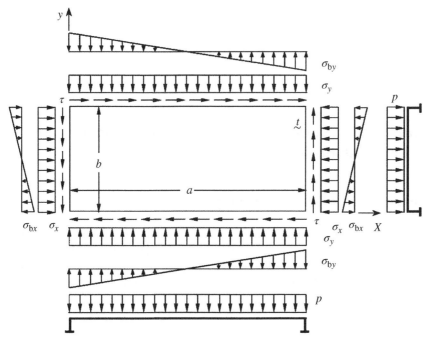

Figure 3.1 A rectangular plate under combined in-plane and lateral pressure loads.

3.2   Geometric and Material Properties

For the convenience of plate buckling analysis, the coordinate system for the plate uses x in the long direction and y in the short direction, as shown in Figure 3.1. The dimensions of the plate are a in length (i.e., in the x direction), b in breadth (i.e., in the y direction), and t in thickness. In this case, the plate aspect ratio, a/b, is always greater than 1 (i.e., $a/b \geq 1$). Young's modulus is E, and Poisson's ratio is v. The elastic shear modulus is $G = E/[2(1+v)]$. The yield stress of the material is σ_Y; $\tau_Y = \sigma_Y/\sqrt{3}$. The plate bending rigidity is $D = Et^3/[12(1-v^2)]$. The plate slenderness ratio is $\beta = (b/t)\sqrt{\sigma_Y/E}$.

3.3   Loads and Load Effects

The plate elements in a continuous plated structure are likely to be subjected to combined in-plane and lateral pressure loads. For plate elements in a complex structure, the load effects (stresses) are calculated by linear elastic finite element analysis (FEA) or by the classical theory of structural mechanics. The individual load components have both local and overall structural effects.

In calculating the load effects, the structure and the associated load effects are often divided into primary, secondary, and tertiary levels. Figure 3.2 illustrates a typical example of these three levels in a ship structure (Paulling 1988). In this case, the primary level is related to the response of the entire ship's hull as a beam under bending or twisting moments.

Figure 3.2 Three structural response levels: primary, secondary, and tertiary.

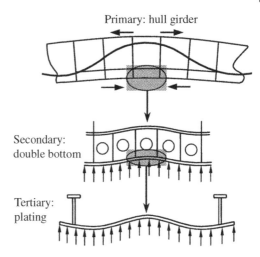

Primary: hull girder

Secondary: double bottom

Tertiary: plating

The secondary level is associated with the load effects of a stiffened panel, such as at the outer bottom plate panels of a double-bottom structure between two adjacent transverse bulkheads. The boundaries of the secondary structure (i.e., stiffened panel) are usually formed by other secondary structures (e.g., side shells or bulkheads). The tertiary level represents the load effects of the individual plating between stiffeners. The boundaries of the tertiary structure (i.e., plating) are formed by the stiffeners of the secondary structure (i.e., stiffened panel) of which it is a part. It is important to realize that the load effect analysis must account for the three responses noted earlier.

These load components are not always applied simultaneously, but more than one normally exists and interacts. Therefore, the buckling strength formulations must account for such combined load effects. In the buckling strength design, the plate element is considered to be subjected to average in-plane stresses, $\sigma_x = \sigma_{xav}$, $\sigma_y = \sigma_{yav}$, σ_{bx}, σ_{by}, $\tau = \tau_{av}$, and lateral pressure, p, or their combinations, as shown in Figure 3.1.

For perforated plates under in-plane loads, the membrane stress distribution may be nonuniform even before buckling, and thus the average values of the applied stresses of a perforated plate may be lower than those of a perfect plate, that is, one without perforations. For practical design purposes of perforated plates, the average stresses of the plates excluding perforations are often used as the characteristic measure of the applied stresses, whereas smaller, partial safety factors related to the load effects are usually adopted in this case.

For convenience in this chapter, axial compressive stress is taken as positive in sign, whereas axial tension is taken as negative in sign unless otherwise specified.

3.4 Boundary Conditions

The plate elements in plated structures are supported by various types of members along the edges, which have finite values of torsional rigidity, in contrast to the idealized simply supported boundary conditions that are often assumed for design purposes.

Depending on the torsional rigidity of support members, the rotation along the plate edges is restrained to some extent. When the rotational restraints are zero, the edge condition corresponds to a simply supported case, whereas the edge condition becomes clamped when the rotational restraints are infinite.

Most current practical design guidelines for the buckling and ultimate strength of plates are based on boundary conditions in which all (four) edges have idealized edge conditions, such as being simply supported or clamped. In real plates of continuous plated structures, such idealized edge conditions rarely if ever occur because of the finite rotational restraints.

For more advanced design of plates against buckling, it is therefore important to better understand the buckling strength characteristics of plates as a function of the rotational restraints associated with the support members along the edges.

This chapter deals with the buckling strength of plates with various edge conditions that are simply supported, clamped, or partially rotation restrained. The first two types of edge conditions are ideal but are often adequate for practical design purposes.

3.5 Linear Elastic Behavior

The behavior of a plate either before buckling or under predominantly axial tensile loading may be linear elastic until buckling occurs or gross yielding is formed. The linear elastic behavior of either perfect plates before buckling or imperfect plates under predominantly axial tensile loading can typically be represented by the relationship between average stresses and strains in a plane stress state as follows:

$$
\left.
\begin{aligned}
\varepsilon_{xav} &= \frac{1}{E}\sigma_{xav} - \frac{\nu}{E}\sigma_{yav} \\
\varepsilon_{yav} &= -\frac{\nu}{E}\sigma_{xav} + \frac{1}{E}\sigma_{yav} \\
\gamma_{av} &= \frac{1}{G}\tau_{av}
\end{aligned}
\right\}
\tag{3.1a}
$$

where ε_{xav}, ε_{yav}, and γ_{av} are the average strain components corresponding to σ_{xav}, σ_{yav}, and τ_{av}, respectively. Equation (3.1a) can be rewritten in the matrix form as follows:

$$
\left\{
\begin{aligned}
\sigma_{xav} \\
\sigma_{yav} \\
\tau_{av}
\end{aligned}
\right\}
= \left[D_\mathrm{p}\right]^{E}
\left\{
\begin{aligned}
\varepsilon_{xav} \\
\varepsilon_{yav} \\
\gamma_{av}
\end{aligned}
\right\}
\tag{3.1b}
$$

where

$$
\left[D_\mathrm{p}\right]^{E} = \frac{E}{1-\nu^2}
\begin{bmatrix}
1 & \nu & 0 \\
\nu & 1 & 0 \\
0 & 0 & (1-\nu)/2
\end{bmatrix}
$$

3.6 Elastic Buckling of Simply Supported Plates Under Single Types of Loads

The elastic buckling stress solutions for a plate under single in-plane loading and common idealized edge conditions are widely available from classic works on the theory of

Table 3.1 Buckling coefficients for a simply supported plate under single types of loads for $a/b \geq 1$.

Load type	σ_E	k
σ_x	$\sigma_{xE,1}$	$k_x = [a/(m_o b) + m_o b/a]^2$ where m_o is the buckling half-wave number for the plate in the x direction, which is the minimum integer that satisfies $a/b \leq \sqrt{[m_o(m_o + 1)]}$. For practical use, the half-wave number m may be taken as $m_o = 1$ for $1 \leq a/b \leq \sqrt{2}$, $m_o = 2$ for $\sqrt{2} < a/b \leq \sqrt{6}$, and $m_o = 3$ for $\sqrt{6} < a/b \leq 3$. If $a/b > 3$, the buckling coefficient can be approximated to $k_x = 4$
σ_y	$\sigma_{yE,1}$	$k_y = \left[1 + (b/a)^2\right]^2$
τ	$\tau_{E,1}$	$k_\tau \approx 4(b/a)^2 + 5.34$ for $a/b \geq 1$ ($k_\tau \approx 5.34(b/a)^2 + 4.0$ for $a/b < 1$)
σ_{bx}	$\sigma_{bxE,1}$	$k_{bx} \approx 23.9$
σ_{by}	$\sigma_{byE,1}$	$k_{by} \approx \begin{cases} 23.9 & \text{for } 1 \leq a/b \leq 1.5 \\ 15.87 + 1.87(a/b)^2 + 8.6(b/a)^2 & \text{for } a/b > 1.5 \end{cases}$

Note: The subscript "1" represents buckling under a single type of load.

elasticity (e.g., Bleich 1952, Timoshenko & Gere 1982, Hughes & Paik 2013). The elastic buckling strength of a plate with $a/b \geq 1$ is typically given in the following form:

$$\sigma_E = k\frac{\pi^2 D}{b^2 t} = k\frac{\pi^2 E}{12(1 - v^2)}\left(\frac{t}{b}\right)^2 \tag{3.2}$$

where σ_E is the plate buckling strength under a single type of load and k is the buckling coefficient for the corresponding load. Values of σ_E and k for various single types of loads are given in Table 3.1.

3.7 Elastic Buckling of Simply Supported Plates Under Two Load Components

3.7.1 Biaxial Compression or Tension

As described in Chapter 4, an analytical solution for the elastic buckling condition of a simply supported plate subject to biaxial loads is given by

$$\frac{m^2}{a^2}\sigma_x + \frac{n^2}{b^2}\sigma_y - \frac{\pi^2 D}{t}\left(\frac{m^2}{a^2} + \frac{n^2}{b^2}\right)^2 = 0 \tag{3.3}$$

where m and n are the buckling half-wave numbers in the x and y directions, respectively.

One half-wave number is normally taken in either the short edge or the direction in which the axial tensile loads are predominant. For a long plate considered in this chapter, that is, with $a/b \geq 1$, $n = 1$ can typically be taken. By holding the applied loading ratio $c = \sigma_y/\sigma_x$ constant, therefore, Equation (3.3) can be rewritten as follows:

$$\sigma_x\left(\frac{m^2}{a^2} + \frac{c}{b^2}\right) - \frac{\pi^2 D}{t}\left(\frac{m^2}{a^2} + \frac{1}{b^2}\right)^2 = 0 \tag{3.4}$$

Because buckling occurs when Equation (3.4) is satisfied, the bifurcation (buckling) stress is obtained by replacing σ_x with σ_{xE} as follows:

$$\sigma_{xE} = \frac{\pi^2 D}{b^2 t} \frac{(1 + m^2 b^2/a^2)^2}{c + m^2 b^2/a^2} = \frac{\pi^2 D}{t} \frac{(m^2/a^2 + 1/b^2)^2}{m^2/a^2 + c/b^2} \tag{3.5a}$$

where σ_{xE} indicates the longitudinal axial buckling stress component of a long plate under combined biaxial loading.

Under axial compression in the x direction alone, the longitudinal axial buckling stress of the plate can be obtained from Equation (3.3) as follows:

$$\sigma_{xE,1} = \frac{\pi^2 D}{b^2 t} \left(\frac{a}{mb} + \frac{mb}{a} \right)^2 \tag{3.5b}$$

As the buckling strength value should be the same at the transition of the buckling mode, the buckling half-wave number, m, in the x direction of the plate under biaxial loads can be predicted from Equation (3.5a) as a minimum integer that satisfies the following condition:

$$\frac{(m^2/a^2 + 1/b^2)^2}{m^2/a^2 + c/b^2} \le \frac{\left[(m+1)^2/a^2 + 1/b^2\right]^2}{(m+1)^2/a^2 + c/b^2} \tag{3.6a}$$

where it is evident that the buckling half-wave number is affected by the applied loading ratio as well as the plate aspect ratio.

When $c = \sigma_y/\sigma_x = 0$, Equation (3.6a) is simplified to the well-known criterion as follows:

$$\frac{a}{b} \le \sqrt{m(m+1)} \tag{3.6b}$$

Because the applied loading ratio, $c = \sigma_y/\sigma_x$, is kept constant, the elastic buckling axial stress in the y direction is obtained by

$$\sigma_{yE} = c\sigma_{xE} \tag{3.7a}$$

where σ_{yE} is the component of the elastic transverse axial buckling stress of the plate under combined biaxial loading.

Under axial compression in the y direction alone, $m = n = 1$ can be taken, and thus the elastic transverse axial buckling strength of the plate can be obtained from Equation (3.3) as follows:

$$\sigma_{yE,1} = \frac{\pi^2 D}{b^2 t} \left[1 + \left(\frac{b}{a} \right)^2 \right]^2 \tag{3.7b}$$

By substituting the buckling half-wave number, m, to be calculated from Equation (3.6a) and $n = 1$ (for the long plate) into Equations (3.5a) and (3.7a), the elastic buckling interaction relationship for a simply supported long plate subject to biaxial loads is obtained.

The benefit of Equation (3.5a) is that it is applicable to the plates under any combination of biaxial loading, for example, axial compressive loading in one direction and axial tensile loading in the other direction as well as axial compressive loading in both directions, whereas an opposite sign convention must be considered for axial tension.

It is evident from Equation (3.5a) that as long as $c = \sigma_y/\sigma_x$ is greater than $-m^2 b^2/a^2$, the buckling phenomenon can occur even if axial tensile loads are applied in one direction.

For a long plate, that is, with $a/b \geq 1$, Equation (3.3) may be rewritten as a function of stress components normalized by the buckling stresses under the corresponding single load component as follows:

$$\frac{(m^2 b^2/a^2)(m_o b/a + a/m_o b)^2}{(m^2 b^2/a^2 + 1)^2} \frac{\sigma_x}{\sigma_{xE,1}} + \frac{(b^2/a^2 + 1)^2}{(m^2 b^2/a^2 + 1)^2} \frac{\sigma_y}{\sigma_{yE,1}} = 1 \tag{3.8}$$

where $\sigma_{xE,1}$ and $\sigma_{yE,1}$ are defined in Equation (3.2) together with Table 3.1, m is defined in Equation (3.6a), and m_o is the buckling half-wave number of the plate under uniaxial compression in the x direction alone, which can be determined from Equation (3.6b).

Figure 3.3 shows the elastic buckling strength interaction curves of a simply supported rectangular plate under biaxial loads for $a/b = 3$ and 5. It is apparent from Figure 3.3 that the buckling half-wave number in the long direction varies with the loading ratio and with the plate aspect ratio.

In practice, a numerical iteration process may be necessary to compute the half-wave number, m, with Equation (3.6a). It is desirable for structural designers to have an approximate closed-form expression for the resulting plate buckling interaction relationship. Based on a series of computations for a variety of aspect ratios and loading ratios, an empirical buckling interaction equation for the plate subjected to biaxial compressive loading may be derived by curve fitting as follows (Ueda et al. 1987):

$$\left(\frac{\sigma_{xE}}{\sigma_{xE,1}}\right)^{\alpha_1} + \left(\frac{\sigma_{yE}}{\sigma_{yE,1}}\right)^{\alpha_2} = 1 \tag{3.9a}$$

where α_1 and α_2 are constants that are functions of the plate aspect ratio. Based on the computed results, the constants may be determined empirically as follows:

$$\alpha_1 = \alpha_2 = 1 \quad \text{for} \quad 1 \leq a/b \leq \sqrt{2} \tag{3.9b}$$

$$\left. \begin{array}{l} \alpha_1 = 0.0293(a/b)^3 - 0.3364(a/b)^2 + 1.5854(a/b) - 1.0596 \\ \alpha_2 = 0.0049(a/b)^3 - 0.1183(a/b)^2 + 0.6153(a/b) + 0.8522 \end{array} \right\} \quad \text{for} \quad a/b > \sqrt{2} \tag{3.9c}$$

Figure 3.4 shows the elastic buckling strength interaction curves for a plate under biaxial compression with varying aspect ratios, as obtained by Equation (3.9).

3.7.2 Longitudinal Axial Compression and Longitudinal In-Plane Bending

The elastic buckling strength interaction relationship of a simply supported plate under combined longitudinal axial compression and longitudinal in-plane bending is typically given by (Ueda et al. 1987)

$$\frac{\sigma_{xE}}{\sigma_{xE,1}} + \left(\frac{\sigma_{bxE}}{\sigma_{bxE,1}}\right)^c = 1 \tag{3.10}$$

where the constant, c, is often taken as $c = 2$ (JWS 1971) or $c = 1.75$ (Hughes & Paik 2013).

(a)

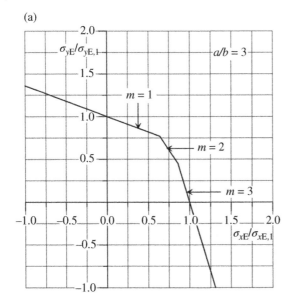

(b)

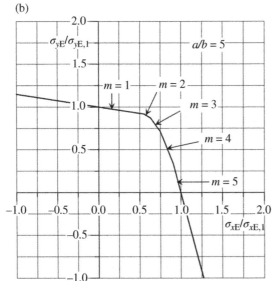

Figure 3.3 Elastic buckling interaction relationship for a plate under biaxial loads, simply supported at all edges: (a) $a/b = 3$ and $n = 1$; (b) $a/b = 5$ and $n = 1$.

3.7.3 Transverse Axial Compression and Longitudinal In-Plane Bending

The elastic buckling strength interaction relationship of a simply supported plate under combined transverse axial compression and longitudinal in-plane bending is typically given by (Ueda et al. 1987)

$$\left(\frac{\sigma_{yE}}{\sigma_{yE,1}}\right)^{\alpha_3} + \left(\frac{\sigma_{bxE}}{\sigma_{bxE,1}}\right)^{\alpha_4} = 1 \tag{3.11a}$$

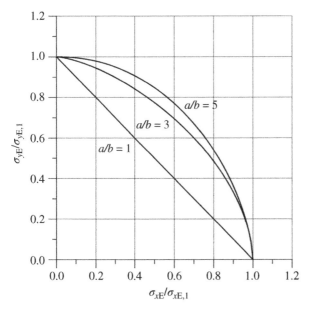

Figure 3.4 Approximate elastic buckling interaction relationships for a plate under biaxial compression with varying aspect ratios, simply supported at all edges, as obtained by Equation (3.9).

where the constants, α_3 and α_4, may be estimated as follows (JWS 1971):

$$\alpha_3 = \alpha_4 = 1.50(a/b) - 0.30 \quad \text{for} \quad 1 \le a/b \le 1.6 \tag{3.11b}$$

$$\left.\begin{array}{l} \alpha_3 = -0.625(a/b) + 3.10 \\ \alpha_4 = 6.25(a/b) - 7.90 \end{array}\right\} \quad \text{for} \quad 1.6 \le a/b \le 3.2 \tag{3.11c}$$

$$\left.\begin{array}{l} \alpha_3 = 1.10 \\ \alpha_4 = 12.10 \end{array}\right\} \quad \text{for} \quad 3.2 \le a/b \tag{3.11d}$$

3.7.4 Longitudinal Axial Compression and Transverse In-Plane Bending

The elastic buckling strength interaction relationship of a simply supported plate under combined longitudinal axial compression and transverse in-plane bending is typically given by (Ueda et al. 1987)

$$\left(\frac{\sigma_{xE}}{\sigma_{xE,1}}\right)^{\alpha_5} + \left(\frac{\sigma_{byE}}{\sigma_{byE,1}}\right)^{\alpha_6} = 1 \tag{3.12a}$$

where the constants, α_5 and α_6, may be estimated by (Ueda et al. 1987)

$$\left.\begin{array}{l} \alpha_5 = 0.930(a/b)^2 - 2.890(a/b) + 3.160 \\ \alpha_6 = 1.20 \end{array}\right\} \quad \text{for} \quad 1 \le a/b \le 2 \tag{3.12b}$$

$$\left.\begin{array}{l} \alpha_5 = 0.066(a/b)^2 - 0.246(a/b) + 1.328 \\ \alpha_6 = 1.20 \end{array}\right\} \quad \text{for} \quad 2 < a/b \le 5 \tag{3.12c}$$

$$\left.\begin{array}{l} \alpha_5 = 1.117(a/b) - 3.837 \\ \alpha_6 = -0.167(a/b) + 2.035 \end{array}\right\} \quad \text{for } 5 < a/b \le 8 \tag{3.12d}$$

$$\left.\begin{array}{l} \alpha_5 = 5.10 \\ \alpha_6 = 0.70 \end{array}\right\} \quad \text{for } 8 < a/b \tag{3.12e}$$

3.7.5 Transverse Axial Compression and Transverse In-Plane Bending

The elastic buckling strength interaction relationship of a simply supported plate under combined transverse axial compression and transverse in-plane bending is typically given by (Ueda et al. 1987)

$$\left(\frac{\sigma_{yE}}{\sigma_{yE,1}}\right)^{\alpha_7} + \left(\frac{\sigma_{byE}}{\sigma_{byE,1}}\right)^{\alpha_8} = 1 \tag{3.13a}$$

where the constants, α_7 and α_8, may be estimated by (Klöppel & Sheer 1960)

$$\left.\begin{array}{l} \alpha_7 = 1.0 \\ \alpha_8 = (14.0 - a/b)/6.5 \end{array}\right\} \quad \text{for } 1 \le a/b \le 7.5 \tag{3.13b}$$

$$\alpha_7 = \alpha_8 = 1.0 \quad \text{for } 7.5 < a/b \tag{3.13c}$$

3.7.6 Biaxial In-Plane Bending

The elastic buckling strength interaction relationship of a simply supported plate under combined biaxial in-plane bending is typically given by (Ueda et al. 1987)

$$\left(\frac{\sigma_{bxE}}{\sigma_{bxE,1}}\right)^{\alpha_9} + \left(\frac{\sigma_{byE}}{\sigma_{byE,1}}\right)^{\alpha_{10}} = 1 \tag{3.14a}$$

where the constants, α_9 and α_{10}, may be estimated by (Ueda et al. 1987)

$$\left.\begin{array}{l} \alpha_9 = 0.050(a/b) + 1.080 \\ \alpha_{10} = 0.268(a/b) - 1.248(b/a) + 2.112 \end{array}\right\} \quad \text{for } 1 \le a/b \le 3 \tag{3.14b}$$

$$\left.\begin{array}{l} \alpha_9 = 0.146(a/b)^2 - 0.533(a/b) + 1.515 \\ \alpha_{10} = 0.268(a/b) - 1.248(b/a) + 2.112 \end{array}\right\} \quad \text{for } 3 < a/b \le 5 \tag{3.14c}$$

$$\left.\begin{array}{l} \alpha_9 = 3.20(a/b) - 13.50 \\ \alpha_{10} = -0.70(a/b) + 6.70 \end{array}\right\} \quad \text{for } 5 < a/b \le 8 \tag{3.14d}$$

$$\left.\begin{array}{l} \alpha_9 = 12.10 \\ \alpha_{10} = 1.10 \end{array}\right\} \quad \text{for } 8 < a/b \tag{3.14e}$$

3.7.7 Longitudinal Axial Compression and Edge Shear

When a plate buckles under edge shear, the deflection pattern is quite complex compared with that under axial compressive loading, and thus a number of terms are normally needed to more properly represent the plate deflection by a Fourier series function. Bleich (1952) studied the buckling of a simply supported rectangular plate subject to longitudinal axial compression and edge shear using the energy method and developed a design chart for plate buckling. Ueda et al. (1987) derived an empirical buckling strength interaction equation for the plate under combined longitudinal axial compression and edge shear by curve fitting based on the results of Bleich:

$$\frac{\sigma_{xE}}{\sigma_{xE,1}} + \left(\frac{\tau_E}{\tau_{E,1}}\right)^{\alpha_{11}} = 1 \tag{3.15a}$$

where the constant, α_{11}, may be given by

$$\alpha_{11} = \begin{cases} -0.160(a/b)^2 + 1.080(a/b) + 1.082 & \text{for } 1 \le a/b \le 3.2 \\ 2.90 & \text{for } a/b > 3.2 \end{cases} \tag{3.15b}$$

Figure 3.5 shows the elastic buckling strength interaction curves of a simply supported plate under combined longitudinal axial compression and edge shear with varying aspect ratios, as obtained by Equation (3.15).

3.7.8 Transverse Axial Compression and Edge Shear

Based on the theoretical results of the buckling strength for rectangular plates subject to combined transverse axial compression and edge shear as obtained by Bleich (1952) and

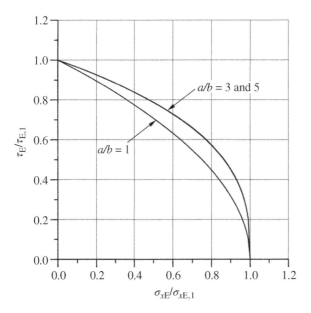

Figure 3.5 Approximate elastic buckling interactions of a plate under longitudinal axial compression and edge shear with varying aspect ratios, simply supported at all edges, as obtained by Equation (3.15).

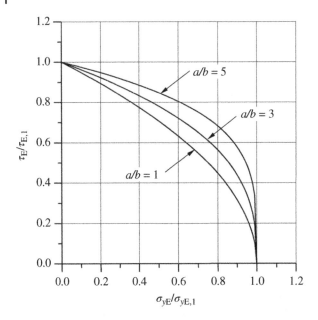

Figure 3.6 Approximate elastic buckling interactions of a plate under transverse axial compression and edge shear with varying aspect ratios, simply supported at all edges, as obtained by Equation (3.16).

Timoshenko and Gere (1982), Ueda et al. (1987) derived the following buckling interaction equation by curve fitting:

$$\frac{\sigma_{yE}}{\sigma_{yE,1}} + \left(\frac{\tau_E}{\tau_{E,1}}\right)^{\alpha_{12}} = 1 \tag{3.16a}$$

where the constant, α_{12}, may be given by

$$\alpha_{12} = \begin{cases} 0.10(a/b) + 1.90 & \text{for } 1 \le a/b \le 2 \\ 0.70(a/b) + 0.70 & \text{for } 2 < a/b \le 6 \\ 4.90 & \text{for } 6 < a/b \end{cases} \tag{3.16b}$$

Figure 3.6 shows the elastic buckling interaction curves of a simply supported plate under combined transverse axial compression and edge shear with varying aspect ratios, as obtained by Equation (3.16).

3.7.9 Longitudinal In-Plane Bending and Edge Shear

The elastic buckling interaction relationship of a simply supported plate under combined longitudinal in-plane bending and edge shear is typically given by (Ueda et al. 1987)

$$\left(\frac{\sigma_{bxE}}{\sigma_{bxE,1}}\right)^c + \left(\frac{\tau_E}{\tau_{E,1}}\right)^c = 1 \tag{3.17}$$

where the constant, c, is sometimes taken as $c = 2$ (JWS 1971).

3.7.10 Transverse In-Plane Bending and Edge Shear

The elastic buckling interaction relationship of a simply supported plate under combined transverse in-plane bending and edge shear is typically given by (Ueda et al. 1987)

$$\left(\frac{\sigma_{byE}}{\sigma_{byE,1}}\right)^c + \left(\frac{\tau_E}{\tau_{E,1}}\right)^c = 1 \tag{3.18}$$

where the constant, c, is sometimes taken as $c = 2$ (JWS 1971).

3.8 Elastic Buckling of Simply Supported Plates Under More than Three Load Components

The elastic buckling interaction relationship of a plate under a combination of three load components, such as longitudinal axial compression, transverse axial compression, and edge shear, is now derived based on three sets of interaction relationships between two load components, such as that between the longitudinal axial compression and the transverse axial compression, that between the longitudinal axial compression and the edge shear, and that between the transverse axial compression and the edge shear.

Figure 3.7 shows a schematic for the development of the buckling interaction equation between the three load components (Ueda et al. 1987). Two sets of interaction relationships between two load components are chosen so that one of the load components is common to both relationships. These relationships are in turn combined to obtain a new relationship between three load components.

Consider that the plate buckles under three load components, which are denoted by σ_{xE}^*, σ_{yE}^*, and τ_E^*. When no transverse axial compression is applied, the interaction among

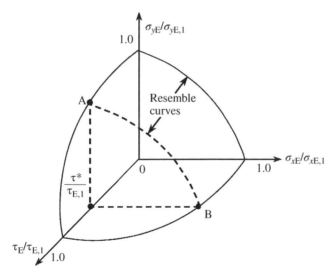

Figure 3.7 A schematic representation for derivation of the buckling interaction relationship among three load components.

σ_{xE}^*, σ_{yE}^*, and τ_E^* corresponds to that of the $\sigma_{xE} - \tau_E$ relationship, as indicated in Equation (3.15). In this case, the critical value, σ_{xE}^*, of the longitudinal axial compression, σ_x, which causes buckling together with τ_E^*, is obtained from Equation (3.15) as follows:

$$\sigma_{xE}^* = \sigma_{xE,1}\left[1 - \left(\frac{\tau_E^*}{\tau_{E,1}}\right)^{\alpha_{11}}\right] \tag{3.19}$$

Similarly, when no longitudinal axial compression is applied, the critical value, σ_{yE}^*, of transverse axial compression, σ_y, which causes the buckling, is obtained from Equation (3.16):

$$\sigma_{yE}^* = \sigma_{yE,1}\left[1 - \left(\frac{\tau_E^*}{\tau_{E,1}}\right)^{\alpha_{12}}\right] \tag{3.20}$$

It is assumed that a relationship similar to that of Equation (3.9a), that is, for the case under biaxial compression, exists between σ_{xE}^* and σ_{yE}^* in any plane of $\tau*/\tau_{E,1}$ = constant, that is, together with any value of edge shear. When we replace $\sigma_{xE,1}$ and $\sigma_{yE,1}$ in Equation (3.9a) with σ_{xE}^* and σ_{yE}^* from Equations (3.19) and (3.20), respectively, the following buckling interaction relationship among σ_{xE}, σ_{yE}, and τ is obtained when $\tau_E = \tau_E^*$:

$$\left\{\frac{\sigma_{xE}}{\sigma_{xE,1}\left[1-(\tau_E/\tau_{E,1})^{\alpha_{11}}\right]}\right\}^{\alpha_1} + \left\{\frac{\sigma_{yE}}{\sigma_{yE,1}\left[1-(\tau_E/\tau_{E,1})^{\alpha_{12}}\right]}\right\}^{\alpha_2} = 1 \tag{3.21}$$

where it is important to realize that the plate can, of course, buckle from edge shear alone, that is, if the values of $[1-(\tau_E/\tau_{E,1})^{\alpha_{11}}]$ or $[1-(\tau_E/\tau_{E,1})^{\alpha_{12}}]$ in the denominator reach zero or become negative.

The buckling strength interaction equations under other sets of the three load components can also be derived in a similar manner to that between σ_x, σ_y, and τ. Also, all five potential in-plane load components—longitudinal axial compression, transverse axial compression, edge shear, longitudinal in-plane bending, and transverse in-plane bending—may be obtained with a similar approach as follows:

$$\left\{\frac{\sigma_{xE}}{C_1 C_4 \sigma_{xE,1}\left[1-(\tau_E/(C_3 C_6 \tau_{E,1}))^{\alpha_{11}}\right]}\right\}^{\alpha_1} + \left\{\frac{\sigma_{yE}}{C_2 C_5 \sigma_{yE,1}\left[1-(\tau_E/(C_3 C_6 \tau_{E,1}))\right]^{\alpha_{12}}}\right\}^{\alpha_2} = 1$$

$$\tag{3.22}$$

where

$$C_1 = 1 - \left(\frac{\sigma_{bxE}}{C_7 \sigma_{bxE,1}}\right)^2, \quad C_2 = \left[1 - \left(\frac{\sigma_{bxE}}{C_7 \sigma_{bxE,1}}\right)^{\alpha_4}\right]^{1/\alpha_3}, \quad C_3 = \left[1 - \left(\frac{\sigma_{bxE}}{C_7 \sigma_{bxE,1}}\right)^2\right]^{0.5},$$

$$C_4 = \left[1 - \left(\frac{\sigma_{byE}}{\sigma_{byE,1}}\right)^{\alpha_6}\right]^{1/\alpha_5}, \quad C_5 = \left[1 - \left(\frac{\sigma_{byE}}{\sigma_{byE,1}}\right)^{\alpha_8}\right]^{1/\alpha_7}, \quad C_6 = \left[1 - \left(\frac{\sigma_{byE}}{\sigma_{byE,1}}\right)^2\right]^{0.5},$$

$$C_7 = \left[1 - \left(\frac{\sigma_{byE}}{\sigma_{byE,1}}\right)^{\alpha_{10}}\right]^{1/\alpha_9}.$$

In Equation (3.22), $c = 2$ has been used for Equations (3.10), (3.17), and (3.18). As may be surmised from Equation (3.22), the plate can, of course, buckle from edge shear alone, that is, when the denominator reaches zero or becomes negative.

3.9 Elastic Buckling of Clamped Plates

3.9.1 Single Types of Loads

The elastic bifurcation buckling stress of plates with clamped edge conditions and under single types of loads may also be calculated from Equation (3.2), but with the use of different buckling coefficients. Table 3.2 indicates the buckling coefficients of clamped plates with $a/b \geq 1$, where the long edges are taken in the x direction and the short edges are taken in the y direction (Bleich 1952).

3.9.2 Combined Loads

For practical purposes, it is often assumed that the elastic buckling interactive relationship between combined loads for a plate with the boundary condition clamped at some or all edges is the same as that for a plate simply supported at all edges but using the corresponding buckling strength components under single loads.

Figures 3.8 and 3.9 show the elastic buckling interaction relationships of rectangular plates clamped at all edges between biaxial compressions and between uniaxial compression and edge shear, respectively, with varying aspect ratios, as those obtained by the eigenvalue analysis using the finite element method. The corresponding buckling interaction equations for plates simply supported at four edges, such as Equations (3.9) and (3.15), are also shown in the figures.

It is apparent from Figures 3.8 and 3.9 that due to the rotational restraints at the edges, the buckling interaction of clamped plates becomes more convex than that of simply supported plates, but the buckling interaction for simply supported plates seems to represent the results fairly well, although slightly on the pessimistic side.

3.10 Elastic Buckling of Partially Rotation Restrained Plates

In a continuous stiffened plate structure, the rotation of the plating at the edges is to some extent restrained depending on the torsional rigidity of support members (stiffeners). This section describes closed-form elastic plate buckling strength formulations that account for the effect of rotational restraints at the plate edges, as originally developed by Paik and Thayamballi (2000).

3.10.1 Rotational Restraint Parameters

The support members of the plate elements have finite values of torsional rigidity, and thus the rotation along the plate edges is to some extent restrained. Essentially, the buckling strength of the plate elements is affected by these rotational restraints.

When the dimensions of support members (stiffeners) are defined as shown in Figure 5.3, the rotational restraint parameters of the longitudinal (x) and transverse (y) support members in a continuous plate structure can be determined as follows:

$$\zeta_L = C_L \frac{GJ_L}{bD}, \quad \zeta_S = C_S \frac{GJ_S}{aD} \tag{3.23}$$

Table 3.2 Elastic buckling coefficients of clamped plates under single types of loads for $a/b \geq 1$.

Load type	σ_E	BC	k	
Uniaxial compression in the x direction, σ_x	$\sigma_{xE,1}$	SSLC	$k_x = \begin{cases} 7.39(a/b)^2 - 19.6(a/b) + 20 \\ 6.98 \end{cases}$	for $1.0 \leq a/b \leq 1.33$ for $1.33 < a/b$
		SCLS	$k_x = \begin{cases} -0.95(a/b)^3 + 6.4(a/b)^2 - 14.86(a/b) + 16.34 \\ 0.2(a/b)^2 - 1.4(a/b) + 6.64 \\ -0.05(a/b) + 4.4 \\ 4.0 \end{cases}$	for $1.0 \leq a/b < 2.0$ for $2.0 \leq a/b < 3.0$ for $3.0 \leq a/b < 8.0$ for $8.0 \leq a/b$
		AC	$k_x = \begin{cases} -1.23(a/b)^3 + 7.9(a/b)^2 - 17.65(a/b) + 21.35 \\ 0.2(a/b)^2 - 1.62(a/b) + 10.35 \\ -0.062(a/b) + 7.476 \\ 6.98 \end{cases}$	for $1.0 \leq a/b < 2.0$ for $3.0 \leq a/b < 3.0$ for $2.0 \leq a/b < 8.0$ for $8.0 \leq a/b$
Uniaxial compression in the y direction, σ_y	$\sigma_{yE,1}$	SSLC	$k_y = [1.0 + (b/a)^2]^2 + 3.01$	for $0.0 < b/a < 1.0$
		SCLS	$k_y = \begin{cases} [1.0 + (b/a)^2]^2 + 0.12 \\ [0.95 + 1.89(b/a)^2]^2 \\ 13.98(b/a) - 6.20 \end{cases}$	for $0.0 < b/a < 0.34$ for $0.34 \leq b/a \leq 0.96$ for $0.96 < b/a \leq 1.0$
		AC	$k_y = \begin{cases} [1.0 + (b/a)^2]^2 + 4.8 \\ [1.92 + 1.305(b/a)^2]^2 \end{cases}$	for $0.0 < b/a < 0.8$ for $0.8 \leq b/a \leq 1.0$
Uniform edge shear, τ	$\tau_{E,1}$	SSLC	$k_\tau = 2.4(b/a)^2 + 1.08(b/a) + 9.0$	for $0.0 < b/a \leq 1.0$
		SCLS	$k_\tau = \begin{cases} 2.25(b/a)^2 + 1.95(b/a) + 5.35 \\ 22.92(b/a)^3 - 33.0(b/a)^2 + 20.43(b/a) + 2.13 \end{cases}$	for $0.0 < b/a \leq 0.4$ for $0.4 < b/a \leq 1.0$
		AC	$k_\tau = 5.4(b/a)^2 + 0.6(b/a) + 9.0$	for $0.0 < b/a \leq 1.0$

Note: AC, all edges clamped; BC, boundary condition; SCLS, short (y) edges clamped and long (x) edges simply supported; SSLC, short (y) edges simply supported and long (x) edges clamped.

Figure 3.8 Elastic buckling interaction relationships of plates between biaxial compressive loads (line: Equation (3.9) for plates simply supported at all edges; symbols: eigenvalue finite element solutions for plates clamped at all edges).

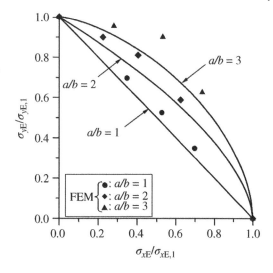

Figure 3.9 Elastic buckling interaction relationships of plates between axial compression and edge shear (line: Equation (3.15) for plates simply supported at all edges; symbols: eigenvalue finite element solutions for plates clamped at all edges).

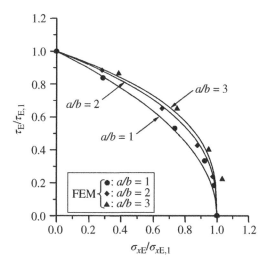

where ζ_L and ζ_S are the rotational restraint parameters for the longitudinal or transverse support member, $J_L = \left(h_{wx}t_{wx}^3 + b_{fx}t_{fx}^3\right)/3$ is the torsional constant of the longitudinal support member, $J_S = \left(h_{wy}t_{wy}^3 + b_{fy}t_{fy}^3\right)/3$ is the torsional constant of the transverse support member, G and D are defined in Section 3.2, and C_L and C_S are constants.

The support members may have welding induced initial distortions or in some cases distort sideways due to axial compression before plate buckling, so that they may not fully contribute to the rotational restraints along the plate edges. C_L and C_S in Equation (3.23) are constants used to account for this effect and are taken as values less than 1.0. For the sake of simplicity, however, $C_L = C_S = 1$ is usually applied; otherwise, they can be determined as the relative torsional rigidity of the support member to the plate part, as follows:

$$C_L = \frac{J_L}{J_{pL}} \le 1.0, \quad C_S = \frac{J_S}{J_{pS}} \le 1.0 \tag{3.24}$$

where

$$J_{pL} = \frac{bt^3}{3}, \quad J_{pS} = \frac{at^3}{3}.$$

3.10.2 Longitudinal Axial Compression

The elastic buckling stress of a plate with partially rotation restrained edge conditions in longitudinal axial compression can also be calculated from Equation (3.2) but using a different buckling coefficient, k_x. In the following, empirical formulas for the buckling coefficient, k_x, which are expressed in terms of the plate aspect ratio and the torsional rigidity of support members, are presented, following Paik and Thayamballi (2000).

3.10.2.1 Partially Rotation Restrained at Long Edges and Simply Supported at Short Edges

$$k_x = \begin{cases} 0.396\zeta_L^3 - 1.974\zeta_L^2 + 3.565\zeta_L + 4.0 & \text{for } 0 \le \zeta_L < 2 \\ 6.951 - 0.881/(\zeta_L - 0.4) & \text{for } 2 \le \zeta_L < 20 \\ 7.025 & \text{for } 20 \le \zeta_L \end{cases} \tag{3.25a}$$

The accuracy of Equation (3.25a) is verified in Figure 3.10 by comparison with the exact theoretical solutions as obtained by direct solution of the characteristic equation.

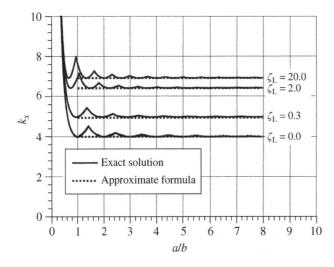

Figure 3.10 Accuracy of Equation (3.25a) for a plate under longitudinal axial compression, partially rotation restrained at the long edges and simply supported at the short edges.

3.10.2.2 Partially Rotation Restrained at Short Edges and Simply Supported at Long Edges

$$k_x = d_1 \zeta_S^4 + d_2 \zeta_S^3 + d_3 \zeta_S^2 + d_4 \zeta_S + d_5 \tag{3.25b}$$

where

$$d_1 = \begin{cases} -1.010(a/b)^4 + 12.827(a/b)^3 - 52.553(a/b)^2 + 67.072(a/b) - 27.585 & \text{for } 0 \le \zeta_S < 0.4 \\ 0.047(a/b)^4 - 0.586(a/b)^3 + 2.576(a/b)^2 - 4.410(a/b) + 1.748 & \text{for } 0.4 \le \zeta_S < 0.8 \\ -0.017(a/b)^2 + 0.099(a/b) - 0.150 & \text{for } 0.8 \le \zeta_S < 2 \\ 0.0 & \text{for } 2 \le \zeta_S \end{cases},$$

$$d_2 = \begin{cases} 0.881(a/b)^4 - 10.851(a/b)^3 + 41.688(a/b)^2 - 43.150(a/b) + 14.615 & \text{for } 0 \le \zeta_S < 0.4 \\ -0.123(a/b)^4 + 1.549(a/b)^3 - 6.788(a/b)^2 + 11.299(a/b) - 3.662 & \text{for } 0.4 \le \zeta_S < 0.8 \\ 0.138(a/b)^2 - 0.793(a/b) + 1.171 & \text{for } 0.8 \le \zeta_S < 2 \\ 0.0 & \text{for } 2 \le \zeta_S \end{cases},$$

$$d_3 = \begin{cases} -0.190(a/b)^4 + 2.093(a/b)^3 - 5.891(a/b)^2 - 2.096(a/b) + 1.792 & \text{for } 0 \le \zeta_S < 0.4 \\ 0.114(a/b)^4 - 1.412(a/b)^3 + 5.933(a/b)^2 - 8.638(a/b) + 0.224 & \text{for } 0.4 \le \zeta_S < 0.8 \\ -0.457(a/b)^2 + 2.571(a/b) - 3.712 & \text{for } 0.8 \le \zeta_S < 2 \\ 0.0 & \text{for } 2 \le \zeta_S \end{cases},$$

$$d_4 = \begin{cases} 0.004(a/b)^4 - 0.007(a/b)^3 - 0.243(a/b)^2 + 0.630(a/b) + 3.617 & \text{for } 0 \le \zeta_S < 0.4 \\ -0.021(a/b)^4 + 0.184(a/b)^3 - 0.126(a/b)^2 - 2.625(a/b) + 6.457 & \text{for } 0.4 \le \zeta_S < 0.8 \\ 0.822(a/b)^2 - 4.516(a/b) + 6.304 & \text{for } 0.8 \le \zeta_S < 2 \\ -0.106(a/b) + 0.176 & \text{for } 2 \le \zeta_S \\ 0.0 & \text{for } 20 \le \zeta_S \end{cases},$$

$$d_5 = \begin{cases} 4.0 & \text{for } 0 \le \zeta_S < 0.4 \\ -0.001(a/b)^4 + 0.033(a/b)^3 - 0.241(a/b)^2 + 0.684(a/b) + 3.539 & \text{for } 0.4 \le \zeta_S < 0.8 \\ -0.148(a/b)^2 + 0.596(a/b) + 3.847 & \text{for } 0.8 \le \zeta_S < 2 \\ -1.822(a/b) + 7.850 & \text{for } 2 \le \zeta_S < 20 \\ 0.041(a/b)^4 - 0.602(a/b)^3 + 3.303(a/b)^2 - 8.176(a/b) + 12.144 & \text{for } 20 \le \zeta_S \end{cases}.$$

In calculating k_x of Equation (3.25b), the following conditions must be satisfied for the approximations to hold: (i) if $4.0 < a/b \le 4.5$ and $\zeta_S \ge 0.2$, then $\zeta_S = 0.2$; (ii) if $a/b > 4.5$ and $\zeta_S \ge 0.1$, then $\zeta_S = 0.1$; (iii) if $a/b \ge 2.2$ and $\zeta_S \ge 0.4$, then $\zeta_S = 0.4$; (iv) if $a/b \ge 1.5$ and $\zeta_S \ge 1.4$, then $\zeta_S = 1.4$; (v) if $8 \le a/b \le 20$, then $\zeta_S = 8$; and (vi) if $a/b \ge 5$, then $a/b = 5$.

Figure 3.11a and b shows the variation of the buckling coefficient, k_x, as a function of the plate aspect ratio and the torsional rigidity of the support members at the short edges. The accuracy of Equation (3.25b) is verified in Figure 3.11b by comparison with the exact theoretical solutions as obtained by direct solution of the characteristic equation described in Paik and Thayamballi (2000).

(a)

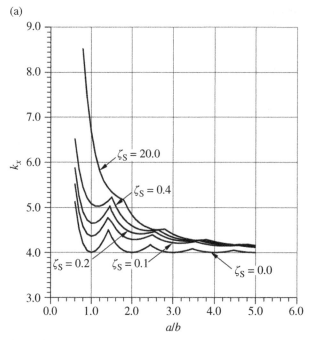

(b)

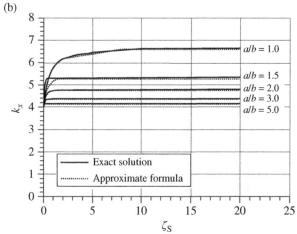

Figure 3.11 (a) Variation of the buckling coefficient, k_x, for a plate under longitudinal axial compression, partially rotation restrained at the short edges and simply supported at the long edges; (b) accuracy of Equation (3.25b) for a plate under longitudinal axial compression, partially rotation restrained at the short edges and simply supported at the long edges.

3.10.2.3 Partially Rotation Restrained at Both Long and Short Edges

For practical design purposes, the partially rotation restrained boundary conditions at both long and short edges may be expressed by a relevant combination of the previous two edge conditions and the simply supported edge condition. Specifically, one can assume

$$k_x = k_{x1} + k_{x2} - k_{x0} \tag{3.25c}$$

where k_x is the buckling coefficient of a plate partially rotation restrained at both long and short edges, k_{x1} is the buckling coefficient of a plate partially rotation restrained at the long edges and simply supported at the short edges, as defined in Equation (3.25a), k_{x2} is the buckling coefficient of a plate partially rotation restrained at the short edges and simply supported at the long edges, as defined in Equation (3.25b), and k_{x0} is the buckling coefficient of a plate simply supported at all edges, as defined in Table 3.1.

3.10.3 Transverse Axial Compression

The elastic buckling stress of a plate with partially rotation restrained edge conditions and under transverse axial compression can also be calculated from Equation (3.2) but using a different buckling coefficient, k_y. In the following, the empirical formulas of the buckling coefficient, k_y, which are expressed in terms of the plate aspect ratio and the torsional rigidity of support members, are presented, following Paik and Thayamballi (2000).

3.10.3.1 Partially Rotation Restrained at Long Edges and Simply Supported at Short Edges

$$k_y = e_1 \zeta_L^2 + e_2 \zeta_L + e_3 \tag{3.26a}$$

where

$$e_1 = \begin{cases} 1.322(b/a)^4 - 1.919(b/a)^3 + 0.021(b/a)^2 + 0.032(b/a) & \text{for } 0 \le \zeta_L < 2 \\ -0.463(b/a)^4 + 1.023(b/a)^3 - 0.649(b/a)^2 + 0.073(b/a) & \text{for } 2 \le \zeta_L < 8, \\ 0.0 & \text{for } 8 \le \zeta_L \end{cases}$$

$$e_2 = \begin{cases} -0.179(b/a)^4 - 3.098(b/a)^3 + 5.648(b/a)^2 - 0.199(b/a) & \text{for } 0 \le \zeta_L < 2 \\ 5.432(b/a)^4 - 11.324(b/a)^3 + 6.189(b/a)^2 - 0.068(b/a) & \text{for } 2 \le \zeta_L < 8 \\ -1.047(b/a)^4 + 2.624(b/a)^3 - 2.215(b/a)^2 + 0.646(b/a) & \text{for } 8 \le \zeta_L < 20 \\ 0.0 & \text{for } 20 \le \zeta_L \end{cases},$$

$$e_3 = \begin{cases} 0.994(b/a)^4 + 0.011(b/a)^3 + 1.991(b/a)^2 + 0.003(b/a) + 1.0 & \text{for } 0 \le \zeta_L < 2 \\ -3.131(b/a)^4 + 4.753(b/a)^3 + 3.587(b/a)^2 - 0.433(b/a) + 1.0 & \text{for } 2 \le \zeta_L < 8 \\ 20.111(b/a)^4 - 43.697(b/a)^3 + 30.941(b/a)^2 - 1.836(b/a) + 1.0 & \text{for } 8 \le \zeta_L < 20 \\ 0.751(b/a)^4 - 0.047(b/a)^3 + 2.053(b/a)^2 - 0.015(b/a) + 4.0 & \text{for } 20 \le \zeta_L \end{cases}.$$

Figure 3.12a and b shows the variation in the buckling coefficient, k_y, for a plate under transverse axial compression, partially rotation restrained at the long edges and simply supported at the short edges, as a function of the plate aspect ratio and the torsional rigidity of the support members at the long edges. The accuracy of Equation (3.26a) is verified in Figure 3.12b by comparison with the exact theoretical solutions as obtained by direct solution of the characteristic equation described in Paik and Thayamballi (2000).

(a)

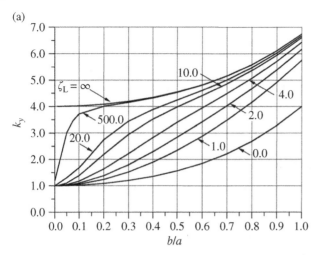

(b)

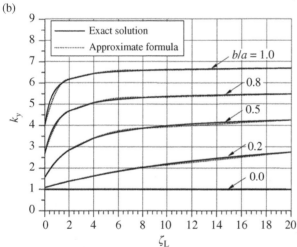

Figure 3.12 (a) Variation of the buckling coefficient, k_y, for a plate under transverse axial compression, partially rotation restrained at the long edges and simply supported at the short edges; (b) accuracy of Equation (3.26a) for a plate under transverse axial compression, partially rotation restrained at the long edges and simply supported at the short edges.

3.10.3.2 Partially Rotation Restrained at Short Edges and Simply Supported at Long Edges

$$k_y = f_1 \zeta_S^2 + f_2 \zeta_S + f_3 \qquad (3.26b)$$

where

$$f_1 = \begin{cases} 0.543(b/a)^4 - 1.297(b/a)^3 + 0.192(b/a)^2 - 0.016(b/a) & \text{for } 0 \leq \zeta_S < 2 \\ -0.347(b/a)^4 + 0.403(b/a)^3 - 0.147(b/a)^2 + 0.016(b/a) & \text{for } 2 \leq \zeta_S < 6, \\ 0.0 & \text{for } 6 \leq \zeta_S \end{cases}$$

$$f_2 = \begin{cases} -1.094(b/a)^4 + 4.401(b/a)^3 - 0.751(b/a)^2 + 0.068(b/a) & \text{for } 0 \le \zeta_S < 2 \\ 2.139(b/a)^4 - 1.76l(b/a)^3 + 0.419(b/a)^2 - 0.030(b/a) & \text{for } 2 \le \zeta_S < 6 \\ -0.199(b/a)^4 + 0.308(b/a)^3 - 0.118(b/a)^2 + 0.013(b/a) & \text{for } 6 \le \zeta_S < 20 \\ 0.0 & \text{for } 20 \le \zeta_S \end{cases},$$

$$f_3 = \begin{cases} 0.994(b/a)^4 + 0.011(b/a)^3 + 1.991(b/a)^2 + 0.003(b/a) + 1.0 & \text{for } 0 \le \zeta_S < 2 \\ -2.031(b/a)^4 + 5.765(b/a)^3 + 0.870(b/a)^2 + 0.102.(b/a) + 1.0 & \text{for } 2 \le \zeta_S < 6 \\ -0.289(b/a)^4 + 7.507(b/a)^3 - 1.029(b/a)^2 + 0.398(b/a) + 1.0 & \text{for } 6 \le \zeta_S < 20 \\ -6.278(b/a)^4 + 17.135(b/a)^3 - 5.026(b/a)^2 + 0.860(b/a) + 1.0 & \text{for } 20 \le \zeta_S \end{cases}.$$

Figure 3.13a and b shows the variation in the buckling coefficient, k_y, for a plate under transverse axial compression, partially rotation restrained at the short edges and simply supported at the long edges, as a function of the plate aspect ratio and the torsional rigidity of the support members at the short edges. The accuracy of Equation (3.26b) is verified in Figure 3.13b by comparison with the exact theoretical solutions as obtained by direct solution of the characteristic equation described in Paik and Thayamballi (2000).

3.10.3.3 Partially Rotation Restrained at Both Long and Short Edges

For a plate partially rotation restrained at both long and short edges under transverse axial compression, the buckling coefficient, k_y, may be expressed by a relevant combination of the previous two edge conditions in addition to the condition in which all edges are simply supported, as follows:

$$k_y = k_{y1} + k_{y2} - k_{y0} \tag{3.26c}$$

where k_y is the buckling coefficient of a plate partially rotation restrained at both long and short edges, k_{y1} is the buckling coefficient of a plate partially rotation restrained at the long edges and simply supported at the short edges, as defined in Equation (3.26a), k_{y2} is the buckling coefficient of a plate partially rotation restrained at the short edges and simply supported at the long edges, as defined in Equation (3.26b), and k_{y0} is the buckling coefficient of a plate simply supported at all edges, as defined in Table 3.1.

3.10.4 Combined Loads

It is often assumed with reasonable certainty that the elastic buckling interactive relationship between combined loads for a plate partially rotation restrained at the edges is similar to that for a plate simply supported at all edges. Therefore, the same buckling interactive relationship of a simply supported plate under combined loads may be used for a plate partially rotation restrained at the edges but replacing the buckling strengths under single types of load components by those for the corresponding edge conditions.

(a)

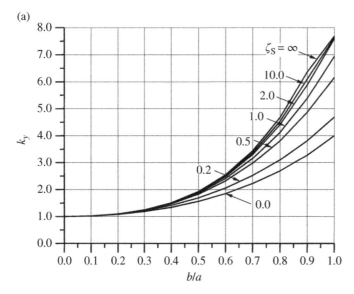

(b)

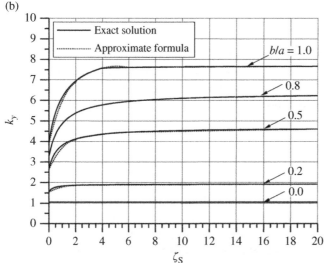

Figure 3.13 (a) Variation of the buckling coefficient, k_y, for a plate under transverse axial compression, partially rotation restrained at the short edges and simply supported at the long edges; (b) accuracy of Equation (3.26b) for a plate under transverse axial compression, partially rotation restrained at the short edges and simply supported at the long edges.

3.11 Effect of Welding Induced Residual Stresses

Welding induced residual stresses reduce the plate buckling strength. For a plate element between stiffeners, the elastic buckling stress may be given by considering that an effective compressive residual stress reduces the buckling strength. Therefore, the elastic buckling stress of a plate with welding induced residual stress in axial compression in the x direction can be calculated from Equation (3.2):

$$\sigma_{xE,1} = k_x \frac{\pi^2 E}{12(1-v^2)} \left(\frac{t}{b}\right)^2 - \sigma_{rex} \tag{3.27a}$$

where $\sigma_{rex} = \sigma_{rcx}(b-2b_t)/b$ is the effective welding induced compressive residual stress in the x direction, σ_{rcx} is the actual welding induced compressive residual stress, and b_t is the breadth of the heat-affected zone in the y direction as described in Section 1.7.3.

In a similar way, the elastic buckling stress of a plate under axial compression in the y direction can be calculated from Equation (3.2) after including the effect of welding residual stress, as follows:

$$\sigma_{yE,1} = k_y \frac{\pi^2 E}{12(1-v^2)} \left(\frac{t}{b}\right)^2 - \sigma_{rey} \tag{3.27b}$$

where $\sigma_{rey} = \sigma_{rcy}(a-2a_t)/a$ is the effective welding induced compressive residual stress in the y direction, σ_{rcy} is the actual welding induced compressive residual stress, and a_t is the breadth of the heat-affected zone in the x direction as described in Chapter 1.

Figure 3.14 shows the influence of welding induced residual stress on the compressive buckling stress for a simply supported plate with a yield stress of 352 MPa. In the calculations indicated in Figure 3.14, the level of residual stresses and the plate slenderness ratio vary. It is evident from Figure 3.14 that welding induced residual stresses can significantly reduce the compressive buckling stress of plates in some cases. The reduction tendency of the buckling stress for thin plates is more significant than that for thick plates, as expected.

It is interesting to note that very thin plates, such as those used to build living quarters in offshore platforms, often buckle under a biaxial compressive residual stress condition alone that arises from welding performed to attach support members in both longitudinal and transverse directions, as described in Chapter 1, even without external mechanical loads. In this case, an optimum design of the plate and support members to prevent buckling can be achieved by applying Equation (3.9) and Section 3.10 associated with welding induced residual stresses and partially rotation restrained edge conditions (Paik & Yi 2016).

3.12 Effect of Lateral Pressure Loads

Plates in a continuous stiffened panel are sometimes subjected to lateral pressure loads. For example, the bottom plates of ships or ship-shaped offshore structures are subjected to lateral pressure loads from cargo and/or water in addition to additional in-plane loads in operation.

As shown in Figure 3.15, the edges of a plate under lateral pressure loads approach the condition of being clamped, depending on the thickness of the plate and the pressures involved. Also, the lateral pressure loading may beneficially disturb the occurrence of the inherent plate buckling pattern. As a result, the elastic buckling strength of a long plate under lateral pressure loading is greater than that without it.

(a)

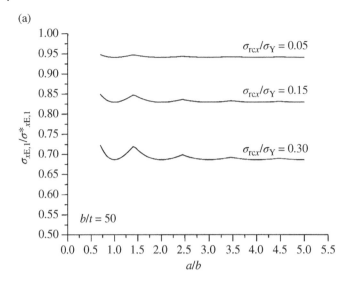

(b)

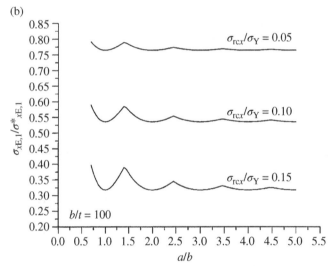

Figure 3.14 Variation of the elastic compressive buckling stress (normalized by the elastic buckling compressive stress without residual stresses) and the magnitude of welding induced residual stress for (a) a relatively thick plate and (b) a relatively thin plate.

For practical design purposes, a correction factor may be used to account for the effect of lateral pressure on the plate buckling strength, with the factor being applied by multiplication to the buckling strength calculated for the plate without lateral pressure loads. In this regard, Fujikubo et al. (1998) proposed plate compressive buckling strength correction factors to account for the effect of lateral pressure by curve fitting based on the finite element solutions for a long plate element in a continuous stiffened panel, as follows:

$$C_{px} = 1 + \frac{1}{576}\left(\frac{pb^4}{Et^4}\right)^{1.6} \quad \text{for } \frac{a}{b} \geq 2 \tag{3.28a}$$

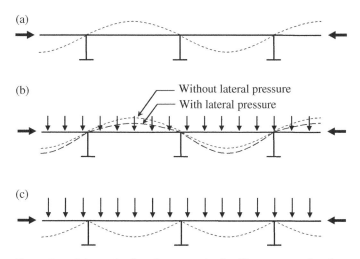

Figure 3.15 Schematic of axial compressive buckling pattern of a plate with or without lateral pressure loads: (a) without lateral pressure; (b) with relatively small amount of lateral pressure; (c) with relatively large amount of lateral pressure (Hughes & Paik 2013).

$$C_{py} = 1 + \frac{1}{160}\left(\frac{b}{a}\right)^{0.95}\left(\frac{pb^4}{Et^4}\right)^{1.75} \quad \text{for } \frac{a}{b} \geq 2 \tag{3.28b}$$

where C_{px} and C_{py} are correction factors of the elastic compressive buckling strength in the x and y directions, respectively, to account for the effect of lateral pressure, and p is the magnitude of net lateral pressure loads.

For a nearly square plate (i.e., with $a/b \approx 1$) under combined axial compression and lateral pressure, one half-wave deflection occurs from the beginning, so the bifurcation buckling phenomenon may not appear as the axial compressive loads increase. In this case, it is beneficial to define an equivalent buckling strength for practical design purposes. It is considered that the increase in the buckling strength due to the rotational restraints and the decrease in the buckling strength due to one half-wave deflection caused by lateral pressure may be offset. For a square plate, therefore, $C_{px} = C_{py} = 1.0$ may approximately be adopted.

The elastic compressive buckling stress of a plate to account for the effects of lateral pressure and welding induced residual stresses can then be calculated from Equations (3.27a) and (3.27b) but using the multiplicative correction factors of Equations (3.28a) and (3.28b) as follows:

$$\sigma_{xE,1} = C_{px}\left[k_x\frac{\pi^2 E}{12(1-\nu^2)}\left(\frac{t}{b}\right)^2 - \sigma_{rex}\right] \tag{3.29a}$$

$$\sigma_{yE,1} = C_{py}\left[k_y\frac{\pi^2 E}{12(1-\nu^2)}\left(\frac{t}{b}\right)^2 - \sigma_{rey}\right] \tag{3.29b}$$

Figure 3.16 plots Equations (3.29a) and (3.29b) for a specific steel plate with $a \times b = 2400 \text{ mm} \times 800 \text{ mm}$ and $E = 210 \text{ GPa}$ as a function of the plate thickness and water head when no welding residual stresses exist. It is apparent from Figure 3.16 that the

(a)

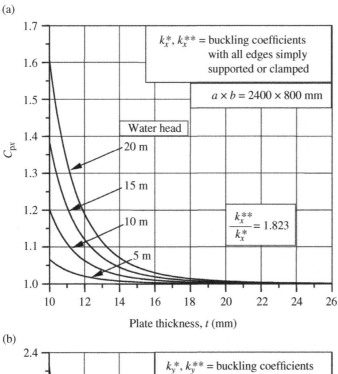

(b)

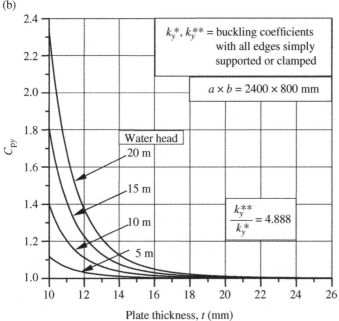

Figure 3.16 Effect of lateral pressure on the plate compressive buckling strength: (a) C_{px} versus plate thickness; (b) C_{py} versus plate thickness.

increasing trend in buckling strength due to lateral pressure for a thin plate is greater than that for a thick plate. It is noted that lateral pressure may not affect the buckling strength of a perforated plate because the perforated plate may not be subjected to lateral pressure loads.

3.13 Effect of Opening

In a plate element of plated structures, an opening (cut out) is sometimes made to create a point of access or to lighten the structure. This perforation reduces the plate's buckling strength. The opening must be included in the buckling strength formulations as a parameter of influence where significant.

Figure 3.17 shows a plate with a centrally located opening. To account for the effect of the opening on the plate buckling strength, one must use a relevant buckling strength reduction factor, which is defined as the ratio of the buckling coefficient of a perforated plate to that of a plate without the hole. In this case, empirical formulations may be derived for the plate buckling strength reduction factors due to opening by curve fitting based on the results of eigenvalue analysis of the finite element method.

In structural analysis and design, it is noted that the load effects (e.g., stresses) in a plate are usually defined for a perfect plate (i.e., one without an opening), although the plate capacity is evaluated by considering the effect of the opening. In this case, the partial safety factor may be adjusted to account for the effect of the opening in applying Equation (1.17). In the following sections, empirical formulations of the elastic buckling strength for a plate with a circular opening, that is, $a_c = b_c = d_c$, are presented.

Chapter 4 describes the ultimate strength of perforated plates. For buckling and ultimate strength of perforated plate panels, interested readers may refer to Narayanan and der Avanessian (1984), Brown et al. (1987), Paik (2007a, 2007b, 2008), Kim et al. (2009), Suneel Kumar et al. (2009), and Wang et al. (2009), among others.

3.13.1 Longitudinal Axial Compression

Figure 3.18 shows the variation of the buckling reduction factor, R_{xE}, of the plate under σ_x varying the size of the opening and the plate aspect ratio, as obtained by the eigenvalue

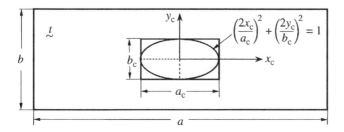

Figure 3.17 A plate with a centrally located opening.

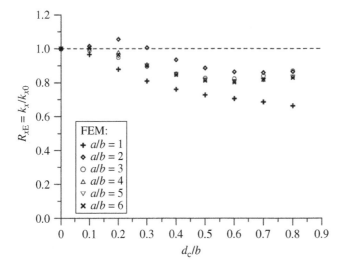

Figure 3.18 Variation of the buckling strength reduction factor of a plate under longitudinal axial compression varying the size of the opening and the plate aspect ratio as obtained by eigenvalue FEA (k_x, k_{x0} = longitudinal compressive buckling coefficients for the plate with and without an opening).

analysis of the finite element method. In this case, R_{xE} may be defined by a cubic equation in terms of the size of the opening and the plate's aspect ratio, as follows:

$$R_{xE} = \alpha_{E1} \left(\frac{d_c}{b}\right)^3 + \alpha_{E2} \left(\frac{d_c}{b}\right)^2 + \alpha_{E3} \frac{d_c}{b} + 1 \tag{3.30a}$$

where

$$\alpha_{E1} = \begin{cases} 0.002(a/b)^{8.238} & \text{for } 1 \le a/b < 2 \\ -1.542(a/b)^2 + 7.232a/b - 7.666 & \text{for } 2 \le a/b < 3, \\ -0.052(a/b)^2 + 0.526a/b - 0.964 & \text{for } 3 \le a/b \le 6 \end{cases}$$

$$\alpha_{E2} = \begin{cases} 0.655 + 1/[4.123(a/b) - 8.922] & \text{for } 1 \le a/b < 2 \\ 1.767(a/b)^2 - 7.937a/b + 7.982 & \text{for } 2 \le a/b < 3, \\ 0.071(a/b)^2 - 0.732a/b + 1.631 & \text{for } 3 \le a/b \le 6 \end{cases}$$

$$\alpha_{E3} = \begin{cases} -0.945 + 1/[-5.661(a/b) + 12.342] & \text{for } 1 \le a/b < 2 \\ -0.248(a/b)^2 + 0.796a/b - 0.565 & \text{for } 2 \le a/b < 3. \\ -0.020(a/b)^2 + 0.199a/b - 0.826 & \text{for } 3 \le a/b \le 6 \end{cases}$$

Figure 3.19 shows the accuracy of Equation (3.30a) by comparison with the eigenvalue finite element buckling solutions. The corresponding elastic plate buckling stress of a plate with a centrally located circular hole can be calculated as follows:

$$\sigma_{xE,1} = R_{xE} k_{x0} \frac{\pi^2 E}{12(1-v^2)} \left(\frac{t}{b}\right)^2 \tag{3.30b}$$

where k_{x0} is the longitudinal compressive buckling coefficient of a plate without an opening.

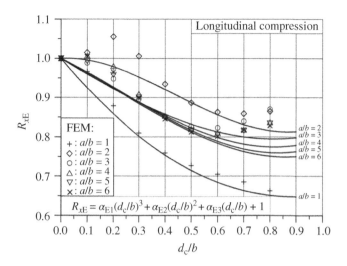

Figure 3.19 Accuracy of Equation (3.30a) for longitudinal axial compressive loading.

3.13.2 Transverse Axial Compression

Figure 3.20 shows the variation of the buckling reduction factor, R_{yE}, of a plate under σ_y varying the size of the opening and the plate aspect ratio, as obtained by the eigenvalue analysis of the finite element method. In this case, R_{yE} may be defined by a cubic equation in terms of the size of the opening and the plate aspect ratio as follows:

$$R_{yE} = \alpha_{E4}\left(\frac{d_c}{b}\right)^2 + \alpha_{E5}\frac{d_c}{b} + 1 \tag{3.31a}$$

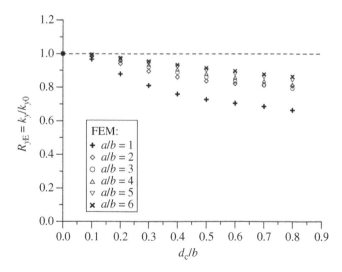

Figure 3.20 Variation of the buckling strength reduction factor of the plate under transverse axial compression varying the size of the opening and the plate aspect ratio as obtained by eigenvalue FEA (k_y, k_{y0} = transverse compressive buckling coefficients for the plate with and without an opening).

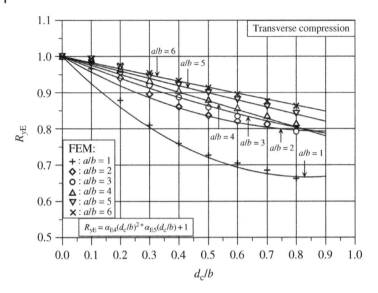

Figure 3.21 The accuracy of Equation (3.31a) for transverse axial compressive loading.

where

$$
\alpha_{E4} = \begin{cases} 0.034(a/b)^2 - 0.327a/b + 0.768 & \text{for } 1 \le a/b < 4 \\ 0.004 & \text{for } 4 \le a/b \le 6 \end{cases},
$$

$$
\alpha_{E5} = -0.008 - 1/[0.967(a/b) + 0.302] \quad \text{for } 1 \le a/b \le 6.
$$

Figure 3.21 shows the accuracy of Equation (3.31a) by comparison with the finite element buckling eigenvalue solutions. The corresponding elastic plate buckling stress of the perforated plate is then calculated as follows:

$$
\sigma_{yE,1} = R_{yE} k_{y0} \frac{\pi^2 E}{12(1 - v^2)} \left(\frac{t}{b}\right)^2 \tag{3.31b}
$$

where k_{y0} is the transverse compressive buckling coefficient of a plate without an opening.

3.13.3 Edge Shear

Figure 3.22 shows the variation of the buckling reduction factor, $R_{\tau E}$, of a plate under τ varying the size of the opening and the plate aspect ratio, as obtained by the eigenvalue analysis of the finite element method. In this case, $R_{\tau E}$ may be defined by a cubic equation in terms of the opening size and the plate aspect ratio as follows:

$$
R_{\tau E} = \alpha_{E6} \left(\frac{d_c}{b}\right)^3 + \alpha_{E7} \left(\frac{d_c}{b}\right)^2 + \alpha_{E8} \frac{d_c}{b} + 1 \tag{3.32a}
$$

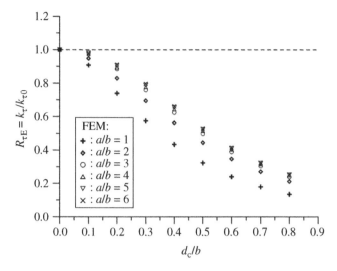

Figure 3.22 Variation of the buckling strength reduction factor of a plate under edge shear varying the size of the opening and the plate aspect ratio as obtained by eigenvalue FEA (k_τ, $k_{\tau 0}$ = shear buckling coefficients for the plate with and without an opening).

where

$$\alpha_{E6} = \begin{cases} 0.094(a/b)^2 + 0.035a/b + 1.551 & \text{for } 1 \le a/b < 3 \\ 2.502 & \text{for } 3 \le a/b \le 6 \end{cases},$$

$$\alpha_{E7} = \begin{cases} -0.039(a/b)^2 - 0.807a/b - 0.405 & \text{for } 1 \le a/b < 3 \\ -3.177 & \text{for } 3 \le a/b \le 6 \end{cases},$$

$$\alpha_{E8} = \begin{cases} -0.053(a/b)^2 + 0.785a/b - 1.875 & \text{for } 1 \le a/b < 3 \\ 0.003 & \text{for } 3 \le a/b \le 6 \end{cases}.$$

Figure 3.23 shows the accuracy of Equation (3.32a) by comparison with the eigenvalue finite element buckling solutions. The corresponding elastic plate buckling stress of the perforated plate is then calculated as follows:

$$\tau_{E,1} = R_{\tau E} k_{\tau 0} \frac{\pi^2 E}{12(1 - v^2)} \left(\frac{t}{b}\right)^2 \tag{3.32b}$$

where $k_{\tau 0}$ is the shear buckling coefficient of a plate without an opening.

3.13.4 Combined Loads

For practical design purposes, it may be assumed that the elastic buckling interaction relationship between combined loads for a perforated plate is the same as that for a plate without an opening, but using the corresponding buckling strength components under

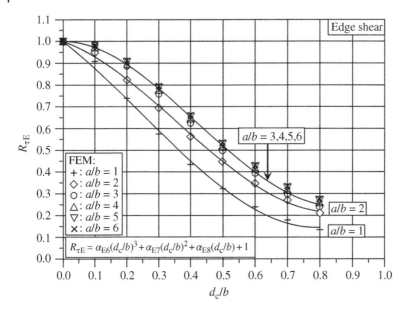

Figure 3.23 The accuracy of Equation (3.32a) for edge shear.

single types of loads. Figure 3.24 shows some selected elastic buckling strength interaction curves of plates with a centrally located circular opening and under combined loads. The finite element eigenvalue solutions are also compared. It is apparent that the assumption of the buckling interaction relationship made earlier is relevant.

3.14 Elastic–Plastic Buckling Strength

3.14.1 Single Types of Loads

A thick plate with a high elastic buckling strength will not buckle in the elastic regime and will reach the ULS with a certain degree of plasticity. Methods to predict the buckling capacity of a plate while accounting for the effect of plasticity differ depending on the presence of an opening.

3.14.1.1 Plates Without Opening

The elastic–plastic buckling strength is obtained by the plasticity correction of the corresponding elastic buckling stress using the Johnson–Ostenfeld formulation method, described in Chapter 2, and it is often termed the "critical" buckling strength. Under single types of loads, the elastic–plastic buckling stress is then approximated by substituting the computed elastic buckling stress into Equation (2.93) as follows:

$$\sigma_{xcr} = \begin{cases} \sigma_{xE,1} & \text{for } \sigma_{xE,1} \le \alpha\sigma_Y \\ \sigma_Y[1 - \sigma_Y/(4\sigma_{xE,1})] & \text{for } \sigma_{xE,1} > \alpha\sigma_Y \end{cases} \qquad (3.33a)$$

(a)

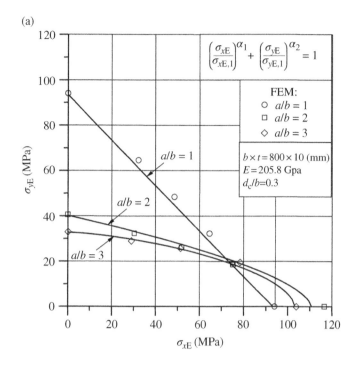

(b)

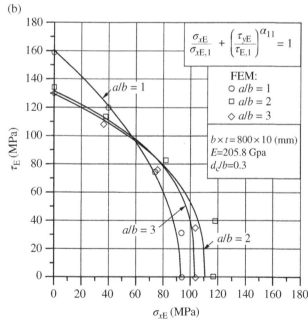

Figure 3.24 Elastic buckling strength interaction of a plate with a centrally located circular hole: (a) under combined longitudinal and transverse axial compression (α_1, α_2 = as defined in Equation (3.9)); (b) under combined longitudinal axial compression and edge shear (α_{11} = as defined in Equation (3.15)); (c) under combined transverse axial compression and edge shear (α_{12} = as defined in Equation (3.16)).

(c)

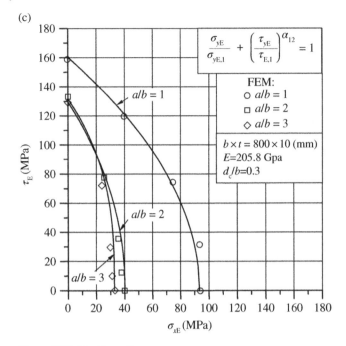

Figure 3.24 (Continued)

$$\sigma_{ycr} = \begin{cases} \sigma_{yE,1} & \text{for } \sigma_{yE,1} \le \alpha\sigma_Y \\ \sigma_Y\left[1 - \sigma_Y/\left(4\sigma_{yE,1}\right)\right] & \text{for } \sigma_{yE,1} > \alpha\sigma_Y \end{cases} \qquad (3.33b)$$

$$\tau_{cr} = \begin{cases} \tau_{E,1} & \text{for } \tau_{E,1} \le \alpha\tau_Y \\ \tau_Y\left[1 - \tau_Y/\left(4\tau_{E,1}\right)\right] & \text{for } \tau_{E,1} > \alpha\tau_Y \end{cases} \qquad (3.33c)$$

where $\tau_Y = \sigma_Y/\sqrt{3}$, $\alpha = 0.5$ or 0.6. In Equation (3.33), the compressive normal stress and shear stress are taken as positive. The critical buckling strengths are often regarded as the corresponding ultimate strengths for practical design purposes.

Figure 3.25 shows the resulting relationships between the critical buckling strength and the elastic buckling strength for steel plates (without opening) with varying edge conditions (Jun 2002). The ultimate strength obtained by the elastic–plastic large-deflection finite element method solutions using the elastic–perfectly plastic material model without the strain-hardening effect is also shown for comparison. It is evident that regardless of the edge conditions, the Johnson–Ostenfeld formulation method predicts fairly well the elastic–plastic or plastic buckling strength of relatively thick steel plates (without an opening) as a function of the elastic buckling strength, albeit on the pessimistic side.

3.14.1.2 Perforated Plates

For a perforated plate, however, Equation (3.33) using the Johnson–Ostenfeld formulation method may lead to inadequate results with an overestimation or underestimation of the critical buckling strength depending on the plate's slenderness ratio and/or the

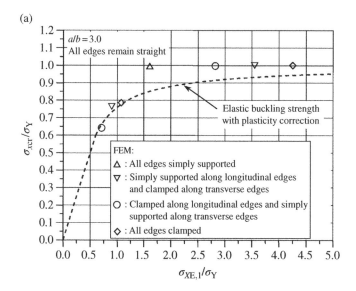

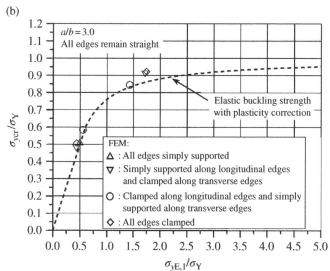

Figure 3.25 (a) The critical buckling stress, σ_{xcr}, versus the elastic bifurcation buckling stress, $\sigma_{xE,1}$, of plates without hole under longitudinal compression alone, $a/b = 3$; (b) the critical buckling stress, σ_{ycr}, versus the elastic bifurcation buckling stress, $\sigma_{yE,1}$, of plates without hole under transverse compression alone, $a/b = 3$; (c) the critical buckling stress, τ_{cr}, versus the elastic bifurcation buckling stress, $\tau_{E,1}$, of plates without hole under edge shear alone, $a/b = 3$.

opening size. Figure 3.26 compares the critical buckling strengths, as obtained from Equation (3.33), with the ultimate strengths, as obtained from nonlinear finite element method, for plates with a centrally located circular opening (Paik 2008). For a thin perforated plate, the critical buckling strength is significantly underestimated compared with the ultimate strength. In contrast, the critical buckling strength of a thick perforated plate is overestimated, specifically when the opening size is large.

(c)

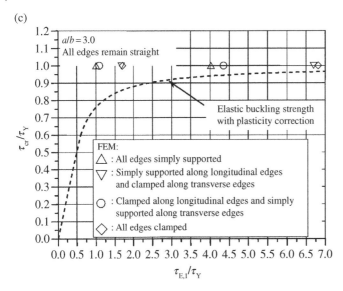

Figure 3.25 (Continued)

It is emphasized that the intention of the critical buckling strength predictions for relatively thick perforated plates is to account for the effect of elastic–plastic buckling. It is important to realize from these figures that the Johnson–Ostenfeld formulation method should not be used to predict the critical bucking strength of a perforated plate and that the ultimate strength described in Chapter 4 must be a better and more consistent basis for structural design. In this case, the ultimate strength of a perforated plate can be determined from Equation (4.86) as follows:

$$
\left.
\begin{aligned}
\sigma_{xcr} &= \sigma_{xu} = R_{xu}\sigma_{xuo} \\
\sigma_{ycr} &= \sigma_{yu} = R_{yu}\sigma_{yuo} \\
\tau_{cr} &= \tau_u = R_{\tau u}\tau_{uo}
\end{aligned}
\right\}
\tag{3.34}
$$

where σ_{xu}, σ_{yu}, and τ_u are the ultimate strengths of a perforated plate; σ_{xuo}, σ_{yuo}, and τ_{uo} are the ultimate strengths of a plate without an opening, as described in Section 4.10; and R_{xu}, R_{yu}, and $R_{\tau u}$ are the ultimate strength reduction factors of Equations (4.87a), (4.87b), and (4.87c), respectively.

3.14.2 Combined Loads

For the buckling strength design of a plate under combined longitudinal compression and tension, σ_x, transverse compression/tension, σ_y, and edge shear, τ, the critical buckling strength interaction function, Γ_B, is often expressed as follows:

$$
\Gamma_B = \left(\frac{\sigma_x}{\sigma_{xcr}}\right)^2 - \alpha\left(\frac{\sigma_x}{\sigma_{xcr}}\right)\left(\frac{\sigma_y}{\sigma_{ycr}}\right) + \left(\frac{\sigma_y}{\sigma_{ycr}}\right)^2 + \left(\frac{\tau}{\tau_{cr}}\right)^2 - 1
\tag{3.35}
$$

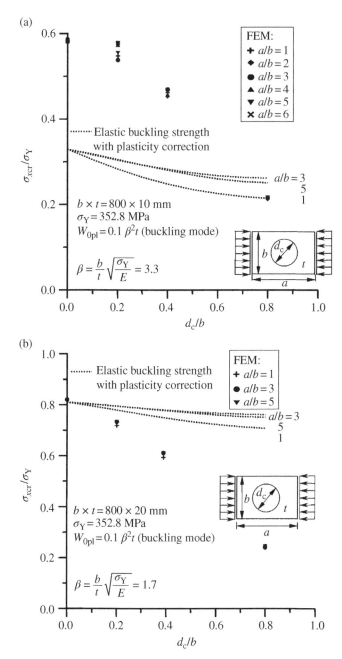

Figure 3.26 (a) A comparison of the critical buckling strength, σ_{xcr}, with the ultimate strength for plates with a centrally located circular hole and under longitudinal axial compression, $\beta = 3.3$; (b) a comparison of the critical buckling strength, σ_{xcr}, with the ultimate strength for plates with a centrally located circular hole and under longitudinal axial compression, $\beta = 1.7$; (c) a comparison of the critical buckling strength, σ_{ycr}, with the ultimate strength for plates with a centrally located circular hole and under transverse axial compression, $\beta = 3.3$; (d) a comparison of the critical buckling strength, σ_{ycr}, with the ultimate strength for plates with a centrally located circular hole and under transverse axial compression, $\beta = 1.7$; (e) a comparison of the critical buckling strength, τ_{cr}, with the ultimate strength for plates with a centrally located circular hole and under edge shear, $\beta = 3.3$; (f) a comparison of the critical buckling strength, τ_{cr}, with the ultimate strength for plates with a centrally located circular hole and under edge shear, $\beta = 1.7$. (β, plate slenderness ratio; w_{opl}, plate initial deflection).

(c)

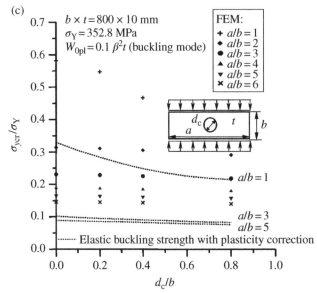

(d)

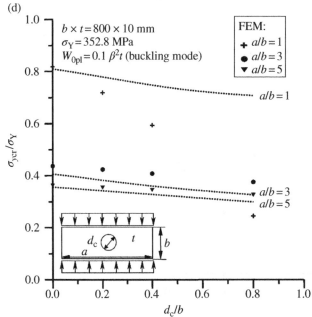

Figure 3.26 (Continued)

(e)

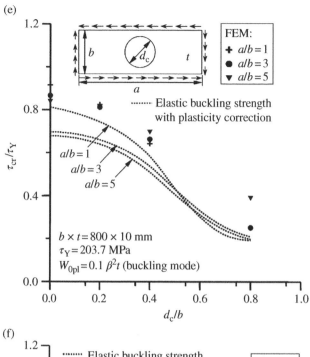

(f)

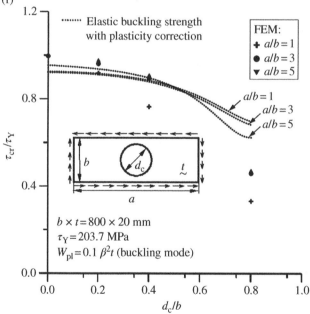

Figure 3.26 (Continued)

where σ_x, σ_y, and τ are the applied stress components and σ_{xcr}, σ_{ycr}, and τ_{cr} are the critical buckling strength components obtained from Equation (3.33) or (3.34); $\alpha = 0$ when both σ_x and σ_y are compressive, whereas $\alpha = 1$ when either σ_x, σ_y, or both are tensile.

In the use of Equation (3.35), a compressive stress takes a negative sign, whereas a tensile stress takes a positive sign. Before buckling, the value of Γ_B is smaller than zero, whereas buckling takes place if Γ_B just reaches zero and buckling has occurred if $\Gamma_B > 0$.

References

Bleich, F. (1952). *Buckling strength of metal structures*. McGraw-Hill, New York.

Brown, C.J., Yettram, A.L. & Burnett, M. (1987). Stability of plates with rectangular holes. *Journal of Structural Engineering*, **113**(5): 1111–1116.

Fujikubo, M., Yao, T., Varghese, B., Zha, Y. & Yamamura, K. (1998). Elastic local buckling strength of stiffened plates considering plate/stiffener interaction and lateral pressure. *Proceedings of the International Offshore and Polar Engineering Conference*, Montreal, IV: 292–299.

Hughes, O.F. & Paik, J.K. (2013). *Ship structural analysis and design*. The Society of Naval Architects and Marine Engineers, Alexandria, VA.

JWS (1971). *Handbook of buckling strength of unstiffened and stiffened plates*. The Plastic Design Committee, The Japan Welding Society, Tokyo (in Japanese).

Kim, U.N., Choe, I.H. & Paik, J.K. (2009). Buckling and ultimate strength of perforated plate panels subject to axial compression: experimental and numerical investigations with design formulations. *Ships and Offshore Structures*, **4**(4): 337–361.

Klöppel, K. & Sheer, J. (1960). *Beulwerte Ausgesteifter Rechteck-platten*. Verlag von Wilhelm Ernst & Sohn, Berlin (in German).

Narayanan, R. & der Avanessian, N.G.V. (1984). Elastic buckling of perforated plates under shear. *Thin-Walled Structures*, **2**: 51–73.

Paik, J.K. (2007a). Ultimate strength of steel plates with a single circular hole under axial compressive loading along short edges. *Ships and Offshore Structures*, **2**(4): 355–360.

Paik, J.K. (2007b). Ultimate strength of perforated steel plates under edge shear loading. *Thin-Walled Structures*, **45**: 301–306.

Paik, J.K. (2008). Ultimate strength of perforated steel plates under combined biaxial compression and edge shear loads. *Thin-Walled Structures*, **46**: 207–213.

Paik, J.K. & Thayamballi, A.K. (2000). Buckling strength of steel plating with elastically restrained edges. *Thin-Walled Structures*, **37**: 27–55.

Paik, J.K. & Yi, M.S. (2016). *Experimental and numerical investigations of welding-induced distortions and stresses in steel stiffened plate structures*. The Korea Ship and Offshore Research Institute, Pusan National University, Busan.

Paulling, J.R. (1988). Strength of ships. Chapter 4 in *Principles of Naval Architecture, Vol. I, Stability and Strength*. The Society of Naval Architects and Marine Engineers, Alexandria, VA.

Suneel Kumar, M., Alagusundaramoorthy, P. & Sunsaravadivelu, R. (2009). Interaction curves for stiffened panel with circular opening under axial and lateral loads. *Ships and Offshore Structures*, **4**(2): 133–143.

Timoshenko, S.P. & Gere, J.M. (1982). *Theory of elastic stability*. Second Edition, McGraw-Hill, London.

Ueda, Y., Rashed, S.M.H. & Paik, J.K. (1987). New interaction equation for plate buckling. *JWRI Transactions, Joining and Welding Research Institute (JWRI), Osaka University, Osaka*, **14**(2): 159–173.

Wang, G., Sun, H.H., Peng, H. & Uemori, R. (2009). Buckling and ultimate strength of plates with openings. *Ships and Offshore Structures*, **4**(1): 43–53.

4

Large-Deflection and Ultimate Strength Behavior of Plates

4.1 Fundamentals of Plate Collapse Behavior

The ultimate strength behavior of a plate under predominantly compressive loads is much more complex involving buckling and plastic collapse than a plate under predominantly axial tensile loads that would fail by gross yielding.

Figures 4.1, 4.2, and 4.3 show the elastic–plastic large-deflection behavior of steel or aluminum plates under longitudinal axial compressive loads with or without initial imperfections until and after the ultimate strength is reached. The plates are 2400 mm long and 800 mm wide and are simply supported at all four edges, keeping them straight. The nonlinear finite element method as described in Chapter 12 is used for the analysis where the elastic–perfectly plastic material model is applied by neglecting the strain-hardening effect.

Figure 4.1 presents the ultimate strength behavior of a steel plate that is made of mild steel with a yield strength of $\sigma_Y = 235$ MPa, an elastic modulus of $E = 205.8$ GPa, and Poisson's ratio of $\nu = 0.3$. Two kinds of plate thicknesses are considered with $t = 15$ mm or the plate slenderness ratio, $\beta = (b/t)\sqrt{\sigma_Y/E} = 1.80$, and with $t = 9$ mm or $\beta = 3.0$. Initial deflection is considered for the thinner plate with $w_{0pl} = 8.1$ mm where w_{0pl} is the maximum initial deflection at the plate center, although the thicker plate is perfectly flat without initial deflection. It is apparent from Figure 4.1 that the ultimate strength behavior of a thick plate is different from that of a thin plate as would be expected and the initially deflected plate does not show the bifurcation buckling but the total deflection is increased progressively from the very beginning of loading.

Figure 4.2 presents the ultimate strength behavior of an aluminum plate that is made of aluminum alloy 5083-O with a yield strength of $\sigma_Y = 125$ MPa, an elastic modulus of $E = 72$ GPa, and Poisson's ratio of $\nu = 0.33$. Similarly, two kinds of plate thicknesses are considered with $t = 13$ mm or $\beta = 2.56$ and with $t = 6$ mm or $\beta = 5.56$. The thinner plate has the initial deflection of $w_{0pl} = 8.0$ mm. It is observed from Figure 4.2 that the ultimate strength behavior of an aluminum plate is similar to that of a steel plate.

It is recognized that the material strength of welded aluminum plates in the softened zone is recovered by natural aging over a period of time (Lancaster 2003), but the plate ultimate strength may be reduced by softening phenomenon as far as the softened zone is not recovered. To illustrate the effects of welding induced residual stresses and softening phenomenon in the heat-affected zone on the ultimate strength behavior of a welded

Ultimate Limit State Analysis and Design of Plated Structures, Second Edition. Jeom Kee Paik.
© 2018 John Wiley & Sons Ltd. Published 2018 by John Wiley & Sons Ltd.

(a)

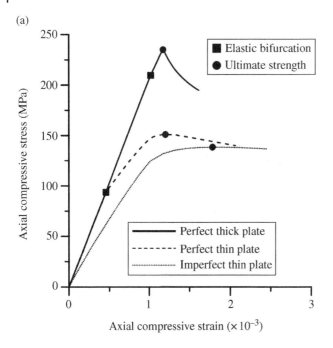

(b)

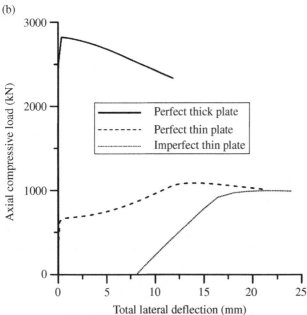

Figure 4.1 The ultimate strength behavior of a steel plate under longitudinal axial compressive loads: (a) axial compressive stress versus axial compressive strain relation; (b) axial compressive loads versus lateral deflection relation; (c) deflected shape at the ultimate limit state of flat thick plate with $t = 15\,mm$; (d) deflected shape at the ultimate limit state of flat thin plate with $t = 9\,mm$; (e) deflected shape at the ultimate limit state of initially deflected thin plate with $t = 9\,mm$ (the deflected shape is amplified by 10 times).

(c)

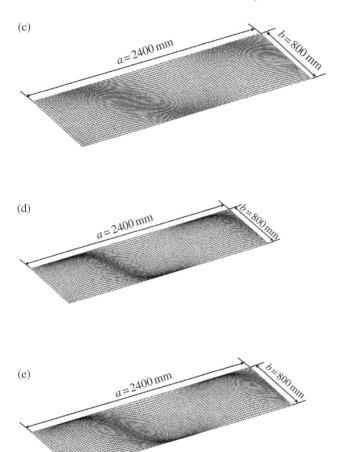

(d)

(e)

Figure 4.1 (Continued)

aluminum plate as shown in Figure 4.3, a specific scenario is considered. The plate is made of aluminum alloy 5083-H116 with a yield strength of $\sigma_Y = 215$ MPa. The plate thickness is $t = 9$ mm or $\beta = 4.86$. A very small initial deflection with $w_{0pl} = 1$ mm is considered, while the plate has an average level of residual stresses with $\sigma_{rcx} = -0.15\sigma_Y$ as per Figure 1.34 where the breadth of the heat-affected zone is $b_p' = 63.2$ mm as per Figure 1.27 and the yield strength in the softened zone is assumed to be 167 MPa ($= 0.777\sigma_Y$). Welding is performed in the longitudinal edges only, and thus neither residual stress nor softening takes place in the transverse direction. It is observed from Figure 4.3 that the welding induced residual stresses significantly reduce the ultimate strength, but the effect of softening phenomenon can be neglected in this specific case.

The behavior of a plate under applied loads may be classified into five regimes: pre-buckling, buckling, post-buckling, ultimate strength, and post-ultimate strength. In the pre-buckling regime, the structural response between loads and displacements is usually linear, and the structural component is stable. As the predominantly compressive stress reaches a critical value, buckling occurs, as described in Chapter 3.

(a)

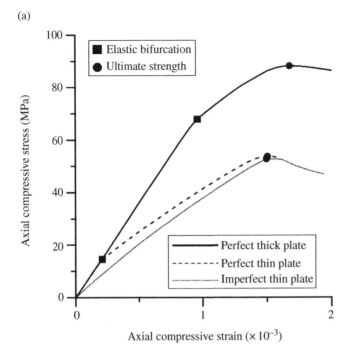

(b)

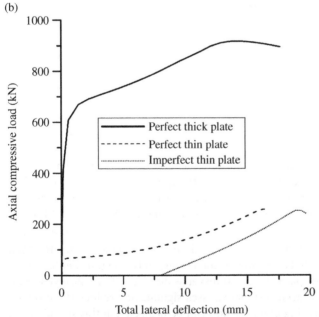

Figure 4.2 The ultimate strength behavior of an aluminum plate under longitudinal axial compressive loads: (a) axial compressive stress versus axial compressive strain relation; (b) axial compressive loads versus lateral deflection relation; (c) deflected shape at the ultimate limit state of flat thick plate with $t = 13$ mm; (d) deflected shape at the ultimate limit state of flat thin plate with $t = 6$ mm; (e) deflected shape at the ultimate limit state of initially deflected thin plate with $t = 6$ mm (the deflected shape is amplified by 10 times).

(c)

(d)

(e)

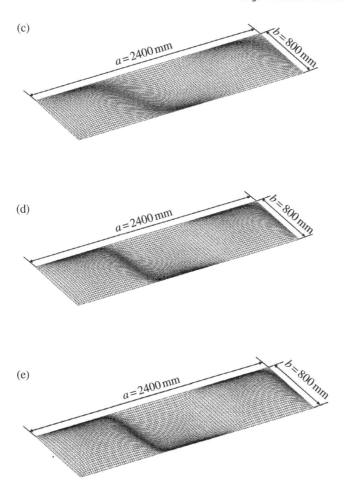

Figure 4.2 (Continued)

Unlike columns, in which buckling is meant to cause collapse, plates that buckle in the elastic regime may have sufficient redundancy to remain stable in the sense that further loading can be sustained until the ultimate strength is reached, even if the in-plane stiffness significantly decreases after the inception of buckling. In this regard, the elastic buckling of a plate between stiffeners may in some design cases be allowed to reduce the structural weight associated with efficient and economic design. Because residual strength is not expected in a plate after buckling occurs in the inelastic regime, however, inelastic buckling is considered to be the plate's ultimate limit state (ULS).

As the applied loads increase, the plate eventually reaches the ULS due to expansion of the yielded region. The in-plane stiffness of the collapsed plate takes a "negative" value in the post-ULS regime, and this insufficient redundancy leads to a high degree of instability. A plate with initial imperfections begins to deflect from the very beginning as the compressive loads increase, so a bifurcation buckling phenomenon does not occur. The ultimate strength of imperfect structures is lower than that of perfect structures.

(a)

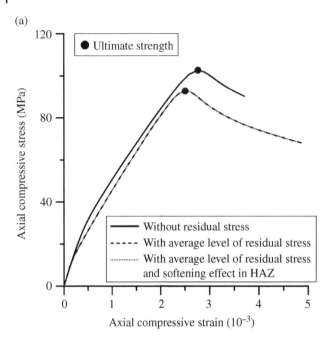

(b)

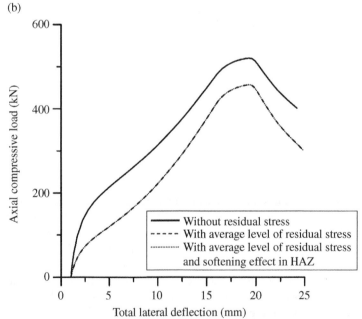

Figure 4.3 Effects of residual stresses and softening on the ultimate strength behavior of a welded aluminum plate under longitudinal axial compressive loads: (a) axial compressive stress versus axial compressive strain relation; (b) axial compressive loads versus lateral deflection relation.

The ultimate strength behavior of a plate depends on a variety of influential factors such as geometric or material properties, loading characteristics, initial imperfections (i.e., initial deflections, residual stresses, or softening in the heat-affected zone of a welded aluminum plate), boundary conditions, and existing local damage related to corrosion, fatigue crack, and denting.

In the ULS design of plate elements using Equation (1.17), the demand indicates the extreme value of the applied stresses, and the capacity represents the ultimate strength. This chapter presents the ultimate strength formulations of plates under combined in-plane and lateral pressure loads accounting for the effects of initial imperfections in the form of initial deflection and welding residual stresses. The effects of openings, corrosion diminution, fatigue cracking, and local denting damage on the plate ultimate strength are described. The average stress–strain relationships of plates are also described until and after the ultimate strength is reached. It is noted that the theories and methodologies described in this chapter can be commonly applied to both steel and aluminum plates.

4.2 Structural Idealizations of Plates

4.2.1 Geometric Properties

Figure 4.4 shows the coordinate of a rectangular plate element between longitudinal stiffeners and transverse frames in a continuous stiffened plate structure. The x axis of the plate is taken in any one reference direction, and the y axis is taken in the direction normal to the x direction. Therefore, one may not always be required to take the plate length to be located along the long edges. One benefit of this type of coordinate system is that computerization of strength calculations is much easier for a large plated structure that is

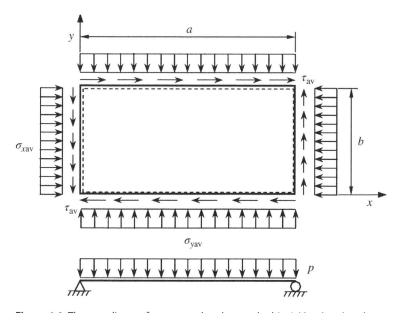

Figure 4.4 The coordinate of a rectangular plate under biaxial loads, edge shear, and lateral pressure.

composed of a number of individual plate elements in which some plate elements are "wide" and others are "long." The plate length and breadth are denoted by a and b, respectively. The plate thickness is t.

4.2.2 Material Properties

The plate elements in a plated structure are made of mild steel, high tensile steel, or aluminum alloy, with a yield strength, σ_Y. Young's modulus and Poisson's ratio are E and ν, respectively. The elastic shear modulus is $G = E/[2(1 + \nu)]$. The plate bending rigidity is denoted by $D = Et^3/[12(1 - \nu^2)]$. The plate slenderness ratio is defined as $\beta = (b/t)\sqrt{(\sigma_Y/E)}$.

4.2.3 Loads and Load Effects

When a continuous plated structure is subjected to external loads, the load effects (e.g., stress, deformation) of the plate elements can be computed by linear elastic finite element analysis or by the classical theory of structural mechanics.

The potential load components that act on a plate element generally include four types (or six load components): biaxial loads (i.e., compression or tension), edge shear, biaxial in-plane bending, and lateral pressure, as described in Chapter 3. When the plate is relatively small compared with the entire plated structure, the influence of in-plane bending effects on the plate ultimate strength may be negligible. In contrast, the effect of in-plane bending on the plate buckling strength may need to be considered, as described in Chapter 3. In this regard, this chapter deals with three types of loads (or four load components): longitudinal compression or tension (σ_{xav}), transverse compression or tension (σ_{yav}), edge shear ($\tau = \tau_{av}$), and lateral pressure loads (p), as shown in Figure 4.4.

In ships and offshore structures, lateral pressure loading arises from water pressure and/or cargo weight. The still-water magnitude of water pressure depends on the vessel's draught, and the still-water value of cargo pressure is determined by the amount and density of the cargo loaded. These still-water pressure values are augmented by wave action and vessel motion at sea. Typically, the larger in-plane loads are caused by longitudinal vessel hull girder bending, both in still-water and in waves at sea.

In this chapter, it is denoted that the compressive stress is negative and the tensile stress is positive, unless otherwise specified. That is, the longitudinal axial load has a negative value when the corresponding load is compressive, and vice versa.

4.2.4 Fabrication Related Initial Imperfections

Welding is normally used to fabricate steel- or aluminum-plated structures, so initial imperfections develop that may in some cases significantly affect (reduce) the structural capacity. In advanced structural design, therefore, strength calculations of plates should accommodate the initial imperfections as parameters of influence. The characteristics of the welding induced initial imperfections are uncertain, and an idealized model is used to represent them as described in Section 1.7.

4.2.5 Boundary Conditions

In a continuous plated structure, the edges of the plate elements are supported by longitudinal stiffeners and transverse frames. The bending rigidities of the boundary support members are usually quite large compared with that of the plate itself, which implies that the relative lateral deflections of the support members to the plate itself are very small, even up to plate collapse. Therefore, it is presumed that the support members at the four plate edges remain in the same plane. The rotational restraints along the plate edges depend on the torsional rigidities of the support members, and these are neither zero nor infinite, as described in Chapter 3.

When predominantly in-plane compressive loads are applied to a continuous plated structure surrounded by support members, the buckling pattern of the plates is expected to be asymmetrical; that is, one plate element tends to buckle up and the adjacent plate element tends to deflect down. In this case (after buckling), the rotational restraints along the plate edges can be considered to be small.

When the plated structure is subjected to combined axial compression and lateral pressure loads, however, the structure's buckling pattern can tend to be symmetrical, at least for sufficiently large pressures, that is, each adjacent plate element may deflect in the direction of lateral pressure loading. In this case, the edge rotational restraints can become so large that they may be considered to correspond to a clamped condition at the beginning of loading. However, if plasticity occurs earlier along the edges, where the larger bending moments are developed, the rotational restraints at the yielded edges decrease as the applied loads increase.

In fact, slender stiffeners prone to torsional buckling may even destabilize the plate in the sense that the overall buckling of the stiffened panel, together with stiffeners, can then occur at a stress level lower than that of a simply supported plate. However, our treatment in this chapter is based on the normal presumption that stiffeners and other support members have been properly designed so that their local instability does not occur before the plating fails. When the stiffeners are very weak, they can buckle together with the plate as part of what is called overall buckling. The design and analysis procedure for the overall buckling of stiffened panels are treated separately, as described in Chapters 5 and 6.

In some cases, specifically under large lateral pressure loading, the plate edges may not remain straight. This is a special case that must be treated separately. However, as long as the stiffeners are sufficiently strong to avoid failure before the plate buckles, which is the case with which this chapter is concerned, the plate will fail locally. In a continuous plated structure for which such a hypothesis can be accepted, the edges of the individual plate elements remain nearly straight due to the relative structural response to the adjacent plate elements until the ULS is reached.

In this chapter, therefore, it is basically presumed that the plate edges are simply supported, with zero deflection and zero rotational restraints along four edges and with all edges kept straight. This contrasts with our more sophisticated treatment of buckling of plates in Chapter 3, wherein the effects of the rotational restraints of the support members are accommodated. Part of the reason is mathematical convenience. The effects of clamped or partially rotation restrained edge conditions are, of course, described in this chapter separately.

Figure 4.5 shows the effects of the straight edge condition on the ultimate strength behavior of a simply supported plate under axial compression or edge shear, as obtained

(a)

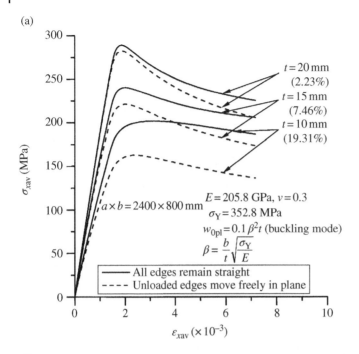

(b)

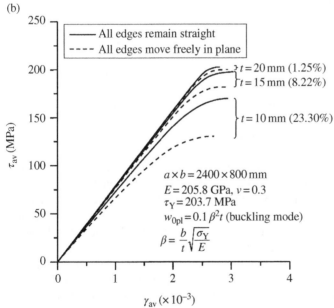

Figure 4.5 Effect of the straight edge condition on the collapse behavior of a simply supported steel plate: (a) under longitudinal compression; (b) under edge shear, as obtained by elastic–plastic large-deflection finite element method.

by the nonlinear finite element method analysis (FEA). As would be expected, the ultimate strength of a plate with the unloaded edges kept straight is greater than that when the unloaded edges move freely in plane. For relatively thick plates with a thickness of 20 mm, the difference in the ultimate strength is very small. However, the difference becomes about 20% for relatively thin plates with a thickness of 10 mm.

Another illustrative example to investigate the effects of the plate boundary condition on collapse behavior is that of a plate under combined axial compression and lateral pressure. Two types of structural idealizations using nonlinear FEA may be relevant: one is a single plate with the condition that all edges are simply supported, as shown in Figure 4.6a, and the other is that a three-bay plate model is taken as the extent of the analysis, as shown in Figure 4.6b. It is supposed that all plate edges are simply supported, keeping them straight. In the FEA, the simply supported condition is applied along the transverse frames, and some rotational restraints automatically develop due to the action of lateral pressure.

Figure 4.7 shows the ultimate strength behavior with varying magnitudes of lateral pressure, as obtained by nonlinear FEA. It is apparent from Figure 4.7 that the ultimate strength of the three-bay plate model is greater than that of the single-bay plate model and that the strength increase tends to grow as the magnitude of the lateral pressure increases, because the plate edges along the transverse frames become clamped due to the action of lateral pressure; the three-bay plate model automatically takes this effect into account. However, the effect of rotational restraints due to lateral pressure is small, and thus the simply supported plate edge condition may be relevant regardless of the lateral pressure load applied.

4.3 Nonlinear Governing Differential Equations of Plates

The post-buckling or large-deflection behavior of plates in the elastic regime can be analyzed by solving the two nonlinear governing differential equations of the large-deflection plate theory: the equilibrium equation and the compatibility equation (Marguerre 1938, Timoshenko & Woinowsky-Krieger 1981):

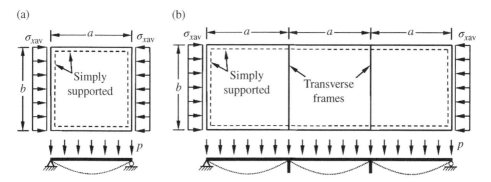

Figure 4.6 (a) A single-bay plate model and (b) a three-bay plate model under combined axial compression and lateral pressure.

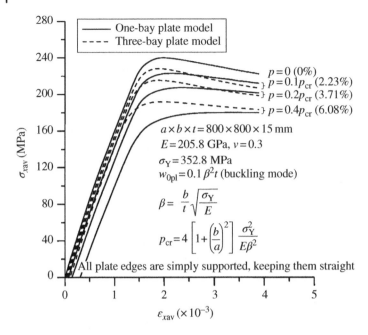

Figure 4.7 The ultimate strength behavior of a plate under combined axial compression and lateral pressure, as obtained by elastic–plastic large-deflection FEA.

$$D\left(\frac{\partial^4 w}{\partial x^4} + 2\frac{\partial^4 w}{\partial x^2 \partial y^2} + \frac{\partial^4 w}{\partial y^4}\right)$$
$$-t\left[\frac{\partial^2 F}{\partial y^2}\frac{\partial^2 (w+w_0)}{\partial x^2} - 2\frac{\partial^2 F}{\partial x \partial y}\frac{\partial^2 (w+w_0)}{\partial x \partial y} + \frac{\partial^2 F}{\partial x^2}\frac{\partial^2 (w+w_0)}{\partial y^2} + \frac{p}{t}\right] = 0 \tag{4.1a}$$

$$\frac{\partial^4 F}{\partial x^4} + 2\frac{\partial^4 F}{\partial x^2 \partial y^2} + \frac{\partial^4 F}{\partial y^4}$$
$$-E\left[\left(\frac{\partial^2 w}{\partial x \partial y}\right)^2 - \frac{\partial^2 w}{\partial x^2}\frac{\partial^2 w}{\partial y^2} + 2\frac{\partial^2 w_0}{\partial x \partial y}\frac{\partial^2 w}{\partial x \partial y} - \frac{\partial^2 w_0}{\partial x^2}\frac{\partial^2 w}{\partial y^2} - \frac{\partial^2 w}{\partial x^2}\frac{\partial^2 w_0}{\partial y^2}\right] = 0 \tag{4.1b}$$

where w and w_0 are the added and initial deflections, respectively, and F is Airy's stress function. When Airy's stress function, F, and the added deflection, w, are known, the stresses inside the plate can be calculated as follows:

$$\sigma_x = \frac{\partial^2 F}{\partial y^2} - \frac{Ez}{1-v^2}\left(\frac{\partial^2 w}{\partial x^2} + v\frac{\partial^2 w}{\partial y^2}\right) \tag{4.2a}$$

$$\sigma_y = \frac{\partial^2 F}{\partial x^2} - \frac{Ez}{1-v^2}\left(\frac{\partial^2 w}{\partial y^2} + v\frac{\partial^2 w}{\partial x^2}\right) \tag{4.2b}$$

$$\tau = -\frac{\partial^2 F}{\partial x \partial y} - \frac{Ez}{2(1+v)}\frac{\partial^2 w}{\partial x \partial y} \tag{4.2c}$$

where z is the coordinate in the plate thickness direction, with $z = 0$ at its mid-thickness.

By solving the governing differential equations subject to the given boundary conditions, load application, and initial imperfections, the stress distribution inside the plate can be calculated, and thus it is possible to examine the elastic large-deflection behavior of the plate. The normal stresses are taken as negative for compression and positive for tension.

To solve Equations (4.1a) and (4.1b), initial and added deflection functions that satisfy the boundary conditions are presumed. Using energy methods such as the Galerkin method (Fletcher 1984), the unknown values in the presumed added deflection function can be determined as a function of the applied loads. It is of course desirable to adopt an added deflection function that can accurately represent the plate's deflection patterns due to the corresponding load applications. This may obviously require a number of unknown values to be included in the presumed added deflection function.

In fact, the solution of the nonlinear governing differential equations of a plate with such a deflection function is difficult for several reasons, and some simplifications are therefore suggested. One reason is that it is difficult to deal with more than two unknown values of the added deflection function in the analytical approach, and thus it is suggested to use a single deflection component that represents the plate behavior. Another reason is that combined load applications usually result in more complex patterns of deflection, and thus a simplified approach using a single deflection component is not always successful. In this case, the load conditions are considered only by a single load component or at most a few components that can be represented by the presumed deflection function with the single deflection component.

4.4 Elastic Large-Deflection Behavior of Simply Supported Plates

To accurately analyze the elastic large-deflection behavior of the simply supported plates under the considerations described in Section 4.2, the initial and added deflection functions that satisfy the simply supported boundary conditions can be expressed by a Fourier series function as follows:

$$w_0 = \sum_{m=1}^{i}\sum_{n=1}^{j} A_{0mn} \sin\frac{m\pi x}{a}\sin\frac{n\pi y}{b} \tag{4.3a}$$

$$w = \sum_{m=1}^{i}\sum_{n=1}^{j} A_{mn} \sin\frac{m\pi x}{a}\sin\frac{n\pi y}{b} \tag{4.3b}$$

where A_{0mn} and A_{mn} are the initial and added deflection amplitudes and i and j are the maximum numbers of deflection components in the x or y direction; as many as possible should be taken. For modeling of the plate initial deflection, Section 1.7 can be referred to.

It is confirmed that Equations (4.3a) and (4.3b) satisfy the simply supported boundary conditions at the four edges of the plate because the deflections and out-of-plane bending moments (i.e., rotations should occur freely) should be zero at the plate edges, namely,

$$w = 0 \quad \text{at} \quad x = 0, a \quad \text{and} \quad y = 0, b \tag{4.4a}$$

$$\frac{\partial^2 w}{\partial x^2} = 0 \quad \text{at} \quad y = 0, b \tag{4.4b}$$

$$\frac{\partial^2 w}{\partial y^2} = 0 \quad \text{at} \quad x = 0, a \tag{4.4c}$$

For the sake of simplicity, the added deflection function and the related initial deflection function are presumed with a single deflection component but with consideration of the best representation of the deflection patterns that correspond to the load applications as described in Section 4.3. It is realized that the deflection patterns due to in-plane loads are quite different from those under lateral pressure loads; these load conditions are thus dealt with separately, whereas their interacting effects shall be considered approximately. Also, the deflection patterns due to edge shear are quite complex, so it cannot be represented by any deflection function with a single deflection component. In this case, numerical computations are more convenient than the analytical approach.

4.4.1 Lateral Pressure Loads

The initial and added deflection functions of a simply supported plate under lateral pressure loads alone can be expressed from Equations (4.3a) and (4.3b), considering one half-wave number in both the x and y directions, as follows:

$$w_0 = A_{01} \sin\frac{\pi x}{a} \sin\frac{\pi y}{b} \tag{4.5a}$$

$$w = A_1 \sin\frac{\pi x}{a} \sin\frac{\pi y}{b} \tag{4.5b}$$

where $A_{01} (= A_{011})$ and $A_1 (= A_{11})$ are the initial and added deflection amplitudes.

The substitution of Equations (4.5a) and (4.5b) into Equation (4.1b) results in

$$\frac{\partial^4 F}{\partial x^4} + 2\frac{\partial^4 F}{\partial x^2 \partial y^2} + \frac{\partial^4 F}{\partial y^4} = -\frac{\pi^4 EA_1(A_1 + 2A_{01})}{2a^2 b^2}\left(\cos\frac{2\pi x}{a} + \cos\frac{2\pi y}{b}\right) \tag{4.6}$$

The particular solution, F_P, of the Airy stress function, F, is obtained by solving Equation (4.6) as follows:

$$F_P = \frac{EA_1(A_1 + 2A_{01})}{32}\left(\frac{a^2}{b^2}\cos\frac{2\pi x}{a} + \frac{b^2}{a^2}\cos\frac{2\pi y}{b}\right) \tag{4.7}$$

The homogeneous solution, F_H, of the stress function, F, that satisfies the loading condition is given by treating the welding induced residual stress as an initial stress parameter, namely,

$$F_H = \sigma_{rx}\frac{y^2}{2} + \sigma_{ry}\frac{x^2}{2} \tag{4.8}$$

where σ_{rx} and σ_{ry} are the welding induced residual stresses, as defined in Section 1.7.3.

The applicable stress function, F, can then be expressed as the sum of the particular solution and the homogeneous solution as follows:

$$F = \sigma_{rx}\frac{y^2}{2} + \sigma_{ry}\frac{x^2}{2} + \frac{EA_1(A_1 + 2A_{01})}{32}\left(\frac{a^2}{b^2}\cos\frac{2\pi x}{a} + \frac{b^2}{a^2}\cos\frac{2\pi y}{b}\right) \tag{4.9}$$

By substituting Equations (4.5a), (4.5b), and (4.9) into Equation (4.1a) and applying the Galerkin method (Fletcher 1984), the following equation is obtained:

$$\int_0^a\int_0^b\left\{D\left(\frac{\partial^4 w}{\partial x^4} + 2\frac{\partial^4 w}{\partial x^2\partial y^2} + \frac{\partial^4 w}{\partial y^4}\right)\right.$$
$$\left. -t\left[\frac{\partial^2 F}{\partial y^2}\frac{\partial^2(w+w_0)}{\partial x^2} - 2\frac{\partial^2 F}{\partial x\partial y}\frac{\partial^2(w+w_0)}{\partial x\partial y} + \frac{\partial^2 F}{\partial x^2}\frac{\partial^2(w+w_0)}{\partial y^2} + \frac{p}{t}\right]\right\} \tag{4.10}$$
$$\times\sin\frac{\pi x}{a}\sin\frac{\pi y}{b}dxdy = 0$$

By performing the integration of Equation (4.10) over the entire plate, a third-order equation with respect to the unknown variable, A_1, is obtained as follows:

$$C_1A_1^3 + C_2A_1^2 + C_3A_1 + C_4 = 0 \tag{4.11}$$

where

$$C_1 = \frac{\pi^2 E}{16}\left(\frac{b}{a^3} + \frac{a}{b^3}\right),$$

$$C_2 = \frac{3\pi^2 EA_{01}}{16}\left(\frac{b}{a^3} + \frac{a}{b^3}\right),$$

$$C_3 = \pi^2 E\frac{A_{01}^2}{8}\left(\frac{b}{a^3} + \frac{a}{b^3}\right) + \frac{b}{a}\sigma_{rex} + \frac{a}{b}\sigma_{rey} + \frac{\pi^2 D}{t}\frac{1}{ab}\left(\frac{b}{a} + \frac{a}{b}\right)^2,$$

$$C_4 = A_{01}\left[\frac{b}{a}\sigma_{rex} + \frac{a}{b}\sigma_{rey}\right] - \frac{16ab}{\pi^4 t}p,$$

$$\sigma_{rex} = \sigma_{rcx} + \frac{2}{b}(\sigma_{rtx} - \sigma_{rcx})\left(b_t - \frac{b}{2\pi}\sin\frac{2\pi b_t}{b}\right) \approx \frac{b - 2b_t}{b}\sigma_{rcx},$$

$$\sigma_{rey} = \sigma_{rcy} + \frac{2}{a}(\sigma_{rty} - \sigma_{rcy})\left(a_t - \frac{a}{2m\pi}\sin\frac{2m\pi a_t}{a}\right) \approx \frac{a - 2a_t}{a}\sigma_{rcy}.$$

The solution of Equation (4.11) can be obtained by the so-called Cardano method as follows:

$$A_m = -\frac{C_2}{3C_1} + k_1 + k_2 \tag{4.12}$$

where

$$k_1 = \left(-\frac{Y}{2} + \sqrt{\frac{Y^2}{4} + \frac{X^3}{27}} \right)^{1/3},$$

$$k_2 = \left(-\frac{Y}{2} - \sqrt{\frac{Y^2}{4} + \frac{X^3}{27}} \right)^{1/3},$$

$$X = \frac{C_3}{C_1} - \frac{C_2^2}{3\,C_1^2},$$

$$Y = \frac{2\,C_2^3}{27\,C_1^3} - \frac{C_2 C_3}{3\,C_1^2} + \frac{C_4}{C_1}.$$

Equation (4.12) can be dealt with numerically using the FORTRAN computer subroutine CARDANO given in the appendices to this book. Once A_1 is determined as a function of lateral pressure loads p and initial imperfections, the lateral deflection and membrane stresses inside the plate can be calculated from Equations (4.3b) and (4.2), respectively. The membrane stress distribution of the plate must be nonuniform, and the maximum and minimum membrane stresses of the plate in the x and y directions are obtained considering the distribution of welding induced residual stresses, as shown in Figure 4.8, as follows:

$$\sigma_{x\,\text{max}} = \left. \frac{\partial^2 F}{\partial y^2} \right|_{x=0,\, y=b_t \text{ or } b-b_t} \tag{4.13a}$$

$$\sigma_{x\,\text{min}} = \left. \frac{\partial^2 F}{\partial y^2} \right|_{x=0,\, y=b/2} \tag{4.13b}$$

$$\sigma_{y\,\text{max}} = \left. \frac{\partial^2 F}{\partial x^2} \right|_{x=a_t \text{ or } a-a_t,\, y=0} \tag{4.13c}$$

$$\sigma_{y\,\text{min}} = \left. \frac{\partial^2 F}{\partial x^2} \right|_{x=a/2,\, y=0} \tag{4.13d}$$

For this case, the maximum and minimum membrane stresses are then computed from Equation (4.13) as follows:

$$\sigma_{x\,\text{max}} = \sigma_{rtx} - \frac{E\pi^2 A_1 (A_1 + 2A_{01})}{8a^2} \cos\frac{2\pi b_t}{b} \tag{4.14a}$$

$$\sigma_{x\,\text{min}} = \sigma_{rcx} + \frac{E\pi^2 A_1 (A_1 + 2A_{01})}{8a^2} \tag{4.14b}$$

$$\sigma_{y\,\text{max}} = \sigma_{rty} - \frac{E\pi^2 A_1 (A_1 + 2A_{01})}{8b^2} \cos\frac{2\pi a_t}{a} \tag{4.14c}$$

$$\sigma_{y\,\text{min}} = \sigma_{rcy} + \frac{E\pi^2 A_1 (A_1 + 2A_{01})}{8b^2} \tag{4.14d}$$

(a)

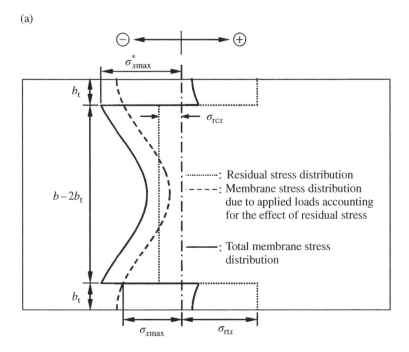

(b)

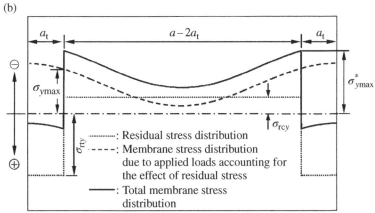

Figure 4.8 A schematic of the total membrane stress distribution inside the plate considering the effect of welding induced residual stress distribution: (a) x direction; (b) y direction.

4.4.2 Combined Biaxial Loads

In this case, the plate behavior is represented by the buckling mode component, denoted by the m half-wave number in the x direction and the n half-wave number in the y direction. The initial and added deflection functions of a plate under biaxial loads may thus be expressed as follows:

$$w_0 = A_{0mn} \sin\frac{m\pi x}{a}\sin\frac{n\pi y}{b} \tag{4.15a}$$

$$w = A_{mn} \sin\frac{m\pi x}{a}\sin\frac{n\pi y}{b} \tag{4.15b}$$

where A_{0mn} and A_{mn} are the initial and added deflection amplitudes and m and n are the buckling half-wave numbers in the x or y directions. For a plate with $a/b \geq 1$, for which the x coordinate is taken in the plate length direction, $n = 1$ is identified, and m is determined as a minimum integer that satisfies the following equation, as described in Chapter 3:

a) When σ_{xav} and σ_{yav} are both nonzero compressive,

$$\frac{(m^2/a^2 + 1/b^2)^2}{m^2/a^2 + c/b^2} \leq \frac{[(m+1)^2/a^2 + 1/b^2]^2}{(m+1)^2/a^2 + c/b^2} \tag{4.16a}$$

where $c = \sigma_{yav}/\sigma_{xav}$.

b) When σ_{xav} is tensile or zero whatever σ_{yav},

$$m = 1 \tag{4.16b}$$

c) When σ_{xav} is compressive and σ_{yav} is tensile or zero,

$$\frac{a}{b} \leq \sqrt{m(m+1)} \tag{4.16c}$$

In a similar way to Equation (4.9), the Airy stress function F can be expressed as follows:

$$F = (\sigma_{xav} + \sigma_{rx})\frac{y^2}{2} + (\sigma_{yav} + \sigma_{ry})\frac{x^2}{2}$$
$$+ \frac{EA_{mn}(A_{mn} + 2A_{0mn})}{32}\left(\frac{n^2 a^2}{m^2 b^2}\cos\frac{2m\pi x}{a} + \frac{m^2 b^2}{n^2 a^2}\cos\frac{2n\pi y}{b}\right) \tag{4.17}$$

The Galerkin method is applied with Equation (4.1a) as follows:

$$\int_0^a \int_0^b \left\{ D\left(\frac{\partial^4 w}{\partial x^4} + 2\frac{\partial^4 w}{\partial x^2 \partial y^2} + \frac{\partial^4 w}{\partial y^4}\right) \right.$$
$$\left. - t\left[\frac{\partial^2 F}{\partial y^2}\frac{\partial^2 (w+w_0)}{\partial x^2} - 2\frac{\partial^2 F}{\partial x \partial y}\frac{\partial^2 (w+w_0)}{\partial x \partial y} + \frac{\partial^2 F}{\partial x^2}\frac{\partial^2 (w+w_0)}{\partial y^2}\right] \right\} \tag{4.18}$$
$$\times \sin\frac{m\pi x}{a}\sin\frac{n\pi y}{b}dxdy = 0$$

By substituting Equations (4.15a), (4.15b), and (4.17) into Equation (4.18) and performing the integration over the entire plate, the following third-order equation with regard to the unknown amplitude A_{mn} is obtained as follows:

$$C_1 A_{mn}^3 + C_2 A_{mn}^2 + C_3 A_{mn} + C_4 = 0 \tag{4.19}$$

where

$$C_1 = \frac{\pi^2 E}{16}\left(\frac{m^4 b}{a^3} + \frac{n^4 a}{b^3}\right),$$

$$C_2 = \frac{3\pi^2 E A_{0mn}}{16} \left(\frac{m^4 b}{a^3} + \frac{n^4 a}{b^3} \right),$$

$$C_3 = \pi^2 E \frac{A_{0mn}^2}{8} \left(\frac{m^4 b}{a^3} + \frac{n^4 a}{b^3} \right) + \frac{m^2 b}{a} (\sigma_{xav} + \sigma_{rex})$$

$$+ \frac{n^2 a}{b} (\sigma_{yav} + \sigma_{rey}) + \frac{\pi^2 D m^2 n^2}{t} \left(\frac{mb}{ab} + \frac{na}{mb} \right)^2,$$

$$C_4 = A_{0mn} \left[\frac{m^2 b}{a} (\sigma_{xav} + \sigma_{rex}) + \frac{n^2 a}{b} (\sigma_{yav} + \sigma_{rey}) \right],$$

$$\sigma_{rex} = \sigma_{rcx} + \frac{2}{b} (\sigma_{rtx} - \sigma_{rcx}) \left(b_t - \frac{b}{2n\pi} \sin \frac{2n\pi b_t}{b} \right) \approx \frac{b - 2b_t}{b} \sigma_{rcx},$$

$$\sigma_{rey} = \sigma_{rcy} + \frac{2}{b} (\sigma_{rty} - \sigma_{rcy}) \left(a_t - \frac{a}{2m\pi} \sin \frac{2m\pi a_t}{a} \right) \approx \frac{a - 2a_t}{a} \sigma_{rcy}.$$

The unknown deflection component (amplitude) A_{mn} can be obtained as a solution of Equation (4.19) using the Cardano method or the FORTRAN computer program CARDANO given in the appendices to this book. The maximum and minimum membrane stresses inside the plate are in this case determined from Equation (4.13) as follows:

$$\sigma_{x\,max} = \sigma_{xav} + \sigma_{rtx} - \frac{E\pi^2 m^2 A_{mn}(A_{mn} + 2A_{0mn})}{8a^2} \cos \frac{2n\pi b_t}{b} \tag{4.20a}$$

$$\sigma_{x\,min} = \sigma_{xav} + \sigma_{rcx} + \frac{E\pi^2 m^2 A_{mn}(A_{mn} + 2A_{0mn})}{8a^2} \tag{4.20b}$$

$$\sigma_{y\,max} = \sigma_{yav} + \sigma_{rty} - \frac{E\pi^2 n^2 A_{mn}(A_{mn} + 2A_{0mn})}{8b^2} \cos \frac{2m\pi a_t}{a} \tag{4.20c}$$

$$\sigma_{y\,min} = \sigma_{yav} + \sigma_{rcy} + \frac{E\pi^2 n^2 A_{mn}(A_{mn} + 2A_{0mn})}{8b^2} \tag{4.20d}$$

For the case of a perfectly flat plate without initial deflections but with welding induced residual stresses, Equation (4.19) is simplified as follows, because $C_2 = C_4 = 0$:

$$A_{mn}\left(C_1 A_{mn}^2 + C_3\right) = 0 \tag{4.21}$$

where

$$C_1 = \frac{\pi^2 E}{16} \left(\frac{m^4 b}{a^3} + \frac{n^4 a}{b^3} \right),$$

$$C_3 = \frac{m^2 b}{a} (\sigma_{xav} + \sigma_{rex}) + \frac{n^2 a}{b} (\sigma_{yav} + \sigma_{rey}) + \frac{\pi^2 D m^2 n^2}{t} \left(\frac{mb}{na} + \frac{na}{mb} \right)^2.$$

The nonzero solution of A_{mn} is defined from Equation (4.21) as follows:

$$A_{mn} = \sqrt{-\frac{C_3}{C_1}} \tag{4.22}$$

Immediately before or after buckling, no deflection must have occurred. That is,

$$A_{mn} = \sqrt{-\frac{C_3}{C_1}} = 0 \quad \text{or} \quad C_3 = 0 \tag{4.23}$$

Equation (4.23) indicates the bifurcation buckling condition of a perfectly flat plate under biaxial loads with welding induced residual stresses, namely,

$$\frac{m^2 b}{a}(\sigma_{xav} + \sigma_{rex}) + \frac{n^2 a}{b}(\sigma_{yav} + \sigma_{rey}) + \frac{\pi^2 D m^2 n^2}{t}\left(\frac{mb}{na} + \frac{na}{mb}\right)^2 = 0 \qquad (4.24)$$

Because $c = \sigma_{yav}/\sigma_{xav}$, the elastic longitudinal compressive buckling strength σ_{xE} of the plate under biaxial loads can be calculated from Equation (4.24) as follows:

$$\sigma_{xE} = -\frac{ab}{m^2 b^2 + cn^2 a^2}\left[\frac{\pi^2 D m^2 n^2}{t}\left(\frac{mb}{na} + \frac{na}{mb}\right)^2 + \frac{m^2 b}{a}\sigma_{rex} + \frac{n^2 a}{b}\sigma_{rey}\right] \qquad (4.25)$$

For a plate with $a/b \geq 1$ and thus $n = 1$, the buckling half-wave number m in the x direction is determined from Equation (4.25) as a minimum integer that satisfies the following condition:

$$\frac{ab}{m^2 b^2 + ca^2}\left[\frac{\pi^2 D m^2}{t}\left(\frac{mb}{a} + \frac{a}{mb}\right)^2 + \frac{m^2 b}{a}\sigma_{rex} + \frac{a}{b}\sigma_{rey}\right]$$

$$\leq \frac{ab}{(m+1)^2 b^2 + ca^2}\left[\frac{\pi^2 D (m+1)^2}{t}\left(\frac{(m+1)b}{a} + \frac{a}{(m+1)b}\right)^2 + \frac{(m+1)^2 b}{a}\sigma_{rex} + \frac{a}{b}\sigma_{rey}\right]$$

$$(4.26a)$$

By neglecting the effect of welding induced residual stresses on the buckling mode, Equation (4.26a) is simplified to

$$\frac{(m^2 b^2 + a^2)^2}{m^2 a^2 b^4 + ca^4 b^2} \leq \frac{\left[(m+1)^2 b^2 + a^2\right]^2}{(m+1)^2 a^2 b^4 + ca^4 b^2} \qquad (4.26b)$$

Equation (4.26b) is equivalent to Equation (4.16a). For uniaxial compression σ_{xav}, Equation (4.26b) is further simplified as $c = 0$:

$$\frac{a}{b} \leq \sqrt{m(m+1)} \qquad (4.26c)$$

In this case, Equation (4.25) is simplified to

$$\sigma_{xE} = -\frac{\pi^2 D}{b^2 t}\left(\frac{mb}{a} + \frac{a}{mb}\right)^2 - \sigma_{rex} - \frac{a^2}{m^2 b^2}\sigma_{rey} \qquad (4.27)$$

For uniaxial compression σ_{yav}, the elastic transverse compressive buckling strength σ_{yE} is determined as follows as $m = n = 1$ for a plate with $a/b \geq 1$:

$$\sigma_{yE} = -\frac{\pi^2 D}{b^2 t}\left[1 + \left(\frac{b}{a}\right)^2\right]^2 - \frac{b^2}{a^2}\sigma_{rex} - \sigma_{rey} \qquad (4.28)$$

4.4.3 Interaction Effect Between Biaxial Loads and Lateral Pressure

The elastic large-deflection behavior of plates under combined biaxial loads and lateral pressure is significantly affected by the amount of lateral pressure loads, among other factors (Hughes & Paik 2013). In fact, it is not possible to analyze large-deflection plate behavior with the deflection functions of Equation (4.5) or (4.15), which have a single deflection component; a greater number of deflection components must be included in these functions to make it possible.

For the sake of simplicity, however, the contribution made by the lateral pressure loads to the nonlinear membrane stresses inside the plate is accounted for in an approximate fashion, where the membrane stresses that arise from only the deflection components of $m = 1$ and $n = 1$ are linearly superposed to those that arise from the biaxial loads. In this case, the coefficient C_4 of Equation (4.19) is redefined as follows (Hughes & Paik 2013):

$$C_4 = A_{0mn} \left[\frac{m^2 b}{a} \left(\sigma_{xav} + \sigma_{rex} \right) + \frac{n^2 a}{b} \left(\sigma_{yav} + \sigma_{rey} \right) \right] - \frac{16ab}{\pi^4 t} p \qquad (4.29)$$

Figure 4.9 confirms the applicability of this approach by comparison with the present method solutions and more refined numerical computations for a simply supported square plate under combined uniaxial compression and lateral pressure loads, with lateral pressure loads of various magnitudes. It is observed from this figure that, due to lateral pressure loads, the deflection increases from the beginning of axial compressive loading; therefore, no bifurcation (buckling) point can be defined because lateral pressure loads are applied. It should be noted, however, that this observation is true only for a square or near-square plate. For a long plate, a bifurcation point in longitudinal

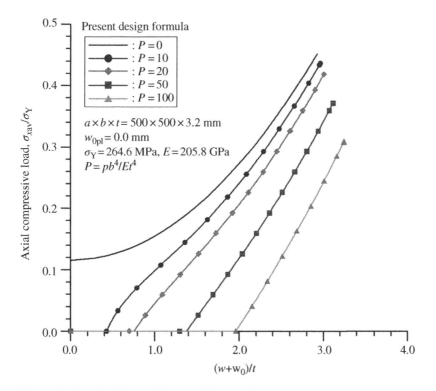

Figure 4.9 The elastic large-deflection behavior of a simply supported square plate element under combined longitudinal axial compression and lateral pressure (Hughes & Paik 2013).

compression may appear even in the presence of lateral pressure loads as long as the magnitude of these loads is not large. In this case, the value of the "elastic" bifurcation load is, however, normally greater than that without the lateral pressure loads.

4.4.4 Interaction Effect Between Biaxial and Edge Shear Loads

The deflection patterns of a plate under edge shear loading is very complex in its geometrical shape, and thus Equation (4.5) or (4.15) cannot represent the plate behavior under edge shear together with biaxial and lateral pressure loads.

Ueda et al. (1984) proposed an approximate method to predict the maximum and minimum membrane stresses inside the plate, where empirical coefficients developed based on numerical computations are introduced with Equations (4.13) and (4.20):

$$\sigma_{x\,max} = \sigma_{xav} + \sigma_{rtx} - \frac{E\pi^2 m^2 A_{mn}(A_{mn} + 2A_{0mn})}{8a^2} \cos\frac{2n\pi b_t}{b}\left[1.3\left(\frac{\tau_{av}}{\tau_E}\right)^c + 1\right]$$

$$+ 1.62\sigma_{xE}\left(\frac{\tau_{av}}{\tau_E}\right)^{2.4}$$

$$(4.30a)$$

$$\sigma_{x\,min} = \sigma_{xav} + \sigma_{rcx} + \frac{E\pi^2 m^2 A_{mn}(A_{mn} + 2A_{0mn})}{8a^2}\left(0.3\frac{\tau_{av}}{\tau_E} + 1\right)$$

$$- 1.3\sigma_{xE}\left(\frac{\tau_{av}}{\tau_E}\right)^{2.1}$$

$$(4.30b)$$

$$\sigma_{y\,max} = \sigma_{yav} + \sigma_{rty} - \frac{E\pi^2 n^2 A_{mn}(A_{mn} + 2A_{0mn})}{8b^2} \cos\frac{2m\pi a_t}{a}\left[1.3\left(\frac{\tau_{av}}{\tau_E}\right)^c + 1\right]$$

$$+ 1.62\sigma_{yE}\left(\frac{\tau_{av}}{\tau_E}\right)^{2.4}$$

$$(4.30c)$$

$$\sigma_{y\,min} = \sigma_{yav} + \sigma_{rcy} + \frac{E\pi^2 n^2 A_{mn}(A_{mn} + 2A_{0mn})}{8b^2}\left(0.3\frac{\tau_{av}}{\tau_E} + 1\right)$$

$$- 1.3\sigma_{yE}\left(\frac{\tau_{av}}{\tau_E}\right)^{2.1}$$

$$(4.30d)$$

where σ_{xE} is the elastic buckling stress under axial compression in the x direction, σ_{yE} is the elastic buckling stress under axial compression in the y direction, τ_E is the elastic buckling stress under edge shear, and $c = 1.5$ for $\tau_{av} \leq \tau_E$ or $c = 1$ for $\tau_{av} > \tau_E$.

Figure 4.10 confirms the applicability of this approach by comparison with more refined numerical computations for a simply supported plate under combined longitudinal compression and edge shear, where SPINE represents the elastic large-deflection behavior obtained by the incremental Galerkin method described in Chapter 11. It is observed from this figure that edge shear amplifies the maximum and minimum membrane stresses inside the plate.

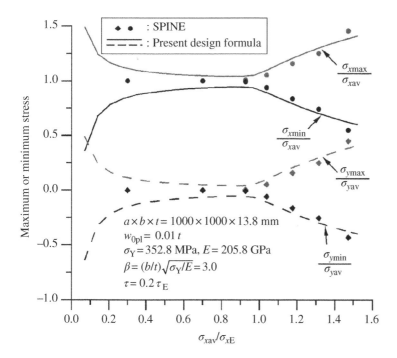

Figure 4.10 Validation of Equation (4.30) for a simply supported plate under combined longitudinal compression and edge shear (SPINE represents the incremental Galerkin method solutions described in Chapter 11).

4.5 Elastic Large-Deflection Behavior of Clamped Plates

When the torsional rigidity of the support members is very strong compared with the bending rigidity of the plate itself, as described in Chapter 3, and/or when the plate element is subjected to predominantly lateral pressure loads, the rotational restraints at the plate edges tend to become large. In this situation, the plate edges may be assumed to be clamped. The elastic large-deflection behavior of a clamped plate is completely different from that of a simply supported plate. In the following, the elastic large-deflection behavior of a plate clamped at four edges is described by solving the nonlinear governing differential equations, Equations (4.1a) and (4.1b). The effects of initial deflection and welding induced residual stress are considered.

4.5.1 Lateral Pressure Loads

In this case, the initial and added deflection functions may be assumed as follows:

$$w_0 = \frac{1}{4}A_{01}\left(1 - \cos\frac{2\pi x}{a}\right)\left(1 - \cos\frac{2\pi y}{b}\right) \tag{4.31a}$$

$$w = \frac{1}{4}A_1\left(1 - \cos\frac{2\pi x}{a}\right)\left(1 - \cos\frac{2\pi y}{b}\right) \tag{4.31b}$$

where A_{01} and A_1 are the initial and added deflection amplitudes.

It is confirmed that Equations (4.31a) and (4.31b) satisfy the clamped conditions at the plate edges, because the deflections and rotations must be zero, as follows:

$$w = 0 \quad \text{at} \quad x = 0, a \quad \text{and} \quad y = 0, b \tag{4.32}$$

$$\frac{\partial w}{\partial x} = 0 \quad \text{at} \quad y = 0, b \tag{4.32}$$

$$\frac{\partial w}{\partial y} = 0 \quad \text{at} \quad x = 0, a \tag{4.32}$$

The substitution of Equations (4.31a) and (4.31b) into Equation (4.1b) gives the Airy stress function F as follows:

$$F = \sigma_{rx}\frac{y^2}{2} + \sigma_{ry}\frac{x^2}{2}$$

$$+ \frac{EA_1(A_1 + 2A_{01})}{512a^2b^2}\left(\begin{array}{c} \left(16a^4\cos\frac{2\pi x}{a} - a^4\cos\frac{4\pi x}{a} + b^4\left(16\cos\frac{2\pi y}{b} - \cos\frac{4\pi y}{b}\right)\right) \\[3mm] \left(\dfrac{\cos\left(\dfrac{2\pi x}{a} - \dfrac{4\pi y}{b}\right)}{(b^2 + 4a^2)^2} - \dfrac{2\cos\left(\dfrac{2\pi x}{a} - \dfrac{2\pi y}{b}\right)}{(b^2 + a^2)^2} + \dfrac{\cos\left(\dfrac{4\pi x}{a} - \dfrac{2\pi y}{b}\right)}{(4b^2 + a^2)^2}\right) \\[5mm] + 8a^4 \\[5mm] \left(-\dfrac{2\cos\left(\dfrac{2\pi x}{a} + \dfrac{2\pi y}{b}\right)}{(b^2 + a^2)^2} + \dfrac{2\cos\left(\dfrac{4\pi x}{a} + \dfrac{2\pi y}{b}\right)}{(4b^2 + a^2)^2} + \dfrac{\cos\left(\dfrac{2\pi x}{a} + \dfrac{4\pi y}{b}\right)}{(b^2 + 4a^2)^2}\right) \end{array}\right)$$

$$\tag{4.33}$$

Similar to Equation (4.10), the application of the Galerkin method leads to

$$\int_0^b \int_0^a \left\{ \begin{array}{c} D\left(\dfrac{\partial^4 w}{\partial x^4} + 2\dfrac{\partial^4 w}{\partial x^2 \partial y^2} + \dfrac{\partial^4 w}{\partial y^4}\right) \\[4mm] -t \left[\begin{array}{c} \dfrac{\partial^2 F}{\partial y^2}\dfrac{\partial^2 (w + w_0)}{\partial x^2} - 2\dfrac{\partial^2 F}{\partial x \partial y}\dfrac{\partial^2 (w + w_0)}{\partial x \partial y} \\[3mm] + \dfrac{\partial^2 F}{\partial x^2}\dfrac{\partial^2 (w + w_0)}{\partial y^2} + \dfrac{p}{t} \end{array}\right] \end{array}\right\} \times \left(1 - \cos\frac{2\pi x}{a}\right)\left(1 - \cos\frac{2\pi y}{b}\right)dxdy = 0$$

$$\tag{4.34}$$

Substituting Equations (4.31a) and (4.31b) into Equation (4.34) and performing the integration over the entire plate, the following equation with regard to the unknown added deflection amplitude A_1 is given by

$$C_1 A_1^3 + C_2 A_1^2 + C_3 A_1 + C_4 = 0 \tag{4.35}$$

where

$$C_1 = \frac{\pi^2 E}{256a^3 b^3}K,$$

$$C_2 = \frac{3\pi^2 EA_{01}}{256a^3b^3}K,$$

$$C_3 = \frac{\pi^2 E A_{01}^2}{128a^3b^3}K + \frac{3b}{4a}\sigma_{rex} + \frac{3a}{4b}\sigma_{rey} + \frac{\pi^2 D}{t}\left(\frac{3b}{a^3} + \frac{3a}{b^3} + \frac{2}{ab}\right),$$

$$C_4 = \frac{3A_{01}}{4ab}\left(b^2\sigma_{rex} + a^2\sigma_{rey}\right) - \frac{ab}{\pi^2 t}p,$$

$$\sigma_{rex} = \sigma_{rcx} + \frac{2}{b}\left(\sigma_{rtx} - \sigma_{rcx}\right)\left(b_t - \frac{b}{2\pi}\sin\frac{2\pi b_t}{b}\right) \approx \frac{b - 2b_t}{b}\sigma_{rcx},$$

$$\sigma_{rey} = \sigma_{rcy} + \frac{2}{b}\left(\sigma_{rty} - \sigma_{rcy}\right)\left(a_t - \frac{a}{2\pi}\sin\frac{2\pi a_t}{a}\right) \approx \frac{a - 2a_t}{a}\sigma_{rcy},$$

$$K = \frac{H}{\left(4a^6 + 21a^4b^2 + 21a^2b^4 + 4b^6\right)^2},$$

$$H = 272a^{16} + 2856a^{14}b^2 + 11273a^{12}b^4 + 23146a^{10}b^6 + 23146a^6b^{10}$$

$$+ 31506a^8b^8 + 11273a^4b^{12} + 2856a^2b^{14} + 272b^{16}.$$

Once A_1 is determined as a function of the lateral pressure loads p using the Cardano method or the FORTRAN computer program CARDANO given in the appendices to this book, the deflection and membrane stresses inside the plate can be obtained.

4.5.2 Combined Biaxial Loads

In this case, the initial and added deflection functions may be assumed as follows:

$$w_0 = \frac{1}{4}A_{0mn}\left(1 - \cos\frac{2m\pi x}{a}\right)\left(1 - \cos\frac{2n\pi y}{b}\right) \tag{4.36a}$$

$$w = \frac{1}{4}A_{mn}\left(1 - \cos\frac{2m\pi x}{a}\right)\left(1 - \cos\frac{2n\pi y}{b}\right) \tag{4.36b}$$

where A_{0mn} and A_{mn} are the initial and added deflection amplitudes and m and n are the buckling half-wave numbers in the x or y direction of the plate. The Airy stress function F is obtained as follows:

$$F = \left(\sigma_{xav} + \sigma_{rx}\right)\frac{y^2}{2} + \left(\sigma_{yav} + \sigma_{ry}\right)\frac{x^2}{2}$$

$$+ \frac{EA_{mn}(A_{mn} + 2A_{0mn})}{512m^2n^2a^2b^2}\left(\begin{array}{l} 16n^4a^4\cos\dfrac{2m\pi x}{a} - n^4a^4\cos\dfrac{4m\pi x}{a} + m^4b^4\left(16\cos\dfrac{2n\pi y}{b} - \cos\dfrac{4n\pi y}{b}\right) \\[2ex] + 8n^4a^4\left(\begin{array}{l}\dfrac{\cos\left(\dfrac{2m\pi x}{a} - \dfrac{4n\pi y}{b}\right)}{(m^2b^2 + 4n^2a^2)^2} - \dfrac{2\cos\left(\dfrac{2m\pi x}{a} - \dfrac{2n\pi y}{b}\right)}{(m^2b^2 + n^2a^2)^2} + \dfrac{\cos\left(\dfrac{4m\pi x}{a} - \dfrac{2n\pi y}{b}\right)}{(4m^2b^2 + n^2a^2)^2} \\[3ex] - \dfrac{2\cos\left(\dfrac{2m\pi x}{a} + \dfrac{2n\pi y}{b}\right)}{(m^2b^2 + n^2a^2)^2} + \dfrac{2\cos\left(\dfrac{4m\pi x}{a} + \dfrac{2n\pi y}{b}\right)}{(4m^2b^2 + n^2a^2)^2} + \dfrac{\cos\left(\dfrac{2m\pi x}{a} + \dfrac{4n\pi y}{b}\right)}{(m^2b^2 + 4n^2a^2)^2}\end{array}\right)\end{array}\right)$$

$$\tag{4.37}$$

The Galerkin method is applied to Equation (4.1a) as follows:

$$\int_0^b \int_0^a \left\{ \begin{aligned} & D\left(\frac{\partial^4 w}{\partial x^4} + 2\frac{\partial^4 w}{\partial x^2 \partial y^2} + \frac{\partial^4 w}{\partial y^4}\right) \\ & -t\left[\begin{aligned} &\frac{\partial^2 F}{\partial y^2}\frac{\partial^2 (w + w_0)}{\partial x^2} - 2\frac{\partial^2 F}{\partial x \partial y}\frac{\partial^2 (w + w_0)}{\partial x \partial y} \\ &+ \frac{\partial^2 F}{\partial x^2}\frac{\partial^2 (w + w_0)}{\partial y^2} \end{aligned}\right] \end{aligned} \right\} \times \left(1 - \cos\frac{2m\pi x}{a}\right)\left(1 - \cos\frac{2n\pi y}{b}\right)dx\,dy = 0$$

$$(4.38)$$

Substituting Equations (4.36a) and (4.36b) into Equation (4.38) and performing the integration over the entire plate, the following equation with regard to the unknown deflection A_{mn} is obtained:

$$C_1 A_{mn}^3 + C_2 A_{mn}^2 + C_3 A_{mn} + C_4 = 0 \qquad (4.39)$$

where

$$C_1 = \frac{\pi^2 E}{256 a^3 b^3}K,$$

$$C_2 = \frac{3\pi^2 E A_{0mn}}{256 a^3 b^3}K,$$

$$C_3 = \frac{\pi^2 E\, A_{0mn}^2}{128 a^3 b^3}K + \frac{3m^2 b}{4a}(\sigma_{xav} + \sigma_{rex}) + \frac{3n^2 a}{4b}(\sigma_{yav} + \sigma_{rey})$$
$$+ \frac{\pi^2 D}{t}\left(\frac{3m^4 b}{a^3} + \frac{3n^4 a}{b^3} + \frac{2m^2 n^2}{ab}\right),$$

$$C_4 = \frac{3A_{0mn}}{4ab}\left[m^2 b^2(\sigma_{xav} + \sigma_{rex}) + n^2 a^2(\sigma_{yav} + \sigma_{rey})\right],$$

$$\sigma_{rex} = \sigma_{rcx} + \frac{2}{b}(\sigma_{rtx} - \sigma_{rcx})\left(b_t - \frac{b}{2n\pi}\sin\frac{2n\pi b_t}{b}\right) \approx \frac{b - 2b_t}{b}\sigma_{rcx},$$

$$\sigma_{rey} = \sigma_{rcy} + \frac{2}{b}(\sigma_{rty} - \sigma_{rcy})\left(a_t - \frac{a}{2m\pi}\sin\frac{2m\pi a_t}{a}\right) \approx \frac{a - 2a_t}{a}\sigma_{rcy},$$

$$K = \frac{H}{\left(4n^6 a^6 + 21m^2 n^4 a^4 b^2 + 21m^4 n^2 a^2 b^4 + 4m^6 b^6\right)^2},$$

$$H = 272n^{16}a^{16} + 2856m^2 n^{14}a^{14}b^2 + 11273m^4 n^{12}a^{12}b^4 + 23146m^6 n^{10}a^{10}b^6 + 23146m^{10}n^6 a^6 b^{10}$$

$$+ 31506m^8 n^8 a^8 b^8 + 11273m^{12}n^4 a^4 b^{12} + 2856m^{14}n^2 a^2 b^{14} + 272m^{16}b^{16}.$$

Again, once A_{mn} is determined as a function of σ_{xav} and σ_{yav} using the Cardano method or the FORTRAN computer program CARDANO given in the appendices to this book, the deflection and membrane stresses inside the plate can be obtained. For particular cases with no initial deflections, that is, $A_{0mn} = 0$, $C_2 = C_4 = 0$. Therefore, A_{mn} is determined from Equation (4.22), whereas C_1 and C_3 are defined in Equation (4.39). In this case, the same expression of Equation (4.23) can be used to determine the elastic buckling strength condition of a clamped plate under combined biaxial loads as follows, because $C_3 = 0$:

$$\frac{3m^2 b}{4a}\left(\sigma_{xav} + \sigma_{rex}\right) + \frac{3n^2 a}{4b}\left(\sigma_{yav} + \sigma_{rey}\right) + \frac{\pi^2 D}{t}\left(\frac{3m^4 b}{a^3} + \frac{3n^4 a}{b^3} + \frac{2m^2 n^2}{ab}\right) = 0 \quad (4.40)$$

When the ratio of biaxial loading, $c = \sigma_{yav}/\sigma_{xav}$, is kept constant, the elastic compressive buckling strength σ_{xE} in the x direction is obtained as follows:

$$\sigma_{xE} = -\frac{4ab}{3m^2 b^2 + 3cn^2 a^2}\left[\frac{3m^2 b}{4a}\sigma_{rex} + \frac{3n^2 a}{4b}\sigma_{rey} + \frac{\pi^2 D}{t}\left(\frac{3m^4 b}{a^3} + \frac{3n^4 a}{b^3} + \frac{2m^2 n^2}{ab}\right)\right]$$
$$(4.41)$$

For a long plate with $a/b \geq 1$, $n = 1$ may be taken. In contrast, the buckling half-wave number m in the x direction can be determined as a minimum integer that satisfies the following condition:

$$\frac{4ab}{3m^2 b^2 + 3ca^2}\left[\frac{3m^2 b}{4a}\sigma_{rex} + \frac{3a}{4b}\sigma_{rey} + \frac{\pi^2 D}{t}\left(\frac{3m^4 b}{a^3} + \frac{3a}{b^3} + \frac{2m^2}{ab}\right)\right]$$
$$\leq \frac{4ab}{3(m+1)^2 b^2 + 3ca^2}\left[\frac{3(m+1)^2 b}{4a}\sigma_{rex} + \frac{3a}{4b}\sigma_{rey} + \frac{\pi^2 D}{t}\left(\frac{3(m+1)^4 b}{a^3} + \frac{3a}{b^3} + \frac{2(m+1)^2}{ab}\right)\right]$$
$$(4.42a)$$

Equation (4.42a) can be simplified by neglecting the effect of welding induced residual stresses as follows:

$$\frac{4ab}{3m^2 b^2 + 3ca^2}\left(\frac{3m^4 b}{a^3} + \frac{3a}{b^3} + \frac{2m^2}{ab}\right) \leq \frac{4ab}{3(m+1)^2 b^2 + 3ca^2}\left[\frac{3(m+1)^4 b}{a^3} + \frac{3a}{b^3} + \frac{2(m+1)^2}{ab}\right]$$
$$(4.42b)$$

For uniaxial compression σ_{xav}, with $\sigma_{yav} = 0$ or $c = 0$, the elastic buckling strength σ_{xE} of a clamped plate is then determined from Equation (4.42) as follows:

$$\sigma_{xE} = -\frac{4a}{3m^2 b}\left[\frac{3m^2 b}{4a}\sigma_{rex} + \frac{3a}{4b}\sigma_{rey} + \frac{\pi^2 D}{t}\left(\frac{3m^4 b}{a^3} + \frac{3a}{b^3} + \frac{2m^2}{ab}\right)\right] \quad (4.43a)$$

If no welding induced residual stresses exist, Equation (4.43a) is simplified to

$$\sigma_{xE} = -\frac{4\pi^2 D}{3t}\frac{3m^4 b^4 + 3a^4 + 2m^2 a^2 b^2}{m^2 a^2 b^4} \quad (4.43b)$$

In this case, the buckling half-wave number m can be determined as a minimum integer that satisfies the following condition:

$$\frac{3m^4 b^4 + 3a^4 + 2m^2 a^2 b^2}{m^2 a^2 b^4} \leq \frac{3(m+1)^4 b^4 + 3a^4 + 2(m+1)^2 a^2 b^2}{(m+1)^2 a^2 b^4} \quad (4.44)$$

For uniaxial compression σ_{yav}, with $\sigma_{xav} = 0$, the elastic buckling strength σ_{yE} of a clamped plate is obtained as follows:

$$\sigma_{yE} = -\frac{4b}{3a}\left[\frac{3b}{4a}\sigma_{rex} + \frac{3a}{4b}\sigma_{rey} + \frac{\pi^2 D}{t}\left(\frac{3b}{a^3} + \frac{3a}{b^3} + \frac{2}{ab}\right)\right] \tag{4.45a}$$

where $m = n = 1$ was taken for a plate with $a/b \geq 1$.

If no welding induced residual stresses exist, Equation (4.45a) is simplified to

$$\sigma_{yE} = -\frac{4\pi^2 D}{3t}\frac{3b^4 + 3a^4 + 2a^2 b^2}{a^4 b^2} \tag{4.45b}$$

4.5.3 Interaction Effect Between Biaxial Loads and Lateral Pressure

Similar to Equation (4.29), the effect of lateral pressure loads is approximately accounted for as follows:

$$C_4 = \frac{3A_{0mn}}{4ab}\left[m^2 b^2\left(\sigma_{xav} + \sigma_{rex}\right) + n^2 a^2\left(\sigma_{yav} + \sigma_{rey}\right)\right] - \frac{ab}{\pi^2 t}p \tag{4.46}$$

Upon calculating the deflection and membrane stresses for a clamped plate under combined in-plane and lateral pressure loads, C_4 in Equation (4.46) is used instead of C_4 in Equation (4.39).

4.6 Elastic Large-Deflection Behavior of Partially Rotation Restrained Plates

The rotational restraints at the plate edges are neither zero nor infinite because the plate edges are supported by longitudinal stiffeners and transverse frames; rather, they depend on the torsional rigidities of the support members, as described in Chapter 3.

The elastic large-deflection behavior of such plates certainly depends on the degree of rotational restraints. Dealing with the elastic large-deflection behavior of a partially rotation restrained plate is not easy. For the sake of simplicity, C_3 and C_4 in Equation (4.19) for a simply supported plate under combined biaxial and lateral pressure loads are modified as follows (Hughes & Paik 2013):

$$C_3 = \frac{\pi^2 E A_{0mn}^2}{8}\left(\frac{m^4 b}{a^3} + \frac{n^4 a}{b^3}\right) + \frac{m^2 b}{a}\left(\sigma_{xav} + \sigma_{rex}\right) + \frac{n^2 a}{b}\left(\sigma_{yav} + \sigma_{rey}\right)$$

$$+ \frac{\pi^2 D m^2 n^2}{t}\left(\frac{mb}{na} + \frac{na}{mb}\right)^2 \sqrt{\frac{k_x k_y}{k_{xo} k_{yo}}}\sqrt{C_{px} C_{py}} \tag{4.47a}$$

$$C_4 = A_{0mn}\left[\frac{m^2 b}{a}\left(\sigma_{xav} + \sigma_{rex}\right) + \frac{n^2 a}{b}\left(\sigma_{yav} + \sigma_{rey}\right)\right] - \frac{16ab}{\pi^4 t}p \tag{4.47b}$$

where k_{xo} and k_{yo} are the buckling coefficients for a simply supported plate under uniaxial compression in the x or y direction as described in Chapter 3, k_x and k_y are the buckling coefficients for a partially rotation restrained plate under uniaxial compression in the

x or y direction as described in Chapter 3, and C_{px} and C_{py} are the coefficients that deal with the effects of lateral pressure loads on the compressive buckling strength in the x or y direction, as defined in Equation (3.28). The plate deflection amplitude A_{mn} is determined as a solution of Equation (4.19), but C_3 and C_4 in this case are defined as those in Equations (4.47a) and (4.47b).

Figure 4.11 shows the variation of the coefficient C_3 in Equation (4.47a) as a function of the parameters of rotational restraints, ζ_L and ζ_S, as defined in Equation (3.23). It is apparent from Figure 4.11 that the coefficient C_3 progressively increases with increase in the rotational restraints. This suggestion is straightforward to apply while giving reasonably accurate solutions as discussed in the following section.

In the following, the applicability of this approach is confirmed by comparison with more refined finite element method solutions for partially rotation restrained plates under uniaxial or biaxial compression (Paik et al. 2012). Tables 4.1 and 4.2 indicate the dimensions of longitudinal stiffeners and transverse frames together with the rotational restraint parameters ζ_L and ζ_S, as defined in Equation (3.23). The plate aspect ratio, dimensions of the support members, and biaxial loading ratio are varied in the illustrative examples, where the plate breadth is $b = 1000$ mm, the plate thickness is $t = 20$ mm, Young's modulus is $E = 205.8$ GPa, and Poisson's ratio is $\nu = 0.3$. The maximum initial deflection of the plate is assumed to be $w_{0pl} = b/200$.

Table 4.3 and Figure 4.12 indicate the boundary condition and mesh modeling of the finite element analysis. It is noted that the finite element method cannot deal with the parameters of rotational restraints in an explicit manner. Instead, a two-bay finite element model as shown in Figure 4.12 is employed in both the x and y directions as the extent of the analysis to automatically take into account the effect of partially rotation restrained edges at the support member locations.

Figure 4.13a shows the mesh modeling. A number of 14 rectangular-type plate–shell elements were used in the y direction of the plating between longitudinal stiffeners, and a number of six rectangular-type plate–shell elements were used for the longitudinal stiffener web in the web height direction. A number of two rectangular plate–shell elements were used for T-type stiffener flange in the flange breadth direction. In the x direction of the panel, the plate–shell elements were assigned so that the element aspect ratio becomes unity. Figure 4.13b and c shows illustrative examples of the plate initial deflection patterns applied, in which the buckling mode shape of the plate initial deflection is considered in the finite element analysis. The buckling mode of a plate can be determined from Equation (4.16) depending on the plate aspect ratio and the loading ratio or condition. Figure 4.14 illustrates three types of loading conditions considered. Under biaxial compression, its ratio between longitudinal compression and transverse compression is varied.

4.6.1 Longitudinal Compression

Figures 4.15, 4.16, and 4.17 present comparisons of the theory and the FEA for a plate under longitudinal axial compression with various sizes of support members in terms of the elastic large-deflection behavior of plates, in which the vertical axis represents the applied average longitudinal compressive stress, σ_{xav}, normalized by the corresponding elastic buckling stress, σ_{xE}, of the simply supported plate and the horizontal axis

(a)

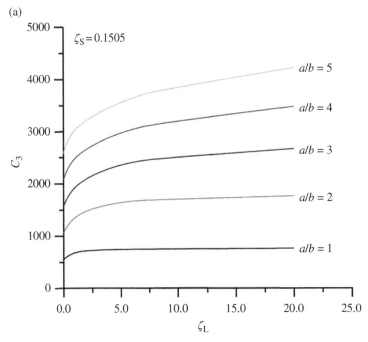

(b)

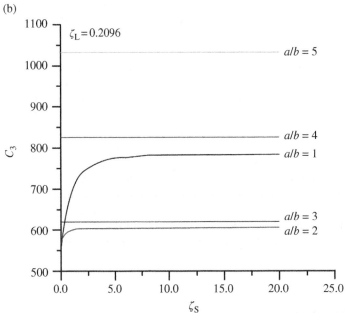

Figure 4.11 Variation of the coefficient C_3 as a function of the rotational restraint parameters with varying the plate aspect ratio: (a) for the longitudinal stiffeners; (b) for the transverse frames.

Table 4.1 Dimensions of the longitudinal stiffeners of the plate.

| Case | Longitudinal stiffener size (mm) | | | | ζ_L | | |
	h_{wx}	t_{wx}	b_{fx}	t_{fx}	$a/b=1$	3	5
I	250	12	150	15		0.1642	
II	400	12	150	15		0.2096	
III	500	12	150	15		0.2398	

Table 4.2 Dimensions of the transverse frames of the plate.

| Case | Transverse stiffener size (mm) | | | | ζ_S | | |
	h_{wy}	t_{wy}	b_{fy}	t_{fy}	$a/b=1$	3	5
A	650	12	150	15	0.2852	0.0951	0.0570
B	1200	12	150	15	0.4515	0.1505	0.0903

Table 4.3 Boundary conditions of the finite element models using the two-bay/double span stiffened panel as indicated in Figure 4.12.

Boundary	Description
$A–A'''$ and $D–D'''$	Symmetric condition with $R_y = R_z = 0$ and uniform displacement in the x direction (U_x = uniform), coupled with the longitudinal stiffener
$A–D$ and $A'''–D'''$	Symmetric condition with $R_x = R_z = 0$ and uniform displacement in the y direction (U_y = uniform), coupled with the transverse frame
$A'–D'$, $A''–D''$, $B–B'$, and $C–C'$	$U_z = 0$

Note: U_x, U_y, and U_z indicate the translational degrees of freedom in the x, y, and z direction, and R_x, R_y, and R_z indicate the rotational degrees of freedom in the x, y, and z direction.

represents the maximum total deflection of the plate including both initial and added deflections. It is apparent from these figures that the behavior of the plate with partially rotation restrained edges is in between those of the plates with simply supported and clamped edges. It is concluded that the solutions obtained with the theory show reasonably good correlation with those of the nonlinear FEA.

4.6.2 Transverse Compression

Figures 4.18, 4.19, and 4.20 present comparisons of the results of the theory and the FEA for a plate under transverse axial compression with various sizes of support members in terms of the elastic large-deflection behavior of plates, in which the vertical axis represents the applied average transverse compressive stress, σ_{yav}, normalized by the

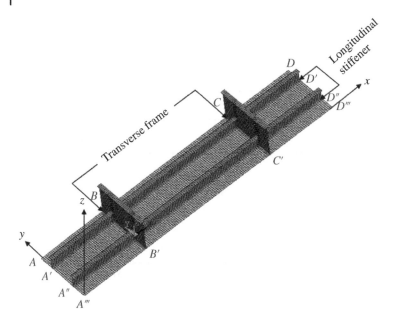

Figure 4.12 Boundary conditions for the finite element model.

corresponding elastic buckling stress, σ_{yE}, of the simply supported plate and the horizontal axis represents the maximum total deflection of the plate including both initial and added deflections. It is apparent from these figures that the behavior of the plate with partially rotation restrained edges is in between those of the plates with simply supported and clamped edges. Again, it is concluded that the solutions obtained with the theory show very good correlation with those of the nonlinear FEA.

4.6.3 Biaxial Compression

Figures 4.21, 4.22, and 4.23 present comparisons of the results of the theory and the FEA for a plate under biaxial compression with various sizes of support members and biaxial loading ratios. In these figures, the vertical axis represents the applied average longitudinal compressive stress, σ_{xav}, normalized by the corresponding elastic buckling stress, σ_{xE}, of the simply supported plate, although the plate is actually subjected to both longitudinal and transverse compressive loads, keeping the biaxial loading ratio constant. It is also concluded that the solutions of the theory show reasonably good correlation with those of the nonlinear FEA.

4.7 Effect of the Bathtub Deflection Shape

For a square or long plate, the plate deflection is normally quite similar to a sinusoidal pattern. For a long plate under predominantly transverse compressive loading, however, the plate deflection may differ somewhat from the sinusoidal pattern; it normally takes

(a)

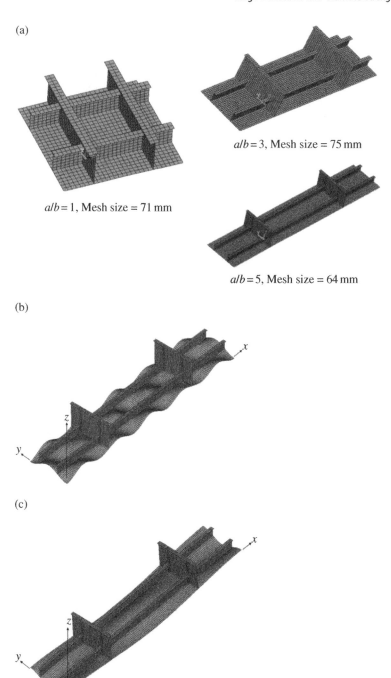

$a/b=3$, Mesh size = 75 mm

$a/b=1$, Mesh size = 71 mm

$a/b=5$, Mesh size = 64 mm

(b)

(c)

Figure 4.13 (a) Mesh modeling for the finite element analysis; (b) Initial deflection pattern applied for a plate with $a/b = 5$ under predominantly longitudinal axial compression (plate initial deflection amplified by 80 times); (c) Initial deflection pattern applied for a plate with $a/b = 5$ under predominantly transverse axial compression (plate initial deflection amplified by 80 times).

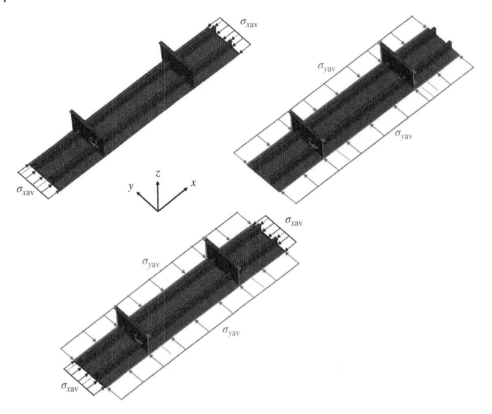

Figure 4.14 Three types of loading conditions considered in the finite element analysis, including longitudinal compression, σ_{xav}, transverse compression, σ_{yav}, and biaxial compression, σ_{xav} and σ_{yav}.

the so-called "bathtub" (or bulb) shape around the plate edges, whereas the deflected shape in the middle part of the plate is nearly flat, as shown in Figure 4.24. Due to the bathtub-type deflection, the rotation and deflection around the plate edges are normally greater than those of the sinusoidal pattern, resulting in larger values of the membrane stresses.

This implies that the presumed deflection functions with one mode term may not be valid further for a plate in compression along the short edges, and thus a more refined deflection function, that is, one with deflection terms of at least more than 2, may be needed. However, it is in this case not straightforward to analytically solve the nonlinear governing differential equations.

As an easier alternative, while maintaining the deflection functions with a single component, the maximum and minimum membrane stresses along the edges of the plate in predominantly transverse compressive loading are approximately corrected by introducing a factor, ρ, to account for the deflection effects of the bathtub shape as follows (Ueda et al. 1984):

$$\rho = \frac{1}{\sqrt{2}}\left(\frac{b}{a} - \sqrt{2}\right) + 2 \tag{4.48}$$

The correction factor ρ of Equation (4.48) is then applied to the maximum and minimum membrane stresses of Equation (4.20) for a simply supported plate under biaxial loads as follows:

$$\sigma_{x\,max} = \sigma_{xav} + \sigma_{rtx} - \rho\frac{E\pi^2 m^2 A_{mn}(A_{mn} + 2A_{0mn})}{8a^2}\cos\frac{2n\pi b_t}{b} \qquad (4.49a)$$

(a)

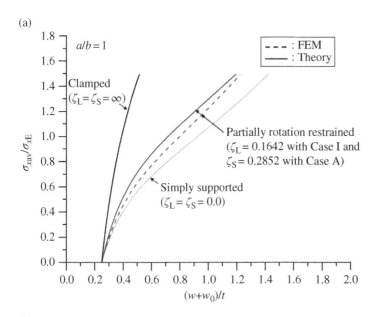

(b)

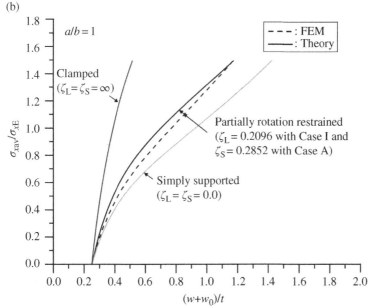

Figure 4.15 Comparison between the theory and the FEA for a plate with $a/b = 1$ under longitudinal compression with the transverse frames of Case A and the longitudinal stiffeners of (a) Case I; (b) Case II and (c) Case III.

(c)

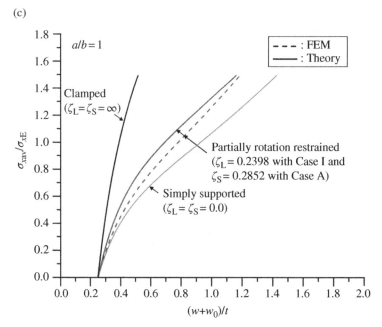

Figure 4.15 (Continued)

$$\sigma_{x\,min} = \sigma_{xav} + \sigma_{rcx} + \rho \frac{E\pi^2 m^2 A_{mn}(A_{mn} + 2A_{0mn})}{8a^2} \tag{4.49b}$$

$$\sigma_{y\,max} = \sigma_{yav} + \sigma_{rty} - \rho \frac{E\pi^2 n^2 A_{mn}(A_{mn} + 2A_{0mn})}{8b^2} \cos \frac{2m\pi a_t}{a} \tag{4.49c}$$

$$\sigma_{y\,min} = \sigma_{yav} + \sigma_{rcy} + \rho \frac{E\pi^2 n^2 A_{mn}(A_{mn} + 2A_{0mn})}{8b^2} \tag{4.49d}$$

For other types of load applications, a similar modification of the maximum and minimum membrane stresses shall be made with the correction factor ρ of Equation (4.48).

4.8   Evaluation of In-Plane Stiffness Reduction Due to Deflection

The membrane stress distribution inside a plate is no longer uniform once buckling or deflection has occurred. Figure 4.25 shows a schematic of the membrane normal stress distribution inside a plate under predominantly longitudinal compressive loading before and after buckling.

It is important to realize that many factors, including buckling, initial deflection, and lateral pressure loading, can cause the membrane stress distribution in the loading (x) direction to become nonuniform as the plate deflects. The membrane stress distribution in the y direction also becomes nonuniform as long as the unloaded plate edges remain

(a)

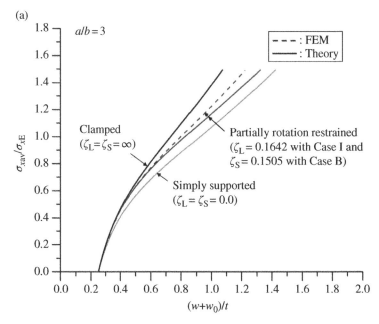

(b)

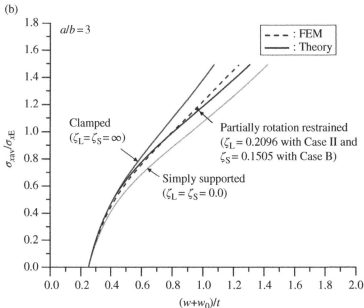

Figure 4.16 Comparison between the theory and the FEA for a plate with $a/b = 3$ under longitudinal compression with the transverse frames of Case B and the longitudinal stiffeners of (a) Case I; (b) Case II; (c) Case III.

(c)

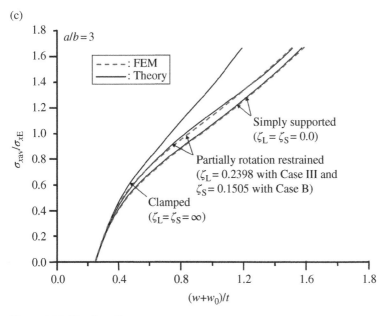

Figure 4.16 (Continued)

straight, whereas no membrane stresses develop in the *y* direction if the unloaded plate edges move freely in plane.

It is apparent from Figure 4.25 that the maximum compressive membrane stresses develop around the plate edges that remain straight, whereas the minimum membrane stresses occur in the middle of the plate, where a membrane tension field is formed by the plate deflection because the plate edges remain straight. The location of the maximum compressive stresses depends on the residual stresses. If there are no residual stresses, the maximum compressive stresses develop along the edges. In contrast, when residual stresses do exist, the maximum compressive stresses are found inside the plate at the limits of the tensile residual stress block breadths from the plate edges, as illustrated in Figure 4.8.

To model the large-deflection behavior of a plate deflected by buckling and/or lateral pressure loading, two concepts are relevant (Paik 2008a):

- The effective width or length concept
- The effective shear modulus concept

The membrane stress distribution inside the deflected plate is nonuniform, but that inside the undeflected plate is uniform. The basic ideas behind the two concepts mentioned earlier are to deal with the deflected plate as an undeflected plate but with the reduced in-plane stiffness. This idealization is beneficial because the theory of linear structural mechanics can still be applied. In the following section, the formulations of the two concepts are derived in detail.

(a)

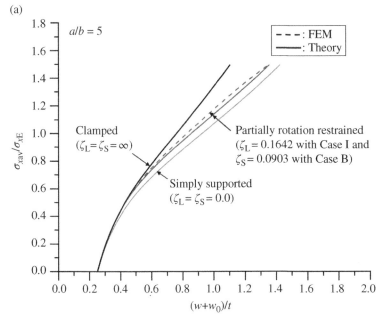

(b)

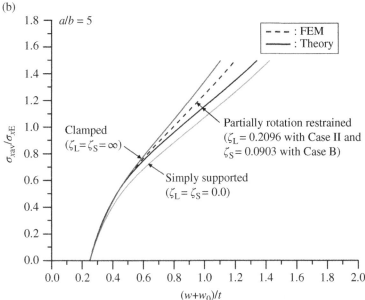

Figure 4.17 Comparison between the theory and the FEA for a plate with $a/b = 5$ under longitudinal compression with the transverse fames of Case B and the longitudinal stiffeners of (a) Case I; (b) Case II and (c) Case III.

(c)

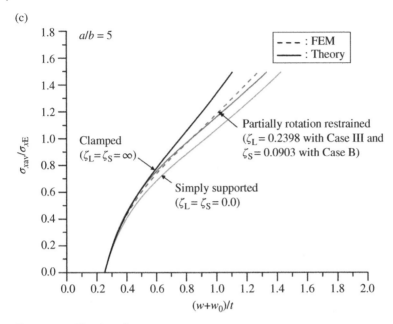

Figure 4.17 (Continued)

4.8.1 Effective Width

The effective width is a virtually reduced width of a plate between two longitudinal stiffeners. The effective width of a plate with initial imperfections under combined in-plane and lateral pressure loads is defined as the ratio of the average stress to the maximum stress as follows:

$$\frac{b_e}{b} = \frac{\sigma_{xav}}{\sigma_{x\,max}} \tag{4.50}$$

where σ_{xmax} is the maximum compressive stress, which is expressed as a function of combined in-plane and lateral pressure loads as well as initial imperfections, and b_e may also be called the "effective breadth" if lateral pressure, p, exists, because in that case, the shear lag effect also develops.

It is of interest to calculate the ultimate effective width, b_{eu}, at the ULS of the plate, which can be obtained from Equation (4.50) when $\sigma_{xav} = \sigma_{xu}$ as follows:

$$\frac{b_{eu}}{b} = \frac{\sigma_{xu}}{\sigma_{xmax}^u} \tag{4.51a}$$

where $\sigma_{xmax}^u = \sigma_{x\,max}$ at $\sigma_{xav} = \sigma_{xu}$ and σ_{xu} is the plate ultimate strength as described in Section 4.9.

Equation (4.50) or (4.51a) explicitly accounts for the influence of the initial imperfections and lateral pressure as parameters of influence. In contrast, the more typical approach in this regard is exemplified by Faulkner (1975), as indicated in Equation (2.19a), who suggested an empirical effective width formula for simply

(a)

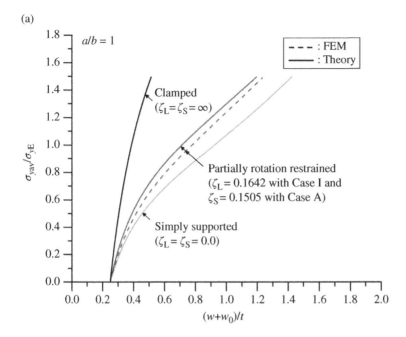

(b)

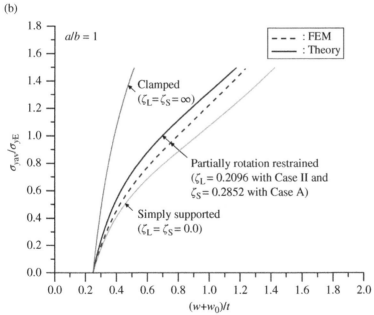

Figure 4.18 Comparison between the theory and the FEA for a plate with $a/b = 1$ under transverse compression with the transverse frames of Case A and the longitudinal stiffeners of (a) Case I; (b) Case II and (c) Case III.

(c)

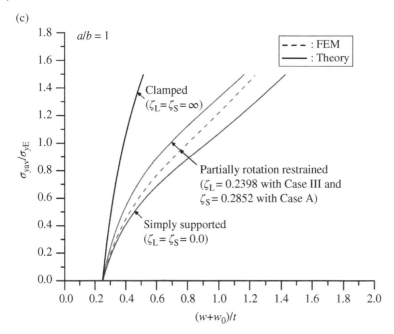

Figure 4.18 (Continued)

supported plates under longitudinal compression alone, that is, without lateral pressure, at the ULS as follows:

$$\frac{b_{eu}}{b} = \begin{cases} 1.0 & \text{for } \beta \le 1 \\ 2/\beta - 1/\beta^2 & \text{for } \beta > 1 \end{cases} \tag{4.51b}$$

where Equation (4.51b) implicitly involves the influence of initial imperfections at an "average" level. In some design codes, the terms 2 and 1 for $\beta > 1$ are changed to 1.8 and 0.9, respectively.

Figure 4.26 plots Equations (4.50) and (4.51a) with increasing σ_{xav}, varying plate slenderness ratios, initial deflections, residual stresses, and lateral pressure. The Faulkner formula, Equation (4.51b), is also shown for comparison. The plate ultimate strength, σ_{xu}, as obtained from Section 4.9, is also plotted. The Faulkner formula corresponds well to the effective width only for relatively thick plates with an "average" level of initial imperfections. It is apparent from Figure 4.26 that the plate effective width varies with the level of initial imperfections as well as applied loads, and Equation (4.50) or (4.51a) therefore better embodies the nature of the plate effective width. It is evident from Figure 4.26c that the lateral pressure is also a significant factor that influences (reduces) the plate effective "breadth," as would be expected.

It is often useful to derive a closed-form expression of the reduced (tangent) effective width to represent the in-plane effectiveness of the buckled plate, namely,

$$\frac{b_e^*}{b} = \left(\frac{\partial \sigma_{x\,max}}{\partial \sigma_{xav}} \right)^{-1} \tag{4.52}$$

where b_e^* is the reduced (tangent) effective width.

(a)

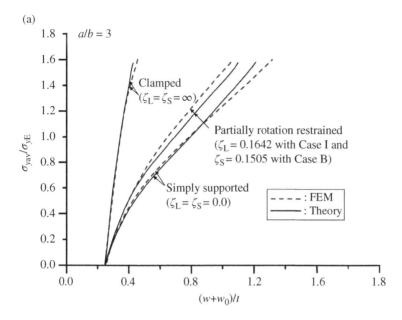

(b)

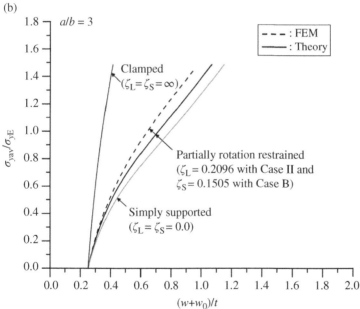

Figure 4.19 Comparison between the theory and the FEA for a plate with $a/b = 3$ under transverse compression with the transverse frames of Case B and the longitudinal stiffeners of (a) Case I; (b) Case II and (c) Case III.

(c)

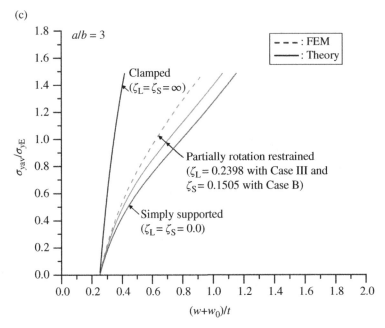

Figure 4.19 (Continued)

For a simply supported plate without initial imperfections and under uniaxial compressive loads in the x direction, Equation (4.20) of the maximum and minimum membrane stresses can be further simplified to

$$\sigma_{x\,max} = a_1\sigma_{xav} + a_2, \quad \sigma_{x\,min} = b_1\sigma_{xav} + b_2, \quad \sigma_{y\,max} = c_1\sigma_{xav} + c_2, \quad \sigma_{y\,min} = d_1\sigma_{xav} + d_2$$

$$(4.53)$$

where

$$a_1 = 1 + \rho\frac{2m^4}{a^4(m^4/a^4 + 1/b^4)}, \quad a_2 = \rho_x\frac{2m^2}{a^2(m^4/a^4 + 1/b^4)}\frac{\pi^2 D}{t}\left(\frac{m^2}{a^2} + \frac{1}{b^2}\right)^2,$$

$$b_1 = 1 - \rho\frac{2m^4}{a^4(m^4/a^4 + 1/b^4)}, \quad b_2 = -\rho_x\frac{2m^2}{a^2(m^4/a^4 + 1/b^4)}\frac{\pi^2 D}{t}\left(\frac{m^2}{a^2} + \frac{1}{b^2}\right)^2,$$

$$c_1 = \rho\frac{2m^2}{a^2 b^2(m^4/a^4 + 1/b^4)}, \quad c_2 = \rho\frac{2}{b^2(m^4/a^4 + 1/b^4)}\frac{\pi^2 D}{t}\left(\frac{m^2}{a^2} + \frac{1}{b^2}\right)^2,$$

$$d_1 = -\rho\frac{2m^2}{a^2 b^2(m^4/a^4 + 1/b^4)}, \quad d_2 = -\rho\frac{2}{b^2(m^4/a^4 + 1/b^4)}\frac{\pi^2 D}{t}\left(\frac{m^2}{a^2} + \frac{1}{b^2}\right)^2.$$

When neither initial imperfections nor lateral pressure is involved, the effective width formula of a simply supported plate under uniaxial compression in the x direction can be

(a)

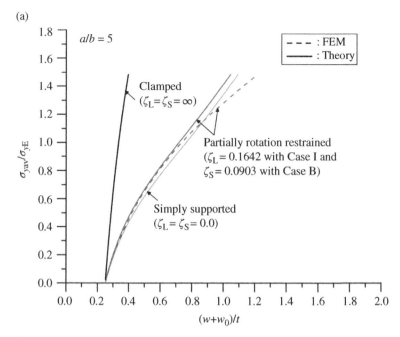

(b)

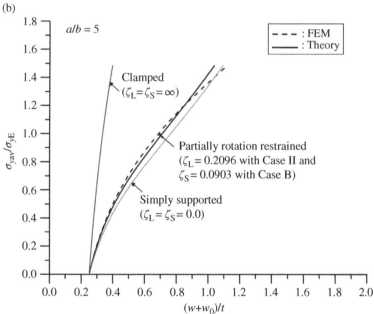

Figure 4.20 Comparison between the theory and the FEA for a plate with $a/b = 5$ under transverse compression with the transverse frames of Case B and the longitudinal stiffeners of (a) Case I; (b) Case II and (c) Case III.

(c)

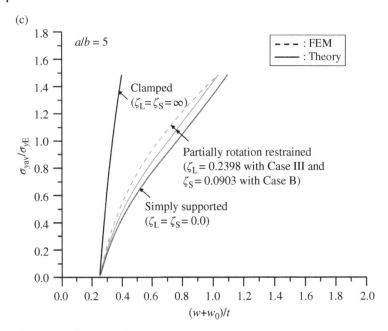

Figure 4.20 (Continued)

expressed from Equation (4.50) as a function of the average stress as follows because $\sigma_{x\max} = a_1\sigma_{xav} + a_2$ from Equation (4.53):

$$\frac{b_e}{b} = \frac{\sigma_{xav}}{a_1\sigma_{xav} + a_2} \tag{4.54a}$$

or as a function of the average strain because $\sigma_{x\max} = a_1\sigma_{xav} + a_2 = E\varepsilon_{xav}$:

$$\frac{b_e}{b} = \frac{1}{a_1}\left(1 - \frac{a_2}{E}\frac{1}{\varepsilon_{xav}}\right) \tag{4.54b}$$

The reduced effective width of a simply supported plate under uniaxial compressive loads in the x direction is obtained from Equation (4.52) when neither initial imperfections nor lateral pressure is involved:

$$\frac{b_e^*}{b} = \frac{1}{a_1} \tag{4.55}$$

4.8.2 Effective Length

The effective "length" is a virtually reduced length of a plate between two transverse frames. Similar to the effective width of Equation (4.50), the effective length of a plate in association with the axial compressive stress σ_{yav} in the y direction can be defined as follows:

$$\frac{a_e}{a} = \frac{\sigma_{yav}}{\sigma_{y\max}} \tag{4.56}$$

(a)

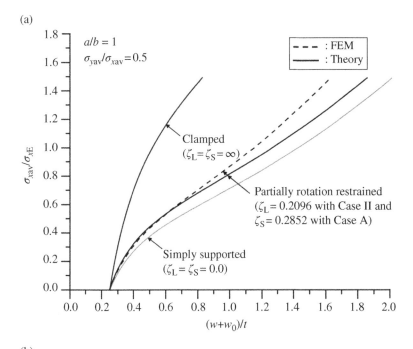

(b)

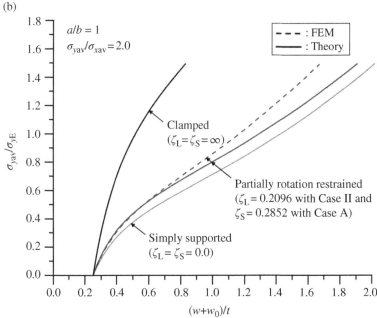

Figure 4.21 Comparison between the theory and the FEA for a plate with $a/b = 1$ under biaxial compression with the longitudinal stiffeners of Case II and the transverse frames of Case A: (a) $\sigma_{yav}/\sigma_{xav} = 0.5$; (b) $\sigma_{yav}/\sigma_{xav} = 2.0$.

(a)

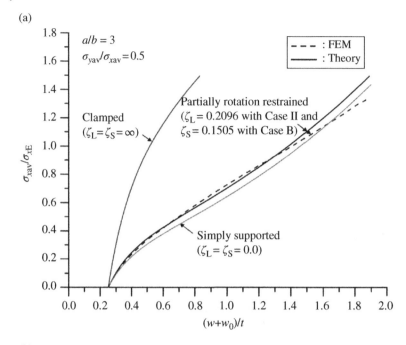

(b)

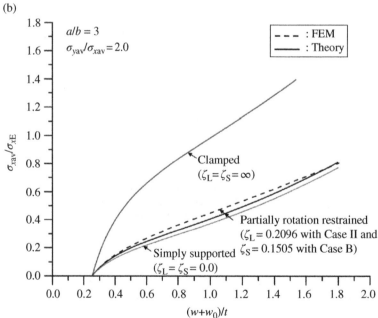

Figure 4.22 Comparison between the theory and the FEA for a plate with $a/b = 3$ under biaxial compression with the longitudinal stiffeners of Case II and the transverse frames of Case B: (a) $\sigma_{yav}/\sigma_{xav} = 0.5$; (b) $\sigma_{yav}/\sigma_{xav} = 2.0$.

(a)

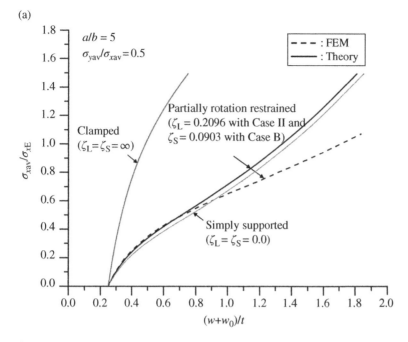

(b)

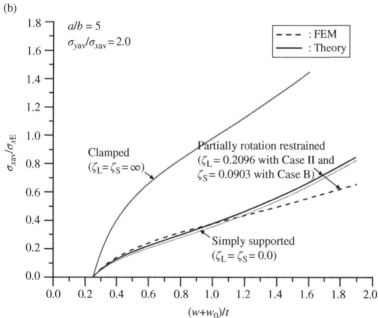

Figure 4.23 Comparison between the theory and the FEA for a plate with $a/b = 5$ under biaxial compression with the longitudinal stiffeners of Case II and the transverse frames of Case B: (a) $\sigma_{yav}/\sigma_{xav} = 0.5$; (b) $\sigma_{yav}/\sigma_{xav} = 2.0$.

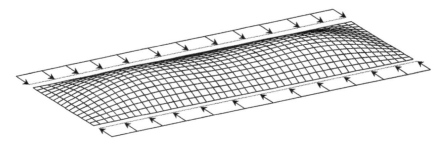

Figure 4.24 The so-called "bathtub" shape of the plate deflection.

(a)

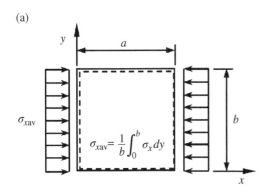

Figure 4.25 Membrane stress distribution inside the plate under predominantly longitudinal compressive loads: (a) before buckling; (b) after buckling, unloaded edges move freely in plane; (c) after buckling, unloaded edges remain straight.

(b)

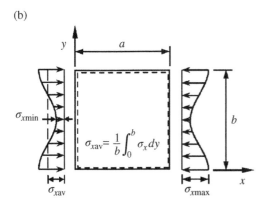

(c)

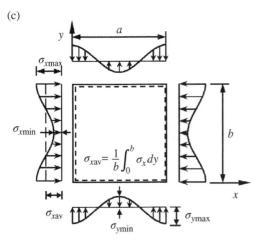

(a)

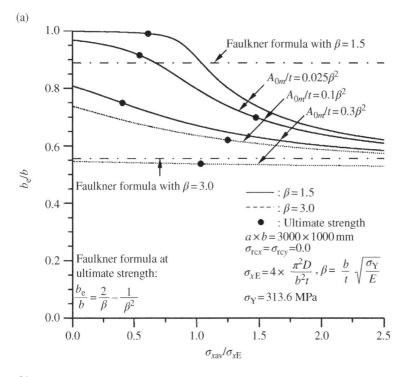

(b)

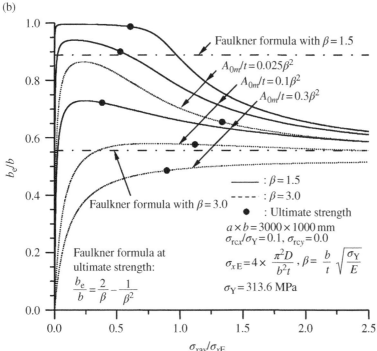

Figure 4.26 Variation of the effective width of a simply supported plate under uniaxial compression: (a) effective of initial deflection, on welding induced residual stresses; (b) effective of initial deflection and welding induced residual stresses: (c) effects of lateral pressure (σ_{xE}, elastic compressive buckling stress).

(c)

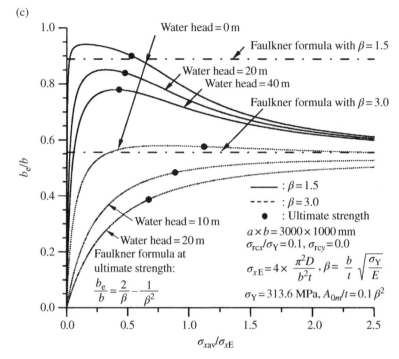

Figure 4.26 (Continued)

where σ_{ymax} is the maximum compressive stress, which is expressed as a function of the combined in-plane and lateral pressure loads and the initial imperfections.

It is again of interest to calculate the ultimate effective length, a_{eu}, at the ULS of the plate, which can be obtained from Equation (4.56) when $\sigma_{yav} = \sigma_{yu}$ as follows:

$$\frac{a_{eu}}{a} = \frac{\sigma_{yu}}{\sigma_{ymax}^u} \tag{4.57a}$$

where $\sigma_{ymax}^u = \sigma_{y\,max}$ at $\sigma_{yav} = \sigma_{yu}$, and σ_{yu} is the plate ultimate strength as described in Section 4.9.

Although Equation (4.57) explicitly accounts for the influence of initial imperfections and lateral pressure as parameters of influence, the more typical approach is that exemplified by Faulkner et al. (1973), who suggested an empirical effective length formula for simply supported plates under transverse compression alone, that is, without lateral pressure, at the ULS as follows:

$$\frac{a_{eu}}{a} = \frac{0.9}{\beta^2} + \frac{b\,1.9}{a\,\beta}\left(1 - \frac{0.9}{\beta^2}\right) \quad \text{for } \frac{a}{b} \geq 1.9 \tag{4.57b}$$

where Equation (4.57b) implicitly involves the influence of initial imperfections at an "average" level.

The reduced tangent effective length of a plate representing the in-plane effectiveness due to buckling or other reasons of deflection is given by

$$\frac{a_e^*}{a} = \left(\frac{\partial \sigma_{y\,max}}{\partial \sigma_{yav}}\right)^{-1} \tag{4.58}$$

where a_e^* is the reduced (tangent) effective length.

For a simply supported plate without initial imperfections and lateral pressure and under uniaxial compressive loads in the y direction, Equation (4.20) of the maximum and minimum membrane stresses can be further simplified to

$$\sigma_{x\,max} = e_1 \sigma_{yav} + e_2, \quad \sigma_{x\,min} = f_1 \sigma_{yav} + f_2, \quad \sigma_{y\,max} = g_1 \sigma_{yav} + g_2, \quad \sigma_{y\,min} = h_1 \sigma_{yav} + h_2 \tag{4.59}$$

where

$$e_1 = \rho \frac{2n^2}{a^2 b^2 (1/a^4 + n^4/b^4)}, \quad e_2 = \rho \frac{2}{a^2(1/a^4 + n^4/b^4)} \frac{\pi^2 D}{t} \left(\frac{1}{a^2} + \frac{n^2}{b^2}\right)^2,$$

$$f_1 = -\rho \frac{2n^2}{a^2 b^2 (1/a^4 + n^4/b^4)}, \quad f_2 = -\rho \frac{2}{a^2(1/a^4 + n^4/b^4)} \frac{\pi^2 D}{t} \left(\frac{1}{a^2} + \frac{n^2}{b^2}\right)^2,$$

$$g_1 = 1 + \rho \frac{2n^4}{b^4 (1/a^4 + n^4/b^4)}, \quad g_2 = \rho \frac{2n^2}{b^2(1/a_4 + n^4/b^4)} \frac{\pi^2 D}{t} \left(\frac{1}{a^2} + \frac{n^2}{b^2}\right)^2,$$

$$h_1 = 1 - \rho \frac{2n^4}{b^4 (1/a^4 + n^4/b^4)}, \quad h_2 = -\rho \frac{2n^2}{b^2(1/a^4 + n^4/b^4)} \frac{\pi^2 D}{t} \left(\frac{1}{a^2} + \frac{n^2}{b^2}\right)^2.$$

For a simply supported plate without both the initial imperfections and lateral pressure loads, Equation (4.56) can be given with $\sigma_{ymax} = g_1 \sigma_{yav} + g_2$ from Equation (4.59):

$$\frac{a_e}{a} = \frac{\sigma_{yav}}{g_1 \sigma_{yav} + g_2} \tag{4.60a}$$

We can recast Equation (4.60a) as a function of the membrane strain because $\sigma_{ymax} = g_1 \sigma_{yav} + g_2 = E \varepsilon_{yav}$ as follows:

$$\frac{a_e}{a} = \frac{1}{g_1} \left(1 - \frac{g_2}{g_1} \frac{1}{E \varepsilon_{yav}}\right) \tag{4.60b}$$

The reduced effective length of a simply supported plate representing the in-plane effectiveness due to buckling under uniaxial compressive loads in the y direction is given from Equation (4.58) when neither the initial imperfections nor lateral pressure is involved:

$$\frac{a_e^*}{a} = \frac{1}{g_1} \tag{4.61}$$

4.8.3 Effective Shear Modulus

Although the effective width is recognized as an efficient approach to evaluate the large-deflection behavior of a plate under predominantly axial compressive loads, the concept of the effective shear modulus, originally suggested by Paik (1995), may be useful to represent the behavior of a plate buckled in edge shear.

The basic concept of the effective shear modulus for a plate buckled in edge shear is now described. In plane stress problems, the relationship between membrane shear stress, τ, and shear strain, γ, is given by

$$\tau = G\gamma \tag{4.62}$$

where $G = E/[2(1 + v)]$ is the shear modulus.

Although the shear strain distribution would be uniform inside the plate before buckling, it is no longer uniform after shear buckling occurs. The shear strain at any point inside the buckled plate may be calculated by accounting for the large-deflection effects as follows:

$$\gamma = \left(\frac{\partial u}{\partial y} + \frac{\partial v}{\partial x}\right) + \left(\frac{\partial w}{\partial x}\frac{\partial w}{\partial y} + \frac{\partial w}{\partial x}\frac{\partial w_0}{\partial y} + \frac{\partial w_0}{\partial x}\frac{\partial w}{\partial y}\right) \tag{4.63}$$

where u and v are the axial displacements in the x and y directions, respectively. The first bracketed term on the right side of the previous equation represents the membrane shear strain component, and the second term indicates the additional shear strain component due to large-deflection effects.

The basic idea of either the effective width or the effective shear modulus concepts is to regard the deflected (buckled) plate as an equivalent "flat" (undeflected) plate, but with a reduced (effective) in-plane stiffness. Therefore, the membrane shear strain component, γ_m, of the buckled plate must in this case be evaluated as follows:

$$\gamma_m = \frac{\partial u}{\partial y} + \frac{\partial v}{\partial x} = \frac{\tau}{G} - \left(\frac{\partial w}{\partial x}\frac{\partial w}{\partial y} + \frac{\partial w}{\partial x}\frac{\partial w_0}{\partial y} + \frac{\partial w_0}{\partial x}\frac{\partial w}{\partial y}\right) \tag{4.64}$$

The membrane shear strain at any point inside the plate can, in a real case, be computed using numerical methods such as the finite element method described in Chapter 12 or the incremental Galerkin method described Chapter 11. The mean membrane shear strain, γ_{av}, may be defined as an average of the shear strains thus computed over the entire plate as follows:

$$\gamma_{av} = \frac{1}{ab}\int_0^a \int_0^b \gamma_m \, dx \, dy \tag{4.65}$$

Because the shear stress at the plate edges may equal the average shear stress, that is, $\tau = \tau_{av}$, the effective shear modulus, G_e, that represents the effectiveness of a plate buckled in edge shear can be defined by

$$G_e = \frac{\tau_{av}}{\gamma_{av}} \tag{4.66}$$

An empirical expression for the effective shear modulus can be developed by curve fitting based on numerical computations and on various influential factors such as the plate aspect ratio and initial imperfections. For instance, the effective shear modulus formula of a simply supported rectangular plate with initial deflections may be empirically derived based on the results of the incremental Galerkin method described in Chapter 11 (Paik 1995):

$$\frac{G_e}{G} = \begin{cases} c_1 V^3 + c_2 V^2 + c_3 V + c_4 & \text{for } V \leq 1.0 \\ d_1 V^2 + d_2 V + d_3 & \text{for } V > 1.0 \end{cases} \tag{4.67}$$

where

$$c_1 = -0.309 W_0^3 + 0.590 W_0^2 - 0.286 W_0,$$

$$c_2 = 0.353 W_0^3 - 0.644 W_0^2 + 0.270 W_0,$$

$$c_3 = -0.072 W_0^3 + 0.134 W_0^2 - 0.059 W_0,$$

$$c_4 = 0.005 W_0^3 - 0.033 W_0^2 + 0.001 W_0 + 1.0,$$

$$d_1 = -0.007 W_0^3 + 0.015 W_0^2 - 0.018 W_0 + 0.015,$$

$$d_2 = -0.022 W_0^3 + 0.006 W_0^2 + 0.075 W_0 - 0.118,$$

$$d_3 = 0.008 W_0^3 + 0.025 W_0^2 - 0.130 W_0 + 1.103.$$

with $V = \tau_{av}/\tau_E$, $W_0 = w_{0pl}/t$, and τ_E is the elastic shear buckling stress of the plate, as defined in Chapter 3.

When the initial deflection is not involved, Equation (4.67) is simplified to

$$\frac{G_e}{G} = \begin{cases} 1.0 & \text{for } \tau_{av}/\tau_E \leq 1 \\ 0.015(\tau_{av}/\tau_E)^2 - 0.118\tau_{av}/\tau_E + 1.103 & \text{for } \tau_{av}/\tau_E > 1 \end{cases} \tag{4.68}$$

Figure 4.27 plots Equations (4.67). It is apparent from Figure 4.27 that the effective shear modulus of a plate decreases after buckling as the edge shear stress increases. The initial deflection also reduces the effective shear modulus, as would be expected.

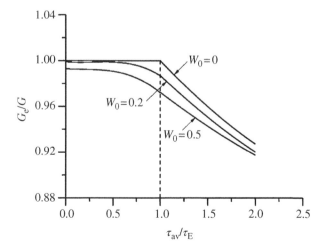

Figure 4.27 Variation of the effective shear modulus for a plate with increase in edge shear.

4.9 Ultimate Strength

The existing analytical methods to calculate the ultimate strength of plates can be categorized into two approaches:

- The rigid-plastic theory method
- The membrane stress-based method

This section describes the ultimate strength formulations for a plate element under combined biaxial loads, edge shear, or lateral pressure, which consider the effects of initial imperfections. The effects of opening or structural damage on the plate ultimate strength are described separately.

4.9.1 Ultimate Strength by Gross Yielding

For a plate element under predominantly axial tensile loading and/or with a very large thickness or a very low plate slenderness ratio, the ultimate strength is reached by gross yielding. In this case, the ultimate strength criterion is typically given by the von Mises yield condition, Equation (1.31c), which can be used as an upper limit of the plate ultimate strength, as follows:

$$\left(\frac{\sigma_{xav}}{\sigma_Y}\right)^2 - \left(\frac{\sigma_{xav}}{\sigma_Y}\right)\left(\frac{\sigma_{yav}}{\sigma_Y}\right) + \left(\frac{\sigma_{yav}}{\sigma_Y}\right)^2 + \left(\frac{\tau_{av}}{\tau_Y}\right)^2 = 1 \tag{4.69}$$

4.9.2 Rigid-Plastic Theory Method

In the classical rigid-plastic theory method (Wood 1961), the kinematically admissible collapse mechanisms of the plate at the ultimate strength are presumed, and the ultimate strength is then determined by applying the classical energy principle so that the internal strain energy is in equilibrium with the external potential energy. To account for the large-deflection effect, the rigid-plastic approach must be combined with the elastic large-deflection theory of the plate. This method is often used to indicate the upper or lower bound solutions of the ultimate strength.

4.9.2.1 Lateral Pressure Loads

Jones (1975) derived the collapse strength of a rectangular plate under lateral pressure loads using the rigid-plastic theory method without accounting for the large-deflection effect:

$$\frac{8M_P}{b^2}\left(1 + \alpha + \alpha^2\right) \leq p_u \leq \frac{24M_P}{b^2}\frac{1}{\left(\sqrt{3 + \alpha^2} - \alpha\right)^2} \quad \text{for a simply supported plate}$$

$$\tag{4.70a}$$

$$\frac{16M_P}{b^2}\left(1 + \alpha^2\right) \leq p_u \leq \frac{48M_P}{b^2}\frac{1}{\left(\sqrt{3 + \alpha^2} - \alpha\right)^2} \quad \text{for a clamped plate} \tag{4.70b}$$

where p_u is the collapse strength, $M_P = \sigma_Y t^2/4$ is the full plastic bending moment, and $\alpha = b/a$.

For a square plate with $\alpha = 1$, Equation (4.70a) or (4.70b) can be simplified to

$$\frac{24M_\mathrm{P}}{b^2} \leq p_\mathrm{u} \leq \frac{24M_\mathrm{P}}{b^2} \quad \text{for a simply supported square plate} \tag{4.71a}$$

$$\frac{32M_\mathrm{P}}{b^2} \leq p_\mathrm{u} \leq \frac{48M_\mathrm{P}}{b^2} \quad \text{for a clamped square plate} \tag{4.71b}$$

It is apparent from Equation (4.71a) that the lower and upper limits coincide for a simply supported plate, whereas they differ significantly, that is, $2:3$, for a clamped plate. In this case, Fox (1974) showed that the collapse load equals $42.85M_\mathrm{P}/b^2$. In this regard, an upper limit, p_cr, of the ultimate lateral pressure load for a simply supported plate may be given as follows:

$$p_\mathrm{cr} = \frac{6t^2\sigma_\mathrm{Y}}{b^2} \frac{1}{\left(\sqrt{3+\alpha^2} - \alpha\right)^2} \tag{4.72}$$

The ultimate lateral pressure load, p_uo, should not be greater than the upper limit, p_cr. It is noted that the rigid-plastic theory formulas noted earlier do not account for the membrane effects; thus, they may predict the critical lateral pressure pessimistically as far as the assumed collapse mechanism is admissible. Interestingly, the so-called permanent deflection of the plate under lateral pressure loads may be defined as the maximum deflection at the ultimate lateral pressure.

4.9.2.2 Axial Compressive Loads

Paik and Pedersen (1996) used the rigid-plastic theory method by taking into account the large-deflection effect to derive the ultimate strength formulation of a plate under axial compressive loads. The effects of both welding induced residual stress and initial deflections with a complex shape are also considered. Figure 4.28 shows a schematic of the Paik–Pedersen method.

In this method, the initial and added deflections of the plate are assumed to be similar to Equation (4.15) as follows:

$$w_0 = A_{0i} \sin\frac{i\pi x}{a}\sin\frac{\pi y}{b} \tag{4.73a}$$

$$w = A_i \sin\frac{i\pi x}{a}\sin\frac{\pi y}{b} \tag{4.73b}$$

where A_{0i} and A_i are the initial or added deflection amplitudes with respect to the half-wave mode i, which is considered to be from 1 to $2m$ in the plate collapse strength calculations in which m is the buckling half-wave number to be taken as an integer that satisfies $a/b \leq \sqrt{m(m+1)}$.

The unknown amplitude A_i can be determined from Equation (4.19) as a function of $\sigma_{x\mathrm{av}}$, which is the average (applied) compressive stress taken as $P_x/(bt)$, where P_x is the axial compressive load in the x direction. According to the rigid-plastic theory, the following equilibrium condition between external work and internal energy associated with

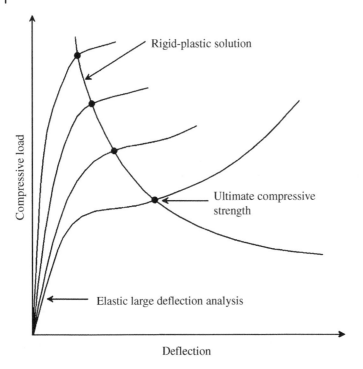

Figure 4.28 A schematic of the Paik–Pedersen method used to calculate the ultimate strength of a plate under axial compression and with a complex shape of initial deflection and welding induced residual stress.

virtual forces, stresses, and strains must be satisfied for a presumed plastic collapse mechanism (Jones 2012), namely,

$$\sigma_{xav}bt\delta u = -\sum_{n=1}^{r}\int_{L_n} N\delta U dL_n + \sum_{n=1}^{s}(M + wN)\delta\theta dL_n \tag{4.74}$$

where L_n is the length of the nth plastic hinge, M is the moment per unit length along the plastic hinge line, N is the axial force per unit length along the plastic hinge line, r is the number of inclined hinge lines, s is the number of horizontal or vertical hinge lines, U is the axial displacement along the plastic hinge line, u is the axial displacement in the x direction, w is the lateral deflection of the plate, and θ is the rotation along the plastic hinge line.

In Equation (4.74), the prefix δ denotes the virtual variable. The left and right terms of Equation (4.74) represent the external virtual-work and the internal virtual-energy dissipation, respectively. The first and second terms on the right side represent the energy contributions due to virtual-axial displacement and virtual rotation along the plastic hinge lines, respectively. In the Paik–Pedersen method, the plate is considered to have three different types of collapse mechanisms, depending on the plate aspect ratio and the deflection shape, among other factors, as shown in Figure 4.29.

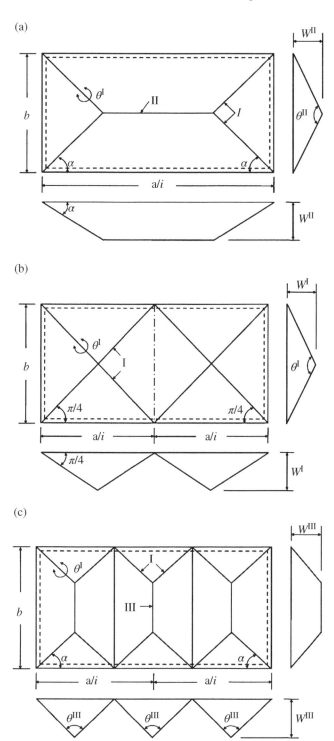

Figure 4.29 Collapse mechanisms for a plate under axial compressive loads: (a) mode I for $(a/ib) > 1$; (b) mode II for $(a/ib) = 1$; (c) mode III for $(a/ib) < 1$.

For each of the three types of presumed collapse mechanisms shown in Figure 4.29, the deflection of the plate can be determined from Equation (4.74) as follows:

a) Mode I for $\dfrac{a}{ib} > 1$

In this mode, as shown in Figure 4.29a, the virtual deflections, w, virtual rotations, $\delta\theta$, and virtual-axial displacements, δu, along plastic hinge lines I and II are determined as follows:

$$w^{\mathrm{I}} = A_i\left(1 - \frac{2\sin\alpha}{b}L_n\right), \quad w^{\mathrm{II}} = A_i, \quad \delta\theta^{\mathrm{I}} = \frac{4A_i\sin^2\alpha}{b\cos\alpha},$$

$$\delta\theta^{\mathrm{II}} = \frac{4A_i}{b}, \quad \delta U^{\mathrm{I}} = \delta u\sin\alpha, \quad \delta U^{\mathrm{II}} = 0 \tag{4.75a}$$

where α is the angle between plastic hinge lines I and II. The superscripted I and II indicate plastic hinge lines I and II.

The axial force and bending moment per unit length along the hinge lines are calculated as follows:

$$N^{\mathrm{I}} = \frac{\sigma_{xav}t}{2}(\cos 2\alpha - 1), \quad N^{\mathrm{II}} = 0,$$

$$M^{\mathrm{I}} = \frac{4(1 - p_x^2)M_{\mathrm{P}}}{\sqrt{16 - 3p_x^2(\cos 2\alpha + 1)^2 - 12p_x^2\sin^2 2\alpha}}, \quad M^{\mathrm{II}} = \frac{2(1 - p_x^2)M_{\mathrm{P}}}{\sqrt{4 - 3p_x^2}} \tag{4.75b}$$

where $p_x = \sigma_{xav}/\sigma_Y$, $M_{\mathrm{P}} = \sigma_Y t^2/4 =$ plastic moment along the plastic hinge lines, and σ_Y is the material yield stress.

The substitution of Equations (4.75a) and (4.75b) into Equation (4.74) yields the following:

$$\sigma_{xav}bt\delta u = -4\int_0^{\frac{b}{2\sin\alpha}}\left(N^{\mathrm{I}}\delta U^{\mathrm{I}} - M^{\mathrm{I}}\delta\theta^{\mathrm{I}} - w^{\mathrm{I}}N^{\mathrm{I}}\delta\theta^{\mathrm{I}}\right)dL_n$$

$$- \int_0^{\frac{a}{i} - b\cot\alpha}\left(N^{\mathrm{II}}\delta U^{\mathrm{II}} - M^{\mathrm{II}}\delta\theta^{\mathrm{II}} - w^{\mathrm{II}}N^{\mathrm{II}}\delta\theta^{\mathrm{II}}\right)dL_n \tag{4.75c}$$

$$= -\sigma_{xav}bt\delta u(\cos 2\alpha - 1) + 8A_i M^{\mathrm{I}}\tan\alpha + 4A_i M^{\mathrm{II}}\frac{1}{b}\left(\frac{a}{i} - b\cot\alpha\right)$$

$$+ 2A_i^2\sigma_{xav}t(\cos 2\alpha - 1)\tan\alpha$$

The angle α can be determined by minimizing the total potential energy, but for the sake of simplicity, $\alpha = \pi/4$ is assumed for all three collapse modes. In this case, Equation (4.75c) gives the maximum deflection, W_i, for the mode I collapse mechanism as follows:

$$W_i = \frac{A_i}{t} = \frac{1 - p_x^2}{p_x}\left[\frac{4}{\sqrt{16 - 15p_x^2}} + \frac{1}{\sqrt{4 - 3p_x^2}}\left(\frac{a}{ib} - 1\right)\right] \tag{4.75d}$$

b) Mode II for $\dfrac{a}{ib} = 1$

Similar to mode I and with $\alpha = \pi/4$, as shown in Figure 4.29b, the virtual deflections, virtual rotations, and virtual in-plane deformations along hinge line I are determined as follows:

$$w^{\mathrm{I}} = A_i\left(1 - \dfrac{\sqrt{2}L_n}{b}\right), \quad \delta\theta^{\mathrm{I}} = \dfrac{2\sqrt{2}A_i}{b}, \quad \delta U^{\mathrm{I}} = \dfrac{\sqrt{2}\delta u}{2} \tag{4.76a}$$

The axial forces and bending moments along hinge line I are determined as follows:

$$N^{\mathrm{I}} = -\dfrac{\sigma_{xav}t}{2}, \quad M^{\mathrm{I}} = \dfrac{4\left(1 - p_x^2\right)M_{\mathrm{P}}}{\sqrt{16 - 15\,p_x^2}} \tag{4.76b}$$

Substitution of Equation (4.76a) and (4.76b) into Equation (4.74) yields the following:

$$\sigma_{xav}bt\delta u = -4\int_0^{\frac{\sqrt{2}b}{2}} \left(N^{\mathrm{I}}\delta U^{\mathrm{I}} - M^{\mathrm{I}}\delta\theta^{\mathrm{I}} - w^{\mathrm{I}}N^{\mathrm{I}}\delta\theta^{\mathrm{I}}\right)dL_n \tag{4.76c}$$

$$= \sigma_{xav}bt\delta u + 8A_iM^{\mathrm{I}} - 2\sigma_{xav}tA_i^2$$

Equation (4.76c) gives the maximum deflection, W_i, for the mode II collapse mechanism as follows:

$$W_i = \dfrac{A_i}{t} = \dfrac{4\left(1 - p_x^2\right)}{p_x\sqrt{16 - 15\,p_x^2}} \tag{4.76d}$$

c) Mode III for $\dfrac{a}{ib} < 1$

In this mode, shown in Figure 4.29c, the virtual deflections, virtual rotations, and virtual in-plane deformations along the hinge lines are determined as follows:

$$w^{\mathrm{I}} = A_i\left(1 - \dfrac{2i\cos\alpha}{a}L_n\right), \quad w^{\mathrm{III}} = A_i, \quad \delta\theta^{\mathrm{I}} = \dfrac{4iA_i\cos^2\alpha}{a\sin\alpha},$$

$$\delta\theta^{\mathrm{III}} = \dfrac{4i}{a}A_i, \quad U^{\mathrm{I}} = \delta u\sin\alpha, \quad U^{\mathrm{III}} = \delta u \tag{4.77a}$$

The axial force and the bending moment along the hinge lines are determined as follows:

$$N^{\mathrm{I}} = \dfrac{\sigma_{xav}t}{2}(\cos 2\alpha - 1),$$

$$N^{\mathrm{III}} = -\sigma_{xav}t,$$

$$M^{\mathrm{I}} = \dfrac{4\left(1 - p_x^2\right)M_{\mathrm{P}}}{\sqrt{16 - 3\,p_x^2(\cos 2\alpha + 1)^2 - 12\,p_x^2\sin^2 2\alpha,}} \tag{4.77b}$$

$$M^{\mathrm{III}} = \left(1 - p_x^2\right)M_{\mathrm{P}}$$

The substitution of Equation (4.77a) and (4.77b) into Equation (4.74) yields the following:

$$
\sigma_{xav} bt\delta u = -4 \int_0^{\dfrac{a}{2i\cos\alpha}} \left(N^{\mathrm{I}}\delta U^{\mathrm{I}} - M^{\mathrm{I}}\delta\theta^{\mathrm{I}} - w^{\mathrm{I}}N^{\mathrm{I}}\delta\theta^{\mathrm{I}}\right) dL_n
$$

$$
- \int_0^{\dfrac{a}{b-\dfrac{}{i\tan\alpha}}} \left(N^{\mathrm{III}}\delta U^{\mathrm{III}} - M^{\mathrm{III}}\delta\theta^{\mathrm{III}} - w^{\mathrm{III}}N^{\mathrm{III}}\delta\theta^{\mathrm{III}}\right) dL_n
$$

$$
= \sigma_{xav} bt\delta u - \frac{\sigma_{xav} at\delta u}{i}\tan\alpha\cos 2\alpha + 8A_i M^{\mathrm{I}}\cot\alpha + 4A_i M^{\mathrm{III}}\left(\frac{ib}{a} - \tan\alpha\right)
$$

$$
+ 2A_i^2\sigma_{xav}t\left[(\cos 2\alpha - 1)\cot\alpha + 2\tan\alpha - \frac{2ib}{a}\right]
$$

(4.77c)

Equations (4.77c) gives the maximum deflection, W_i, for the mode III collapse mechanism as follows:

$$
W_i = \frac{A_i}{t} = \frac{a}{2ib-a}1 - \frac{p_x^2}{p_x}\left(\frac{4}{\sqrt{16-15\,p_x^2}} + \frac{ib}{2a} - \frac{1}{2}\right)
$$

(4.77d)

The ultimate strength of the plate is now determined as the intersection between A_i and W_i with varying values for i, which may be taken as $1-2 \times m$ (two times the buckling half-wave number), as shown in Figure 4.28. The real plate ultimate strength minimum is taken as the minimum value among the three values obtained for ultimate strength.

4.9.3 Membrane Stress-Based Method

In this method, the membrane stresses inside the plate are computed by solving the nonlinear governing differential equations of elastic large-deflection plate theory, and it is considered that the plate will collapse if the membrane stress reaches a critical value (e.g., the yield stress) or if any relevant criterion in terms of membrane stresses is satisfied.

With an increase in the plate deflection, the upper and/or lower fibers inside the middle of the plate will initially yield by the action of bending. However, as long as it is possible to redistribute the applied loads to the straight plate boundaries by membrane action, the plate will not collapse. Collapse will then occur when the most stressed boundary locations yield, because the plate can no longer keep the boundaries straight, resulting in a rapid increase of lateral plate deflection.

4.9.3.1 Ultimate Strength Conditions

Because of the nature of combined membrane axial stresses in the x and y directions, three possible locations at the edges—plate corners, longitudinal edges, and transverse edges—that could initially yield are generally considered, as shown in Figure 4.30. The stress status for the two edge locations, that is, at each longitudinal or transverse edge,

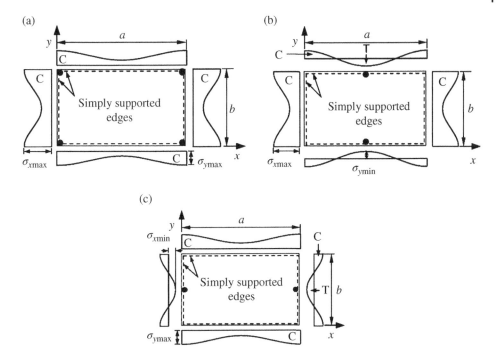

Figure 4.30 Three possible locations for the initial plastic yield at the plate edges under combined loads: (a) plasticity at corners; (b) plasticity at longitudinal edges; (c) plasticity at transverse edges (•, expected yielding locations; C, compression; T, tension).

can be expected to be the same as long as the longitudinal or transverse axial stresses are uniformly applied, that is, without in-plane bending. Depending on the predominant half-wave mode in the long direction, the location of the possible plasticity can vary at the long edges because the location of the minimum membrane stresses can differ, whereas it is always located at the mid-edges in the short direction.

The occurrence of yielding can be assessed using the von Mises yield criterion. The three resulting ultimate strength criteria for the most probable yield locations are as follows:

1) Yielding at corners:

$$\left(\frac{\sigma_{x\,max}}{\sigma_Y}\right)^2 - \left(\frac{\sigma_{x\,max}}{\sigma_Y}\right)\left(\frac{\sigma_{y\,max}}{\sigma_Y}\right) + \left(\frac{\sigma_{y\,max}}{\sigma_Y}\right)^2 + \left(\frac{\tau_{av}}{\tau_Y}\right)^2 = 1 \qquad (4.78a)$$

2) Yielding at longitudinal edges:

$$\left(\frac{\sigma_{x\,max}}{\sigma_Y}\right)^2 - \left(\frac{\sigma_{x\,max}}{\sigma_Y}\right)\left(\frac{\sigma_{y\,min}}{\sigma_Y}\right) + \left(\frac{\sigma_{y\,min}}{\sigma_Y}\right)^2 + \left(\frac{\tau_{av}}{\tau_Y}\right)^2 = 1 \qquad (4.78b)$$

3) Yielding at transverse edges:

$$\left(\frac{\sigma_{x\,min}}{\sigma_Y}\right)^2 - \left(\frac{\sigma_{x\,min}}{\sigma_Y}\right)\left(\frac{\sigma_{y\,max}}{\sigma_Y}\right) + \left(\frac{\sigma_{y\,max}}{\sigma_Y}\right)^2 + \left(\frac{\tau_{av}}{\tau_Y}\right)^2 = 1 \qquad (4.78c)$$

Although the maximum or minimum membrane stresses of a deflected plate under simple types of load applications such as uniaxial compression or combined uniaxial compression and lateral pressure loads may be calculated as described in Sections 4.4–4.7, the calculation of the maximum and minimum membrane stresses of a plate under more complex load applications such as combined biaxial loads, edge shear, and lateral pressure is not straightforward.

As an easier alternative approach, Equation (4.78) may be used to develop the plate ultimate strength formulations under simpler load applications, and a relevant combination of such strength formulations as obtained for various simpler load cases may be adopted to derive the strength formula under all potential load applications.

4.9.3.2 Lateral Pressure Loads

The ultimate strength, p_{u0}, of a plate under lateral pressure alone may be calculated as the lowest value of the three lateral pressures, as obtained by satisfying the three conditions of Equation (4.78) when $\sigma_{xav} = \sigma_{yav} = \tau_{av} = 0$.

Figure 4.31 compares the present method, denoted by ALPS/ULSAP (2017), to the corresponding mechanical collapse test results from Yamamoto et al. (1970) and the incremental Galerkin method solutions as described in Chapter 11, denoted by SPINE, for a plate with $a/b = 3$ under combined longitudinal axial compression and lateral pressure loads.

4.9.3.3 Combined Longitudinal Axial Loads and Lateral Pressure

The maximum and minimum membrane stresses are in this case calculated in terms of σ_{xav} and p together with initial imperfections. Under the present type of load application,

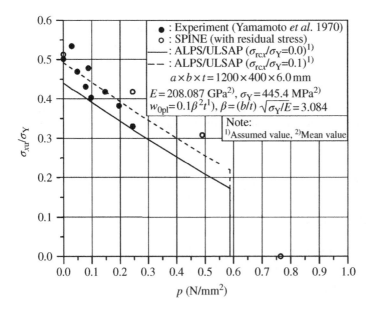

Figure 4.31 Comparison of ALPS/ULSAP, with the collapse test results of Yamamoto et al. and the incremental Galerkin method solutions, denoted by SPINE, for plating under combined longitudinal axial compression and lateral pressure loads.

the initial yield location at the plate edges may be the longitudinal edges by the nature of the von Mises yield condition, Equation (1.31c).

By substituting the maximum and minimum membrane stresses into Equation (4.78b), the ultimate longitudinal axial strength, σ_{xu}, is obtained as the solution of the following equation with regard to σ_{xav}, regarding p as a secondary constant load:

$$\left(\frac{\sigma_{x\,max}}{\sigma_Y}\right)^2 - \left(\frac{\sigma_{x\,max}}{\sigma_Y}\right)\left(\frac{\sigma_{y\,min}}{\sigma_Y}\right) + \left(\frac{\sigma_{y\,min}}{\sigma_Y}\right)^2 = 1 \tag{4.79a}$$

where σ_{xmax} and σ_{ymin} are the maximum and minimum membrane stresses in the x and y directions.

When the lateral pressure is not involved, the ultimate strength, σ_{xu}, is calculated by letting $p = 0$. It is interesting to note that when the unloaded edges move freely in plane, no membrane stresses develop in the y direction, as shown in Figure 4.25b. In this case, the ultimate strength formulation, Equation (4.79a), can be simplified to

$$\sigma_{x\,max} = \sigma_Y \tag{4.79b}$$

Alternatively, using the effective width approach, σ_{xu} is simply given by

$$\sigma_{xu} = \sigma_Y \frac{b_{eu}}{b} \tag{4.79c}$$

where b_{eu} is the effective width at the ultimate strength.

When the plate is subjected to predominantly axial tensile loads, the plate ultimate strength may approximately be taken as $\sigma_{xu} = \sigma_Y$, whereas the ultimate strength of the plate under combined longitudinal axial tension and lateral pressure loads can also be calculated from Equation (4.79a).

Figure 4.32 compares the theoretical results of Equation (4.79a), denoted by ALPS/ULSAP, with the mechanical collapse tests and the nonlinear FEA for simply supported long plates with different plate aspect ratios and under longitudinal axial compressive loads. Whereas Equation (4.79a) deals with initial imperfections as direct parameters of influence, the mechanical collapse tests involve various uncertain levels of both initial deflections and residual stresses. For more details of the test data, readers should refer to Ellinas et al. (1984). The FEA has two types of unloaded plate edge condition: (i) the unloaded plate edges move freely in plane, and (ii) they are kept straight. For the FEAs, an "average" level of initial deflection is considered, and the welding induced residual stresses are not included. The finite element method solutions with edge condition (i) are smaller than those with edge condition (ii), as would be expected.

4.9.3.4 Combined Transverse Axial Loads and Lateral Pressure

The maximum and minimum membrane stresses are in this case calculated in terms of σ_{yav} and p together with initial imperfections. For this type of load application, the initial yield location at the plate edges may be the transverse edges by the nature of the von Mises yield condition, Equation (1.31c).

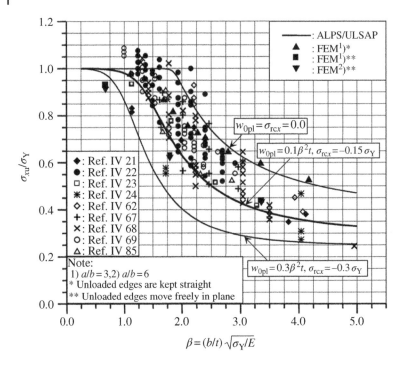

Figure 4.32 Comparison of the present method, denoted by ALPS/ULSAP, with the test database (reference numbers for test data are taken from Ellinas et al. (1984)).

By substituting the maximum and minimum membrane stresses into Equation (4.78c), the ultimate transverse axial strength, σ_{yu}, is obtained as the solution of the following equation with regard to σ_{yav}, regarding p as a secondary constant load:

$$\left(\frac{\sigma_{x\,min}}{\sigma_Y}\right)^2 - \left(\frac{\sigma_{x\,min}}{\sigma_Y}\right)\left(\frac{\sigma_{y\,max}}{\sigma_Y}\right) + \left(\frac{\sigma_{y\,max}}{\sigma_Y}\right)^2 = 1 \tag{4.80a}$$

where σ_{xmin} and σ_{ymax} are the minimum and maximum membrane stresses in the x and y directions.

When the lateral pressure is not involved, the ultimate strength, σ_{yu}, is of course calculated by letting $p = 0$. When the unloaded edges move freely in plane, no membrane stresses develop in the x direction. Therefore, Equation (4.80a) can in this case be simplified as follows:

$$\sigma_{y\,max} = \sigma_Y \tag{4.80b}$$

Alternatively, using the effective width approach, σ_{yu} may be given by

$$\sigma_{yu} = \sigma_Y \frac{a_{eu}}{a} \tag{4.80c}$$

where a_{eu} is the effective length at the ultimate strength.

When the plate is subjected to predominantly axial tensile loads, the plate ultimate strength may approximately be taken as $\sigma_{yu} = \sigma_Y$, and the ultimate strength of the plate

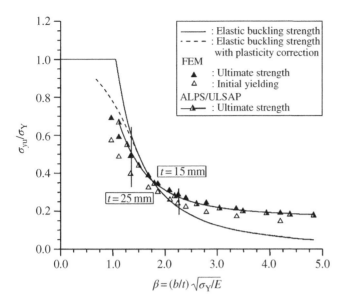

Figure 4.33 Variation of the ultimate transverse compressive strength of a long plate as a function of the reduced slenderness ratio, $a/b = 3$.

under combined transverse axial tension and lateral pressure loads can also be calculated from Equation (4.80a). Figure 4.33 compares Equation (4.80a), denoted by ALPS/ULSAP, with nonlinear finite element method solutions for simply supported plates.

4.9.3.5 Edge Shear

Because the deflection pattern of a plate is quite complex after buckling under predominantly edge shear, the analytical approach by solving the nonlinear governing differential equations may not be straightforward for evaluation of the membrane stress distribution inside the plate buckled in edge shear. In this case, a nonlinear numerical method is more convenient.

In such a case, a series of elastic–plastic large-deflection finite element method solutions for plates under edge shear alone were carried out by Paik et al. (2001) by varying the plate slenderness ratio, the plate aspect ratio, the boundary condition, and the magnitude of initial deflections where the plate edges remained straight. By curve fitting based on the computed results, the following empirical formula for the ultimate strength, τ_{u0}, of a plate under edge shear alone was derived:

$$\frac{\tau_{u0}}{\tau_Y} = \begin{cases} 1.324(\tau_E/\tau_Y) & \text{for } 0 < \tau_E/\tau_Y \leq 0.5 \\ 0.039(\tau_E/\tau_Y)^3 - 0.274(\tau_E/\tau_Y)^2 + 0.676(\tau_E/\tau_Y) + 0.388 & \text{for } 0.5 < \tau_E/\tau_Y \leq 2.0 \\ 0.956 & \text{for } \tau_E/\tau_Y > 2.0 \end{cases}$$

$$(4.81)$$

where τ_E is the elastic shear buckling stress of the plate.

Figure 4.34 shows the variation of the plate ultimate edge shear strength for simply supported plates plotted versus the elastic shear buckling stress for various plate aspect ratios.

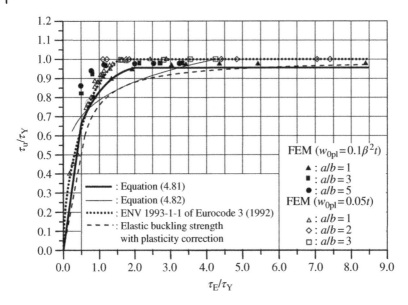

Figure 4.34 The ultimate strength versus the elastic buckling stress of the plate under edge shear.

The dotted line represents the elastic shear buckling strength with plasticity correction using the Johnson–Ostenfeld formulation method as described in Chapter 2. Equation (4.81) has been subdivided into three equations representing the ultimate edge shear strength of thin plates, medium-thickness plates, and thick plates, respectively. It is apparent from Figure 4.34 that Equation (4.81) covers a wider range of the plate slenderness ratio with reasonable accuracy. It shows a modeling error characterized by a mean bias of 0.931 and a coefficient of variation of 0.075 when compared with the nonlinear finite element method solutions for plates by varying the aspect ratios and initial deflections.

Figure 4.35 shows the effect of the plate aspect ratio on the plate ultimate shear strength. As the plate aspect ratio increases, the plate ultimate shear strength tends to decrease. However, it is apparent from Figure 4.35 that the ultimate shear strength depends weakly on the plate aspect ratio, especially for relatively thick plates.

In the treatment, the ultimate strength of a stiffened panel in edge shear is approximately taken as that of the plating between the stiffeners in edge shear. Any strength reserve due to tension field action, where a developing diagonal tension is anchored by the adjoining stiffening, is not included, so the approach is somewhat pessimistic. Also, implicit in the approach is the (usually reasonable) assumption that the stiffeners of a stiffened panel are normally designed so that they remain straight until the panel buckles in edge shear. Corrections are necessary if this is not the case, as described in Chapter 7.

Alternatively, ENV 1993-1-1 (1992) of Eurocode 3 suggests an empirical formula for the plate ultimate shear strength, as described in Chapter 7. Also, the plate ultimate edge shear strength is often predicted by the plasticity correction of the elastic shear buckling strength using the Johnson–Ostenfeld formulation method, as described in Chapter 2. Nara et al. (1988) proposed an empirical closed-form expression of the ultimate shear

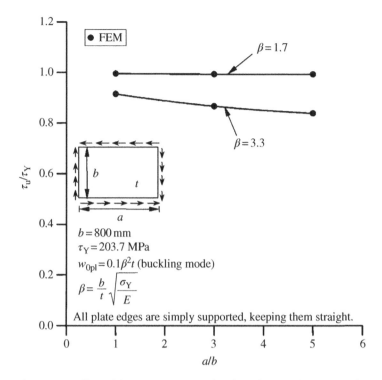

Figure 4.35 Effect of the aspect ratio on the plate ultimate shear strength.

strength of plating obtained by curve fitting based on the nonlinear finite element method solutions as follows:

$$\frac{\tau_u}{\tau_Y} = \left(\frac{0.486}{\lambda}\right)^{0.333} \leq 1.0 \ \ \text{for} \ \ 0.486 \leq \lambda \leq 2.0 \tag{4.82}$$

where $\lambda = \sqrt{\tau_Y/\tau_E}$ and τ_E is the elastic shear buckling stress. In Figure 4.34, Equation (4.82) is compared with Equation (4.81) and with more refined nonlinear finite element method solutions.

4.9.3.6 Combined Edge Shear Loads and Lateral Pressure

Although the effects of lateral pressure loads on the ultimate edge shear strength are typically neglected in most current design procedures of plated structures, the lateral pressure loads may in some cases affect (reduce) the plate ultimate edge shear strength.

Figure 4.36 shows the ultimate strength interactive relationship for a plate under combined edge shear and lateral pressure based on the incremental Galerkin method solutions described in Chapter 11, denoted by SPINE. It is apparent from Figure 4.36 that the ultimate strength interaction between edge shear and lateral pressure is significant and thus cannot be ignored.

From the limited results, it is also observed that their interacting effect tends to become moderate with an increase in the plate aspect ratio. As a pessimistic measure, the plate

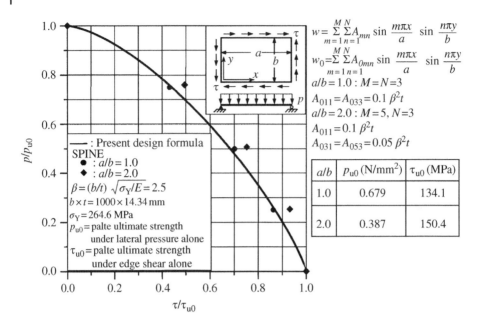

$$w = \sum_{m=1}^{M} \sum_{n=1}^{N} A_{mn} \sin \frac{m\pi x}{a} \sin \frac{n\pi y}{b}$$

$$w_0 = \sum_{m=1}^{M} \sum_{n=1}^{N} A_{0mn} \sin \frac{m\pi x}{a} \sin \frac{n\pi y}{b}$$

$a/b = 1.0 : M = N = 3$

$A_{011} = A_{033} = 0.1\,\beta^2 t$

$a/b = 2.0 : M = 5, N = 3$

$A_{011} = 0.1\,\beta^2 t$

$A_{031} = A_{053} = 0.05\,\beta^2 t$

a/b	p_{u0} (N/mm²)	τ_{u0} (MPa)
1.0	0.679	134.1
2.0	0.387	150.4

——— : Present design formula
SPINE
● : $a/b = 1.0$
◆ : $a/b = 2.0$
$\beta = (b/t)\sqrt{\sigma_Y/E} = 2.5$
$b \times t = 1000 \times 14.34$ mm
$\sigma_Y = 264.6$ MPa
$p_{u0} =$ palte ultimate strength
under lateral pressure alone
$\tau_{u0} =$ palte ultimate strength
under edge shear alone

Figure 4.36 Ultimate strength interaction relationship for a simply supported plate subjected to edge shear and lateral pressure loads.

ultimate strength interaction equation between edge shear and lateral pressure may be derived by curve fitting based on the interaction curve of square plates (i.e., with $a/b = 1$) as follows:

$$\left(\frac{\tau}{\tau_{u0}}\right)^{1.5} + \left(\frac{p}{p_{u0}}\right)^{1.2} = 1 \tag{4.83a}$$

where τ_{u0} is the plate ultimate strength under edge shear alone as defined in Equation (4.81) and p_{u0} is the plate ultimate strength under lateral pressure loads alone.

The ultimate edge shear strength, τ_u, of a plate under combined τ_{av} and p is then obtained as the solution of Equation (4.83a) with regard to τ_{av}, treating p as a secondary constant load parameter, as follows:

$$\tau_u = \tau_{u0} \left[1 - \left(\frac{p}{p_{u0}}\right)^{1.2} \right]^{1/1.5} \tag{4.83b}$$

4.9.3.7 Combined Biaxial Loads, Edge Shear Loads, and Lateral Pressure

The ultimate strength formulation under all of the load components involved can now be derived by a relevant combination. Although various types of plate ultimate strength interactive relationships with biaxial compression have been suggested in the literature, most of them may be generalized to the following form:

$$\left(\frac{\sigma_{xav}}{\sigma_{xu}}\right)^{c_1} + \alpha \left(\frac{\sigma_{xav}}{\sigma_{xu}}\right)\left(\frac{\sigma_{yav}}{\sigma_{yu}}\right) + \left(\frac{\sigma_{yav}}{\sigma_{yu}}\right)^{c_2} = 1 \tag{4.84}$$

where σ_{xu} and σ_{yu} are the ultimate strength under σ_{xav} and σ_{yav} and α, c_1, and c_2 are coefficients.

Some examples of the constants used in Equation (4.84) by different investigators are indicated in Table 4.4. Figure 4.37 plots Equation (4.84) with the various constants indicated in Table 4.4. Figure 4.38 compares the ultimate strength interaction curves using the constants of Paik et al. (2001) with nonlinear finite element method results for a simply supported plate under biaxial compression or tension, where Equation (4.84) is denoted by ALPS/ULSAP (2017).

Table 4.4 Examples of the constants used in Equation (4.84) for biaxial compressive loading.

Reference	Constants used in Equation (4.84)
BS 5400 (2000)	$c_1 = c_2 = 2$, $\alpha = 0$; both σ_{xav} and σ_{yav} are compressive
Valsgård (1980)	$c_1 = 1$, $c_2 = 2$, $\alpha = -0.25$ for $a/b = 3$; both σ_{xav} and σ_{yav} are compressive
Dier and Dowling (1980)	$c_1 = c_2 = 2$, $\alpha = 0.45$; both σ_{xav} and σ_{yav} are compressive
Stonor et al. (1983)	$c_1 = c_2 = 1.5$, $\alpha = 0$ (lower bound)
	$c_1 = c_2 = 2$, $\alpha = -1$ (upper bound)
	Both σ_{xav} and σ_{yav} are compressive
Paik et al. (2001)	$c_1 = c_2 = 2$, $\alpha = 0$; both σ_{xav} and σ_{yav} are compressive (negative)
	$c_1 = c_2 = 2$, $\alpha = -1$; either σ_{xav} or σ_{yav} or both are tensile (positive)

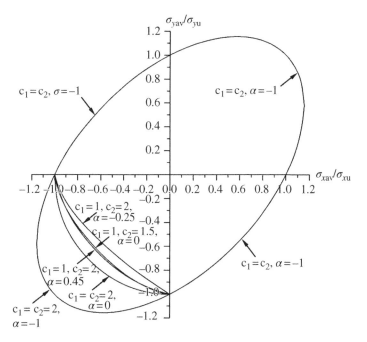

Figure 4.37 Various types of the plate ultimate strength interaction curves under biaxial loads.

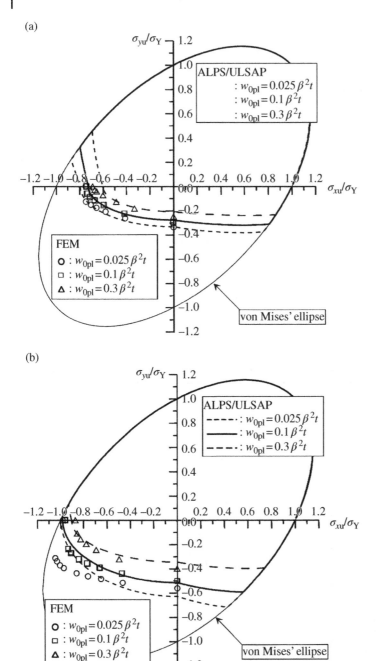

Figure 4.38 Ultimate strength interaction relationship between biaxial compression or tension for a thin plate: $a/b = 3$, $b = 1000$ mm, $E = 205.8$ GPa, $\sigma_Y = 235.2$ MPa: (a) $t = 15$ mm; (b) $t = 25$ mm.

In general, the plate elements that make up a plated structure can sometimes be subjected to axial tension in one direction, while axial compression is applied in the other direction. By the nature of the von Mises yield condition, Equation (1.31c), the biaxial compressive loading condition is not always the most critical, but in some cases the loading condition under axial tension in one direction and axial compression in the other direction could be more important. This implies that the plate ultimate strength interactive relationships should in principle be established by considering any possible combination of axial loads (tensile or compressive) together with edge shear loads.

Based on the insights developed by a series of nonlinear numerical solutions in which the loading ratio and the plate aspect ratio vary, for instance, the following ultimate strength interaction relationship between biaxial compression or tension, edge shear, and lateral pressure may be proposed:

$$\left(\frac{\sigma_{xav}}{\sigma_{xu}}\right)^{c_1} + \alpha\left(\frac{\sigma_{xav}}{\sigma_{xu}}\right)\left(\frac{\sigma_{yav}}{\sigma_{yu}}\right) + \left(\frac{\sigma_{yav}}{\sigma_{yu}}\right)^{c_2} + \left(\frac{\tau_{av}}{\tau_{u}}\right)^{c_3} = 1 \tag{4.85}$$

where σ_{xu}, σ_{yu}, and τ_{u} are the ultimate strength under σ_{xav}, σ_{yav}, and τ_{av} by taking into account the effect of lateral pressure loads. The coefficients of Equation (4.85) may be taken as $c_1 = c_2 = c_3 = 2$, whereas $\alpha = 0$ when both σ_{xav} and σ_{yav} are compressive (negative) and $\alpha = -1$ when either σ_{xav} or σ_{yav}, or both, is tensile (positive).

Figure 4.39 shows the ultimate strength interaction curves of a plate under combined longitudinal compression and edge shear by comparison with Equation (4.85) and more refined method solutions, where SPINE indicates the solutions of the incremental Galerkin method described in Chapter 11.

4.10 Effect of Opening

An opening in a plate can reduce its ultimate strength as well as buckling strength. As described in Chapter 3, it is cautioned that the Johnson–Ostenfeld formulation method is inadequate to predict the "critical" bucking strength of a perforated plate that is regarded as the maximum load-carrying capacity, because it may overestimate the strength for relatively thick plates with opening. The ultimate strength is a better basis to evaluate the load-carrying capacity of perforated plates.

This section describes empirical formulations to predict the ultimate strength of plates with a centrally located opening as shown in Figure 3.17, where the length of hole in the x direction is denoted by a_c and the breadth of hole in the y direction is denoted by b_c. The elastic buckling strength of perforated plates is described in Chapter 3. For detailed description on the ultimate strength of perforated plate panels, Chapter 10 is referred to. Interested readers may also refer to Narayanan and der Avanessian (1984), Brown et al. (1987), Paik (2007a, 2007b, 2008b), Kim et al. (2009), Suneel Kumar et al. (2009), and Wang et al. (2009b), among others.

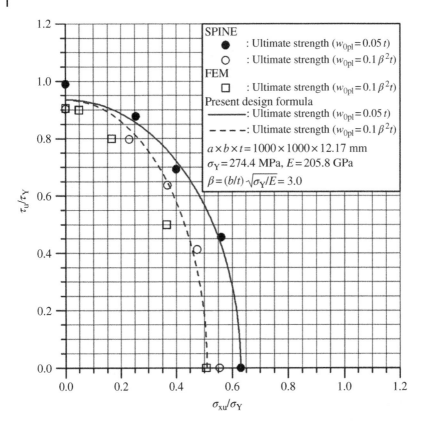

Figure 4.39 Ultimate strength interaction relationship between longitudinal compression and edge shear for a square plate.

4.10.1 Single Types of Loads

The ultimate strength of a plate with an opening can be predicted using the strength reduction factors:

$$\sigma_{xu} = R_{xu}\sigma_{xu0} \tag{4.86a}$$

$$\sigma_{yu} = R_{yu}\sigma_{yu0} \tag{4.86b}$$

$$\tau_u = R_{\tau u}\tau_{u0} \tag{4.86c}$$

where σ_{xu}, σ_{yu}, and τ_u are the ultimate strength of a perforated plate; σ_{xu0}, σ_{yu0}, and τ_{u0} are the ultimate strength of a plate without opening; and R_{xu}, R_{yu} and $R_{\tau u}$ are the strength reduction factors.

The ultimate strength reduction factors in Equation (4.86) are defined as follows:

$$R_{xu} = c_1\left(\frac{b_c}{b}\right)^2 + c_2\frac{b_c}{b} + 1.0 \tag{4.87a}$$

$$R_{yu} = c_3\left(\frac{a_c}{b}\right)^2 + c_4\frac{a_c}{b} + 1.0 \tag{4.87b}$$

$$R_{\tau u} = c_5 \left(\frac{d_c}{b}\right)^2 + c_6 \frac{d_c}{b} + 1.0 \tag{4.87c}$$

where

$$c_1 = -0.700, \; c_2 = -0.365,$$

$$c_3 = \begin{cases} -0.177(a/b)^2 + 1.088a/b - 1.671 & \text{for } 1 \le a/b \le 3 \\ 0.0 & \text{for } 3 \le a/b \le 6 \end{cases},$$

$$c_4 = \begin{cases} -0.048(a/b)^2 + 0.252a/b - 0.386 & \text{for } 1 \le a/b \le 3 \\ -0.062 & \text{for } 3 \le a/b \le 6 \end{cases},$$

$$c_5 = -0.009 \left(\frac{a}{b}\right)^2 - 0.068 \left(\frac{a}{b}\right) - 0.415,$$

$$c_6 = -0.025 \left(\frac{a}{b}\right)^2 + 0.309 \left(\frac{a}{b}\right) - 0.787,$$

$$d_c = \frac{a_c + b_c}{2}.$$

In the previous equations, both elliptical and rectangular holes are approximated as circular holes, but the diameter of the circular hole is represented by the breadth of the hole in the transverse direction, b_c, for longitudinal compressive loads, by the length of the hole in the longitudinal direction, a_c, for transverse compressive loads, or by an average of the hole size, $d_c = (a_c + b_c)/2$, for edge shear. Figure 4.40 confirms the accuracy of Equation (4.86) with Equation (4.87) by comparison with nonlinear finite element method solutions for a simply supported plate with an opening (Paik 2008b).

4.10.2 Biaxial Compression

Figure 4.41 shows a perforated plate where a circular hole with a diameter of d_c is located at the center of the plate. It is considered that the plate has an average level of initial deflection $w_{0pl} = 0.1\beta^2 t$ in the plate buckling mode that can routinely occur during welding fabrication of stiffened plate structures. Figure 4.42 shows a sample of the finite element mesh modeling for a perforated plate with an aspect ratio of $a/b = 3$. Figure 4.43 presents the ultimate strength behavior of perforated plates under biaxial compression (Paik 2008b). From the computed results, the ultimate strength interaction relation of a perforated plate may be represented by the following equation:

$$\left(\frac{\sigma_{xav}}{\sigma_{xu}}\right)^{c_1} + \left(\frac{\sigma_{yav}}{\sigma_{yu}}\right)^{c_2} = 1 \tag{4.88}$$

where σ_{xu} and σ_{yu} are the ultimate strengths of a perforated plate under σ_{xav} or σ_{yav} alone that can be determined from Equation (4.86a) or (4.86b) and c_1 and c_2 are constants that may be taken as $c_1 = 1$ and $c_2 = 7$. In Figure 4.43, Equation (4.88) with $c_1 = c_2 = 2$, which are constants well adopted for plates without opening, is also compared. The ultimate strength interaction relation of a perforated plate under biaxial compression differs from that of a plate without opening.

(a)

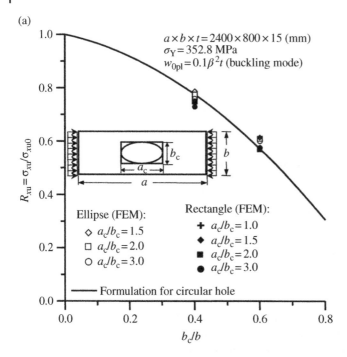

(b)

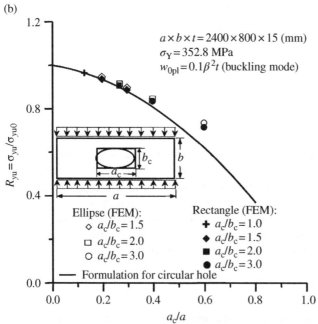

Figure 4.40 Ultimate strength of a perforated plate: (a) under longitudinal compression; (b) under transverse compression; (c) under edge shear.

(c)

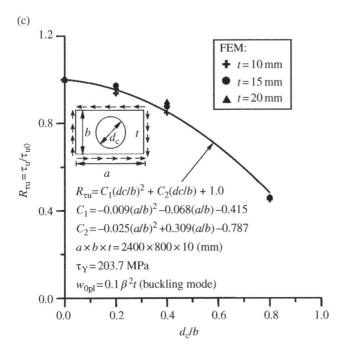

$R_{\tau u} = C_1(dc/b)^2 + C_2(dc/b) + 1.0$

$C_1 = -0.009(a/b)^2 - 0.068(a/b) - 0.415$

$C_2 = -0.025(a/b)^2 + 0.309(a/b) - 0.787$

$a \times b \times t = 2400 \times 800 \times 10$ (mm)

$\tau_Y = 203.7$ MPa

$w_{0pl} = 0.1\,\beta^2 t$ (buckling mode)

Figure 4.40 (Continued)

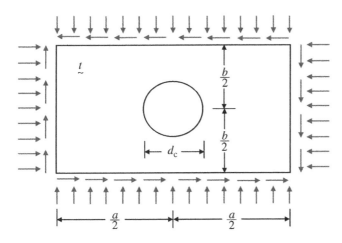

Figure 4.41 A plate with a centrally located circular hole under combined biaxial compression and edge shear loads.

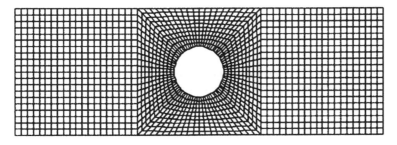

Figure 4.42 A sample of the finite element mesh modeling for a plate with a centrally located circular hole, $a/b = 3$.

(a)

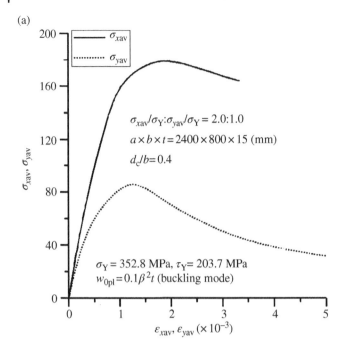

(b)

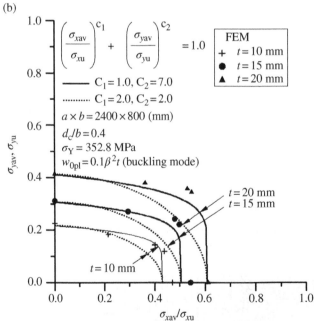

Figure 4.43 (a) The ultimate strength behavior of a perforated plate under biaxial compression; (b) the ultimate strength interaction relation of a perforated plate under biaxial compression, obtained by nonlinear FEA for $a/b = 3$, $\beta = 2.2$, $d_c/b = 0.4$.

4.10.3 Combined Longitudinal Compression and Edge Shear

Figure 4.44 shows the ultimate strength behavior of a plate with a centrally located circular hole under longitudinal compression edge shear (Paik 2008b). It is apparent from this figure that the ultimate strength interaction relation of a perforated plate under combined longitudinal compression and edge shear can be represented as the following equation:

$$\left(\frac{\sigma_{xav}}{\sigma_{xu}}\right)^2 + \left(\frac{\tau_{av}}{\tau_u}\right)^2 = 1 \tag{4.89}$$

where σ_{xu} and τ_u are the ultimate strength of a perforated plate under σ_{xav} or τ_{av} alone that can be determined from Equation (4.86a) or (4.86c).

4.10.4 Combined Transverse Compression and Edge Shear

Figure 4.45 shows the ultimate strength behavior of a plate with a centrally located circular hole under transverse compression edge shear (Paik 2008b). It is apparent from this figure that the ultimate strength interaction relation of a perforated plate under combined transverse compression and edge shear can be represented as the following equation:

$$\left(\frac{\sigma_{yav}}{\sigma_{yu}}\right)^2 + \left(\frac{\tau_{av}}{\tau_u}\right)^2 = 1 \tag{4.90}$$

where σ_{yu} and τ_u are the ultimate strength of a perforated plate under σ_{xav} or τ_{av} alone that can be determined from Equation (4.86b) or (4.86c).

4.11 Effect of Age Related Structural Deterioration

Two primary parameters of age related structural degradation are corrosion and fatigue cracks (Paik & Melchers 2008). As corrosion damage or fatigue cracking can reduce the plate ultimate strength, the evaluation of the capacity associated with Equation (1.17) needs to account for age related damage as a parameter of influence.

4.11.1 Corrosion Damage

For general (uniform) corrosion that uniformly reduces the plate thickness, the plate ultimate strength calculations can be carried out by excluding the thickness lost to corrosion. For localized corrosion, such as pitting or grooving, the strength calculation procedure can be more complex (Paik et al. 2003a, 2004), but for a simplified pessimistic treatment, the corroded plates may also be idealized using an equivalent general corrosion. The ultimate strength of plates with corrosion damage may be determined from Equation (4.86), but the strength reduction factors can be defined as follows:

$$R_{xu} = \frac{A_{xo} - A_{xw}}{A_{xo}} \tag{4.91a}$$

(a)

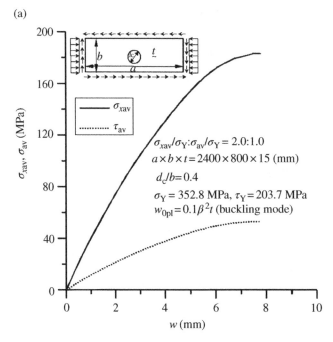

(b)

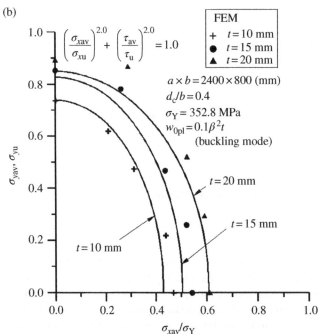

Figure 4.44 (a) The ultimate strength behavior of a plate with a centrally located circular hole under combined longitudinal compression and edge shear; (b) the ultimate strength interaction relation of a perforated plate under combined longitudinal compression and edge shear, as obtained by nonlinear FEA for $a/b = 3$, $\beta = 2.2$, $d_c/b = 0.4$.

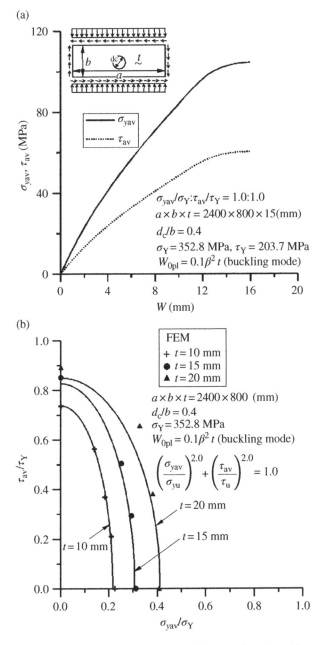

Figure 4.45 (a) The ultimate strength behavior of a plate with a centrally located circular hole under combined transverse compression and edge shear; (b) the ultimate strength interaction relation of a perforated plate under combined transverse compression and edge shear, as obtained by nonlinear FEA for $a/b = 3$, $\beta = 2.2$, $d_c/b = 0.4$.

$$R_{yu} = \frac{A_{yo} - A_{yw}}{A_{yo}} \tag{4.91b}$$

$$R_{ru} = \begin{cases} 1.0 & \text{for } \alpha \le 1.0 \\ 1.0 - 0.18\ell n(\alpha) & \text{for } \alpha > 1.0 \end{cases} \tag{4.91c}$$

where A_{xw} and A_{yw} are the total cross-sectional areas associated with all pits (corrosion wear) at the cross section of the largest number of pits in the x or y direction of the plate; A_{xo} and A_{yo} are the total cross-sectional areas of the original (intact) plate without pits in the x or y direction; $\alpha = \left[\sum_{i=1}^{n} V_{pi} \times 100 \right] / (abt)(\%)$ is the volumetric degree of pits; V_{pi} is the volume of the ith pit, which may be determined as $V_{pi} = \pi d_{wi} d_{wi}^2 / 4$; d_{di} is the depth of the ith pit; d_{wi} is the diameter of the ith pit; and n is the total number of pits.

4.11.2 Fatigue Cracking Damage

As cracking damage can reduce the ultimate strength of plated structures, the capacity associated with Equation (1.17) should be evaluated by accounting for the effect of cracking damage. The ultimate strength of plates with cracking damage may also be predicted from Equation (4.86), but the strength reduction factors are in this case given, as described in Section 9.7, as follows:

$$R_{xu} = \frac{A_{xo} - A_{xc}}{A_{xo}} \tag{4.92a}$$

$$R_{yu} = \frac{A_{yo} - A_{yc}}{A_{yo}} \tag{4.92b}$$

$$R_{ru} = \frac{1}{2} \left(\frac{A_{xo} - A_{xc}}{A_{xo}} + \frac{A_{yo} - A_{yc}}{A_{yo}} \right) \tag{4.92c}$$

where A_{xc} and A_{yc} are the cross-sectional areas associated with crack damage, projected to the x or y direction, and A_{xo} and A_{yo} are the cross-sectional areas of the intact (uncracked) plate in the x or y direction.

For detailed description on the ultimate strength of cracked plates, Chapter 9 is referred to. Interested readers may also refer to Paik et al. (2005), Paik (2008c, 2008d, 2009) and Wang et al. (2009a, 2015), Shi and Wang (2012), Rahbar-Ranji and Zarookian (2014), Underwood et al. (2015), Cui et al. (2016, 2017), and Shi et al. (2017), among others.

4.12 Effect of Local Denting Damage

As local denting damage can reduce the ultimate strength of plated structures, the capacity associated with Equation (1.17) should be evaluated by accounting for the effect of local denting damage. The ultimate strength of plates with local denting damage as shown in Figure 10.33 may also be predicted from Equation (4.86), but the strength reduction factors are in this case given by (Paik et al. 2003b, Paik 2005).

$$R_{xu} = \left[c_1 \ell n \left(\frac{D_d}{t} \right) + c_2 \right] c_3 \tag{4.93a}$$

$$R_{yu} = \left[c_4 \ell n \left(\frac{D_d}{t} \right) + c_5 \right] c_6 \tag{4.93b}$$

$$R_{\tau u} = \begin{cases} \left[1.0 + c_7 \left(\frac{D_d}{t} \right)^2 - c_8 \frac{D_d}{t} \right] & \text{for } 1 < \frac{D_d}{t} \le 10 \\ 1.0 + 100c_7 - 10c_8 & \text{for } \frac{D_d}{t} > 10 \end{cases} \tag{4.93c}$$

where

$$c_1 = -0.042 \left(\frac{d_d}{b} \right)^2 - 0.105 \frac{d_d}{b} + 0.015,$$

$$c_2 = -0.138 \left(\frac{d_d}{b} \right)^2 - 0.302 \frac{d_d}{b} + 1.042,$$

$$c_3 = \begin{cases} -1.44 \left(\frac{h}{b} \right)^2 + 1.74 \frac{h}{b} + 0.49 & \text{for } h \le \frac{b}{2} \\ -1.44 \left(\frac{b-h}{b} \right)^2 + 1.74 \frac{b-h}{b} + 0.49 & \text{for } h > \frac{b}{2} \end{cases},$$

$$c_4 = -0.042 \left(\frac{d_d}{a} \right)^2 - 0.105 \frac{d_d}{a} + 0.015,$$

$$c_5 = -0.138 \left(\frac{d_d}{a} \right)^2 - 0.302 \frac{d_d}{a} + 1.042,$$

$$c_6 = \begin{cases} -1.44 \left(\frac{s}{a} \right)^2 + 1.74 \frac{s}{a} + 0.49 & \text{for } s \le \frac{a}{2} \\ -1.44 \left(\frac{a-s}{a} \right)^2 + 1.74 \frac{a-s}{a} + 0.49 & \text{for } s > \frac{a}{2} \end{cases},$$

$$c_7 = 0.0129 \left(\frac{d_d}{b} \right)^{0.26} - 0.0076,$$

$$c_8 = 0.1888 \left(\frac{d_d}{b} \right)^{0.49} - 0.07.$$

D_d, d_d, h, and s are defined in Figure 10.33.

For detailed description on the ultimate strength of dented plates, Chapter 10 is referred to. Interested readers may also refer to Saad-Eldeen et al. (2015, 2016), Raviprakash et al. (2012), Xu and Guedes Soares (2013, 2015), and Li et al. (2014, 2015), among others.

4.13 Average Stress–Average Strain Relationship of Plates

In this section, the relationship between the average stress and the average strain of a plate element with initial imperfections is analytically derived.

4.13.1 Pre-buckling or Undeflected Regime

In the linear elastic regime without lateral deflection, the relationship between average stresses and average strains in a plane stress state is represented as described in Section 3.5:

$$\left.\begin{array}{l} \varepsilon_{xav} = \dfrac{1}{E}\sigma_{xav} - \dfrac{\nu}{E}\sigma_{yav} \\[2mm] \varepsilon_{yav} = -\dfrac{\nu}{E}\sigma_{xav} + \dfrac{1}{E}\sigma_{yav} \\[2mm] \gamma_{av} = \dfrac{1}{G}\tau_{av} \end{array}\right\} \tag{4.94a}$$

where ε_{xav}, ε_{yav}, and γ_{av} are the average strain components corresponding to σ_{xav}, σ_{yav}, and τ_{av}, respectively. Equation (4.94a) can be rewritten in the matrix form as follows:

$$\left\{\begin{array}{c} \sigma_{xav} \\ \sigma_{yav} \\ \tau_{av} \end{array}\right\} = \left[D_p\right]^E \left\{\begin{array}{c} \varepsilon_{xav} \\ \varepsilon_{yav} \\ \gamma_{av} \end{array}\right\} \tag{4.94b}$$

where

$$\left[D_p\right]^E = \frac{E}{1-\nu^2}\begin{bmatrix} 1 & \nu & 0 \\ \nu & 1 & 0 \\ 0 & 0 & (1-\nu)/2 \end{bmatrix}.$$

4.13.2 Post-buckling or Deflected Regime

For an imperfect plate element under combined biaxial and lateral pressure loads, the average stress–strain relationship can be derived as long as the unloaded edges remain straight:

$$\varepsilon_{xav} = \frac{1}{E}\sigma_{xmax} = \frac{1}{E}\frac{b}{b_e}\sigma_{xav} \quad \text{or} \quad \sigma_{xav} = \frac{b_e}{b}E\varepsilon_{xav} \tag{4.95a}$$

$$\varepsilon_{yav} = \frac{1}{E}\sigma_{y\,max} = \frac{1}{E}\frac{a}{a_e}\sigma_{yav} \quad \text{or} \quad \sigma_{yav} = \frac{a_e}{a}E\varepsilon_{yav} \tag{4.95b}$$

The incremental form of Equation (4.95) is given by

$$\Delta\varepsilon_{xav} = \frac{1}{E}\left(\frac{\partial\sigma_{xmax}}{\partial\sigma_{xav}}\right)\Delta\sigma_{xav} \quad \text{or} \quad \Delta\sigma_{xav} = \left(\partial\frac{\sigma_{xmax}}{\partial\sigma_{xav}}\right)^{-1} E\Delta\varepsilon_{xav} \tag{4.96a}$$

$$\Delta\varepsilon_{yav} = \frac{1}{E}\left(\frac{\partial\sigma_{y\,max}}{\partial\sigma_{yav}}\right)\Delta\sigma_{yav} \quad \text{or} \quad \Delta\sigma_{yav} = \left(\frac{\partial\sigma_{y\,max}}{\partial\sigma_{yav}}\right)^{-1} E\Delta\varepsilon_{yav} \tag{4.96b}$$

where the prefix, Δ, represents the increment of the variable (throughout this chapter). The numerical approach is often more pertinent for the computation of $\partial\sigma_{xmax}/\partial\sigma_{xav}$ with infinitesimal stress variations around σ_{xav} or for the computation of $\partial\sigma_{ymax}/\partial\sigma_{yav}$ with infinitesimal stress variations around σ_{yav}.

For a simply supported plate without both initial imperfections and lateral pressure loads and under uniaxial loads in each direction, Equations (4.95a) and (4.95b) can be simplified to

$$\varepsilon_{xav} = \frac{1}{E}(a_1\sigma_{xav} + a_2) \quad \text{or} \quad \sigma_{xav} = \frac{1}{a_1}(E\varepsilon_{xav} - a_2) \tag{4.97a}$$

$$\varepsilon_{yav} = \frac{1}{E}(g_1\sigma_{yav} + g_2) \quad \text{or} \quad \sigma_{yav} = \frac{1}{g_1}(E\varepsilon_{yav} - g_2) \tag{4.97b}$$

The incremental forms of Equations (4.97a) and (4.97b) are then given by

$$\Delta\sigma_{xav} = \frac{E}{a_1}\Delta\varepsilon_{xav} \tag{4.98a}$$

$$\Delta\sigma_{yav} = \frac{E}{g_1}\Delta\varepsilon_{yav} \tag{4.98b}$$

In Equation (4.98), E/a_1 or E/g_1 is the effective Young's modulus (tangent modulus) after buckling for a perfectly flat plate under uniaxial compressive loads in the x or y direction, namely,

$$E_x = \frac{E}{a_1} = E / \left(1 + \rho\frac{2m^4}{m^4 + a^4/b^4}\right) \tag{4.99a}$$

$$E_y = \frac{E}{a_1} = E / \left(1 + \rho\frac{2n^4}{n^4 + b^4/a^4}\right) \tag{4.99b}$$

where Equation (4.99a) or (4.99b) represents that the tangent modulus of the buckled plate does not change with the applied loads, as it is a function of the plate aspect ratio. Figure 4.46 shows the relationship between average stress and average strain for a simply supported plate without initial imperfections under uniaxial compressive loads in the x direction. Figure 4.47 shows the variation of the tangent modulus of the buckled plate as a function of the plate aspect ratio where $E^* = E_x$. It is interesting to note that the effective tangent modulus varies in a cyclic pattern with regard to a mean equal to $E^*/E = 0.5$, and for a shorter plate, the effect of the aspect ratio is more significant.

The membrane strain components of a deflected or buckled plate element under combined biaxial loads, edge shear, and lateral pressure can now be given by

$$\varepsilon_{xav} = \frac{1}{E}(\sigma_{x\,max} - v\sigma_{yav}) \tag{4.100a}$$

$$\varepsilon_{yav} = \frac{1}{E}(-v\sigma_{xav} + \sigma_{y\,max}) \tag{4.100b}$$

$$\gamma_{av} = \frac{\tau_{av}}{G_e} \tag{4.100c}$$

where σ_{xmax} and σ_{ymax} are the maximum membrane stresses in the x or y direction described in Sections 4.4–4.7 and G_e is the effective shear modulus described in Section 4.8.3.

Because σ_{xmax}, σ_{ymax}, and G_e are nonlinear functions with regard to the corresponding average stress components, Equation (4.100) indicates a set of nonlinear relationships between membrane stresses and strains. The incremental form of the membrane

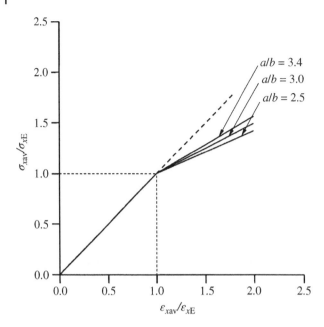

Figure 4.46 The average stress–strain curves of a perfect plate under uniaxial compression in the elastic regime (ε_{xE} = average axial compressive strain at $\sigma_{xav} = \sigma_{xE}$).

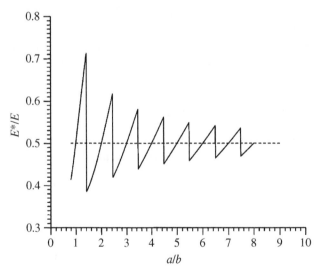

Figure 4.47 Variation of the reduced tangent modulus for a perfect plate after buckling as a function of the plate aspect ratio.

stress–strain relationship is relevant by differentiating Equation (4.100) with regard to the corresponding average stress components as follows:

$$\Delta \varepsilon_{xav} = \frac{1}{E} \left[\frac{\partial \sigma_{x\,max}}{\partial \sigma_{xav}} \Delta \sigma_{xav} + \left(\frac{\partial \sigma_{x\,max}}{\partial \sigma_{yav}} - v \right) \Delta \sigma_{yav} \right] \tag{4.101a}$$

$$\Delta \varepsilon_{yav} = \frac{1}{E} \left[\left(\frac{\partial \sigma_{y\,max}}{\partial \sigma_{xav}} - v \right) \Delta \sigma_{xav} + \frac{\partial \sigma_{y\,max}}{\partial \sigma_{yav}} \Delta \sigma_{yav} \right] \tag{4.101b}$$

$$\Delta \gamma_{av} = \frac{1}{G_e} \left(1 - \frac{\tau_{av}}{G_e} \frac{\partial G_e}{\partial \tau_{av}} \right) \Delta \tau_{av} \tag{4.101c}$$

The differentiation of maximum membrane stresses may often be carried out numerically with infinitesimal stress variations around the corresponding average stress. Equation (4.101) can then be rewritten in the matrix form as follows:

$$\left\{ \begin{array}{c} \Delta \sigma_{xav} \\ \Delta \sigma_{yav} \\ \Delta \tau_{av} \end{array} \right\} = [D_p]^B \left\{ \begin{array}{c} \Delta \varepsilon_{xav} \\ \Delta \varepsilon_{yav} \\ \Delta \gamma_{av} \end{array} \right\} \tag{4.102}$$

where

$$[D_p]^B = \frac{1}{A_1 B_2 - A_2 B_1} \begin{bmatrix} B_2 & -A_2 & 0 \\ -B_1 & A_1 & 0 \\ 0 & 0 & 1/C_1 \end{bmatrix}$$

is the stress–strain matrix of the plate in the post-buckling or deflected regime, with

$$A_1 = \frac{1}{E} \frac{\partial \sigma_{x\,max}}{\partial \sigma_{xav}}, \quad A_2 = \frac{1}{E} \left(\frac{\partial \sigma_{x\,max}}{\partial \sigma_{yav}} - v \right),$$

$$B_1 = \frac{1}{E} \left(\frac{\partial \sigma_{y\,max}}{\partial \sigma_{xav}} - v \right), \quad B_2 = \frac{1}{E} \frac{\partial \sigma_{y\,max}}{\partial \sigma_{yav}},$$

$$C_1 = \frac{1}{G_e} \left(1 - \frac{\tau_{av}}{G_e} \frac{\partial G_e}{\partial \tau_{av}} \right).$$

When no deflections have occurred or under biaxial tensile loading, the differentiations in Equation (4.102) are simplified to

$$\frac{\partial \sigma_{x\,max}}{\partial \sigma_{xav}} = \frac{\partial \sigma_{y\,max}}{\partial \sigma_{yav}} = 1 \quad \text{and} \quad \frac{\partial \sigma_{x\,max}}{\partial \sigma_{yav}} = \frac{\partial \sigma_{y\,max}}{\partial \sigma_{xav}} = \frac{\partial G_e}{\partial \tau_{av}} = 0 \tag{4.103}$$

In this case, Equation (4.103) becomes Equation (4.94b) in the linear elastic regime.

4.13.3 Post-ultimate Strength Regime

In the post-ultimate strength regime, the internal stress decreases as long as the axial compressive displacements continue to increase. In this case, the average membrane stress components may be calculated in terms of the plate effective width or length as follows:

$$\sigma_{xav} = \frac{b_e}{b}\sigma^u_{x\,max} \tag{4.104a}$$

$$\sigma_{yav} = \frac{a_e}{a}\sigma^u_{y\,max} \tag{4.104b}$$

where $\sigma^u_{x\,max}$ and $\sigma^u_{y\,max}$ are the maximum membrane stresses of the plate in the x or y direction, immediately after the ultimate strength is reached, that is, $\sigma^u_{x\,max} = \sigma_{x\,max}$ at $\sigma_{xav} = \sigma_{xu}$ and $\sigma^u_{y\,max} = \sigma_{y\,max}$ at $\sigma_{yav} = \sigma_{yu}$.

The effective width or length of the plate in the post-ultimate strength regime may be defined as follows:

$$\frac{b_e}{b} = \frac{\sigma^*_{xav}}{\sigma^*_{xmax}} \tag{4.105a}$$

$$\frac{a_e}{a} = \frac{\sigma^*_{yav}}{\sigma^*_{ymax}} \tag{4.105b}$$

where the asterisk represents a value of the plate in the post-ultimate strength regime. For a plate without initial imperfections, σ^*_{xmax} or σ^*_{ymax} can be obtained in a simpler form as follows:

$$\sigma^*_{xmax} = E\varepsilon_{xav} = 2\sigma^*_{xav} - \sigma_{xE} \tag{4.106a}$$
$$\sigma^*_{ymax} = E\varepsilon_{yav} = 2\sigma^*_{yav} - \sigma_{yE} \tag{4.106b}$$

where σ_{xE} and σ_{yE} are the elastic compressive buckling stresses in the x or y direction.

By substituting Equation (4.106) into Equation (4.105), the plate effective width or length can be expressed in terms of strain components as follows:

$$\frac{b_e}{b} = \frac{1}{2}\left(1 + \frac{\sigma_{xE}}{E\varepsilon_{xav}}\right) \tag{4.107a}$$

$$\frac{a_e}{a} = \frac{1}{2}\left(1 + \frac{\sigma_{yE}}{E\varepsilon_{yav}}\right) \tag{4.107b}$$

When the effects of initial imperfections are neglected in the post-ultimate strength regime, the average stress–average strain relationship can be derived by substituting Equation (4.107) into Equation (4.104) as follows:

$$\sigma_{xav} = \frac{1}{2}\left(1 + \frac{\sigma_{xE}}{E\varepsilon_{xav}}\right)\sigma^u_{x\,max} \tag{4.108a}$$

$$\sigma_{yav} = \frac{1}{2}\left(1 + \frac{\sigma_{yE}}{E\varepsilon_{yav}}\right)\sigma^u_{y\,max} \tag{4.108b}$$

The incremental form of Equations (4.108) is then given by

$$\Delta\sigma_{xav} = -\frac{\sigma^u_{x\,max}}{2}\frac{\sigma_{xE}}{E\,\varepsilon^2_{xav}}\Delta\varepsilon_{xav} \tag{4.109a}$$

$$\Delta\sigma_{yav} = -\frac{\sigma^{u}_{y\,max}}{2}\frac{\sigma_{yE}}{E\,\varepsilon^{2}_{yav}}\Delta\varepsilon_{yav} \tag{4.109b}$$

In contrast, the average shear stress–average shear strain relationship in the post-ultimate strength regime is given by

$$\Delta\tau_{av} = G^{*}_{e}\Delta\gamma_{av} \tag{4.109c}$$

where G^{*}_{e} is the tangent shear modulus in the post-ultimate strength regime, which is often assumed to be $G^{*}_{e}=0$ when the unloading behavior due to shear is not very significant.

In combined load cases, the average stress–average strain relationship of the plate in the post-ultimate strength regime is therefore given from the combination of all stress–strain relationships derived earlier as follows:

$$\begin{Bmatrix} \Delta\sigma_{xav} \\ \Delta\sigma_{yav} \\ \Delta\tau_{av} \end{Bmatrix} = \begin{bmatrix} D_{p} \end{bmatrix}^{U} \begin{Bmatrix} \Delta\varepsilon_{xav} \\ \Delta\varepsilon_{yav} \\ \Delta\gamma_{av} \end{Bmatrix} \tag{4.110}$$

where

$$\begin{bmatrix} D_{p} \end{bmatrix}^{U} = \begin{bmatrix} A_{1} & 0 & 0 \\ 0 & A_{2} & 0 \\ 0 & 0 & A_{3} \end{bmatrix}$$

is the stress–strain matrix of the plate in the post-ultimate strength regime, with

$$A_{1} = -\frac{\sigma^{u}_{x\,max}}{2}\frac{\sigma_{xE}}{E\,\varepsilon^{2}_{xav}}, \quad A_{2} = -\frac{\sigma^{u}_{y\,max}}{2}\frac{\sigma_{yE}}{E\,\varepsilon^{2}_{yav}}, \quad A_{3} = G^{*}_{e}.$$

References

ALPS/ULSAP (2017). *A computer program for the ultimate strength analysis of plates and stiffened panels*. MAESTRO Marine LLC, Stevensville, MD.

Brown, C.J., Yettram, A.L. & Burnett, M. (1987). Stability of plates with rectangular holes. *Journal of Structural Engineering*, **113**(5): 1111–1116.

BS 5400 (2000). *Steel, concrete and composite bridges. Part 3 code of practice for design of steel bridges*. British Standards Institution, London.

Cui, C., Yang, P., Li, C. & Xia, T. (2017). Ultimate strength characteristics of cracked stiffened plates subjected to uniaxial compression. *Thin-Walled Structures*, **113**: 27–39.

Cui, C., Yang, P., Xia, T. & Du, J. (2016). Assessment of residual ultimate strength of cracked steel plates under longitudinal compression. *Ocean Engineering*, **121**: 174–183.

Dier, A.F. & Dowling, P.J. (1980). *Strength of ship's plating—plates under combined lateral loading and biaxial compression*. CESLIC Report **SP8**, Imperial College, London.

Ellinas, C.P., Supple, W.J. & Walker, A.C. (1984). *Buckling of offshore structures: a state-of-the-art review*. Gulf Publishing, Houston.

ENV 1993-1-1 (1992). *Eurocode 3: design of steel structures, part 1.1 general rules and rules for buildings.* British Standards Institution, London.

Faulkner, D. (1975). A review of effective plating for use in the analysis of stiffened plating in bending and compression. *Journal of Ship Research*, **19**(1): 1–17.

Faulkner, D., Adamchak, J.C., Snyder, G.J. & Vetter, M.F. (1973). Synthesis of welded grillages to withstand compression and normal loads. *Computers & Structures*, **3**: 221–246.

Fletcher, C.A.J. (1984). *Computational Galerkin method.* Springer-Verlag, New York.

Fox, E.N. (1974). Limit analysis for plates: the exact solution for a clamped square plate of isotropic homogeneous material obeying the square yield criterion and loaded by a uniform pressure. *Philosophical Transactions of the Royal Society of London Series A (Mathematical and Physical Sciences)*, **277**: 121–155.

Hughes, O.F. & Paik, J.K. (2013). *Ship structural analysis and design.* The Society of Naval Architects and Marine Engineers, Alexandria, VA.

Jones, N. (1975). Plastic behavior of beams and plates. Chapter 23 in *Ship structural design concepts*, Edited by Harvey Evans, J., Cornell Maritime Press, Cambridge, MD, 747–778.

Jones, N. (2012). *Structural impact.* Second Edition, Cambridge University Press, Cambridge.

Kim, U.N., Choe, I.H. & Paik, J.K. (2009). Buckling and ultimate strength of perforated plate panels subject to axial compression: experimental and numerical investigations with design formulations. *Ships and Offshore Structures*, **4**(4): 337–361.

Lancaster, J. (2003). *Handbook of structural welding: processes, materials and methods used in the welding of major structures, pipelines and process plant.* Abington Publishing, Cambridge.

Li, Z.G., Zhang, M.Y., Liu, F., Ma, C.S., Zhang, J.H., Hu, Z.M., Zhang, J.Z. & Zhao, Y.N. (2014). Influence of dent on residual ultimate strength of 2024-T3 aluminum alloy plate under axial compression. *Transactions of Nonferrous Metals Society of China*, **24**(10): 3084–3094.

Li, Z.G., Zhang, D.N., Peng, C.L., Ma, C.S., Zhang, J.H., Hu, Z.M., Zhang, J.Z. & Zhao, Y.N. (2015). The effect of local dents on the residual ultimate strength of 2024-T3 aluminum alloy plate used in aircraft under axial tension tests. *Engineering Failure Analysis*, **48**: 21–29.

Marguerre, K. (1938). Zur Theorie der gekreummter Platte grosser Formaenderung. *Proceedings of the 5th International Congress for Applied Mechanics*, Cambridge.

Nara, S., Deguchi, Y. & Fukumoto, Y. (1988). Ultimate strength of steel plate panels with initial imperfections under uniform shearing stress. *Proceedings of the Japan Society of Civil Engineers*, **392/I-9**: 265–271 (in Japanese).

Narayanan, R. & der Avanessian, N.G.V. (1984). Elastic buckling of perforated plates under shear. *Thin-Walled Structures*, **2**: 51–73.

Paik, J.K. (1995). A new concept of the effective shear modulus for a plate buckled in shear. *Journal of Ship Research*, **39**(1): 70–75.

Paik, J.K. (2005). Ultimate strength of dented steel plates under edge shear loads. *Thin-Walled Structures*, **43**: 1475–1492.

Paik, J.K. (2007a). Ultimate strength of steel plates with a single circular hole under axial compressive loading along short edges. *Ships and Offshore Structures*, **2**(4): 355–360.

Paik, J.K. (2007b). Ultimate strength of perforated steel plates under edge shear loading. *Thin-Walled Structures*, **45**: 301–306.

Paik, J.K. (2008a). Some recent advances in the concepts of plate-effectiveness evaluation. *Thin-Walled Structures*, **46**: 1035–1046.

Paik, J.K. (2008b). Ultimate strength of perforated steel plates under combined biaxial compression and edge shear loads. *Thin-Walled Structures*, **46**: 207–213.

Paik, J.K. (2008c). Residual ultimate strength of steel plates with longitudinal cracks under axial compression: experiments. *Ocean Engineering*, **35**: 1775–1783.

Paik, J.K. (2008d). Residual ultimate strength of steel plates with longitudinal cracks under axial compression: nonlinear finite element method investigations. *Ocean Engineering*, **36**: 266–276.

Paik, J.K. (2009). Residual ultimate strength of steel plates with longitudinal cracks under axial compression: nonlinear finite element method investigations. *Ocean Engineering*, **36**(3–4): 266–276.

Paik, J.K., Kim, D.K., Lee, H. & Shim, Y.L. (2012). A method for analyzing elastic large deflection behavior of perfect and imperfect plates with partially rotation restrained edges. *Journal of Offshore Mechanics and Arctic Engineering*, **134**: 021603.1–021603.12.

Paik, J.K., Lee, J.M. & Ko, M.J. (2003a). Ultimate compressive strength of plate elements with pit corrosion wastage. *Journal of Engineering for the Maritime Environment*, **217**(M4): 185–200.

Paik, J.K., Lee, J.M. & Ko, M.J. (2004). Ultimate shear strength of plate elements with pit corrosion wastage. *Thin-Walled Structures*, **42**(8): 1161–1176.

Paik, J.K., Lee, J.M. & Lee, D.H. (2003b). Ultimate strength of dented steel plates under axial compressive loads. *International Journal of Mechanical Sciences*, **45**: 433–448.

Paik, J.K. & Melchers, R.E. (2008). *Condition assessment of aged structures.* CRC Press, New York.

Paik, J.K. & Pedersen, P.T. (1996). A simplified method for predicting the ultimate compressive strength of ship panels. *International Shipbuilding Progress*, **43**(434): 139–157.

Paik, J.K., Satish Kumar, Y.V. & Lee, J.M. (2005). Ultimate strength of cracked plate elements under axial compression or tension. *Thin-Walled Structures*, **43**: 237–272.

Paik, J.K., Thayamballi, A.K. & Kim, B.J. (2001). Advanced ultimate strength formulations for ship plating under combined biaxial compression/tension, edge shear and lateral pressure loads. *Marine Technology*, **38**(1): 9–25.

Rahbar-Ranji, A. & Zarookian, A. (2014). Ultimate strength of stiffened plates with a transverse crack under uniaxial compression. *Ships and Offshore Structures*, **10**(4): 416–425.

Raviprakash, A.V., Prabu, B. & Alagumurthi, N. (2012). Residual ultimate compressive strength of dented square plates. *Thin-Walled Structures*, **58**: 32–39.

Saad-Eldeen, S., Garbatov, Y. & Guedes Soares, C. (2015). Stress-strain analysis of dented rectangular plates subjected to uni-axial compressive loading. *Engineering Structures*, **99**: 78–91.

Saad-Eldeen, S., Garbatov, Y. & Guedes Soares, C. (2016). Ultimate strength analysis of highly damaged plates. *Marine Structures*, **45**: 63–85.

Shi, G.J. & Wang, D.Y. (2012). Residual ultimate strength of open box girders with cracked damage. *Ocean Engineering*, **43**: 90–101.

Shi, X.H., Zhang, J. & Guedes Soares, C. (2017). Experimental study on collapse of cracked stiffened plate with initial imperfections under compression. *Thin-Walled Structures*, **114**: 39–51.

Stonor, R.W.P., Bradfield, C.D., Moxham, K.E. & Dwight, J.B. (1983). Tests on plates under biaxial compression. Report CUED/D-Struct/TR98, Engineering Department, Cambridge University, Cambridge.

Suneel Kumar, M., Alagusundaramoorthy, P. & Sunsaravadivelu, R. (2009). Interaction curves for stiffened panel with circular opening under axial and lateral loads. *Ships and Offshore Structures*, **4**(2): 133–143.

Timoshenko, S.P. & Woinowsky-Krieger, S. (1981). *Theory of plates and shells*. Second Edition, McGraw-Hill, London.

Ueda, Y., Rashed, S.M.H. & Paik, J.K. (1984). Buckling and ultimate strength interactions of plates and stiffened plates under combined loads (1st Report): in-plane biaxial and shearing forces. *Journal of the Society of Naval Architects of Japan*, **156**: 377–387 (in Japanese).

Underwood, J.M., Sobey, A.J., Blake, J.I.R. & Shenoi, R.A. (2015). Ultimate collapse strength assessment of damaged steel plated grillages. *Engineering Structures*, **99**: 517–535.

Valsgård, S. (1980). Numerical design prediction of the capacity of plates in in-plane compression. *Computers & Structures*, **12**: 729–739.

Wang, F., Cui, W.C. & Paik, J.K. (2009a). Residual ultimate strength of structural members with multiple crack damage. *Thin-Walled Structures*, **47**: 1439–1446.

Wang, F., Paik, J.K., Kim, B.J., Cui, W.C., Hayat, T. & Ahmad, B. (2015). Ultimate shear strength of intact and cracked stiffened panels. *Thin-Walled Structures*, **88**: 48–57.

Wang, G., Sun, H.H., Peng, H. & Uemori, R. (2009b). Buckling and ultimate strength of plates with openings. *Ships and Offshore Structures*, **4**(1): 43–53.

Wood, R.H. (1961). *Plastic and elastic design of slabs and plates*. Ronald Press, New York.

Xu, M.C. & Guedes Soares, C. (2013). Assessment of residual ultimate strength for side dented stiffened panels subjected to compressive loads. *Engineering Structures*, **49**: 316–328.

Xu, M.C. & Guedes Soares, C. (2015). Effect of a central dent on the ultimate strength of narrow stiffened panels under axial compression. *International Journal of Mechanical Sciences*, **100**: 68–79.

Yamamoto, Y., Matsubara, N. & Murakami, T. (1970). Buckling strength of rectangular plates subjected to edge thrusts and lateral pressure (2nd report). *Journal of the Society of Naval Architects of Japan*, **127**: 171–179 (in Japanese).

5

Elastic and Inelastic Buckling Strength of Stiffened Panels and Grillages

5.1 Fundamentals of Stiffened Panel Buckling

As compressive or edge shear loads increase, a stiffened panel, as an assembly of plating and stiffeners, can buckle if the applied load (or for convenience, stress) reaches a critical value. The buckling patterns of the stiffened panel can normally be categorized into two major groups—overall buckling and local buckling—the latter of which is associated with the buckling of either plating or stiffeners. Figure 5.1 shows typical patterns of stiffened panel buckling, with the focus on the buckling patterns induced by axial compressive loading. Shear buckling patterns occur where tension field actions are formed, as described in Chapter 7.

When the stiffeners are "small," a stiffened panel can buckle together with the plating, in a mode that may be termed overall buckling, as shown in Figure 5.1a. In contrast, when the stiffeners are relatively strong, they remain straight until the plating between them buckles locally, as shown in Figure 5.1b. If the height of the stiffener web is large or the web thickness is small, the stiffener web can buckle locally, much like a plate element, as shown in Figure 5.1c. When the torsional rigidity of the stiffener is not strong enough, the stiffener can twist sideways, in a mode called lateral-torsional buckling (also called tripping), as shown in Figure 5.1d. Although Figure 5.1 illustrates each buckling pattern separately for convenience, in some cases, some buckling modes may interact and occur almost simultaneously.

Unlike columns that buckling is meant to collapse, a stiffened panel can normally sustain further applied loads even after buckling occurs locally, and the stiffened panel ultimate strength is eventually reached by excessive plasticity in the plate field and/or failure of the stiffener. However, any occurrence of elastic overall buckling results in significant instability of the entire structure. In structural design, therefore, the order of the buckling modes for a stiffened panel or grillage (cross-stiffened panel) needs to be controlled so that the overall buckling mode is prevented before local buckling of the plating between stiffeners.

The elastic buckling of a stiffened panel is a good indication of the required panel strength with regard to the serviceability limit state (SLS) design. To better understand the ultimate limit state (ULS) design procedure, it is essential to have basic knowledge of the stiffened panel's buckling strength.

In the SLS or ULS design of stiffened panels using Equation (1.17), the design load effects (i.e., stresses) are calculated with the classical theory of structural mechanics

Ultimate Limit State Analysis and Design of Plated Structures, Second Edition. Jeom Kee Paik.
© 2018 John Wiley & Sons Ltd. Published 2018 by John Wiley & Sons Ltd.

(a)

(b)

(c)

(d)

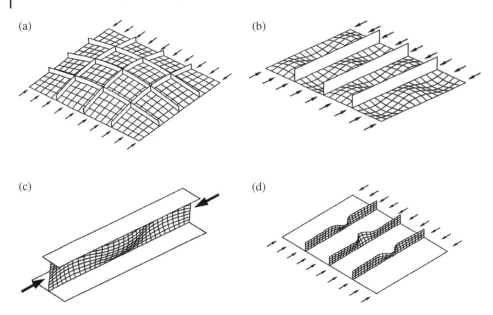

Figure 5.1 Schematics of various types of stiffened panel buckling under predominantly axial compressive loading: (a) overall grillage buckling; (b) local buckling of plating between stiffeners; (c) local buckling of stiffeners; (d) lateral-torsional buckling of stiffener.

or with linear elastic finite element analysis, whereas the design capacity can be determined by relevant buckling strength formulations.

This chapter presents the fundamentals and useful elastic buckling strength design formulations for a stiffened panel under combined loads and under single types of loads. The elastic–plastic buckling strength in such a case may, as usual, be estimated by correcting the elastic buckling strength using the Johnson–Ostenfeld formulation method, Equation (2.93), so that the influence of plasticity is approximately considered. It is also noted that the theories and methodologies described in this chapter can be commonly applied to both steel- and aluminum-stiffened panels.

5.2 Structural Idealizations of Stiffened Panels

5.2.1 Geometric Properties

Figure 5.2 shows a typical stiffened panel surrounded by heavy longitudinal girders and heavy transverse frames in a continuous stiffened plate structure. The stiffened panel usually has a number of stiffeners in one direction, that is, in the longitudinal or long direction. In some cases, the stiffened panel has stiffeners in both directions, which is termed a cross-stiffened panel or grillage.

The length and breadth of the stiffened panel are denoted by L and B, respectively. The thickness of the plate is t. The numbers of x and y stiffeners are n_{sx} and n_{sy}, respectively. The stiffeners are considered to be arranged with the same spacing in any given direction,

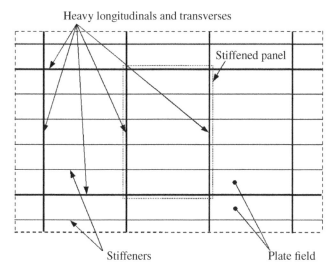

Figure 5.2 A continuous stiffened plate structure.

and the spacing of the stiffeners is denoted by a between y stiffeners, that is, $a = L/(n_{sy} + 1)$, and b between x stiffeners, that is, $b = B/(n_{sx} + 1)$.

Figure 5.3 presents the typical geometry of the stiffeners in the x or y direction. Stiffeners are placed to one side of the panel, that is, on the positive side of the z direction. The geometry of the stiffeners in each direction of the panel is considered to be identical.

5.2.2 Material Properties

The elastic modulus and Poisson's ratio of both the plating and the stiffeners are E and v, respectively. The elastic shear modulus is thus $G = E/[2(1 + v)]$. The bending rigidity of the plating between stiffeners is denoted by $D = Et^3/[12(1 - v^2)]$. The material yield stress is σ_{Yp} for the plating and σ_{Ys} for the stiffeners. When the material of the stiffened flange or web is different from that of the plating, an equivalent yield stress is identified as per Table 2.1. The slenderness ratio of the plating between longitudinal stiffeners is denoted by $\beta = (b/t)\sqrt{\sigma_{Yp}/E}$.

For a stiffened panel that may buckle in the overall mode and reach the ultimate strength primarily by excessive plasticity of the plating, it may be idealized as an "orthotropic plate." In this case, it is proposed that an equivalent yield stress, σ_{Yeq}, for the entire stiffened panel may approximately be defined as follows:

$$\sigma_{Yeq} = \begin{cases} \sigma_{Yx} & \text{for a longitudinally stiffened panel} \\ \sigma_{Yy} & \text{for a transversely stiffened panel} \\ (\sigma_{Yx} + \sigma_{Yy})/2 & \text{for a cross-stiffened panel} \end{cases} \tag{5.1}$$

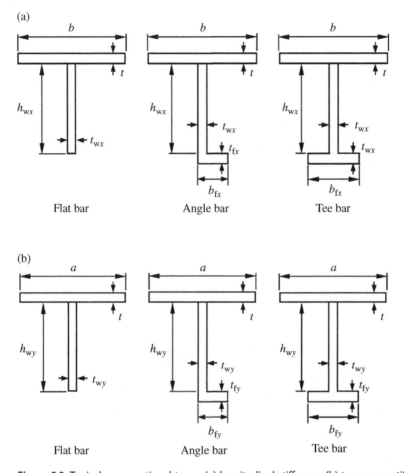

Figure 5.3 Typical cross-sectional types: (a) longitudinal stiffeners; (b) transverse stiffeners.

where

$$\sigma_{Yx} = \frac{Bt\sigma_{Yp} + n_{sx}A_{sx}\sigma_{Ys}}{Bt + n_{sx}A_{sx}},$$

$$\sigma_{Yy} = \frac{Lt\sigma_{Yp} + n_{sy}A_{sy}\sigma_{Ys}}{Lt + n_{sy}A_{sy}},$$

$$A_{sx} = h_{wx}t_{wx} + b_{fx}t_{fx},$$

$$A_{sy} = h_{wy}t_{wy} + b_{fy}t_{fy}.$$

5.2.3 Loads and Load Effects

When a continuous stiffened plate structure is subjected to external loads, the load effects (e.g., stresses, deformations) can be typically analyzed by the linear elastic finite element method or the classical theory of structural mechanics. The structural responses associated with the structure's primary, secondary, and tertiary levels must be accounted for, as described in Section 3.3, in determination of the characteristic measure of the load effects in a stiffened panel.

Figure 5.4 illustrates the potential stresses that act on stiffened panels, which are generally of six types as follows:

- Longitudinal axial stress
- Transverse axial stress
- Edge shear stress
- Longitudinal in-plane bending stress
- Transverse in-plane bending stress
- Lateral pressure-related stress

In this chapter (and Chapter 6), it is taken that compressive stresses are negative and tensile stresses are positive, unless otherwise specified. When a stiffened panel is simultaneously subjected to combined in-plane loads and lateral pressure, the latter is normally considered to be applied first in our treatment, and the other in-plane load components will then be taken to be applied afterward.

In the overall buckling mode, the plating typically deflects together with stiffeners. In this case, the average values of the applied axial stresses are often used as the characteristic measure of the load effects, thus neglecting the influence of in-plane bending as follows:

$$\sigma_{xav} = \frac{\sigma_{x1} + \sigma_{x2}}{2}, \quad \sigma_{yav} = \frac{\sigma_{y1} + \sigma_{y2}}{2} \tag{5.2}$$

where σ_{x1}, σ_{x2}, σ_{y1}, and σ_{y2} are defined in Figure 5.4. In a local buckling mode, the axial compressive stresses applied at the location of the most highly stressed stiffeners may be used as the stress parameters for the local buckling analysis of the stiffeners or plating. The values of the highest applied stresses at the x and y stiffeners are denoted by σ_{xM} and σ_{yM}, respectively, and the edge shear stress and uniform lateral pressure loads are denoted by τ_{av} and p, respectively.

5.2.4 Boundary Conditions

The edges of the stiffened panel extent are usually supported by strong beam members (e.g., girders or frames). The bending rigidities of the boundary support members are normally quite large compared with that of the plating itself, which implies that the normal displacements of the support members in the direction of panel deflections are very small, even up to panel collapse. The rotational restraints along the panel edges depend on the torsional rigidities of the longitudinal girders or transverse frames, which are neither zero nor infinite.

When predominantly in-plane compressive loads are applied to a continuous plated structure surrounded by such support members, the panel's buckling pattern is expected to be asymmetrical; that is, one panel tends to buckle up and the adjacent panel tends to deflect down. In this case, rotational restraints along the panel edges can be considered small.

(a)

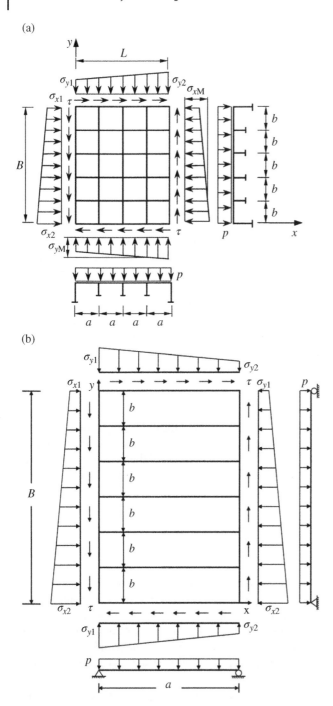

(b)

Figure 5.4 Types of load effects in a stiffened panel: (a) a cross-stiffened panel; (b) a longitudinally stiffened panel.

When the plated structure is subjected predominantly to lateral pressure loads, however, the structure's buckling pattern tends to be symmetrical, at least for large enough pressures, that is, each adjacent panel may deflect in the direction of the lateral pressure load. In this case, the edge rotational restraints can eventually become sufficiently large that they correspond to a clamped condition from the beginning of loading in some cases. However, if plasticity occurs earlier along the panel edges where the large bending moments develop, the rotational restraints at the yielded edges will then decrease as the applied loads increase.

In a continuous plated structure, the edges of individual stiffened panels are considered to remain almost straight because the structural response is relative to the adjacent panels even if the panel deflects. In this regard, an idealized condition, that is, one with no rotational restraints along the panel edges, has been widely used for practical purposes of analysis.

In this chapter (and Chapter 6), it is also assumed that the panel edges are simply supported, with zero deflection and zero rotational restraints along the four edges, and that all edges are kept straight. In most practical situations, this approximation will lead to adequate results. In contrast, the influence of rotational restraints along the junctions of plate–stiffener and/or stiffener web–flange may need to be considered in the calculations of local buckling of either the plating between the stiffeners or the stiffener web, as described in Chapters 3 and 4.

5.2.5   Fabrication Related Initial Imperfections

Although fabrication related initial imperfections in a stiffened panel are described in Section 1.7, it is assumed for the purposes of this chapter that initial distortions do not exist, mostly because a bifurcation buckling phenomenon may not appear in a panel that has initial deformations or curvature. The effects of welding induced residual stresses are dealt with. As described in Chapter 6, however, the influence of the initial imperfections on the panel ultimate strength needs to be accounted for.

5.3   Overall Buckling Versus Local Buckling

Because any occurrence of elastic overall panel buckling leads to significant instability of the entire plated structure, the order of the panel buckling modes is typically controlled such that the overall panel buckling mode is prevented before local buckling of the plating between stiffeners.

Under a multiaxial compressive loading condition, the following criterion must be satisfied so that overall panel buckling may not take place before local buckling of the plating between stiffeners, namely,

$$K_{\mathrm{OB}} \le K_{\mathrm{LB}} \tag{5.3a}$$

where K_{LB} and K_{OB} denote the characteristic measures of local plate buckling and overall panel buckling, respectively, which may be given in terms of the applied stresses and the corresponding buckling strength components. For instance, K_{LB} and K_{OB} may be given for a stiffened panel under biaxial compressive loads as follows:

$$K_{OB} = \sqrt{\left(\frac{\sigma_{xav}}{\sigma_{xEO}}\right)^2 + \left(\frac{\sigma_{yav}}{\sigma_{yEO}}\right)^2}, \quad K_{LB} = \sqrt{\left(\frac{\sigma_{xM}}{\sigma_{xEL}}\right)^2 \left(\frac{\sigma_{yM}}{\sigma_{yEL}}\right)^2} \qquad (5.3b)$$

where σ_{xav} and σ_{yav} are the average compressive stresses in the x or y direction, σ_{xM} and σ_{yM} are the highest compressive stresses in the x or y direction, σ_{xEO} and σ_{yEO} are the elastic overall compressive buckling stresses in the x or y direction, and σ_{xEL} and σ_{yEL} are the elastic compressive buckling stresses of the plating between the stiffeners in the x or y direction.

For uniaxial compression, the criterion is readily obtained from Equations (5.3a) and (5.3b) so that overall buckling will not occur before local buckling of the plating between the stiffeners as long as the following condition is satisfied:

$$\frac{\sigma_{xav}}{\sigma_{xEO}} \le \frac{\sigma_{xM}}{\sigma_{xEL}} \quad \text{or} \quad \frac{\sigma_{yav}}{\sigma_{yEO}} \le \frac{\sigma_{yM}}{\sigma_{yEL}} \qquad (5.3c)$$

In this regard, Equations (5.3a), (5.3b), and (5.3c) can be used to control the order of the buckling mode for a stiffened panel once the elastic buckling strength components in the local and overall modes are known.

5.4 Elastic Overall Buckling Strength

This section presents the elastic overall buckling strength formulations of a stiffened panel under combined loads and single types of loads.

5.4.1 Longitudinal Axial Compression

5.4.1.1 Longitudinally Stiffened Panels
The overall buckling of a panel with only longitudinal stiffeners and under longitudinal axial compression may approximately be represented by the column buckling of a plate–stiffener combination model, as described in Chapter 2. In this case, the plate–stiffener combination model as representative of the panel is supposed to be simply supported at both ends.

5.4.1.2 Transversely Stiffened Panels
The overall buckling strength of the panel with only transverse stiffeners and under longitudinal axial compression may be approximately predicted from the corresponding plate buckling strength formula, as described in Chapter 3, for a wide plate between two stiffeners. It is cautioned that the panel's coordinate system must be rotated to use the buckling strength formulations described in Chapter 3, which are available for only long plates, that is, $L/B \ge 1$.

5.4.1.3 Cross-Stiffened Panels (Grillages)
When the panel has stiffeners in both the x and y directions and it buckles in the overall mode, the panel's elastic overall buckling strength can be calculated by solving the

nonlinear governing differential equations derived from the elastic large-deflection orthotropic plate theory, as described in Chapter 6. An analytical solution of the elastic overall buckling for a simply supported orthotropic plate under uniaxial compressive loads in the x direction is given by

$$\sigma_{xEO,1} = -\frac{\pi^2}{B^2 t}\left(D_x\frac{m^2 B^2}{L^2} + 2Hn^2 + D_y\frac{n^4 L^2}{m^2 B^2}\right) \equiv -k_{xO}\frac{\pi^2 D}{B^2 t} \tag{5.4}$$

where

$$k_{xO} = \frac{1}{D}\left(D_x\frac{m^2 B^2}{L^2} + 2Hn^2 + D_y\frac{n^4 L^2}{m^2 B^2}\right)$$

is the elastic overall buckling strength coefficient for longitudinal axial compression and D_x, D_y, and H are defined in Chapter 6. The subscript "1" indicates that the number of load components is one. m and n are the overall buckling half-wave numbers of the panel in the x and y directions, respectively. As long as longitudinal compressive loads are applied, the buckling half-wave number of the orthotropic plate in the direction of the short edge may be taken as 1, which is similar to that of an isotropic plate, namely,

$$n = 1 \quad \text{for a long orthotropic plate with } L/B \geq 1 \tag{5.5a}$$

$$m = 1 \quad \text{for a wide orthotropic plate with } L/B < 1 \tag{5.5b}$$

When the cross-stiffened panel surrounded by heavy longitudinal girders and transverse frames is relatively wide, therefore, $m = 1$ may sometimes be used for practical purposes. In this case, the buckling half-wave number in the y direction may also be taken as $n = 1$ because no axial compressive loads are applied in the y direction. In contrast, m must be determined for a long orthotropic plate as a minimum integer that satisfies the following condition because $n = 1$, namely,

$$\left(\frac{L}{B}\right)^4 \leq \frac{D_x}{D_y}m^2(m+1)^2 \tag{5.6a}$$

From this, it is evident that the buckling half-wave number of the orthotropic plate is affected by the structural orthotropy and by the panel aspect ratio. For an isotropic plate, Equation (5.6a) becomes

$$\frac{L}{B} \leq \sqrt{m(m+1)} \tag{5.6b}$$

because $D_x = D_y$.

5.4.2 Transverse Axial Compression

5.4.2.1 Longitudinally Stiffened Panels

The overall buckling strength of a panel with only longitudinal stiffeners and under transverse axial compression may approximately be predicted from the corresponding plate buckling strength formula, as described in Chapter 3, for a wide plate between two stiffeners. It is cautioned that the panel's coordinate system must be rotated to use the plate buckling formula of Chapter 3 because the plate buckling strength formulations described in Chapter 3 consider only long plates, that is, $L/B \geq 1$.

5.4.2.2 Transversely Stiffened Panels

The overall buckling of a panel with only transverse stiffeners, and under transverse axial compression, may approximately be represented by the column buckling of the plate–stiffener combination model with the simply supported end condition, as described in Chapter 2.

5.4.2.3 Cross-Stiffened Panels (Grillages)

The large-deflection orthotropic plate theory can be applied to calculate the elastic overall buckling of a cross-stiffened panel under axial compression in the y direction, as described in Chapter 6.

In this case, the solution of the nonlinear governing differential equations of the large-deflection orthotropic plate theory under the boundary conditions of simple supports at all edges gives the elastic overall buckling strength of a stiffened panel under uniaxial compressive loads in the y direction as follows:

$$\sigma_{yEO,1} = -\frac{\pi^2}{B^2 t}\left(D_x\frac{m^4 B^4}{n^2 L^4} + 2H\frac{m^2 B^2}{L^2} + D_y n^2\right) = -k_{yO}\frac{\pi^2 D}{B^2 t} \tag{5.7}$$

where

$$k_{yO} = \frac{1}{D}\left(D_x\frac{m^4 B^4}{n^2 L^4} + 2H\frac{m^2 B^2}{L^2} + D_y n^2\right)$$

is the elastic overall buckling strength coefficient for transverse axial compression and D_x, D_y, and H are defined in Chapter 6. m and n are the buckling half-wave numbers of the panel in the x and y directions, respectively. $m = n = 1$ may be taken for a long orthotropic plate (i.e., with $L/B \geq 1$), because no axial compressive loads are applied in the x direction. For a wide orthotropic plate (i.e., with $L/B < 1$), the buckling half-wave number in the y direction may be determined as a minimum integer that satisfies the following condition because $m = 1$:

$$\left(\frac{B}{L}\right)^4 \leq \frac{D_y}{D_x}n^2(n+1)^2 \tag{5.8a}$$

For an isotropic plate, Equation (5.8a) will be simplified to

$$\frac{B}{L} \leq \sqrt{n(n+1)} \tag{5.8b}$$

because $D_x = D_y$.

5.4.3 Edge Shear

The elastic overall buckling of a stiffened panel under edge shear may be determined with the orthotropic plate theory. The elastic overall buckling strength for a simply supported orthotropic plate in edge shear was obtained by Seydel (1933) and is given by

$$\tau_{EO,1} = k_\tau \frac{\pi^2}{B^2 t}D_x^{1/4}D_y^{3/4} \tag{5.9}$$

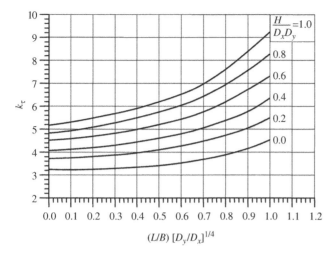

Figure 5.5 Overall buckling coefficient of stiffened panels in edge shear (Allen & Bulson 1980).

where k_τ is the shear buckling coefficient, which may be determined from Figure 5.5 as a function of the panel aspect ratio and the various structural orthotropy parameters. It is apparent from Figure 5.5 that $k_\tau \approx 9.34$ when $D_x = D_y = H = D$ and $L/B = 1$, which corresponds to the elastic shear buckling coefficient of an isotropic square plate.

As long as the stiffeners are not very weak and they remain straight, that is the case when no significant membrane tension field develops, the elastic overall shear buckling strength of a stiffened panel can be approximately taken as the elastic shear buckling strength of a simply supported plate between stiffeners with k_τ, as described in Chapter 3, as follows:

$$\tau_{EO,1} = k_\tau \frac{\pi^2}{12(1 - v^2)} \left(\frac{t}{b}\right)^2 \qquad (5.10)$$

5.4.4 Combined Biaxial Compression or Tension

The elastic overall post-buckling behavior of a stiffened panel under combined biaxial compression or tension can be calculated by analytical solution of the nonlinear governing differential equation of the large-deflection orthotropic plate theory, as described in Chapter 6. Immediately before bifurcation buckling occurs, the panel lateral deflection must be zero. This requirement results in an overall panel buckling criterion under combined biaxial loads as follows:

$$\frac{m^2 B}{L}\sigma_{xav} + \frac{n^2 L}{B}\sigma_{yav} + \frac{\pi^2}{t}\left(D_x \frac{m^4 B}{L^3} + 2H\frac{m^2 n^2}{LB} + D_y \frac{n^4 L}{B^3}\right) = 0 \qquad (5.11)$$

where σ_{xav} and σ_{yav} are the applied axial stresses in the x or d y direction and m and n are the buckling half-wave numbers in the x or y direction.

It is apparent from Equation (5.11) that Equation (5.4) or (5.7), which is applicable to the corresponding uniaxial compressive load cases, may be obtained by setting the other stress component to zero. By holding the loading ratio, $c = \sigma_{yav}/\sigma_{xav}$, constant, the elastic overall buckling strength components, σ_{xEO} or σ_{yEO}, are in this case calculated by the solution of Equation (5.11) as follows:

$$\sigma_{xEO} = -\frac{\pi^2}{t(m^2 B/L + cn^2 L/B)}\left(D_x \frac{m^4 B^2}{n^2 L^4} + 2H \frac{m^2}{L^2} + D_y \frac{n^2}{B^2}\right) \tag{5.12a}$$

$$\sigma_{yEO} = \begin{cases} -\dfrac{\pi^2}{t[m^2 B/(cL) + n^2 L/B]}\left(D_x \dfrac{m^4 B^2}{n^2 L^4} + 2H \dfrac{m^2}{L^2} + D_y \dfrac{n^2}{B^2}\right) & \text{for } \sigma_{xav} \neq 0 \\[4mm] -\dfrac{\pi^2 B}{n^2 L t}\left(D_x \dfrac{m^4 B^2}{n^2 L^4} + 2H \dfrac{m^2}{L^2} + D_y \dfrac{n^2}{B^2}\right) & \text{for } \sigma_{xav} = 0 \end{cases}$$

$$\tag{5.12b}$$

For a long panel, that is, with $L/B \geq 1$, the buckling half-wave number m in the x direction is determined as a minimum integer that satisfies the following condition because $n = 1$ in the direction of the short edges:

$$\frac{D_x(m^4 B^4/L^4) + 2H(m^2/L^2) + D_y(1/B^2)}{m^2 B/L + cL/B}$$

$$\leq \frac{D_x[(m+1)^4 B^2/L^4] + 2H[(m+1)^2/L^2] + D_y(1/B^2)}{(m+1)^2 B/L + cL/B} \tag{5.13}$$

where it is evident that the buckling half-wave number of the panel under biaxial loads is affected by the biaxial loading ratio and by the structural orthotropy and aspect ratio.

For a wide panel, that is, with $L/B < 1$, the buckling half-wave number n in the y direction can be determined as the minimum integer that satisfies the following condition because $m = 1$ in the direction of the short edges:

$$\frac{D_x(B^2/n^2 L^4) + 2H(1/L^2) + D_y(n^2/B^2)}{B/L + cn^2 L/B}$$

$$\leq \frac{D_x[B^2/(n+1)^2 L^4] + 2H(1/L^2) + D_y[(n+1)^2/B^2]}{B/L + c(n+1)^2 L/B} \tag{5.14}$$

5.4.5 Combined Uniaxial Compression and Edge Shear

Following the isotropic buckling strength interaction relationship described in Chapter 3, the elastic overall buckling strength interaction equations for a stiffened panel under combined uniaxial compression and edge shear are sometimes used for practical design purposes as follows:

$$\frac{\sigma_{xav}}{\sigma_{xEO,1}} + \left(\frac{\tau_{av}}{\tau_{EO,1}}\right)^2 = 1 \tag{5.15a}$$

$$\frac{\sigma_{yav}}{\sigma_{yEO,1}} + \left(\frac{\tau_{av}}{\tau_{EO,1}}\right)^2 = 1 \tag{5.15b}$$

5.5   Elastic Local Buckling Strength of Plating Between Stiffeners

When the stiffeners become stiff, they may remain straight until the plating between them buckles locally. In this case, the local buckling strength of the plating can be calculated with the methods for the bare plate element between stiffeners, taking into account the influences of various parameters, as described in Chapter 3.

5.6   Elastic Local Buckling Strength of Stiffener Web

The web of a stiffener in a stiffened panel can locally buckle, a possibility that must usually be considered for built-up sections. This failure mode is termed "stiffener web buckling," which can sometimes be a quite sudden phenomenon that results in unloading of the stiffened panel, particularly with the use of deep webs or flat-bar stiffeners. In this case, once such stiffener web buckling occurs, the buckled or collapsed plating is left with essentially no stiffening, and thus overall stiffened panel collapse may follow with little increase in the loading.

The local buckling of the stiffener web and the buckling or collapse of the plating between the stiffeners normally interact and can take place in any order, depending on the dimensions of plating and stiffener. Clearly, buckling of the stiffener web before the inception of buckling in the plating between the stiffeners is normally an undesirable failure mode.

It is necessary in design that the stiffener web should resist buckling until the plating between the stiffeners buckles or collapses. However, the local buckling strength of the stiffener web depends significantly on the torsional rigidities of the adjacent members to which they are attached, among other factors. Because the rotational restraints along any plate–stiffener intersection can be decreased by the collapse of the plating involved, the stiffener web buckling strength should be calculated by accounting for this effect. In the following, an exact solution of the stiffener web buckling is described, following Paik et al. (1998).

5.6.1   Governing Differential Equation

Figure 5.6 shows a schematic representation of the loading and boundary conditions for a plate–stiffener combination model with the attached effective plating between two adjacent transverse frames in a continuous stiffened plate structure. In this case, the elastic buckling strength for the stiffener web can be analyzed by solving the governing differential equations (Bleich 1952).

The elastic buckling strength of a stiffener web, which is regarded as a very long plate (strip) under the appropriate edge conditions, may be analytically addressed as a characteristic value problem. The boundary condition of the flat-bar stiffener along one edge is free, but for angle-type or Tee stiffeners, it may be assumed that the stiffener flange will not buckle until the inception of local buckling in the stiffener web, whereas the attached plating of the stiffened panel can itself buckle. This would imply that the rotational restraint at the web–flange junction is fully effective until the stiffener web buckles, whereas that at the web-plating junction takes a value derived from an effective cross section.

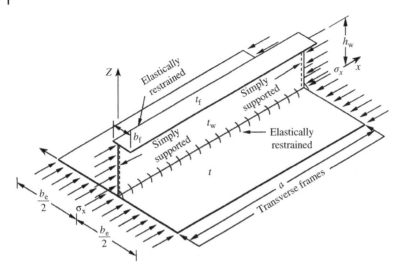

Figure 5.6 Schematic representation of loading and boundary conditions for the plate–stiffener combination model with the attached effective plating between two adjacent transverse frames under axial compression.

The fundamental differential equation for the out-of-plane (i.e., sideways) deflection of the stiffener web with zero initial deflection under axial compressive loads can be derived under the assumption that the deflection is small when compared with the thickness of the stiffener web. The applicable equation for the stiffener web was given by Bleich (1952) as

$$D_w\left(\frac{\partial^4 v}{\partial x^4} + 2\frac{\partial^4 v}{\partial^2 x\partial^2 z} + \frac{\partial^4 v}{\partial z^4}\right) + t_w\sigma_x\frac{\partial^2 v}{\partial x^2} = 0 \tag{5.16}$$

where v is the sideways deflection of the stiffener web, $D_w = Et_w^3/[12(1-v^2)]$ is the bending rigidity of stiffener web, and t_w is the thickness of the stiffener web.

The solution of the aforementioned equation for v will provide the deflected form of the stiffener web under compression σ_x, such as $\sigma_x = \sigma_{xM}$ when in-plane bending is applied, or $\sigma_x = \sigma_{xav}$ for uniform compression, which represents equilibrium but in an unstable position. The buckling strength is defined by the load at a bifurcation point where, in addition to the plane equilibrium from $v = 0$, a deflected but unstable form of equilibrium occurs.

5.6.2 Exact Web Buckling Characteristic Equation

To solve Equation (5.16), edge conditions for the stiffener web that are consistent with its support characteristics should be prescribed. The loaded edges of the stiffener web at $x = 0$ and a are normally supported by transverse frames and can (somewhat pessimistically) be assumed to be simply supported as follows:

$$v = 0 \quad \text{at} \quad x = 0 \quad \text{and} \quad a \tag{5.17a}$$

$$M_z = 0 \quad \text{at} \quad x = 0 \quad \text{and} \quad a \tag{5.17b}$$

where M_z is the bending moment per unit length of stiffener web about the z axis.

In practical cases, because the bending rigidity of the plating in the x–y plane is normally comparatively very large, the deflection (sideways movement) of the stiffener web along the lower edge, that is, $z = 0$, relative to transverse frames, can be assumed to be zero, and hence

$$v = 0 \quad \text{at} \quad z = 0 \tag{5.18}$$

The edge of the stiffener web along $z = 0$, where it joins the plating, cannot be assumed to rotate freely during buckling. Hence, we consider that the lower edge of the stiffener web is rotationally restrained, with the magnitude of the restraint depending on the torsional rigidity of the plating. Along the edge at $z = 0$, the bending moments that appear during buckling of the stiffener web must be equal and opposite to the rate of change of the twisting moments of plating, which gives us the condition that

$$M_x = -\frac{\partial m_x}{\partial x} \quad \text{at} \quad z = 0 \tag{5.19}$$

where M_x is the bending moment per unit length of the stiffener web about the x axis, which is taken as follows:

$$M_x = -D_w\left(\frac{\partial^2 v}{\partial z^2} + v\frac{\partial^2 v}{\partial x^2}\right) \tag{5.20}$$

m_x is the twisting moment per unit length of plating about the x axis, which may be approximated by neglecting the warping rigidity of the plating as

$$m_x = -GJ_p\frac{\partial^2 v}{\partial x \partial z} \tag{5.21}$$

$J_p = b_e t^3/3$ is the torsion constant of attached effective plating, b_e is the effective width of the attached plating, and t is the thickness of the attached plating.

The edge condition for the stiffener web at $z = h_w$ depends on both the bending and the torsional rigidities of the stiffener flange, where h_w is the height of the stiffener web. For flat-bar stiffeners that do not have a stiffener flange, the deflection and rotation along the edge at $z = h_w$ occurs freely. In contrast, for angle or Tee-section stiffeners, the stiffener web is partially rotation restrained by the stiffener flange. Hence, the deflection (sideways movement) along this edge is not zero, but equals the deflection of the stiffener flange. The general condition for deflection along the edge at $z = h_w$ can be expressed by

$$EI_f\frac{\partial^4 v}{\partial x^4} = D_w\left[\frac{\partial^3 v}{\partial z^3} + (2-v)\frac{\partial^3 v}{\partial x^2 \partial z}\right] \quad \text{at} \quad z = h_w \tag{5.22}$$

where I_f is the moment of inertia of the stiffener flange at $z = h_w$ with regard to the z axis, which is taken as $I_f = b_f^3 t_f/3$ in the case of an angle section and $I_f = b_f^3 t_f/12$ in the case of a Tee section.

The rotation along the edge at $z = h_w$ is restrained due to the stiffener flange's torsional rigidity, and the bending moment of the stiffener web equals the rate of change of the twisting moment of the stiffener flange, namely,

$$M_x = -\frac{\partial m_x}{\partial x} \quad \text{at} \quad z = h_w \tag{5.23}$$

In Equation (5.23), m_x is given by

$$m_x = -GJ_f \frac{\partial^2 v}{\partial x \partial z} \tag{5.24}$$

where $J_f = b_f t_f^3/3$ is the torsion constant of the stiffener flange, b_f is the breadth of the stiffener flange, and t_f is the thickness of the stiffener flange.

For flat-bar stiffeners, that is, with $I_f = 0$ or $J_f = 0$, Equation (5.23) corresponds to the condition that the rotation at $z = h_w$ occurs freely so that no moments develop there.

The general solution of Equation (5.16), which satisfies the conditions of simple support at $x = 0$ and a, as expressed by Equation (5.17), can be assumed to have the following form:

$$v = Z(z) \sin \frac{m\pi x}{a} \tag{5.25}$$

where $Z(z)$ indicates a function of z and m represents the number of primary buckling half waves for the stiffener web along the x direction.

Substituting Equation (5.25) into Equation (5.16), an ordinary differential equation of the fourth order is obtained upon replacing σ_x by σ_E^W as follows:

$$\frac{\partial^4 Z}{\partial z^4} - 2\left(\frac{m\pi}{a}\right)^2 \frac{\partial^2 Z}{\partial z^2} + \left(\frac{m\pi}{a}\right)^4 (1 - \mu^2) Z = 0 \tag{5.26}$$

where

$$\mu = \frac{a}{m h_w} \sqrt{k_w}, \quad k_w = \sigma_E^W \frac{h_w^2 t_w}{\pi^2 D_w}.$$

and σ_E^W is the elastic local buckling stress of stiffener web.

The general solution of Equation (5.26) is given by

$$Z(z) = C_1 e^{-\alpha_1 z} + C_2 e^{\alpha_1 z} + C_3 \cos \alpha_2 z + C_4 \sin \alpha_2 z \tag{5.27}$$

where

$$\alpha_1 = \frac{m\pi}{a} \sqrt{\mu + 1}, \quad \alpha_2 = \frac{m\pi}{a} \sqrt{\mu - 1}.$$

From Equations (5.18) and (5.19), the following relationships between the constants in Equation (5.27) can be obtained:

$$C_1 = \frac{C_3 (\alpha_1^2 + \alpha_2^2 - \alpha_1 \alpha_3)}{2 \alpha_1 \alpha_3} + \frac{C_4 \alpha_2}{2 \alpha_1}, \quad C_2 = -\frac{C_3 (\alpha_1^2 + \alpha_2^2 - \alpha_1 \alpha_3)}{2 \alpha_1 \alpha_3} - \frac{C_4 \alpha_2}{2 \alpha_1} \tag{5.28}$$

where

$$\alpha_3 = \frac{GJ_p}{D_w} \left(\frac{m\pi}{a}\right)^2 = h_w \zeta_p \left(\frac{m\pi}{a}\right)^2, \quad \zeta_p = \frac{GJ_p}{h_w D_w}$$

Upon substitution of Equation (5.28) into Equation (5.27), we obtain

$$Z(z) = C_3 \left(\cos \alpha_2 z - \cosh \alpha_1 z - \frac{\alpha_1^2 + \alpha_2^2}{\alpha_1 \alpha_3} \sinh \alpha_1 z \right) + C_4 \left(\sin \alpha_2 z - \frac{\alpha_2}{\alpha_1} \sinh \alpha_1 z \right)$$

$$\tag{5.29}$$

Substitution of Equation (5.29) into Equation (5.25) yields

$$v = \left[C_3 \left(\cos \alpha_2 z - \cosh \alpha_1 z - \frac{\alpha_1^2 + \alpha_2^2}{\alpha_1 \alpha_3} \sinh \alpha_1 z \right) \right.$$
$$\left. + C_4 \left(\sin \alpha_2 z - \frac{\alpha_2}{\alpha_1} \sinh \alpha_1 z \right) \right] \sin \frac{m\pi x}{a} \tag{5.30}$$

Using the boundary conditions, that is, Equations (5.23) and (5.24), along the edge of the stiffener web at $z = h_w$, the unknown constants, C_3 and C_4, are determined from

$$\begin{bmatrix} A_{11} & A_{12} \\ A_{21} & A_{22} \end{bmatrix} \begin{pmatrix} C_3 \\ C_4 \end{pmatrix} = 0 \tag{5.31}$$

where

$$A_{11} = h_w \gamma_f \left(\frac{m\pi}{a} \right)^4 (\cos \alpha_2 h_w - \cosh \alpha_1 h_w - S \sin \alpha_1 h_w)$$
$$- \left(\alpha_2^3 \sin \alpha_2 h_w - \alpha_1^3 \sinh \alpha_1 h_w - S \alpha_1^3 \cosh \alpha_1 h_w \right)$$
$$- (2 - v) \left(\frac{m\pi}{a} \right)^2 (\alpha_2 \sin \alpha_2 h_w + \alpha_1 \sinh \alpha_1 h_w + S \alpha_1 \cosh \alpha_1 h_w),$$

$$A_{12} = h_w \gamma_f \left(\frac{m\pi}{a} \right)^4 \left(\sin \alpha_2 h_w - \frac{\alpha_2}{\alpha_1} \sinh \alpha_1 h_w \right) + \left(\alpha_2^3 \cos \alpha_2 h_w + \alpha_1^2 \alpha_2 \cosh \alpha_1 h_w \right)$$
$$+ (2 - v) \left(\frac{m\pi}{a} \right)^2 (\alpha_2 \cos \alpha_2 h_w + \alpha_2 \cosh \alpha_1 h_w),$$

$$A_{21} = h_w \zeta_f \left(\frac{m\pi}{a} \right)^2 (\alpha_2 \sin \alpha_2 h_w + \alpha_1 \sinh \alpha_1 h_w + S \alpha_1 \cosh \alpha_1 h_w)$$
$$+ \left(\alpha_2^2 \cos \alpha_2 h_w + \alpha_1^2 \cosh \alpha_1 h_w + S \alpha_1^2 \sinh \alpha_1 h_w \right)$$
$$+ v \left(\frac{m\pi}{a} \right)^2 (\cos \alpha_2 h_w - \cos \alpha_1 h_w - S \sinh \alpha_1 h_w),$$

$$A_{22} = - h_w \zeta_f \left(\frac{m\pi}{a} \right)^2 (\alpha_2 \cos \alpha_2 h_w - \alpha_2 \cosh \alpha_1 h_w) + \left(\alpha_2^2 \sin \alpha_2 h_w + \alpha_1 \alpha_2 \sinh \alpha_1 h_w \right)$$
$$+ v \left(\frac{m\pi}{a} \right)^2 \left(\sin \alpha_2 h_w - \frac{\alpha_2}{\alpha_1} \sinh \alpha_1 h_w \right),$$

$$S = \frac{\alpha_1^2 + \alpha_2^2}{\alpha_1 \alpha_3}, \quad \gamma_f = \frac{EI_f}{h_w D_w}, \quad \zeta_f = \frac{GJ_f}{h_w D_w}.$$

Equation (5.31) has solutions different from zero only if the determinant, Δ, of the coefficient matrix vanishes. The condition for determinant $\Delta = 0$ of Equation (5.31) yields

$$\Delta = \begin{vmatrix} A_{11} & A_{12} \\ A_{21} & A_{22} \end{vmatrix} = 0 \ \text{ or } \ \Delta = A_{11} A_{22} - A_{12} A_{21} = 0 \tag{5.32}$$

Equation (5.32) is the characteristic equation for elastic buckling of the stiffener web with or without a stiffener flange, the latter being a flat-bar stiffener. The solution of Equation (5.32) would then provide the value of the buckling coefficient, k_w, for stiffener web buckling. The stiffener web's elastic local buckling strength can be obtained by solving the characteristic equation, with compressive stress taking a negative sign as follows:

$$\sigma_E^W = -k_w \frac{\pi^2 E}{12(1-v^2)} \left(\frac{t_w}{h_w}\right)^2 \tag{5.33}$$

where σ_E^W is the elastic buckling strength of stiffener web and k_w is the elastic buckling strength coefficient of the stiffener web.

To account for the effect of welding residual stress, the web buckling stress computed from Equation (5.33) may be reduced by the compressive residual stress in the stiffener web.

5.6.3 Closed-Form Expressions of Stiffener Web Buckling Strength

The task of solving Equation (5.32) in any specific case is not always straightforward, and so it is most desirable for a designer to have a closed-form expression to predict the local buckling strength of the stiffener web more readily.

An empirical expression is particularly useful to predict the buckling strength of the stiffener web in terms of the relevant torsional rigidities of the plating and the stiffener flange. One such expression for the buckling coefficient, k_w, in Equation (5.33) was given by Paik et al. (1998) as follows:

$$k_w = \begin{cases} C_1 \zeta_p + C_2 & \text{for } 0 \le \zeta_p \le \eta_w \\ C_3 - 1/(C_4 \zeta_p + C_5) & \text{for } \eta_w < \zeta_p \le 60 \\ C_3 - 1/(60 C_4 + C_5) & \text{for } 60 < \zeta_p \end{cases} \tag{5.34}$$

where

$$\eta_w = -0.444 \zeta_f^2 + 3.333 \zeta_f + 1.0,$$

$$C_1 = -0.001 \zeta_f + 0.303,$$

$$C_2 = 0.308 \zeta_f + 0.427,$$

$$C_3 = \begin{cases} -4.350 \zeta_f^2 + 3.965 \zeta_f + 1.277 & \text{for } 0 \le \zeta_f \le 0.2 \\ -0.427 \zeta_f^2 + 2.267 \zeta_f + 1.460 & \text{for } 0.2 < \zeta_f \le 1.5 \\ -0.133 \zeta_f^2 + 1.567 \zeta_f + 1.850 & \text{for } 1.5 < \zeta_f \le 3.0 \\ 5.354 & \text{for } 3.0 < \zeta_f \end{cases},$$

$$C_4 = \begin{cases} -6.70 \zeta_f^2 + 1.40 & \text{for } 0 \le \zeta_f \le 0.1 \\ 1/(5.10 \zeta_f + 0.860) & \text{for } 0.1 < \zeta_f \le 1.0 \\ 1/(4.0 \zeta_f + 1.814) & \text{for } 1.0 < \zeta_f \le 3.0 \\ 0.0724 & \text{for } 3.0 < \zeta_f \end{cases},$$

$$
C_5 = \begin{cases}
-1.135\zeta_f + 0.428 & \text{for } 0 \leq \zeta_f \leq 0.2 \\
-0.299\zeta_f^3 + 0.803\zeta_f^2 - 0.783\zeta_f + 0.328 & \text{for } 0.2 < \zeta_f \leq 1.0 \\
-0.016\zeta_f^3 + 0.117\zeta_f^2 - 0.285\zeta_f + 0.235 & \text{for } 1.0 < \zeta_f \leq 3.0 \\
0.001 & \text{for } 3.0 < \zeta_f
\end{cases}
$$

For flat-bar stiffeners, Equation (5.34) will become much simpler because $\zeta_f = 0$, and the computed results are well approximated by

$$
k_w = \begin{cases}
0.303\zeta_p + 0.427 & \text{for } 0 \leq \zeta_p \leq 1 \\
1.277 - 1/(1.40\zeta_p + 0.428) & \text{for } 1 < \zeta_p \leq 60 \\
1.2652 & \text{for } 60 < \zeta_p
\end{cases} \tag{5.35}
$$

Figure 5.7a shows the variation in the elastic buckling coefficient for a flat-bar stiffener web as a function of the web aspect ratio, a/h_w, and the torsional rigidity of the plating. It can be seen from this figure that an increase in the torsional rigidity of the plating will result in a significant increase in the web buckling coefficient. Therefore, accounting for such effects can be important, particularly in cases in which stiffener web buckling is a possibility. The effects of the web aspect ratio on the stiffener web's buckling strength can be ignored in most practical cases.

Figure 5.7b and c shows the variations in the elastic buckling strength coefficient for an angle or Tee-section stiffener web as a function of three parameters: the aspect ratio of the stiffener web, the torsional rigidity of the plating, and the torsional rigidity of the stiffener flange.

The results are shown in the practical ranges of the parameters applicable to the stiffened panels in merchant ships. With an increase in the torsional rigidities of the stiffener flange and/or plating, the elastic buckling strength of the stiffener web increases significantly, whereas the influence of the web aspect ratio on the buckling strength of the stiffener web can again be ignored for most practical purposes.

The dotted lines in Figure 5.7 represent the approximate solutions for the web buckling coefficient, as given in Equation (5.34) or (5.35). These may be compared with the solid lines in the figure, which represent the results computed by direct solution of the characteristic buckling equation, Equation (5.32). A combination of Equation (5.33) and Equation (5.34) or (5.35) provides reasonably accurate predictions for the elastic buckling strength of the stiffener web with or without a stiffener flange.

5.7 Elastic Local Buckling Strength of Stiffener Flange

Local buckling of the stiffener flange before buckling of the plating between the stiffeners is an undesirable failure mode. In design, stiffener flange buckling must be prevented until the stiffened panel reaches its ultimate strength.

For practical pessimistic assessment, the elastic local compressive buckling strength of the stiffener flange, denoted by σ_E^F, can be estimated with a plate idealization in which

(a)

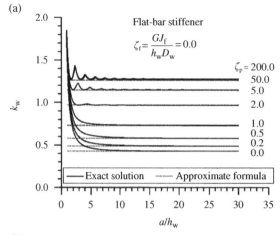

(b)

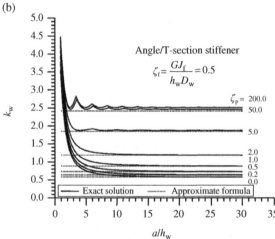

(c)

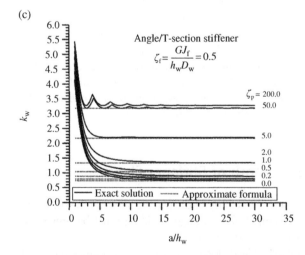

Figure 5.7 Variation of the elastic buckling strength coefficient: (a) flat-bar stiffener web as a function of the web aspect ratio and the torsional rigidity of the plating; (b, c) angle or Tee-section stiffener web as a function of the web aspect ratio and torsional rigidities of the plating or stiffener flange.

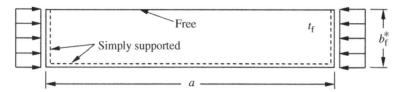

Figure 5.8 A stiffener flange with three simply supported edges and one free edge.

three edges are simply supported and one edge is free, as shown in Figure 5.8, with the results as follows:

$$\sigma_E^F = k_f \frac{\pi^2 E}{12(1-v^2)} \left(\frac{t_f}{b_f^*}\right)^2 \tag{5.36}$$

where $k_f = 0.425 + \left(b_f^*/a\right)^2, b_f^* = b_f$ for asymmetric angle stiffeners and $b_f^* = 0.5b_f$ for symmetric Tee stiffeners.

5.8 Lateral-Torsional Buckling Strength of Stiffeners

5.8.1 Fundamentals of Lateral-Torsional Buckling

Lateral-torsional buckling (also called tripping) of stiffeners is a phenomenon in which the failure of a stiffened panel occurs after the stiffener twists sideways about the edge of the stiffener web attached to the plating. When the stiffener's torsional rigidity is low or the stiffener flange is weak, this phenomenon is more likely to occur.

Like stiffener web buckling, lateral-torsional buckling can be a relatively sudden phenomenon that results in unloading of the stiffened panel. Once tripping occurs, the buckled or collapsed plating is left with essentially much reduced stiffening and thus overall collapse may follow. A plate–stiffener combination model, as described in Chapter 2, with the attached effective plating under combined axial compression and lateral line loads, is typically considered to collapse if lateral-torsional buckling occurs after the plating between the stiffeners collapses.

In a continuous stiffened panel, lateral-torsional buckling may generally involve a coupling of sideways and vertical deflection and rotation of the stiffener web together with local buckling of the attached plating, as shown in Figure 5.9a. Unlike an ordinary beam–column in framed structures, the attached plating of a plate–stiffener combination model in plated structures is restricted from deflecting sideways, whereas the beam (stiffener) flange is relatively free to deflect sideways and vertically.

For asymmetric section profiles (e.g., angle section), vertical bending, sideways bending, and torsion are typically coupled, whereas for symmetric section profiles (e.g., Tee section), only sideways bending and torsion are normally coupled, which implies that the overall flexural Euler buckling and lateral-torsional buckling can sometimes be closely coupled for plate–stiffener combination models.

Many researchers have examined the lateral-torsional buckling of stiffeners in theoretical, numerical, and experimental studies. Very earlier studies that used the classical theory of thin-walled bars were summarized by Bleich (1952). During the 1970s and 1980s,

(a)

(b)

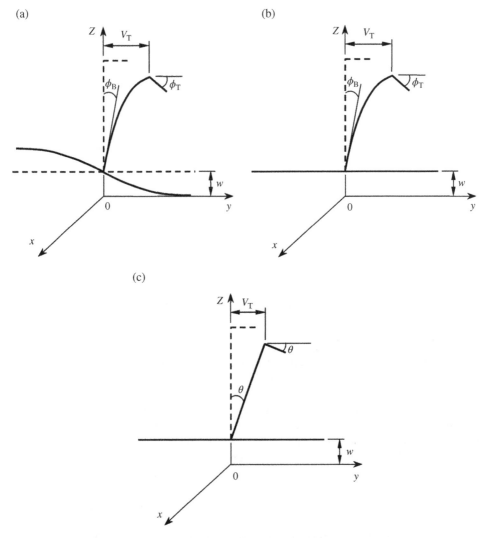

(c)

Figure 5.9 General and idealized tripping deformations of a plate–stiffener combination model: (a) general deformations of flexible stiffener web with plate rotational restraints; (b) simplified deformations of flexible stiffener web without plate rotational restraints; (c) simplified deformations of rigid stiffener web without plate rotational restraints.

further studies were undertaken by Faulkner et al. (1973), Smith (1976), Adamchak (1979), and Faulkner (1975, 1987), among others. Hughes and Paik (2013) reviewed and summarized some of these studies. During the 1990s, in addition to the lateral-torsional buckling problem under axial compression alone (Danielson et al. 1990, Danielson 1995, Hu et al. 1997), the effect of combined axial compression and lateral loads was studied by Hughes and Ma (1996a, 1996b) and by Hu et al. (2000), among others.

Although nonlinear finite element methods can accurately analyze the lateral-torsional buckling behavior in any specific case, it is not straightforward to derive theoretical solutions for the lateral-torsional buckling strength of a plate–stiffener combination model, as described in Chapter 2, considering the general section deformations depicted in Figure 5.9a. For practical design purposes, however, it would be more desirable to use a closed-form expression of the lateral-torsional buckling strength that is based on the corresponding analytical solutions.

Related to this problem, different idealizations of the tripping deformations may be made as approximations to the most general case of tripping deformations, as shown in Figure 5.9a, which all potentially account for the coupling effect between the flexural column buckling and lateral-torsional buckling. Three possible idealizations are as follows:

- Flexible web without the plate rotational restraint (see Figure 5.9b)
- Rigid web with the plate rotational restraint
- Rigid web without the plate rotational restraint (see Figure 5.9c)

Although the rotational restraints between the stiffener web and the attached plating generally play an important role in the lateral-torsional buckling behavior or local buckling of the stiffener web, the rotational restraint effects from the plating may be ignored if the plating between the stiffeners buckles before lateral-torsional buckling occurs, which implies that the contribution of the attached plating to restrict the rotation of the stiffener web about the plate–web junction is small, and thus the stiffener and the attached plating may be considered to be pin-joined.

This assumption will arguably result in a lower bound solution of the lateral-torsional buckling strength because plate rotational restraints will always exist to some extent, whereas the effect of buckled plating may approximately be incorporated within the effective plate width in the sectional properties of the plate–stiffener combination model without needing to change the axis of rotation. Solutions that include such an approach may typically be used when the ratio of the stiffener web height to the web thickness, h_w/t_w, is less than 20 (Hughes & Ma 1996a).

In contrast, as the height of the stiffener web increases, the stiffener web is likely to deflect sideways, and local buckling of the stiffener web can occur in some cases. Because this type of failure is in principle considered in Section 5.6, it is assumed for the lateral-torsional buckling analysis herein that the cross section of the stiffener web does not deflect locally, which is similar to the assumption of the ordinary beam–column theory, whereas it can twist sideways. This assumption may result in an optimistic strength prediction, particularly when the ratio of the stiffener web height to the web thickness is very large, which allows local web buckling to take place. The lateral-torsional buckling strength of a flat-bar stiffener may be considered to equal the local web buckling strength of the stiffener, as described in Section 5.6.

5.8.2 Closed-Form Expressions of Lateral-Torsional Buckling Strength

To derive a closed-form analytical solution of the lateral-torsional buckling strength, therefore, the rigid-web case without the plate rotational restraints may be adopted, as shown in Figure 5.9c.

The elastic lateral-torsional buckling strength of angle or Tee-type stiffeners under combined axial compression, σ_x ($= \sigma_{xM}$ or σ_{xav}), and uniform lateral pressure line load, $q = pb$ (a line force obtained by multiplying the uniform lateral pressure, p, and the breadth, b, of the plating between the stiffeners, as shown in Figure 2.5) in the x direction, can be calculated by applying the principle of minimum potential energy.

The strain energy, U, of the plate–stiffener combination model, as defined in Equation (2.68), with the attached effective plating stored during the lateral-torsional buckling can in this case be given with $v_w = z\theta$ ($=$ sideways deflection of the web), $v_T = h_w\theta$ ($=$ maximum value of v_w), and $\phi_B = \phi_T = \theta$, resulting in

$$U = \frac{E}{2}\int_0^a I_e\left(\frac{\partial^2 w}{\partial x^2}\right)^2 dx + \frac{E}{2}\int_0^a I_z h_w^2\left(\frac{\partial^2 \theta}{\partial x^2}\right)^2 dx + \frac{E}{2}\int_0^a 2I_{zy}h_w\frac{\partial^2 w}{\partial x^2}\frac{\partial^2 \theta}{\partial x^2}dx$$
$$+ \frac{G}{2}\int_0^a (J_w + J_f)\left(\frac{\partial\theta}{\partial x}\right)^2 dx$$

(5.37)

where I is the stiffener's moment of inertia and the attached effective width of the plating with regard to the y axis, given by

$$I_e = b_e\int_{-t/2}^{t/2} (z_p - z)^2 dz + t_w\int_{-h_w/2}^{h_w/2}\left(z_p - \frac{t}{2} - \frac{h_w}{2} - z\right)^2 dz$$
$$+ b_f\int_{-t_f/2}^{t_f/2}\left(z_p - \frac{t}{2} - h_w - \frac{t_f}{2} - z\right)^2 dz$$

(5.38a)

where I_z is the moment of inertia of the panel with respect to the z axis, given by

$$I_z = t\int_0^{b_e} y_0^2 dy + h_w\int_0^{t_w} y_0^2 dz + t_f\int_0^{b_f} (y_0 - y)^2 dz$$

(5.38b)

I_{zy} is the product of inertia of the panel with respect to the yz plane, given by

$$I_{zy} = \int_0^{b_e}\int_{-t/2}^{t/2} (z_p - z)y_0 dzdy + \int_0^{t_w}\int_{-h_w/2}^{h_w/2}\left(z_p - \frac{t}{2} - \frac{h_w}{2} - z\right)y_0 dzdy$$
$$+ \int_0^{b_f}\int_{-t_f/2}^{t_f/2}\left(z_p - \frac{t}{2} - h_w - \frac{t_f}{2} - z\right)(y_0 - y)dzdy$$

(5.38c)

J_w is the torsion constant for the web, given by

$$J_w = \frac{1}{3}t_w^3 h_w\left(1 - \frac{192}{\pi^5}\frac{t_w}{h_w}\sum_{n=1,3,5}^{\infty}\frac{1}{n^5}\tanh\frac{n\pi h_w}{2t_w}\right)$$

(5.38d)

J_f is the torsion constant for the flange, given by

$$J_f = \frac{1}{3}t_f^3 b_f\left(1 - \frac{192}{\pi^5}\frac{t_f}{b_f}\sum_{n=1,3,5}^{\infty}\frac{1}{n^5}\tanh\frac{n\pi b_f}{2t_f}\right)$$

(5.38e)

z_p is the distance from the middle plane of the attached plating to the elastic horizontal neutral axis with attached effective plating, and y_0 is the distance from the middle plane of the web to the elastic vertical neutral axis with attached effective plating.

In Equation (5.37), w is the deflection of the stiffener in the z direction, and θ is the rotation of the stiffener with regard to the x axis. The moments of inertia and other geometric constants used in Equation (5.37) are calculated for the stiffener with attached effective plating, which approximately accommodates the effects of plate buckling.

In contrast, the external potential energy, W, generated during lateral-torsional buckling is given by

$$
W = -\frac{1}{2}\int_{A_p}\sigma_p\int_0^a\left(\frac{\partial w}{\partial x}\right)^2 dxdA - \frac{1}{2}\int_{A_w}\sigma_w\int_0^a\left[\left(\frac{\partial w}{\partial x}\right)^2 + z^2\left(\frac{\partial \theta}{\partial x}\right)^2\right]dxdA
$$
$$
-\frac{1}{2}\int_{A_f}\sigma_f\int_0^a\left[h_w^2\left(\frac{\partial \theta}{\partial x}\right)^2 + \left(\frac{\partial w}{\partial x}\right)^2 - 2y\frac{\partial w}{\partial x}\frac{\partial \theta}{\partial x} + y^2\left(\frac{\partial \theta}{\partial x}\right)^2\right]dxdA
$$
$$(5.39)$$

where σ_p, σ_w, and σ_f are the axial stresses in the plate, web, and flange, respectively. $\int_{A_p}()dA$, $\int_{A_w}()dA$, and $\int_{A_f}()dA$ indicate the integration for the corresponding section area of plating (i.e., $A_p = b_e t$), web (i.e., $A_w = h_w t_w$), and flange (i.e., $A_f = b_f t_f$), respectively.

The total potential energy, Π, is obtained from the sum of U in Equation (5.37) and W in Equation (5.39) as follows:

$$\Pi = U + W \tag{5.40}$$

When the plate–stiffener combination model, as described in Chapter 2, which is assumed to be simply supported at both ends, is subjected to combined axial compressive stress, σ_x, and uniform lateral line load, $q = pb$, the axial stresses at the plate (i.e., σ_p), web (i.e., σ_w), and flange (i.e., σ_f) can be calculated for the stiffener with the associated effective plating as follows:

$$
\sigma_p = \sigma_x - \frac{q}{I_e}z_p\frac{x(L-x)}{2},
$$
$$
\sigma_w = \sigma_x - \frac{q}{I_e}(z_p-z)\frac{x(a-x)}{2}, \tag{5.41}
$$
$$
\sigma_f = \sigma_x - \frac{q}{I_e}(z_p-h_w)\frac{x(a-x)}{2}
$$

The displacement functions of the stiffener due to tripping are assumed (because the stiffener web does not deflect locally and the end conditions are simply supported) as follows:

$$
w = \sum_{m=1}A_m\sin\frac{m\pi x}{a}, \quad \theta = \sum_{m=1}B_m\sin\frac{m\pi x}{a} \tag{5.42}
$$

where A_m and B_m are unknown constants.

In plated structures, a stiffened panel is usually a multi-bay structure supported by transverse frames or brackets. Therefore, Equation (5.42) may be simplified further by taking only the predominant tripping half-wave number of the stiffener between two transverse frames or brackets for the plating's post-buckling behavior. Usually, the primary lateral-torsional buckling half-wave number is not known beforehand, but it can be determined so that the resulting lateral-torsional buckling strength must be the lowest of those obtained for potential half-wave numbers.

Substituting Equation (5.42) into the total potential energy equation, Π, Equation (5.40), and applying the principle of minimum potential energy, the unknown constants A_m and B_m can be determined as follows:

$$\frac{\partial \Pi}{\partial A_m} = 0, \quad \frac{\partial \Pi}{\partial B_m} = 0 \tag{5.43}$$

The characteristic equation for elastic flexural-torsional buckling of the stiffener can turn out as a bifurcation condition. The elastic flexural-torsional buckling stress, σ_E^T, of the plate–stiffener combination model is obtained by solution of the characteristic equation with regard to the axial compressive stress, whereas the lateral pressure is regarded as a given constant load. Interested readers may refer to Hughes and Ma (1996a, 1996b).

5.8.2.1 Elastic Flexural-Torsional Buckling Strength of Asymmetric Angle Stiffeners

A closed-form expression of the elastic flexural-torsional buckling strength of asymmetric angle stiffeners can be obtained with compressive stress taken as negative as follows:

$$\sigma_E^T = (-1) \min_{m=1,2,3,\dots} \left| \frac{C_2 + \sqrt{C_2^2 - 4C_1 C_3}}{2C_1} \right| \tag{5.44}$$

where

$$C_1 = (b_e t + h_w t_w + b_f t_f) I_p - S_f^2,$$

$$C_2 = -I_p \left[EI \left(\frac{m\pi}{a} \right)^2 - \frac{qa^2 S_1}{12 I_e} \left(1 - \frac{3}{m^2 \pi^2} \right) \right] - (b_e t + h_w t_w + b_f t_f)$$

$$\times \left[G(J_w + J_f) + EI_z h_w^2 \left(\frac{m\pi}{a} \right)^2 - \frac{qa^2 S_2}{12 I_e} \left(1 - \frac{3}{m^2 \pi^2} \right) \right]$$

$$+ 2S_f \left[EI_{zy} h_w \left(\frac{m\pi}{a} \right)^2 - \frac{qa^2 S_3}{12 I_e} \left(1 - \frac{3}{m^2 \pi^2} \right) \right],$$

$$C_3 = \left[EI \left(\frac{m\pi}{a} \right)^2 - \frac{qa^2 S_1}{12 I_e} \left(1 - \frac{3}{m^2 \pi^2} \right) \right]$$

$$\times \left[G(J_w + J_f) + EI_z h_w^2 \left(\frac{m\pi}{a} \right)^2 - \frac{qa^2 S_2}{12 I_e} \left(1 - \frac{3}{m^2 \pi^2} \right) \right]$$

$$- \left[EI_{zy} h_w \left(\frac{m\pi}{a} \right)^2 - \frac{qa^2 S_3}{12 I_e} \left(1 - \frac{3}{m^2 \pi^2} \right) \right]^2,$$

$$S_f = -\frac{t_f b_f^2}{2},$$

$$S_1 = -\left(z_p - h_w \right) t_f b_f - b_e t z_p - h_w t_w \left(z_p - \frac{h_w}{2} \right),$$

$$S_2 = -\left(z_p - h_w \right) t_f \left(h_w^2 b_f + \frac{b_f^3}{3} \right) - h_w^3 t_w \left(\frac{1}{3} z_p - \frac{h_w}{4} \right),$$

$$S_3 = (z_p - h_w)\frac{b_f^2 t_f}{2},$$

$$I_e = \frac{b_e t^3}{12} + A_p z_p^2 + \frac{t_w h_w^2}{12} + A_w\left(z_p - \frac{t}{2} - \frac{h_w}{2}\right)^2 + \frac{b_f t_f^3}{12} + A_f\left(z_p - \frac{t}{2} - h_w - \frac{t_f}{2}\right)^2,$$

$$I_z = A_p y_0^2 + A_w y_0^2 + A_f\left(y_0^2 - b_f y_0 + \frac{b_f^2}{3}\right),$$

$$I_{zy} = A_p z_p y_0 + A_w\left(z_p - \frac{t}{2} - \frac{h_w}{2}\right)y_0 + A_f\left(z_p - \frac{t}{2} - h_w - \frac{t_f}{2}\right)\left(y_0 - \frac{b_f}{2}\right).$$

I_p is the polar moment of inertia of stiffener about the toe, given by

$$I_p = \frac{t_w h_w^3}{3} + \frac{t_w^3 h_w}{3} + \frac{b_f^3 t_f}{3} + \frac{b_f t_f^3}{3} + A_f h_w^2$$

$$z_p = \frac{0.5 A_w(t + h_w) + A_f(0.5t + h_w + 0.5t_f)}{b_e t + h_w t_w + b_f t_f}$$

$$y_0 = \frac{b_f^2 t_f}{2(b_e t + h_w t_w + b_f t_f)}$$

q is the equivalent line pressure ($q = pb$), and m is the tripping half-wave number of the stiffener.

5.8.2.2 Elastic Flexural-Torsional Buckling Strength of Symmetric Tee Stiffeners

A closed-form expression of the elastic flexural-torsional buckling strength of symmetric Tee stiffeners can be obtained with compressive stress taken as negative as follows:

$$\sigma_E^T = (-1)\ \min_{m=1,2,3,\ldots}\left|\frac{-a^2 G(J_w + J_f) + EI_f\, h_w^2 m^2 \pi^2}{I_p a^2} + \frac{qa^2}{12}\frac{S_4}{I_e I_p}\left(1 - \frac{3}{m^2 \pi^2}\right)\right| \qquad (5.45)$$

where

$$S_4 = -(z_p - h_w)t_f\left(\frac{h_w^2 b_f + b_f^3}{12}\right) - h_w^3 t_w\left(\frac{1}{3}z_p - \frac{h_w}{4}\right),$$

$$I_p = \frac{t_w h_w^3}{3} + \frac{t_w^3 h_w}{12} + \frac{b_f t_f^3}{3} + \frac{b_f^3 t_f}{12} + A_f h_w^2, \quad I_f = \frac{b_f^3 t_f}{12}.$$

5.8.2.3 Elastic Flexural-Torsional Buckling Strength of Flat-Bar Stiffeners

As previously noted, the elastic flexural-torsional buckling strength of flat-bar stiffeners is taken as being approximately equal to the local buckling strength, σ_E^W, of the stiffener web, as defined in Equation (5.33) with Equation (5.35), that is, $\sigma_E^T = \sigma_E^W$.

5.8.2.4 Effect of Welding Induced Residual Stresses

To account for the effects of welding residual stresses, the flexural-torsional buckling stress computed earlier is typically reduced by an effective compressive residual stress, σ_{rs}^*. Danielson (1995) suggested an empirical formula of σ_{rs}^* as follows:

$$\sigma_{rs}^* = \sigma_{rc}\left(1 + \frac{2\pi^2 I}{b^3 t}\right) \tag{5.46}$$

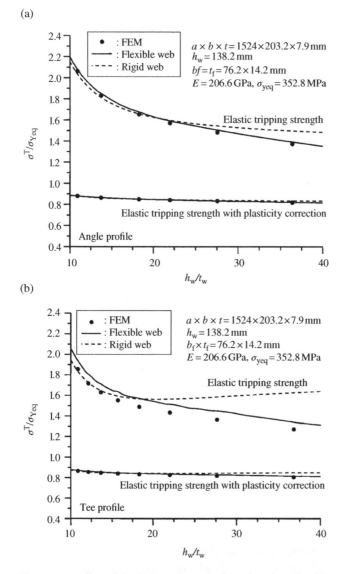

(a)

(b)

Figure 5.10 Effect of the h_w/t_w ratio on the lateral-torsional buckling strength of a plate and flange–stiffener combination without considering the plate rotational restraints (σ_{Yeq} is the equivalent yield strength of the plate–stiffener combination): (a) angle section stiffener with attached effective plating; (b) Tee-section stiffener with attached effective plating.

where σ_{rc} is the compressive residual stress in the stiffener web and I is the moment of inertia for the full section of the plate–stiffener combination model.

Figure 5.10 confirms the validity of the lateral-torsional buckling strength formulations by comparison with nonlinear finite element method solutions. The effect of the h_w/t_w ratio on the lateral-torsional buckling strength for a particular plate–stiffener combination model is investigated in this figure. The two types of idealizations, that is, one for a flexible web as given by Hughes and Ma (1996a) using a simplified numerical method and the other for a rigid web as predicted by Equations (5.44) and (5.45), both without plate rotational restraints, are considered in addition to more refined eigenvalue finite element method solutions. It is apparent from Figure 5.10 that the effects of stiffener web deflection can be ignored when the h_w/t_w ratio is low, but the rigid-web approximation, which neglects the effects of stiffener web deflection, can result in overestimation of the elastic lateral-torsional buckling strength for higher h_w/t_w ratios. The elastic lateral-torsional buckling strength increases further because of the rotational restraint effect along the web–plate intersection (Hu et al. 2000).

5.9 Elastic–Plastic Buckling Strength

The stocky stiffened panel will buckle in the inelastic regime with a certain degree of plasticity. Although the nonlinear finite element method can deal accurately with this behavior, it is a computing- and labor-intensive process. For practical design purposes, an easier alternative to account for the influence of plasticity is to make the plasticity correction of the elastic buckling strength using the Johnson–Ostenfeld formulation method, as described in Equation (2.93).

The applicability of the Johnson–Ostenfeld formulation method in predicting the inelastic lateral-torsional buckling strength is confirmed in Figure 5.10. It is interesting to note from this figure that the inelastic flexural-torsional buckling strength may not be significantly affected by the stiffener web deflection and the Johnson–Ostenfeld formulation method is useful for approximate prediction of the inelastic lateral-torsional buckling strength, which is regarded as the ultimate strength for practical design purposes (Adamchak 1979, Hughes & Paik 2013).

References

Adamchak, J.C. (1979). *Design equations for tripping of stiffeners under in-plane and lateral loads*. DTNSRDC-79/064, Naval Surface Warfare Center, Washington, DC, October.

Allen, H.G. & Bulson, P.S. (1980). *Background to buckling*. McGraw-Hill, London.

Bleich, F. (1952). *Buckling strength of metal structures*. McGraw-Hill, New York.

Danielson, D.A. (1995). Analytical tripping loads for stiffened plates. *International Journal of Solids and Structures*, **32**(8/9): 1317–1328.

Danielson, D.A., Kihl, D.P. & Hodges, D.H. (1990). Tripping of thin-walled plating stiffeners in axial compression. *Thin-Walled Structures*, **10**(2): 121–142.

Faulkner, D. (1975). Compression strength of welded grillages. Chapter 21 in *Ship structural design concepts*, Edited by Harvey Evans, J., Cornell Maritime Press, Cambridge, MD, 633–712.

Faulkner, D. (1987). Toward a better understanding of compression induced tripping. In *Steel and aluminium structures*, Edited by Narayanan, R., Elsevier Applied Science, Barking, 159–175.

Faulkner, D., Adamchak, J.C., Snyder, G.J. & Vetter, M.R. (1973). Synthesis of welded grillages to withstand compression and normal loads. *Computers & Structures*, **3**: 221–246.

Hu, S.Z., Chen, Q., Pegg, N. & Zimmerman, T.J.E. (1997). Ultimate collapse tests of stiffened plate ship structural units. *Marine Structures*, **10**: 587–610.

Hu, Y., Chen, B. & Sun, J. (2000). Tripping of thin-walled stiffeners in the axially compressed stiffened panel with lateral pressure. *Thin-Walled Structures*, **37**: 1–26.

Hughes, O.F. & Ma, M. (1996a). Elastic tripping analysis of asymmetrical stiffeners. *Computers & Structures*, **60**(3): 369–389.

Hughes, O.F. & Ma, M. (1996b). Inelastic analysis of panel collapse by stiffener buckling. *Computers & Structures*, **61**(1): 107–117.

Hughes, O.F. & Paik, J.K. (2013). *Ship structural analysis and design*. The Society of Naval Architects and Marine Engineers, Alexandria, VA.

Paik, J.K., Thayamballi, A.K. & Park, Y.I. (1998). Local buckling of stiffeners in ship plating. *Journal of Ship Research*, **42**(1): 56–67.

Seydel, E. (1933). Uber das Ausbeulen von rechteckigen Isotropen oder orthogonal-anisotropen Platten bei Schubbeanspruchung. *Ingenieur-Archiv*, **4**: 169 (in German).

Smith, C.S. (1976). Compressive strength of welded steel ship grillages. *Transactions of the Royal Institution of Naval Architects*, **118**: 325–359.

6

Large-Deflection and Ultimate Strength Behavior of Stiffened Panels and Grillages

6.1 Fundamentals of Stiffened Panel Ultimate Strength Behavior

A stiffened panel is an assembly of plating and stiffeners (support members). Even if the stiffened panel or its parts initially buckle in the elastic or even inelastic regime, the stiffened panel will normally be able to sustain further applied loads. The ultimate strength of the stiffened panel is eventually reached by excessive plasticity and/or by stiffener failure.

A method for the prediction of a stiffened panel's ultimate strength has its own level of accuracy. In addition to the inherent and modeling uncertainties in the structural properties and failure phenomena involved, the following four aspects are the primary reasons for such differences:

- The numerous collapse modes involved, the manner in which they are idealized for consideration, and the effects of failure mode interactions that must be considered
- Differences in the evaluation of the ineffectiveness in the in-plane stiffness of the plating between the stiffeners
- Consideration of welding induced initial imperfections and existing structural damage
- Consideration of rotational restraints between the plating and the stiffeners and/or between the stiffener web and the flange

First, not all theoretically possible collapse modes are usually considered in the development of any specific design-oriented method of predicting strength. Second, it is important to accurately evaluate the ineffectiveness of the plating or stiffeners after local buckling and/or with large deflection. As the loads increase, the effective width or breadth of the buckled or deflected plating would, by definition, vary because it is a function of the applied stresses. However, most simplified methods assume that the effective width of the plating does not depend on the applied loads, and the ultimate effective width of the plating is instead used as a convenient "constant." Third, fabrication related initial deformations and residual stresses are not always treated as the parameters of influence in development of the method. Most methods account for the influence of initial imperfections in the plating between stiffeners, but only some of them include initial imperfection effects for the stiffeners. Finally, stiffeners have some rotational restraints at their line of attachment to the plating and/or along the stiffener web–flange intersection. Such restraints affect the failure of a stiffener, but most methods neglect this effect.

Ultimate Limit State Analysis and Design of Plated Structures, Second Edition. Jeom Kee Paik.
© 2018 John Wiley & Sons Ltd. Published 2018 by John Wiley & Sons Ltd.

In the ultimate limit state (ULS) design of stiffened panels using Equation (1.17), the capacity indicates the ultimate strength that may be determined by relevant ultimate strength formulations, whereas the demand represents the extreme value of the load effects (stresses), which are determined with the classical theory of structural mechanics or linear elastic finite element analysis once the overall loads are known.

This chapter describes the ultimate strength formulations for stiffened panels and grillages. The formulations presented are designed to be more sophisticated than previous theoretically based simplified procedures. It is noted that the theories and methodologies described in this chapter can be commonly applied to both steel- and aluminum-stiffened panels.

6.2   Classification of Panel Collapse Modes

When subjected to predominantly axial tension, a stiffened panel may fail by gross yielding. In contrast, a stiffened panel under predominantly compressive loads may potentially show a variety of failure modes until the ultimate strength is reached, as shown in Figure 6.1, with the focus on the collapse patterns induced by axial compressive loading for illustrative purposes. The primary modes of overall failure for a stiffened panel are categorized into the following six types:

- Mode I: Overall collapse of plating and stiffeners as a unit
  - Mode I-1: Mode I for uniaxially stiffened panels (see Figure 6.1a)
  - Mode I-2: Mode I for cross-stiffened panels (grillages) (see Figure 6.1b)
- Mode II: Plate collapse without distinct failure of stiffener (see Figure 6.1c)
- Mode III: Beam–column collapse (see Figure 6.1d)
- Mode IV: Collapse by local web buckling of stiffener (see Figure 6.1e)
- Mode V: Collapse by lateral-torsional buckling of stiffener (see Figure 6.1f)
- Mode VI: Gross yielding

Mode I depicts the typical collapse pattern when the stiffeners are relatively weak. In this case, the stiffeners can buckle together with the plating as a unit, and the overall buckling behavior is initially elastic. The stiffened panel can normally sustain further loading even after overall buckling occurs in the elastic regime, and the ultimate strength is eventually reached by the formation of a large yield region inside the panel and/or along the panel edges. In mode I, the collapse behavior of a uniaxially stiffened panel, termed mode I-1, differs slightly from that of a cross-stiffened panel (grillage), termed mode I-2. The former is in fact initiated by beam–column failure, whereas the latter failure resembles that of an "orthotropic plate."

Mode II represents the collapse pattern in which the panel collapses by yielding along the plate–stiffener intersection at the panel edges, with no stiffener failure. This type of collapse can be important in some cases in which the panel is subjected predominantly to biaxial compressive loads.

Mode III indicates a failure pattern in which the ultimate strength is reached by yielding of the plate–stiffener combination at the mid-span. Mode III failure typically occurs when the dimensions of the stiffeners are intermediate, that is, neither weak nor very strong.

Modes IV and V typically arise from stiffener-induced failure when the ratio of the stiffener web height to the stiffener web thickness is large and/or when the type of stiffener flange is unable to remain straight so that the stiffener web buckles or twists sideways. Mode IV represents a failure pattern in which the panel collapses by local compressive buckling of the stiffener web, whereas mode V can occur when the ultimate strength is reached after lateral-torsional buckling (also called tripping) of the stiffener.

Mode VI typically takes place when the panel slenderness is very low (i.e., the panel is very stocky) and/or when the panel is subjected predominantly to axial tensile loading so that neither local nor overall buckling occurs until the panel cross section yields in large regions or entirely.

Although Figure 6.1 illustrates each collapse pattern separately, some collapse modes may in some cases interact and occur simultaneously. It is also important to realize that the division of the behavior of a stiffened panel as illustrated earlier is (i) artificial and (ii) not necessarily completely descriptive of all anticipated actual behavior, although based on insights and experiences, such division is thought to be adequate for design

(a)

(b)

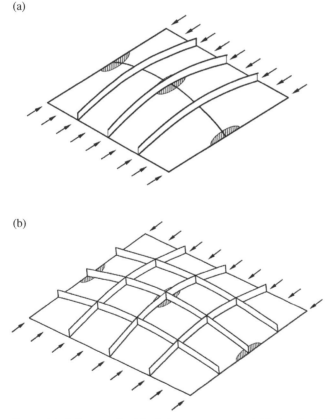

Figure 6.1 Failure modes (shaded areas represent yielded region): (a) mode I-1: overall collapse of a uniaxially stiffened panel; (b) mode I-2: overall collapse of a cross-stiffened panel; (c) mode II: plate collapse without distinct failure of stiffener; (d) mode III: beam–column collapse; (e) mode IV: collapse by local web buckling of stiffener; (f) mode V: collapse by flexural-torsional buckling of stiffener.

(c)

(d)

(e)

(f)

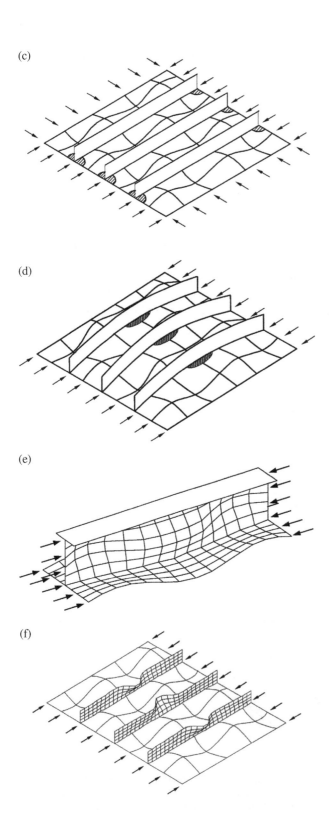

Figure 6.1 (Continued)

purposes in plated structures. Further, even accepting these idealizations of behavior, the calculation of the ultimate strength of the stiffened panel under combined loads is not straightforward because of the interplay of the various factors, such as geometric and material properties, loading, boundary conditions, welding induced initial imperfections, and existing structural damages.

For practical design purposes, therefore, it is typically considered that the collapse of the stiffened panels occurs at the lowest value among the various ultimate loads calculated when considering each of the previously mentioned six collapse patterns separately.

6.3 Structural Idealizations of Stiffened Panels

The structural idealizations of the stiffened panels are described in Section 5.2. The characteristics of the fabrication related initial imperfections are idealized as described in Section 1.7.

A stiffened panel is surrounded by strong support members such as longitudinal girders and transverse frames, and it has stiffeners in the x and y directions. The stiffeners are attached to one side of the panel, that is, they are placed on the positive side in the z direction, as shown in Figure 5.3. The length and breadth of the cross-stiffened panel (or grillage) are denoted by L and B, respectively, as shown in Figure 5.4a. The panel's stiffeners are identical in terms of geometry and material, with the same spacing. The number of stiffeners in the x or y direction is n_{sx} or n_{sy}, and thus the stiffener spacing in the x or y direction is $a = L/(n_{sy} + 1)$ or $b = B/(n_{sx} + 1)$. In many cases in a continuous stiffened plate structure, the stiffeners in the stiffened panel are positioned in only one direction. In such a case, the panel length is denoted by a in which the coordinate of the panel is taken so that the stiffeners are positioned in the x direction, as shown in Figure 5.4b.

The thickness of the plating between the stiffeners is t. In some panels, the thicknesses of the individual plates between the stiffeners may not be the same. In this case, the plate thickness t used in the ultimate strength formulations of the panels is represented by an equivalent plate thickness that is approximated as follows:

$$t = \frac{b}{B} \sum_{i=1}^{n_{sx}+1} t_i \tag{6.1}$$

where t_i is the thickness of the ith plate.

The dimensions of the stiffeners in the x or y directions are defined in Figure 5.3. The plating and stiffeners are of the same material, although their yield stresses may differ. The plating and stiffeners have the same value of either Young's modulus or Poisson's ratio, defined by E and ν, respectively, and the elastic shear modulus is defined by $G = E/[2(1+\nu)]$. The material yield stress of the plating is σ_{Yp}. The slenderness ratio and the flexural (bending) rigidity of the plating between the stiffeners are given by $\beta = (b/t)\sqrt{\sigma_{Yp}/E}$ and $D = Et^3/[12(1-\nu^2)]$, respectively.

In some plate panels, the material yield stress of the plating differs from that of the stiffeners. In a steel-stiffened panel, for example, the plating may be made of mild steel, whereas the stiffeners are made of high-tensile steel. In an aluminum-stiffened panel, the material yield stress of the stiffeners is sometimes greater than that of the plating, with different tempers of aluminum alloys as indicated in Tables 1.3 and 1.4. The yield stress is σ_{Yw} for the stiffener web and σ_{Yf} for the stiffener flange. In this case, an equivalent yield stress σ_{Yeq} can be defined to represent the yield stress of the entire panel:

$$\sigma_{Yeq} = \frac{Bt\sigma_{Yp} + n_{sx}(h_{wx}t_{wx}\sigma_{Yw} + b_{fx}t_{fx}\sigma_{Yf})}{Bt + n_{sx}(h_{wx}t_{wx} + b_{fx}t_{fx})} \tag{6.2a}$$

for a uniaxially stiffened panel in the x direction

$$\sigma_{Yeq} = \frac{1}{2}\left[\frac{Bt\sigma_{Yp} + n_{sx}(h_{wx}t_{wx}\sigma_{Yw} + b_{fx}t_{fx}\sigma_{Yf})}{Bt + n_{sx}(h_{wx}t_{wx} + b_{fx}t_{fx})} + \frac{Lt\sigma_{Yp} + n_{sy}(h_{wy}t_{wy}\sigma_{Yw} + b_{fy}t_{fy}\sigma_{Yf})}{Lt + n_{sy}(h_{wy}t_{wy} + b_{fy}t_{fy})}\right]$$

for a cross-stiffened panel

$$\tag{6.2b}$$

The equivalent yield stress of a plate–stiffener combination model indicated in Table 2.1 is defined as follows:

$$\sigma_{Yeq} = \frac{bt\sigma_{Yp} + h_{wx}t_{wx}\sigma_{Yw} + b_{fx}t_{fx}\sigma_{Yf}}{bt + h_{wx}t_{wx} + b_{fx}t_{fx}} \tag{6.2c}$$

for the plate-stiffener combination model in the x direction

$$\sigma_{Yeq} = \frac{at\sigma_{Yp} + h_{wy}t_{wy}\sigma_{Yw} + b_{fy}t_{fy}\sigma_{Yf}}{at + h_{wy}t_{wy} + b_{fy}t_{fy}} \tag{6.2d}$$

for the plate-stiffener combination model in the y direction

Because the stiffened panel is supported by strong members, the rotational restraints along the panel edges depend on the relative values of the torsional rigidities of the support members to the flexural rigidity of the panel, and these values are neither zero nor infinite. For the sake of simplicity, however, it is often assumed that the stiffened panel edges are simply supported, with zero deflection and zero rotational restraints along the four edges and with all edges kept straight. In engineering practice, this approximation is considered adequate. As described in Chapters 3, 4, and 5, however, the effects of rotational restraints at the four edges of the plating and at the junctions of the plate and the stiffener or the stiffener web–flange are considered in calculations of the local buckling and ultimate strength of either the plating between the stiffeners or the stiffener web, the latter being associated with collapse modes IV and V.

Six potential stress components act on a stiffened panel—longitudinal axial stress, transverse axial stress, edge shear, longitudinal in-plane bending, transverse in-plane bending, and lateral pressure—as described in Section 5.2.3. To develop the ultimate strength formulations for the panel, this chapter simplifies some of the applied load components that depend on the collapse mode: σ_{xav} is the average axial stress in the x direction, σ_{yav} is the average axial stress in the y direction, τ_{av} is the average edge shear stress, and p is the lateral pressure.

The average stress components applied in a stiffened panel are then defined as follows, where in the x direction, σ_{x2} is always larger than σ_{x1}, or in the y direction, σ_{y2} is always larger than σ_{y1}:

6.3.1 Collapse Modes I and VI

The effect of in-plane bending moments is neglected over the stiffened panel, and the following four load components are defined in this case:

$$\sigma_{xav} = \frac{\sigma_{x1} + \sigma_{x2}}{2}, \quad \sigma_{yav} = \frac{\sigma_{y1} + \sigma_{y2}}{2}, \quad \tau_{av}, \quad p \tag{6.3a}$$

6.3.2 Collapse Modes II, III, IV, and V

The most highly stressed plating between the stiffeners is considered to determine the panel's ultimate strength.

$$\sigma_{xM} = \sigma_{x2} - \frac{b}{2B}(\sigma_{x2} - \sigma_{x1}), \quad \sigma_{yM} = \sigma_{y2} - \frac{a}{2L}(\sigma_{y2} - \sigma_{y1}), \quad \tau_{av}, \quad p \tag{6.3b}$$

It is considered that the compressive stress is negative and that the tensile stress is positive. That is, the axial load has a negative value when the corresponding load is compressive, and vice versa.

6.4 Nonlinear Governing Differential Equations of Stiffened Panels

The nonlinear governing differential equations of a stiffened panel can be divided into two groups depending on the buckling modes: those for overall panel buckling and those for local plate buckling. The former type of post-buckling behavior is analyzed with the large-deflection orthotropic plate theory, whereas the latter type is analyzed with the large-deflection isotropic plate theory.

6.4.1 Large-Deflection Orthotropic Plate Theory

If a stiffened panel has several small stiffeners, it may buckle in the overall grillage buckling mode under compressive loads, as shown in Figure 5.1a. In this case, the panel may be idealized as an orthotropic plate, in which the stiffeners are in a sense smeared into the plating.

The orthotropic plate approach implies that the stiffeners are relatively numerous and small so that they deflect together with the plating as a unit and that the physical stiffeners remain stable through the ranges of orthotropic plate behavior. It is recognized that application of the orthotropic plate theory to cross-stiffened panels must be restricted to those with more than three stiffeners in each direction and that the stiffeners in a given direction must be similar (Smith 1966, Troitsky 1976, Mansour 1977). For an approximation, however, the post-buckling behavior of a plate panel with numerous small stiffeners in one direction may also be analyzed with the orthotropic plate theory.

The overall buckling behavior of the panel can then be analyzed by solving the two nonlinear governing differential equations of the large-deflection orthotropic plate theory: the equilibrium equation and the compatibility equation (e.g., Troitsky 1976). By accounting for the effects of the initial deflections, the two governing differential equations for the cross-stiffened panel (i.e., one with stiffeners in both the x and y directions) can be written as follows:

$$D_x \frac{\partial^4 w}{\partial x^4} + 2H \frac{\partial^4 w}{\partial x^2 \partial y^2} + D_y \frac{\partial^4 w}{\partial y^4}$$

$$- t \left[\frac{\partial^2 F}{\partial y^2} \frac{\partial^2 (w + w_0)}{\partial x^2} - 2 \frac{\partial^2 F}{\partial x \partial y} \frac{\partial^2 (w + w_0)}{\partial x \partial y} + \frac{\partial^2 F}{\partial x^2} \frac{\partial^2 (w + w_0)}{\partial y^2} + \frac{p}{t} \right] = 0 \tag{6.4a}$$

$$\frac{1}{E_y} \frac{\partial^4 F}{\partial x^4} + \left(\frac{1}{G_{xy}} - 2 \frac{v_x}{E_x} \right) \frac{\partial^4 F}{\partial x^2 \partial y^2} + \frac{1}{E_x} \frac{\partial^4 F}{\partial y^4}$$

$$- \left[\left(\frac{\partial^2 w}{\partial x \partial y} \right)^2 - \frac{\partial^2 w}{\partial x^2} \frac{\partial^2 w}{\partial y^2} + 2 \frac{\partial^2 w_0}{\partial x \partial y} \frac{\partial^2 w}{\partial x \partial y} - \frac{\partial^2 w_0}{\partial x^2} \frac{\partial^2 w}{\partial y^2} - \frac{\partial^2 w}{\partial x^2} \frac{\partial^2 w_0}{\partial y^2} \right] = 0 \tag{6.4b}$$

where w_0 and w are the initial and added deflection functions for the orthotropic plate, respectively; F is Airy's stress function; E_x and E_y are the elastic moduli of the orthotropic plate in the x and y directions, respectively; G_{xy} is the elastic shear modulus of the orthotropic plate; D_x and D_y are the flexural rigidities of the orthotropic plate in the x and y directions, respectively; and H is the effective torsional rigidity of the orthotropic plate.

Once Airy's stress function, F, and the added deflection, w, are known, the stresses inside the panel can be calculated as follows:

$$\sigma_x = \frac{\partial^2 F}{\partial y^2} - \frac{E_x z}{1 - v_x v_y} \left(\frac{\partial^2 w}{\partial x^2} + v_y \frac{\partial^2 w}{\partial y^2} \right) \tag{6.5a}$$

$$\sigma_y = \frac{\partial^2 F}{\partial x^2} - \frac{E_y z}{1 - v_x v_y} \left(\frac{\partial^2 w}{\partial y^2} + v_x \frac{\partial^2 w}{\partial x^2} \right) \tag{6.5b}$$

$$\tau = - \frac{\partial^2 F}{\partial x \partial y} - 2 G_{xy} z \frac{\partial^2 w}{\partial x \partial y} \tag{6.5c}$$

where σ_x and σ_y are the normal stresses in the x and y direction, respectively; τ is the shear stress; and the z axis is along the plate's thickness, with $z = 0$ at the mid-thickness.

The reliability of the orthotropic plate theory analysis depends significantly on various elastic constants that must be determined when a stiffened panel is replaced with an equivalent orthotropic plate. In the following, the large-deflection orthotropic plate theory constants developed by Paik et al. (2001) are introduced.

An isotropic plate has two independent elastic constants: the elastic modulus, E, and the Poisson ratio, v. For the orthotropic plate, four elastic constants—E_x, E_y, v_x, and v_y—are required to describe the orthotropic stress–strain relationship of the plate. In real stiffened panels, the anisotropy in the two mutually perpendicular directions arises from different properties of the geometry rather than from different properties of the material, which itself is inherently isotropic. In this case, the corresponding orthotropic constants of the elastic moduli can be approximately given by

$$E_x = E \left(1 + \frac{n_{sx} A_{sx}}{Bt} \right) \tag{6.6a}$$

$$E_y = E \left(1 + \frac{n_{sy} A_{sy}}{Lt} \right) \tag{6.6b}$$

$$G_{xy} = \frac{E_x E_y}{E_x + \left(1 + 2\sqrt{v_x v_y}\right) E_y} \approx \frac{\sqrt{E_x E_y}}{2\left(1 + \sqrt{v_x v_y}\right)} \tag{6.6c}$$

The flexural and torsional rigidities of the orthotropic plate are determined as follows:

$$D_x = \frac{E t^3}{12\left(1 - v_x v_y\right)} + \frac{E t\, z_{0x}^2}{1 - v_x v_y} + \frac{E I_x}{b} \tag{6.7a}$$

$$D_y = \frac{E t^3}{12\left(1 - v_x v_y\right)} + \frac{E t\, z_{0y}^2}{1 - v_x v_y} + \frac{E I_y}{a} \tag{6.7b}$$

$$H = \frac{1}{2}\left(v_y D_x + v_x D_y + G_{xy}\frac{t^3}{3}\right) \tag{6.7c}$$

where

$$I_x = \frac{t_{wx} h_{wx}^3}{12} + t_{wx} h_{wx}\left(\frac{h_{wx}}{2} + \frac{t}{2} - z_{0x}\right)^2 + \frac{b_{fx} t_{fx}^3}{12} + b_{fx} t_{fx}\left(\frac{t_{fx}}{2} + h_{wx} + \frac{t}{2} - z_{0x}\right)^2,$$

$$I_y = \frac{t_{wy} h_{wy}^3}{12} + t_{wy} h_{wy}\left(\frac{h_{wy}}{2} + \frac{t}{2} - z_{0y}\right)^2 + \frac{b_{fy} t_{fy}^3}{12} + b_{fy} t_{fy}\left(\frac{t_{fy}}{2} + h_{wy} + \frac{t}{2} - z_{0y}\right)^2,$$

$$z_{0x} = \frac{h_{wx} t_{wx}\left(h_{wx}/2 + t/2\right) + b_{fx} t_{fx}\left(t_{fx}/2 + h_{wx} + t/2\right)}{bt + h_{wx} t_{wx} + b_{fx} t_{fx}},$$

$$z_{0y} = \frac{h_{wy} t_{wy}\left(h_{wy}/2 + t/2\right) + b_{fy} t_{fy}\left(t_{fy}/2 + h_{wy} + t/2\right)}{at + h_{wy} t_{wy} + b_{fy} t_{fy}}.$$

For an isotropic plate, the flexural rigidities will of course simplify into the following well-known expression:

$$D_x = D_y = H = D = \frac{E t^3}{12(1 - v^2)} \tag{6.8}$$

To determine the various elastic constants indicated earlier, Poisson's ratios v_x and v_y due to structural orthotropy, which are not material properties but rather elastic constants that correspond to the given geometrical configuration, should be known in advance. Based on Betti's reciprocity theorem, the following two requirements are then pertinent:

$$v_x E_y = v_y E_x, \quad v_x D_y = v_y D_x \tag{6.9}$$

Substitution of Equations (6.6) and (6.7) into Equation (6.9) leads to

$$\left[\frac{E I_x}{b}\left(\frac{E_y}{E_x}\right)^2 - \frac{E I_y}{a}\left(\frac{E_y}{E_x}\right)\right] v_x^3$$

$$- \left[\frac{E_y}{E_x}\left(\frac{E t^3}{12} + E t z_{0x}^2 + \frac{E I_x}{b}\right) - \frac{E t^3}{12} - E t z_{0y}^2 - \frac{E I_y}{a}\right] v_x = 0 \tag{6.10}$$

The solution of Equation (6.10) together with Equation (6.9) results in the effective Poisson ratios in the x and y directions, namely,

$$v_x = c \left[\frac{\left(E_y/E_x\right)\left(Et^3/12 + Et\, z_{0x}^2 + EI_x/b\right) - Et^3/12 - Et\, z_{0y}^2 - EI_y/a}{\left(EI_x/b\right)\left(E_y/E_x\right)^2 - \left(EI_y/a\right)\left(E_y/E_x\right)} \right]^{0.5} \tag{6.11a}$$

$$v_y = \frac{E_y}{E_x} v_x$$

$$= c\frac{E_y}{E_x} \left[\frac{\left(E_y/E_x\right)\left(Et^3/12 + Et\, z_{0x}^2 + EI_x/b\right) - Et^3/12 - Et\, z_{0y}^2 - EI_y/a}{\left(EI_x/b\right)\left(E_y/E_x\right)^2 - \left(EI_y/a\right)\left(E_y/E_x\right)} \right]^{0.5} \tag{6.11b}$$

where c is a correction factor to correlate Poisson's ratios with $v_x = v_y = v$ for an isotropic plate, which may approximately be taken as $c = v/0.86$. Note also that $v_x = v_y = v$ if

$$\frac{EI_x}{b}\left(\frac{E_y}{E_x}\right)^2 = \frac{EI_y}{a}\left(\frac{E_y}{E_x}\right) \quad \text{or} \quad \frac{E_x}{E_y} = \frac{a I_x}{b I_y} \tag{6.11c}$$

For an orthotropic plate theory application, it is assumed that the effect of welding induced residual stresses can be neglected because the effects of both tensile and compressive residual stresses may be offset in the panel's overall deflection response. For an orthotropic plate under predominantly compressive loads, the membrane stress distribution inside the plate is not uniform, as illustrated in Figure 6.2. In this case, the maximum and minimum membrane stresses of the plate are determined at $z = 0$ as follows:

$$\sigma_{x\max} = \sigma_x\big|_{x=0,\,y=0} \tag{6.12a}$$

$$\sigma_{x\min} = \sigma_y\big|_{x=0,\,y=B/2} \tag{6.12b}$$

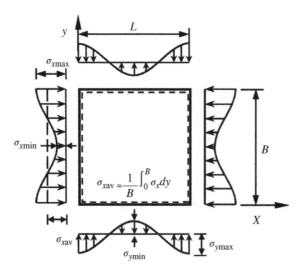

Figure 6.2 Membrane stress distribution inside an orthotropic plate under predominantly longitudinal axial compressive loads.

$$\sigma_{y\,\max} = \sigma_y\big|_{x=0,\,y=0} \tag{6.12c}$$

$$\sigma_{y\,\min} = \sigma_y\big|_{x=L/2,\,B=0} \tag{6.12d}$$

6.4.2 Large-Deflection Isotropic Plate Theory

When the stiffeners have sufficient strength, they will not fail before the plating between them buckles. In this case, the large-deflection behavior (including buckling) of the plating between stiffeners is of primary concern and can be analyzed by solving the nonlinear governing differential equations of the large-deflection isotropic plate theory, as described in Chapter 4.

6.5 Elastic Large-Deflection Behavior After Overall Grillage Buckling

In a manner similar to the isotropic plate theory method described in Chapter 4, the governing differential equations of the orthotropic plate, Equations (6.4a) and (6.4b), can be solved with the Galerkin method. In this case, the initial and added deflection functions of the orthotropic plate that satisfy the simply supported conditions at the panel edges can be presumed as follows:

$$w_0 = A_{0mn}\sin\frac{m\pi x}{L}\sin\frac{n\pi y}{B} \tag{6.13a}$$

$$w = A_{mn}\sin\frac{m\pi x}{L}\sin\frac{n\pi y}{B} \tag{6.13b}$$

where A_{0mn} and A_{mn} are the amplitudes of the initial and added deflection functions, respectively, and m and n are the buckling half-wave numbers in the x and y directions, respectively. For modeling the panel initial deflection, Section 1.7 can be referred to.

6.5.1 Lateral Pressure Loads

For an orthotropic plate under lateral pressure loads alone, the initial and added plate deflection functions are assumed as Equation (6.13) but with $m = n = 1$. In this case, the unknown amplitude $A_1 = A_{11}$ with $A_{01} = A_{011}$ can be determined as the solution to the following equation:

$$C_1 A_1^3 + C_2 A_1^2 + C_3 A_1 + C_4 = 0 \tag{6.14}$$

where $C_1 = \dfrac{\pi^2}{16}\left(E_x\dfrac{B}{a^3}+E_y\dfrac{L}{B^3}\right)$, $C_2 = \dfrac{3\pi^2 A_{01}}{16}\left(E_x\dfrac{B}{a^3}+E_y\dfrac{L}{B^3}\right)$, $C_3 = \dfrac{\pi^2 A_{01}^2}{8}\left(E_x\dfrac{B}{a^3}+E_y\dfrac{L}{B^3}\right)$

$+\dfrac{\pi^2}{t}\left(D_x\dfrac{B}{a^3}+2H\dfrac{1}{aB}+D_y\dfrac{L}{B^3}\right)$, $C_4 = -\dfrac{16aB}{\pi^4 t}p.$

The solution of Equation (6.14) can be obtained using the Cardano method or the FORTRAN computer subroutine CARDANO given in the appendices to this book.

The maximum and minimum membrane stresses inside the orthotropic plate can then be obtained from Equation (6.12) as follows:

$$\sigma_{x\,max} = -\frac{\pi^2 E_x A_1 (A_1 + 2A_{01})}{8L^2} \tag{6.15a}$$

$$\sigma_{x\,min} = \frac{\pi^2 E_x A_1 (A_1 + 2A_{01})}{8L^2} \tag{6.15b}$$

$$\sigma_{y\,max} = -\frac{\pi^2 E_y A_1 (A_1 + 2A_{01})}{8B^2} \tag{6.15c}$$

$$\sigma_{y\,min} = \frac{\pi^2 E_y A_1 (A_1 + 2A_{01})}{8B^2} \tag{6.15d}$$

6.5.2 Combined Biaxial Loads

In this case, the initial and added deflection functions are assumed as Equations (6.13a) and (6.13b), and the unknown amplitude A_{mn} can be determined by solving the following equation:

$$C_1 A_{mn}^3 + C_2 A_{mn}^2 + C_3 A_{mn} + C_4 = 0 \tag{6.16}$$

where

$$C_1 = \frac{\pi^2}{16} \left(E_x \frac{m^4 B}{L^3} + E_y \frac{n^4 L}{B^3} \right),$$

$$C_2 = \frac{3\pi^2 A_{0mn}}{16} \left(E_x \frac{m^4 B}{L^3} + E_y \frac{n^4 L}{B^3} \right),$$

$$C_3 = \pi^2 \frac{A_{0mn}^2}{8} \left(E_x \frac{m^4 B}{L^3} + E_y \frac{n^4 L}{B^3} \right) + \frac{m^2 B}{L} \sigma_{xav} + \frac{n^2 L}{B} \sigma_{yav}$$

$$+ \frac{\pi^2}{t} \left(D_x \frac{m^4 B}{L^3} + 2H \frac{m^2 n^2}{LB} + D_y \frac{n^4 L}{B^3} \right),$$

$$C_4 = A_{0mn} \left(\frac{m^2 B}{L} \sigma_{xav} + \frac{n^2 L}{B} \sigma_{yav} \right).$$

The solution of Equation (6.16) can be obtained using the Cardano method or the FORTRAN computer subroutine CARDANO given in the appendices to this book. It is interesting to determine the elastic bifurcation buckling strength of the orthotropic plate without initial deflections and under biaxial compressive loads in a manner similar to that of an isotropic plate described in Chapter 4. For an orthotropic plate with $L/B \geq 1$, $n = 1$ can be adopted. In this case, the following condition must be satisfied immediately before or after buckling as $A_{0mn} = 0$:

$$A_{m1} = \sqrt{-\frac{C_3}{C_1}} = 0 \tag{6.17}$$

where

$$C_1 = \frac{\pi^2}{16}\left(E_x\frac{m^4 B}{L^3} + E_y\frac{L}{B^3}\right),$$

$$C_3 = \frac{m^2 B}{L}\sigma_{xav} + \frac{L}{B}\sigma_{yav} + \frac{\pi^2}{t}\left(D_x\frac{m^4 B}{L^3} + 2H\frac{m^2}{LB} + D_y\frac{L}{B^3}\right).$$

The solution of Equation (6.17) results in either $C_3 = 0$ or the following equation:

$$\frac{m^2 B}{L}\sigma_{xav} + \frac{L}{B}\sigma_{yav} + \frac{\pi^2}{t}\left(D_x\frac{m^4 B}{L^3} + 2H\frac{m^2}{LB} + D_y\frac{L}{B^3}\right) = 0 \tag{6.18}$$

Equation (6.18) is the elastic bifurcation buckling condition of the orthotropic plate under biaxial compressive loads. When the loading ratio between biaxial compression is defined as $c = \sigma_{yav}/\sigma_{xav}$, the elastic overall buckling strength, σ_{xEO}, of the orthotropic plate in longitudinal compressive loads σ_{xav} is determined from Equation (6.18) as follows:

$$\sigma_{xEO} = -\frac{LB}{m^2 B^2 + cL^2}\frac{\pi^2}{t}\left(D_x\frac{m^2}{L^2} + 2H\frac{1}{B^2} + D_y\frac{L^2}{m^2 B^4}\right) \tag{6.19}$$

In this case, the bucking half-wave number m can be determined as a minimum integer that satisfies the following condition:

$$\frac{LB}{m^2 B^2 + cL^2}\left(D_x\frac{m^2}{L^2} + 2H\frac{1}{B^2} + D_y\frac{L^2}{m^2 B^4}\right)$$
$$\leq \frac{LB}{(m+1)^2 B^2 + cL^2}\left[D_x\frac{(m+1)^2}{L^2} + 2H\frac{1}{B^2} + D_y\frac{L^2}{(m+1)^2 B^4}\right] \tag{6.20}$$

For a uniaxial compression in the x direction with $c = \sigma_{yav}/\sigma_{xav} = 0$, Equation (6.19) is simplified to

$$\sigma_{xEO} = -\frac{\pi^2}{t}\left(D_x\frac{m^2}{L^2} + 2H\frac{1}{B^2} + D_y\frac{L^2}{m^2 B^4}\right) \tag{6.21}$$

In this case, the buckling half-wave number m is determined as a minimum integer that satisfies the following condition:

$$D_x\frac{m^2}{L^2} + 2H\frac{1}{B^2} + D_y\frac{L^2}{m^2 B^4} \leq D_x\frac{(m+1)^2}{L^2} + 2H\frac{1}{B^2} + D_y\frac{L^2}{(m+1)^2 B^4} \tag{6.22a}$$

or, more simply,

$$\left(\frac{L}{B}\right)^4 \leq \frac{D_x}{D_y}m^2(m+1)^2 \tag{6.22b}$$

It is evident from Equation (6.22b) that the buckling mode depends on both the plate aspect ratio and the structural orthotropy. For an isotropic plate under σ_{xav} in compression, Equation (6.22b) will simplify to the well-known condition because $D_x = D_y$ as follows:

$$\frac{L}{B} \leq \sqrt{m(m+1)} \tag{6.22c}$$

For an orthotropic plate with $L/B \geq 1$, the elastic bifurcation buckling strength under transverse compression σ_{yav} is obtained from Equation (6.18) as follows because $\sigma_{xav} = 0$:

$$\sigma_{yEO} = -\frac{\pi^2}{t}\left(D_x\frac{B^2}{L^4} + 2H\frac{1}{L^2} + D_y\frac{1}{B^2}\right) \tag{6.23}$$

Once A_m is determined as the solution of Equation (6.16) for the given values of σ_{xav} and σ_{yav}, the maximum or minimum membrane stresses in the x and y direction can be determined from Equation (6.12) as follows:

$$\sigma_{x\max} = \sigma_{xav} - \frac{m^2\pi^2 E_x A_m(A_m + 2A_{0mn})}{8L^2} \tag{6.24a}$$

$$\sigma_{x\min} = \sigma_{xav} + \frac{m^2\pi^2 E_x A_{mn}(A_{mn} + 2A_{0mn})}{8L^2} \tag{6.24b}$$

$$\sigma_{y\max} = \sigma_{yav} - \frac{\pi^2 E_y A_{mn}(A_{mn} + 2A_{0mn})}{8B^2} \tag{6.24c}$$

$$\sigma_{y\min} = \sigma_{yav} + \frac{\pi^2 E_y A_{mn}(A_{mn} + 2A_{0mn})}{8B^2} \tag{6.24d}$$

6.5.3 Effect of the Bathtub Deflection Shape

In a manner similar to that of the isotropic plate theory method associated with the bathtub-shaped deflection as described in Chapter 4, the maximum and minimum membrane stresses in Equation (6.24) are amplified by multiplying a correction factor (Paik et al. 2001) as follows:

$$\sigma_{x\max} = \sigma_{xav} - \rho\frac{m^2\pi^2 E_x A_{mn}(A_{mn} + 2A_{0mn})}{8L^2} \tag{6.25a}$$

$$\sigma_{x\min} = \sigma_{xav} + \rho\frac{m^2\pi^2 E_x A_{mn}(A_{mn} + 2A_{0mn})}{8L^2} \tag{6.25b}$$

$$\sigma_{y\max} = \sigma_{yav} - \rho\frac{\pi^2 E_y A_{mn}(A_{mn} + 2A_{0mn})}{8B^2} \tag{6.25c}$$

$$\sigma_{y\min} = \sigma_{yav} + \rho\frac{\pi^2 E_y A_{mn}(A_{mn} + 2A_{0mn})}{8B^2} \tag{6.25d}$$

The correction factor ρ in Equation (6.25) may be given as follows:

$$\rho = \begin{cases} \rho_c & \text{for } (L/B)^4 \geq D_y/(4D_x) \\ 2\rho_c & \text{for } (L/B)^4 < D_y/(4D_x) \end{cases} \tag{6.26}$$

where

$$\rho_c = \begin{cases} 1.0 & \text{for } H/D < 1.3569 \\ 0.0894(H/D - 1.3569) + 1.0 & \text{for } H/D \geq 1.3569 \end{cases}.$$

Because ρ_c is always greater than 1.0, the large-deflection-related terms of the maximum and minimum membrane stresses are amplified because of the bathtub-shaped deflection. When tensile loads are applied, $\rho = 1$ is used.

6.5.4 Interaction Effect Between Biaxial Loads and Lateral Pressure

In a manner similar to that of the isotropic plate theory method described in Chapter 4, C_4 in Equation (6.16) is modified to superpose the effects of lateral pressure loads p as follows:

$$C_4 = A_{0mn}\left(\frac{m^2 B}{L}\sigma_{xav} + \frac{n^2 L}{B}\sigma_{yav}\right) - \frac{16LB}{\pi^4 t}p \tag{6.27}$$

6.6 Ultimate Strength

The ultimate strength formulations for the stiffened panel under combined in-plane and lateral pressure loads are now presented for all potential collapse modes noted in Section 6.2. The minimum value among the six ultimate strengths so obtained for the six collapse modes is then taken as the real ultimate strength.

6.6.1 Mode I: Overall Collapse

In collapse mode I, the stiffened panel is idealized as an orthotropic plate under combined σ_{xav}, σ_{yav}, τ_{av}, and p. In this case, the ultimate strength interaction relationship of the stiffened panel is given by

$$\left(\frac{\sigma_{xav}}{\sigma_{xu}^I}\right)^{c_1} - \alpha\left(\frac{\sigma_{xav}}{\sigma_{xu}^I}\right)\left(\frac{\sigma_{yav}}{\sigma_{yu}^I}\right) + \left(\frac{\sigma_{yav}}{\sigma_{yu}^I}\right)^{c_2} + \left(\frac{\tau_{av}}{\tau_u^I}\right)^{c_3} = 1 \tag{6.28}$$

where σ_{xu}^I, σ_{yu}^I, and τ_u^I are the ultimate strengths of the collapse mode I stiffened panel for σ_{xav}, σ_{yav}, and τ_{av}, respectively, taking into account the effects of lateral pressure loads p, $c_1 - c_3$ = coefficients, as defined in Equation (4.85), which may be taken as $c_1 = c_2 = c_3 = 2$, whereas $\alpha = 0$ when both σ_{xav} and σ_{yav} are compressive (negative) and $\alpha = -1$ when σ_{xav}, σ_{yav}, or both are tensile (positive). In the following, the formulations for calculation of σ_{xu}^I, σ_{yu}^I, and τ_u^I are described.

6.6.1.1 Calculation of σ_{xu}^I

σ_{xu}^I is the maximum longitudinal-axial-load carrying capacity when the stiffened panel is subjected to combined σ_{xav} and p. In this case, collapse will occur when the most stressed boundary locations yield, because the longitudinal panel boundaries can no longer be kept straight, resulting in a rapid increase in lateral deflection, as shown in Figure 6.3. Therefore, σ_{xu}^{I} can be determined as the solution of the following equation with regard to σ_{xav}:

$$\left(\frac{\sigma_{x\,max}}{\sigma_{Yeq}}\right)^2 - \left(\frac{\sigma_{x\,max}}{\sigma_{Yeq}}\right)\left(\frac{\sigma_{y\,min}}{\sigma_{Yeq}}\right) + \left(\frac{\sigma_{y\,min}}{\sigma_{Yeq}}\right)^2 = 1 \tag{6.29}$$

where $\sigma_{x\,max}$ and $\sigma_{y\,min}$ are the maximum and minimum membrane stresses in the x and y directions, respectively, as a function of σ_{xav} and p.

6.6.1.2 Calculation of σ_{yu}^{I}

σ_{yu}^{I} is the maximum transverse-axial-load carrying capacity when the stiffened panel is subjected to combined σ_{yav} and p. In this case, collapse will occur when the most stressed boundary locations yield, because the transverse panel boundaries can no longer be kept straight, resulting in a rapid increase in lateral deflection, as shown in Figure 6.4. There-fore, σ_{yu}^{I} can be determined as the solution of the following equation with regard to σ_{yav}:

$$\left(\frac{\sigma_{x\,min}}{\sigma_{Yeq}}\right)^2 - \left(\frac{\sigma_{x\,min}}{\sigma_{Yeq}}\right)\left(\frac{\sigma_{y\,max}}{\sigma_{Yeq}}\right) + \left(\frac{\sigma_{y\,max}}{\sigma_{Yeq}}\right)^2 = 1 \tag{6.30}$$

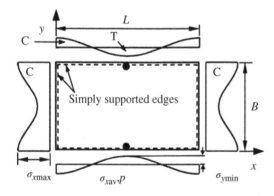

Figure 6.3 Plasticity at panel longitudinal edges for a combined σ_{xav} and p (●, expected yielding locations; C, compression; T, tension).

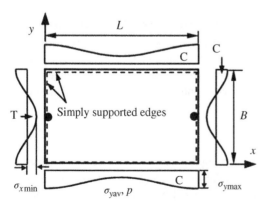

Figure 6.4 Plasticity at panel transverse edges for a combined σ_{yav} and p (●, expected yielding locations; C, compression; T, tension).

where $\sigma_{x\,min}$ and $\sigma_{y\,max}$ are the minimum and maximum membrane stresses in the x and y directions, respectively, as a function of σ_{yav} and p.

6.6.1.3 Calculation of τ_u^I

τ_u^I is the maximum edge-shear-load carrying capacity when the stiffened panel is subjected to combined τ_{av} and p. In this case, Equations (4.83b) can be used as follows:

$$\tau_u^I = \tau_{u0} \left[1 - \left(\frac{p}{p_{u0}} \right)^{1.2} \right]^{1/1.5} \tag{6.31}$$

where p_{u0} is the ultimate lateral pressure load, as the minimum value among three solutions determined from Equation (4.78) with the maximum or minimum membrane stresses defined in Equation (6.25).

τ_{u0} is the ultimate strength of a cross-stiffened panel under edge shear. Mikami et al. (1989) proposed an empirical ultimate shear strength formula of a cross-stiffened panel that was originally intended for the design of the plate girder web under shearing force as follows:

$$\frac{\tau_{u0}}{\tau_Y} = \begin{cases} 1.0 & \text{for } \lambda \le 0.6 \\ 1.0 - 0.614(\lambda - 0.6) & \text{for } 0.6 < \lambda \le \sqrt{2} \\ 1/\lambda^2 & \text{for } \lambda > \sqrt{2} \end{cases} \tag{6.32}$$

with $\lambda = \sqrt{(\tau_E/\tau_Y)}$, τ_E is the elastic shear buckling stress, and $\tau_Y = \sigma_{Yeq}/\sqrt{3}$.

6.6.2 Mode II: Plate Collapse Without Distinct Failure of Stiffener

In collapse mode II, the stiffened panel reaches the ultimate strength via collapse of the plating between the stiffeners under σ_{xM}, σ_{yM}, τ_{av}, and p. The collapse of the plating is considered to occur when the most highly stressed plate corners yield. In this case, the ultimate strength interaction relationship of the stiffened panel is given by

$$\left(\frac{\sigma_{xM}}{\sigma_{xu}^{II}} \right)^{c_1} - \alpha \left(\frac{\sigma_{xM}}{\sigma_{xu}^{II}} \right) \left(\frac{\sigma_{yM}}{\sigma_{yu}^{II}} \right) + \left(\frac{\sigma_{yM}}{\sigma_{yu}^{II}} \right)^{c_2} + \left(\frac{\tau_{av}}{\tau_u^{II}} \right)^{c_3} = 1 \tag{6.33}$$

where σ_{xu}^{II}, σ_{yu}^{II}, and τ_u^{II} are the ultimate strengths of the collapse mode II stiffened panel for σ_{xM}, σ_{yM}, and τ_{av}, respectively, taking into account the effects of lateral pressure loads p, and $c_1 - c_3$ are coefficients, as defined in Equation (6.28). In the following, the formulations for calculation of σ_{xu}^{II}, σ_{yu}^{II}, and τ_u^{II} are described.

6.6.2.1 Calculation of σ_{xu}^{II}

The ultimate longitudinal axial strength, σ_{xu}^{II}, based on mode II is considered for the most highly stressed plating between stiffeners under combined σ_{xM} and p when $\sigma_{yM} = \tau = 0$. The applicable maximum and minimum membrane stress components for the plating between stiffeners can be calculated by solving the nonlinear governing differential equations of the large-deflection isotropic plate theory, as described in Chapter 4. The stiffened panel is then considered to collapse in mode II if the plate corner yields, which results in the condition in Equation (4.78a). In this case, σ_{xmax} and σ_{ymax} are functions of σ_{xM} and p, as well as the initial imperfections.

The panel ultimate longitudinal axial strength, σ_{xu}^{II}, based on mode II is then obtained by solving Equation (4.78a) with regard to σ_{xM} by substituting the maximum and minimum membrane stress components upon replacing σ_{xav} with σ_{xM}. The approach used is quite similar to that for mode I based on the initial yield at the panel boundaries, except for (i) the inclusion of both welding induced residual stresses and initial deflections and (ii) the consideration of yielding at the plate corners.

6.6.2.2 Calculation of σ_{yu}^{II}

For combined σ_{yM} and p, the panel ultimate transverse axial strength, σ_{xu}^{II}, based on mode II is obtained by solving Equation (4.78a) with regard to σ_{yM}, but substituting the maximum and minimum membrane stress components into Equation (4.78a). As before, these membrane stress components are derived by solving the nonlinear governing differential equations of the large-deflection isotropic plate theory, as described in Chapter 4. In this case, σ_{xmax} and σ_{ymax} are functions of σ_{yM} and p, as well as initial imperfections.

6.6.2.3 Calculation of τ_u^{II}

When the stiffeners are relatively strong (or the plating fails before the stiffeners), τ_u^{II} is determined from Equation (6.31), but in this case, τ_{u0} and p_{u0} are considered for the plating between the longitudinal and transverse stiffeners, as described in Section 4.9.3.6.

6.6.3 Mode III: Beam–Column Collapse

In collapse mode III, it is considered that the stiffened panel reaches the ultimate strength if the most highly stressed stiffener in the x or y direction under the four stress components (σ_{xM}, σ_{yM}, τ_{av}, and p) fails as a beam–column, which can be idealized by the plate–stiffener combination model as described in Chapter 2. In this case, the ultimate strength interaction relationship of the stiffened panel is given by

$$\left(\frac{\sigma_{xM}}{\sigma_{xu}^{III}}\right)^{c_1} - \alpha\left(\frac{\sigma_{xM}}{\sigma_{xu}^{III}}\right)\left(\frac{\sigma_{yM}}{\sigma_{yu}^{III}}\right) + \left(\frac{\sigma_{yM}}{\sigma_{yu}^{III}}\right)^{c_2} + \left(\frac{\tau_u}{\tau_u^{III}}\right)^{c_3} = 1 \tag{6.34}$$

where σ_{xu}^{III}, σ_{yu}^{III}, and τ_u^{III} are the ultimate strengths of the collapse mode II stiffened panel for σ_{xM}, σ_{yM}, and τ_{av}, taking into account the effects of lateral pressure loads p, $c_1 - c_3$ = coefficients, as defined in Equation (6.28). In the following, the formulations for calculation of σ_{xu}^{III}, σ_{yu}^{III}, and τ_u^{III} are described.

6.6.3.1 Calculation of σ_{xu}^{III}

σ_{xu}^{III} is the maximum longitudinal-axial-load capacity of the stiffened panel under combined σ_{xM} and p. When the stiffened panel is subjected to axial compressive loads alone, σ_{xu}^{III} can be determined with the Johnson–Ostenfeld formulation method, the Perry–Robertson formulation method, or the empirical formulation method, as described in Section 2.9.5. When the stiffened panel is subjected to combined longitudinal axial loads and lateral pressure, σ_{xu}^{III} can be determined with the modified Perry–Robertson formulation method. Upon using the plate–stiffener combination model, the effective width or breadth of the attached plating should be considered, as described in Section 4.8.

In the original or modified Perry–Robertson formulation method, the ultimate strength is considered to have been reached if the extreme fibers of the cross section (at the mid-span in the simply supported case) yield, that is, when the axial stress at the outmost section reaches the yield stress either on the stiffener or the plate side, with the former being called "stiffener-induced failure" and the latter being called "plate-induced failure." It is recognized that the stiffener-induced failure mode predictions are too pessimistic in comparison with the actual test data or nonlinear finite-element method solutions when the stiffeners are relatively small.

Although the idea of the Perry–Robertson formulation method assumes that stiffener-induced failure occurs if the tip of the stiffener yields, plasticity may grow into the stiffener web as long as lateral-torsional buckling or stiffener web buckling does not occur, so the stiffener may resist further loading even after the first yielding occurs at the extreme fiber of the stiffener. In such cases, one may exclude the stiffener-induced failure condition (i.e., yielding at the tip of the stiffener) from the mode III panel ultimate strength calculations. The possibility of the stiffener-induced failure mode due to local buckling of the stiffener web or lateral-torsional buckling of the stiffener is dealt with in modes IV and V, respectively.

6.6.3.2 Calculation of σ_{yu}^{III}

σ_{yu}^{III} is the maximum transverse-axial-load capacity of the stiffened panel under combined σ_{yM} and p, and it can be calculated in a manner similar to that under combined σ_{xM} and p using a representative of the plate–stiffener combination model. When transverse stiffeners are not present, σ_{yu}^{III} is determined as the ultimate strength of the plating between two longitudinal stiffeners.

6.6.3.3 Calculation of τ_{u}^{III}

τ_{u}^{III} is obtained as $\tau_{u}^{III} = \tau_{u}^{II}$ where the ultimate lateral load, p_{u0}, and the ultimate shear strength, τ_{u0}, are defined as the same as those of mode II.

6.6.4 Mode IV: Collapse by Local Web Buckling of Stiffener

If the height of the stiffener web increases in comparison with its thickness, the stiffener web is likely to deform, and local buckling can in some cases occur. Once web buckling occurs, the buckled or collapsed plating may be left with essentially little stiffening after overall grillage collapse. The welding induced initial imperfections in the plating between the stiffeners are included as parameters of influence.

In mode IV, the stiffened panel under σ_{xM}, σ_{yM}, τ_{av}, and p is considered to have reached the ultimate strength if local buckling of the stiffener web takes place at its most highly stressed location. In this case, the ultimate strength interaction relationship of the stiffened panel is given by

$$\left(\frac{\sigma_{xM}}{\sigma_{xu}^{IV}}\right)^{c_1} - \alpha\left(\frac{\sigma_{xM}}{\sigma_{xu}^{IV}}\right)\left(\frac{\sigma_{yM}}{\sigma_{yu}^{IV}}\right) + \left(\frac{\sigma_{yM}}{\sigma_{yu}^{IV}}\right)^{c_2} + \left(\frac{\tau_{av}}{\tau_{u}^{IV}}\right)^{c_3} = 1 \tag{6.35}$$

where σ_{xu}^{IV}, σ_{yu}^{IV}, and τ_u^{IV} are the ultimate strength of the collapse mode II stiffened panel for σ_{xM}, σ_{yM}, and τ_{av}, taking into account the effect of lateral pressure loads p, $c_1 - c_3 =$ coefficients, as defined in Equation (6.28). In the following, the formulations for calculation of σ_{xu}^{IV}, σ_{yu}^{IV}, and τ_u^{IV} are described.

6.6.4.1 Calculation of σ_{xu}^{IV}

σ_{xu}^{IV} is the maximum longitudinal-axial-load capacity of the stiffened panel under σ_{xM} and p when the longitudinal stiffener web buckles. In this case, the ultimate strength of the stiffened panel is approximated as a weighted average of the ultimate strengths of the stiffener and the associated plating. The intention behind the averaging proposed is to avoid an overly pessimistic estimate of the stiffened panel ultimate strength as follows:

$$\sigma_{xu}^{IV} = \frac{bt\,\sigma_{xu}^P + h_{wx}t_{wx}\,\sigma_{xu}^w - b_{fx}t_{fx}\sigma_{Yf}}{bt + h_{wx}t_{wx} + b_{fx}t_{fx}} \tag{6.36}$$

where σ_{xu}^P is the ultimate strength of the plating between the longitudinal stiffeners under σ_{xM} and p, which can be determined as described in Section 4.9, and σ_{xu}^w is the ultimate strength of the longitudinal stiffener, which can be approximately determined with the Johnson–Ostenfeld formulation method with the elastic buckling strength of the stiffener web, as described in Section 5.6. The negative sign in the numerator of Equation (6.36) indicates the compressive stress associated with the contribution of the stiffener flange.

6.6.4.2 Calculation of σ_{yu}^{IV}

Similar to σ_{xu}^{IV}, σ_{yu}^{IV} is the maximum longitudinal-axial-load capacity of the stiffened panel under σ_{yM} and p when the transverse stiffener web buckles. In this case, σ_{yu}^{IV} can be determined as follows:

$$\sigma_{yu}^{IV} = \frac{at\,\sigma_{yu}^P + h_{wy}t_{wy}\,\sigma_{yu}^w - b_{fy}t_{fy}\sigma_{Yf}}{at + h_{wy}t_{wy} + b_{fy}t_{fy}} \tag{6.37a}$$

where σ_{yu}^P is the ultimate strength of the plating between the transverse stiffeners under σ_{yM} and p, which can be determined as described in Section 4.9, and σ_{yu}^w is the ultimate strength of the transverse stiffener, which can be approximately determined with the Johnson–Ostenfeld formulation method with the elastic buckling strength of the stiffener web, as described in Section 5.6. For longitudinally stiffened panels, Equation (6.37a) is simplified to

$$\sigma_{yu}^{IV} = \sigma_{yu}^P \tag{6.37b}$$

6.6.4.3 Calculation of τ_u^{IV}

τ_u^{IV} is obtained as $\tau_u^{IV} = \tau_u^{III} = \tau_u^{II}$, where the ultimate lateral load, p_{u0}, and the ultimate shear strength, τ_{u0}, are defined as the same as in mode II.

6.6.5 Mode V: Collapse by Lateral-Torsional Buckling of Stiffener

Lateral-torsional buckling (also called tripping) of stiffeners is a phenomenon in which a stiffened panel fails after the stiffener twists sideways about the edge of the stiffener web attached to the plating. When the torsional rigidity of the stiffener is low or the stiffener flange is weak, this phenomenon is more likely to occur.

Like the stiffener web buckling previously described, lateral-torsional buckling can be a relatively sudden phenomenon that results in unloading of the stiffened panel. Once lateral-torsional buckling occurs, the buckled or collapsed plating is left with little stiffening, and thus overall collapse may follow. In mode V, the stiffened panel is considered to collapse if lateral-torsional buckling occurs.

Local buckling of the stiffener web is treated in mode IV, and thus for the purposes of mode V, we consider a type of tripping in which the cross section of the stiffener web does not deform locally, consistent with a similar assumption used in ordinary beam–column theory. It therefore follows that the lateral-torsional buckling strength of the flat-bar type of stiffener equals the local buckling of the stiffener web, and such a case is treated as part of mode IV and not mode V.

Similar to mode IV, it is idealized that the most highly stressed stiffener under consideration is subjected to combined σ_{xM}, σ_{yM}, τ_{av}, and p. The welding induced initial imperfections in the plating between the stiffeners are included as parameters of influence. In this case, the ultimate strength interaction relationship of the stiffened panel is given by

$$\left(\frac{\sigma_{xM}}{\sigma_{xu}^V}\right)^{c_1} - \alpha\left(\frac{\sigma_{xM}}{\sigma_{xu}^V}\right)\left(\frac{\sigma_{yM}}{\sigma_{yu}^V}\right) + \left(\frac{\sigma_{yM}}{\sigma_{yu}^V}\right)^{c_2} + \left(\frac{\tau_{av}}{\tau_u^V}\right)^{c_3} = 1 \qquad (6.38)$$

where σ_{xu}^V, σ_{yu}^V, and τ_u^V are the ultimate strengths of the collapse mode II stiffened panel for σ_{xM}, σ_{yM}, and τ_{av}, taking into account the effect of lateral pressure loads p, $c_1 - c_3 =$ coefficients, as defined in Equation (6.28). In the following, the formulations for calculation of σ_{xu}^V, σ_{yu}^V, and τ_u^V are described.

6.6.5.1 Calculation of σ_{xu}^V

σ_{xu}^V is the maximum longitudinal-axial-load capacity of the stiffened panel under σ_{xM} and p when the longitudinal stiffener fails by lateral-torsional buckling. Similar to the stiffener web buckling in mode IV, the ultimate strength of the stiffened panel is approximated as a weighted average of the ultimate strengths of the stiffener and the associated plating as follows:

$$\sigma_{xu}^V = \frac{bt\,\sigma_{xu}^P + h_{wx}t_{wx}\,\sigma_{xu}^W - b_{fx}t_{fx}\sigma_{Yf}}{bt + h_{wx}t_{wx} + b_{fx}t_{fx}} \qquad (6.39)$$

where σ_{xu}^P is the ultimate strength of the plating between the longitudinal stiffener under σ_{xM} and p, which can be determined as described in Section 4.9, and σ_{xu}^W is the ultimate strength of the longitudinal stiffener, which can be approximately determined by the Johnson–Ostenfeld formulation method with the elastic flexural-torsional buckling strength, as described in Section 5.8. The negative sign in the numerator of Equation (6.39) indicates the compressive stress associated with the contribution of the stiffener flange.

6.6.5.2 Calculation of σ_{yu}^V

Similar to σ_{xu}^V, σ_{yu}^V is the maximum longitudinal-axial-load capacity of the stiffened panel under σ_{yM} and p when the transverse stiffener fails by flexural-torsional buckling. In this case, σ_{yu}^V can be determined as follows:

$$\sigma_{yu}^V = \frac{at\,\sigma_{yu}^P + h_{wy}t_{wy}\,\sigma_{yu}^w - b_{fy}t_{fy}\sigma_{Yf}}{at + h_{wy}t_{wy} + b_{fy}t_{fy}} \tag{6.40a}$$

where σ_{yu}^P is the ultimate strength of the plating between the transverse stiffener under σ_{yM} and p, which can be determined as described in Section 4.9, and σ_{yu}^w is the ultimate strength of the transverse stiffener, which can be approximately determined by the Johnson–Ostenfeld formulation method with the elastic lateral-torsional buckling strength, as described in Section 5.8. For longitudinally stiffened panels, Equation (6.40a) is simplified to

$$\sigma_{yu}^V = \sigma_{yu}^P \tag{6.40b}$$

6.6.5.3 Calculation of τ_u^V

τ_u^V is obtained as $\tau_u^V = \tau_u^{IV} = \tau_u^{III} = \tau_u^{II}$, where the ultimate lateral load, p_{uo}, and the ultimate shear strength, τ_{uo}, are defined as the same as those of mode II.

6.6.6 Mode VI: Gross Yielding

In mode VI, the stiffened panel reaches the ultimate strength by gross yielding of the cross section with neither local nor overall (grillage) buckling. The applicable ultimate strength interaction relationship for a stiffened panel under combined loads is in this case similar in form to the von Mises yield condition:

$$\left(\frac{\sigma_{xM}}{\sigma_{xu}^{VI}}\right)^{c_1} - \left(\frac{\sigma_{xM}}{\sigma_{xu}^{VI}}\right)\left(\frac{\sigma_{yM}}{\sigma_{yu}^{VI}}\right) + \left(\frac{\sigma_{yM}}{\sigma_{yu}^{VI}}\right)^{c_2} + \left(\frac{\tau_{av}}{\tau_u^{VI}}\right)^{c_3} = 1 \tag{6.41}$$

where $\sigma_{xu}^{VI} = \pm\sigma_{Yeq}$ (+ for σ_{xM} in tension, – for σ_{xM} in compression), $\sigma_{yu}^{VI} = \pm\sigma_{Yeq}$ (+ for σ_{yM} in tension, – for σ_{yM} in compression), and $\tau_u^{VI} = \sigma_{Yeq}/\sqrt{3}$, σ_{Yeq} is the equivalent yield stress, as defined in Equation (6.2). The coefficients c_1–c_3 are taken as those of Equation (6.28).

6.6.7 Determination of the Real Ultimate Strength

The six modes of the stiffened panel collapse have been considered separately, but some modes may interact and take place simultaneously. For the sake of simplicity, however, it is considered that a stiffened panel will reach the ULS if the dominant collapse mode occurs first among the six types of collapse patterns.

The ultimate strengths of the stiffened panel are therefore computed separately for each of the six collapse modes, and the smallest value among the computed strengths is taken as the real ultimate strength of the stiffened panel. For the purposes of automated computations, it is considered that the stiffened panel reaches the ULS if any one condition among the six ultimate strength characteristic conditions, that is, Equations (6.28), (6.33), (6.34), (6.35), (6.38), and (6.41), is satisfied first as the applied loads are increased. This method is also of benefit because the dominant collapse mode of the stiffened panel can be recognized in an explicit form so that the safety design to prevent the computed collapse pattern can be developed more efficiently.

6.7 Effects of Age Related and Accident Induced Damages

Age related and accident induced damages may significantly reduce the ultimate strength of stiffened panels and they should be dealt with as parameters of influence. For general corrosion that uniformly reduces the panel thickness, the ultimate strength or effectiveness of the panel can be evaluated by excluding the corrosion diminution (reduction in thickness). For stiffened panels with premised cracking damage, the strength reduction factor method may be used as described in Chapters 4 and 9. The buckling and ultimate strength of stiffened panels with accident induced damage such as local denting can be evaluated as described in Chapters 4 and 10.

6.8 Benchmark Studies

The ultimate strength formulations for stiffened panels described in this chapter are automated within the computer program ALPS/ULSAP (2017). In this section, the benchmark studies by ALPS/ULSAP are presented with other methods, including the nonlinear finite element method (Paik et al. 2011, ISSC 2012).

Table 6.1 indicates the candidate methods used for the benchmark studies. Stiffened panels surrounded by longitudinal girders and transverse frames are selected as the target structure of the benchmark studies as shown in Figure 6.5. Two types of stiffened panels are considered, that is, panels taken from the bottom panels of a bulk carrier

Table 6.1 Candidate methods used for the benchmark studies (Paik et al. 2011, ISSC 2012).

Method/tool	Symbol	Working organization
ALPS/ULSAP	ALPS/ULSAP (PNU)	Pusan National University
BV Advanced Buckling (BV 2011)	BV Advanced Buckling (BV)	Bureau Veritas
DNV/PULS	DNV/PULS (DNV)	Det Norske Veritas
ABAQUS	ABAQUS (NTUA)	National Technical University of Athens
	ABAQUS (DNV)	Det Norske Veritas
ANSYS	ANSYS (ULG)	University of Liege
	ANSYS (IRS)	Indian Register of Shipping
	ANSYS (PNU)	Pusan National University
MSC/MARC	MSC/MARC (OU)	Osaka University

(a)

• Yield stress of plate, $\sigma_{Yp} = 313.6$ MPa

• Yield stress of stiffener, $\sigma_{Ys} = 313.6$ MPa

• Elastic modulus, $E = 205800$ MPa

• Poisson's ratio, $v = 0.3$

• Plate length, $a = 2550$ mm

• Plate breath, $b = 850$ mm

• Plate thickness, $t_p = 9.5, 11, 13, 16, 22, 33$ mm

• Number of stiffeners: 2 stiffeners in a panel

(b)

• Yield stress of plate, $\sigma Yp = 313.6$ MPa

• Yield stress of stiffener, $\sigma Ys = 313.6$ MPa

• Elastic modulus, $E = 205800$ MPa

• Poisson's ratio, $v = 0.3$

• Plate length, $a = 4750$ mm

• Plate breath, $b = 950$ mm

• Plate thickness, $t_p = 11, 12.5, 15, 18.5, 25, 37$ mm

• Number of the stiffeners: 8 stiffeners in a panel

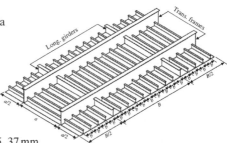

Figure 6.5 Target stiffened panels selected from (a) a bulk carrier and (b) a double-hulled oil tanker.

and the deck panels of a very large double-hulled crude oil tanker. The stiffened panels have only longitudinal stiffeners. Table 6.2 indicates the types and dimensions of the stiffeners. The yield stresses of the stiffener web, the flange, and the plating are the same. Modeling techniques for nonlinear finite element analysis are described in Chapter 12.

Although the effects of welding induced residual stresses are not considered, three types of initial distortions—plate initial deflection, the column-type initial distortion

Table 6.2 Types and dimensions of the stiffeners (in mm).

Size	Flat bar, $h_w \times t_w$	Angle bar, $h_w \times b_f \times t_w/t_f$	Tee bar, $h_w \times b_f \times t_w/t_f$
Size 1	150×17	$138 \times 90 \times 9/12$	$138 \times 90 \times 9/12$
Size 2	250×25	$235 \times 90 \times 10/15$	$235 \times 90 \times 10/15$
Size 3	350×35	$383 \times 100 \times 12/17$	$383 \times 100 \times 12/17$
Size 4	550×35	$580 \times 150 \times 15/20$	$580 \times 150 \times 15/20$

Table 6.3 Coefficients of initial plate deflections and stiffener distortions.

Method/tool	A_{0m}	B_0	C_0
ALPS/ULSAP	$0.1\beta^2 t_p$	$0.0015a$	$0.0015a$
DNV/PULS	$b/200$	$0.001a$	$0.001a$
ABAQUS	$0.1\beta^2 t_p$	$0.0015a$	$0.0015a$
ANSYS	$0.1\beta^2 t_p$	$0.0015a$	$0.0015a$
MSC/MARC	$0.1\beta^2 t_p$	$0.0015a$	$0.0015a$

of the stiffener, and the sideways initial distortion of the stiffener—are considered, which can be expressed as described in Section 1.7:

- Buckling mode initial deflection of plating: $w_{0pl} = A_{0m} \sin\dfrac{m\pi x}{a}\sin\dfrac{\pi y}{b}$
- Column-type distortion of stiffener: $w_{0c} = B_0 \sin\dfrac{\pi x}{a}\sin\dfrac{\pi y}{B}$
- Sideways initial distortion of stiffener: $w_{0s} = C_0 \dfrac{z}{h_w}\sin\dfrac{\pi x}{a}$

where m is the buckling mode of the plate and A_{0m}, B_0, and C_0 are the coefficients of the initial distortions, which are presumed as indicated in Table 6.3, where $\beta = \left(b/t_p\right)\sqrt{\sigma_{Yp}/E}$, t_p is the thickness of the plating, σ_{Yp} is the yield stress of the plating, E is the elastic

(a)

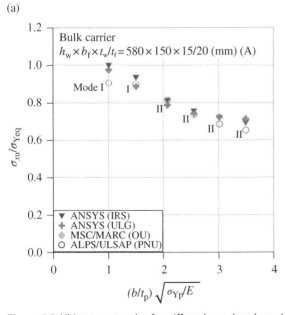

Figure 6.6 Ultimate strength of a stiffened panel under uniaxial compression: (a) bulk carrier panel with Tee bars (size 4); (b) bulk carrier panel with angle bars (size 3); (c) tanker panel with angle bars (size 4); (d) tanker panel with flat bars (size 2); (e) tanker panel with Tee bars (size 4).

(b)

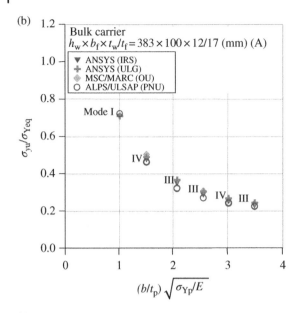

(c)

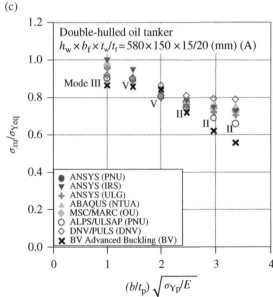

Figure 6.6 (Continued)

modulus, a is the panel length or spacing between the transverse frames, b is the breadth of the plating or the spacing between the longitudinal stiffeners, and h_w is the height of the stiffener web.

Figure 6.6a–e presents the ultimate strengths of stiffened panels under longitudinal or transverse compression with various types or dimensions. Figure 6.7a–e presents the ultimate strength interaction relationships of the stiffened panels under biaxial compression. Figure 6.8a and b presents the ultimate strength interaction relationships of the stiffened panels under combined axial compression and lateral pressure loads. As would be

(d)

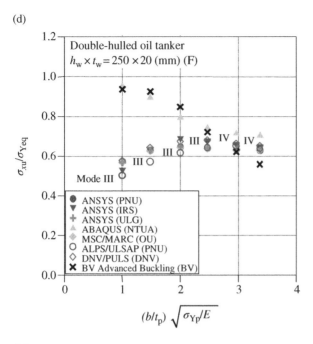

(e)

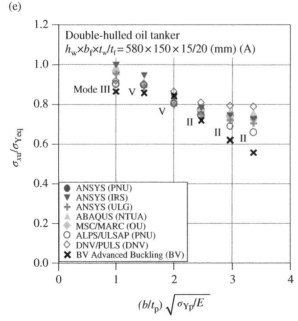

Figure 6.6 (Continued)

expected, the candidate methods provide different results, implying that many modeling uncertainties associated with the ultimate strength calculations are involved. The ultimate strength formulations of the stiffened panels described in this chapter are considered to be comparable with more refined methods, such as the nonlinear finite element method.

(a)

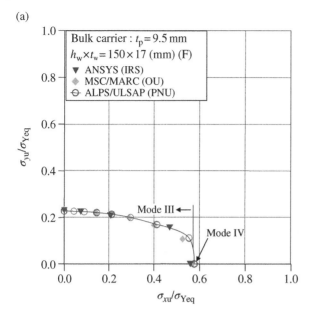

(b)

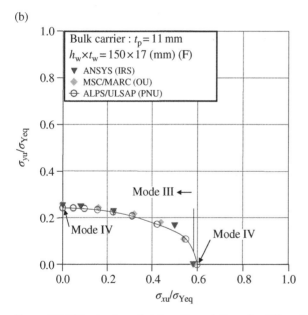

Figure 6.7 Ultimate strength interaction relation of a stiffened panel under biaxial compression: (a) bulk carrier panel with flat bars (size 1), $t_p = 9.5$ mm; (b) bulk carrier panel with flat bars (size 1), $t_p = 11$ mm; (c) bulk carrier panel with angle bars (size 3), $t_p = 13$ mm; (d) bulk carrier panel with Tee bars (size 3), $t_p = 13$ mm; (e) tanker panel with Tee bars (size 4), $t_p = 18.5$ mm.

(c)

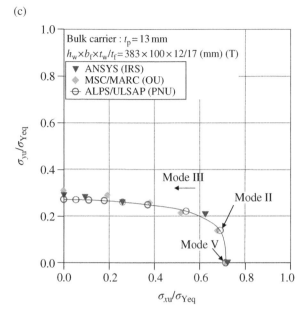

(d)

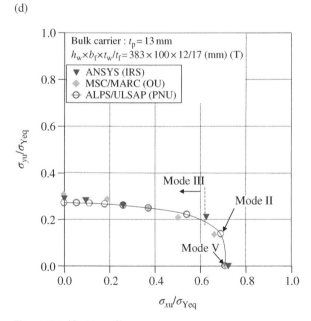

Figure 6.7 (Continued)

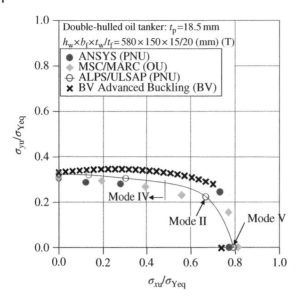

Figure 6.7 (Continued)

(a)

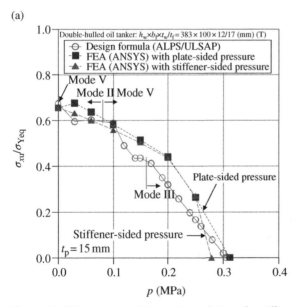

Figure 6.8 Ultimate strength interaction relation of a stiffened panel under combined axial compression and lateral pressure loads: (a) tanker panel with Tee bars (size 3), $t_p = 15$ mm; (b) tanker panel with Tee bars (size 3), $t_p = 18.5$ mm.

(b)

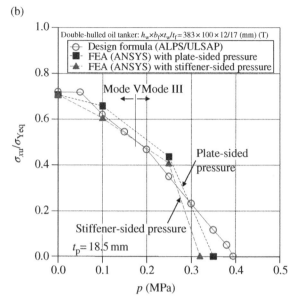

Figure 6.8 (Continued)

References

ALPS/ULSAP (2017). *A computer program for the ultimate strength analysis of plates and stiffened panels*. MAESTRO Marine LLC, Stevensville, MD.

BV (2011). Rules for the classification of steel ships. NR 467.B2 DT R05 E, Bureau Veritas, Paris.

ISSC (2012). Ultimate strength. In *Report of technical committee III.1, 18th international ship and offshore structures congress*, Edited by Fricke, W. & Bronsart, R., Schiffbautechnische Gesellschaft, Hamburg.

Mansour, A.E. (1977). *Gross panel strength under combined loading*, **SSC-270**, Ship Structure Committee, Washington, DC.

Mikami, I., Kimura, T. & Yamazato, Y. (1989). Prediction of ultimate strength of plate girders for design. *Journal of Structural Engineering*, **35A**: 511–522 (in Japanese).

Paik, J.K., Kim, S.J., Kim, D.H., Kim, D.C., Frieze, P.A., Abbattista, M., Vallascas, M. & Hughes, O.F. (2011). Benchmark study on use of ALPS/ULSAP method to determine plate and stiffened panel ultimate strength. *Proceedings of MARSTRUCT 2011 Conference*, Hamburg.

Paik, J.K., Thayamballi, A.K. & Kim, B.J. (2001). Large deflection orthotropic plate approach to develop ultimate strength formulations for stiffened panels under combined biaxial compression/tension and lateral pressure. *Thin-Walled Structures*, **39**: 215–246.

Smith, C.S. (1966). Elastic analysis of stiffened plating under lateral loading. *Transactions of the Royal Institution of Naval Architects*, **108**(2): 113–131.

Troitsky, M.S. (1976). *Stiffened panels: bending, stability and vibrations*. Elsevier Scientific Publishing Company, Amsterdam.

7

Buckling and Ultimate Strength of Plate Assemblies

Corrugated Panels, Plate Girders, Box Columns, and Box Girders

7.1 Introduction

Units or blocks of plate assemblies often constitute the primary strength parts in building plated structures. These include corrugated plate panels, plate girders, box columns, and box girders. This chapter is a first principle-based treatment of the buckling and ultimate strength of plate assemblies. In usual design practice, the design of such assemblies relies largely on structural codes and classification society rules, which contain a wealth of varying approaches to the strength treatment of these members, some based on classical theory and others based on the results of numerical computation and structural model testing.

Corrugated panels are typically seen at the transverse bulkheads of merchant ships that carry bulk cargoes such as iron ore or coal. In civil engineering structures, the webs of plate girders are sometimes constructed of corrugated plate panels. In the transverse bulkheads of merchant ships, corrugated panels are likely to be subjected to lateral pressure and axial loads, whereas, in civil engineering structures, they are typically subjected to axial loads and shearing forces.

Welded plate girders, which are one major type of plate assembly of interest, are used as the primary strength members in industrial buildings, bridges, ships, and offshore platforms. They are typically used to resist bending about their strong axis. The flanges of a plate girder are designed to effectively sustain bending stresses, whereas the web is designed to resist stresses due to shearing forces.

In buildings, especially for the fabrication of portal frames in the competitive industrial sector, the web of plate girders is sometimes not stiffened, except at the bearings or at locations at which point loads are applied. A slender web may buckle in the elastic regime. To improve the load-carrying capacity of plate girders, therefore, the web should be stiffened in the longitudinal and/or transverse directions to the extent required, with longitudinal stiffeners located in the compression zone of the web.

In the past, when the linear theory of plate buckling was used for the design of large plate girders, longitudinal stiffeners were invariably used, but unstiffened webs could be more economical than stiffened webs in many cases (Maquoi 1992). Plate girder webs without stiffeners are not necessarily stocky in the field of buildings. For instance, many industrial buildings used as warehouses are now erected using deep girders with unstiffened slender webs with slenderness ratios (i.e., the ratio of the girder depth to the web thickness) sometimes as high as 300.

Ultimate Limit State Analysis and Design of Plated Structures, Second Edition. Jeom Kee Paik.
© 2018 John Wiley & Sons Ltd. Published 2018 by John Wiley & Sons Ltd.

The webs of plate girders are normally welded to each flange with a single-sided butt weld. When the web has stiffeners in both the longitudinal and transverse directions, the longitudinal stiffeners may be welded on one side of the web, with the transverse stiffeners located on the other side. Because of extensive welding processes used during the fabrication of plate girders, initial imperfections, including initial deflection and residual stress or softening in the heat-affected zone of welded aluminum structures, normally develop and may affect the ultimate strength.

Welded built-up box columns or girders with relatively large sizes are commonly used in offshore structures, building frames, and other civil engineering structures. Box columns are predominantly subjected to axial compressive loads, whereas box girders are used to sustain predominantly bending moments. Box girders can have a variety of cross-sectional shapes, ranging from a deep narrow box to a wide shallow box. The flanges of box girders are normally wider and more slender than those of plate girders, whereas the slenderness of the webs of box girders is comparable with those of plate girders. These are, however, generalizations that may not always hold true in specific cases. The relative distribution of material in a plated beam or column unit is, however, usually driven by cost and arrangement considerations.

This chapter deals with the ultimate strength formulations for selected types of plate assemblies. With the load effects computed with the usual linear elastic finite element method or the classical theory of structural mechanics, the ultimate limit state (ULS) design of such plate assemblies is undertaken so that Equation (1.17) is satisfied. It is noted that the theories and methodologies described in this chapter can be commonly applied to steel- and aluminum-plated structures.

7.2 Ultimate Strength of Corrugated Panels

This section presents the ultimate strength formulations of a corrugated panel under some typical types of load application. Figure 7.1 is a schematic of the corrugated panel considered.

7.2.1 Ultimate Strength Under Axial Compression

The ultimate strength of the corrugated panel under axial compressive loads may be obtained as the sum of the ultimate compressive strengths of the individual walls of the corrugations, that is, the flanges and webs. In this case, all four edges of each plate may be assumed to be simply supported. The ultimate strength formulations of the individual plate elements presented in Chapter 4 can be used for this purpose. Alternatively, a simpler approach may be applied with the elastic plate buckling strength by the plasticity correction using the Johnson–Ostenfeld formulation method described in Equation (2.93).

7.2.2 Ultimate Strength Under Shearing Force

Due to corrugations, the shear strength of a corrugated panel is greater than that of a similar flat plate with the same thickness and overall dimensions. Under shearing force,

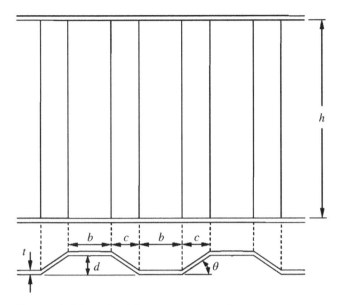

Figure 7.1 Schematic of the corrugated panel.

two distinct buckling modes are normally relevant (Maquoi 1992): (i) shear buckling that occurs locally in the largest plane wall element of the folds and is restricted to that region only and (ii) global shear buckling that generally involves several folds that may occur with snap-through and causes yield lines to cross these folds, resulting in an appropriate change in the panel configuration.

In contrast to plane webs, corrugated webs do not usually exhibit a significant strength reserve after shear buckling. Hence, the corrugated panel may be considered to reach the ULS if the wall of the corrugation buckles. For this case, the elastic local shear buckling strength may be calculated for the corrugation flange (plate) with the condition of simple supports at all four edges. As shown in Figure 7.1, the length, breadth, and thickness of the plate are denoted by h, b, and t, respectively. The critical local shear buckling strength, τ_L, may be estimated with the plasticity correction of the corresponding elastic buckling stress using the Johnson–Ostenfeld formulation method described in Equation (2.93).

The elastic global shear buckling strength, τ_G^*, of the corrugated panel may be given by (Maquoi 1992)

$$\tau_G^* = \frac{36}{h^2 t} \sqrt[4]{D_x D_y^3} \tag{7.1}$$

where

$$D_x = \frac{Et^3}{12(1-\nu^2)} \frac{b+c}{b+c/\cos\theta}, \quad D_y = \frac{Ed^2 t}{12(1-\nu^2)} \frac{3b+c/\cos\theta}{b+c},$$

E is Young's modulus and ν is Poisson's ratio.

The critical global shear buckling strength, τ_G, is then estimated by the plasticity correction of τ_G^* using the Johnson–Ostenfeld formulation method. Corrugated panel collapse may involve both buckling modes. In this case, Maquoi (1992) suggests the following simple expression for the ultimate shear strength, τ_u, of the corrugated panel, taking into account the interactive effects between the buckling modes, namely,

$$\tau_u = 1.3 \frac{\tau_L \tau_G}{\tau_L + \tau_G} \tag{7.2}$$

where τ_u should not be greater than either τ_L or τ_G.

Equation (7.2) does not have a physical meaning but serves to interpolate between the local and global buckling strengths.

7.2.3 Ultimate Strength Under Lateral Pressure

The corrugated transverse bulkheads of merchant ships such as bulk carriers are often arranged to efficiently sustain large lateral pressure loads. Because of the importance of such members to the integrity of the ships that carry dense bulk cargoes, the ultimate strength of corrugated panels under lateral pressure loads has been studied by many investigators (Caldwell 1955, Paik et al. 1997, Ji et al. 2001, among others).

One of the important insights from these experiments (Caldwell 1955, Paik et al. 1997) is that each corrugation of the corrugated panel deforms similarly (or can normally be designed to deform similarly) under a similar distribution of pressure, which implies that the behavior of a single central corrugation could be considered nearly representative of an entire corrugated panel.

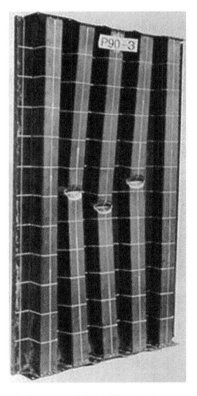

Figure 7.2 represents a typical collapse pattern of a corrugated panel under uniform lateral pressure as obtained by an experiment (Paik et al. 1997). Given such behavior, the ultimate strength of a corrugated panel under lateral pressure loads, p, may be estimated for an equivalent beam with the single corrugation cross section and under a line load, that is, $q = p(b + c)$, which is multiplied by p with the breadth (i.e., $b + c$) of the beam.

For a single corrugation beam simply supported at both ends and under a triangular type of lateral line loading, as indicated in Figure 2.17c, for instance, the ultimate strength is given by replacing the plastic bending moment with the ultimate bending moment in the same figure:

$$q_u = \frac{9\sqrt{3}}{h^2} M_u \tag{7.3}$$

where M_u is the ultimate bending moment of the single corrugation beam.

For other types of end conditions or line load applications, a similar procedure can be applied

Figure 7.2 A typical collapse pattern of the corrugated panel under lateral pressure (Paik et al. 1997).

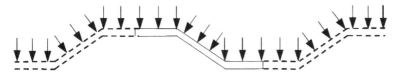

Figure 7.3 A single corrugation under lateral pressure load.

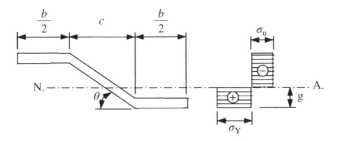

Figure 7.4 Idealized stress distribution at a plastic hinge in a single corrugation cross section.

by replacing the plastic bending capacity with the relevant ultimate bending capacity, as described in Section 2.8.

When a single corrugation is subjected to lateral pressure loads, as shown in Figure 7.3, M_u in Equation (7.3) may in the limit be estimated by considering a relevant bending stress distribution under a plastic hinge condition, representing that all parts in compression reach the ultimate compressive stress, whereas all parts in tension reach the yield stress, as shown in Figure 7.4. This accommodates local buckling in the compressed part of the corrugation. In this case, M_u is given by (Paik et al. 1997)

$$M_u = \sigma_Y \left(A_f g + A_W \frac{g^2}{d} \sin \theta \right) + \sigma_u (d - g) \left(A_f + A_W \frac{d-g}{d} \sin \theta \right) \tag{7.4}$$

where $A_w = ct/\cos \theta$, $A_f = bt$, $g = d[2\sigma_u A_w \sin \theta - (\sigma_Y - \sigma_u)A_f]/[2(\sigma_u + \sigma_Y)A_w \sin \theta]$, σ_Y is the yield stress, t, d is defined in Figure 7.1, and σ_u is the ultimate compressive stress of the corrugation flange to account for buckling.

7.3 Ultimate Strength of Plate Girders

Figure 7.5 shows a plate girder with transverse stiffeners under combined bending moments and shearing forces. This section describes the ultimate strength formulations of such plate girders under bending moment, shearing force, patch load, and combinations thereof. For the ULS design of plate girders, interested readers may refer to Maquoi (1992) and Kitada and Dogaki (1997), among others. For plate girders with perforated web under axial compression and bending moment, Zhao et al. (2015) may be referred to.

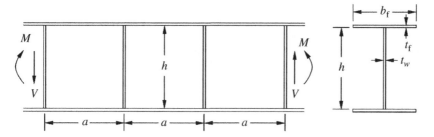

Figure 7.5 A plate girder under longitudinal bending and shearing forces.

7.3.1 Ultimate Strength Under Shearing Force

A stocky plate girder web under shearing force may not fail until the following upper limit of the load-carrying capacity is reached, namely,

$$V_P = h t_w \tau_Y \tag{7.5}$$

where $\tau_Y = \sigma_{Yw}/\sqrt{3}$ is the shear yield stress, σ_{Yw} is the web yield stress, and V_P is the plastic shear strength.

However, a slender web can buckle before it reaches its ultimate strength. Shear buckling is considered to take place if the following criteria are satisfied (ENV 1993-1-1 1992):

$$\frac{h}{t_w} = 69\varepsilon \text{ for an unstiffened web} \tag{7.6a}$$

$$\frac{h}{t_w} = 30\varepsilon\sqrt{k_s} \text{ for a stiffened web} \tag{7.6b}$$

where $\varepsilon = \sqrt{(235/\sigma_{Yw})}$, σ_{Yw} is the web yield stress (N/mm^2), and k_s is the web shear buckling coefficient (which is given as 5.34 for webs with transverse stiffeners at the supports but no intermediate transverse stiffeners); $k_s = 4.0 + 5.34(h/a)^2$ for webs with transverse stiffeners at the supports and intermediate transverse stiffeners with $a/h < 1$; $k_s = 5.34 + 4.0(h/a)^2$ for webs with transverse stiffeners at the supports and intermediate transverse stiffeners with $a/h \geq 1$.

Equation (7.6a) implies that all webs with h/t_w greater than 69ε may be designed to have transverse stiffeners at the supports. The shear buckling strength of a web depends on the h/t_w ratio and the spacing, a, of any intermediate web stiffeners. The shear buckling strength may also be affected by the anchorage of the tension fields associated with the end stiffeners or flanges. The anchorage provided by flanges will normally be reduced by longitudinal stresses due to bending moment and axial load.

For shear buckling strength estimation of plate girder webs without intermediate transverse stiffeners or of webs with transverse stiffeners only, the following two methods are useful (ENV 1993-1-1 1992):

1) The simple post-elastic critical buckling strength method, which can be used for the webs of plate girders, with or without intermediate transverse stiffeners, provided that the web has transverse stiffeners at the supports, but only for webs for which $a/h \geq 3.0$. It has been found that the simple post-critical method tends to underestimate the strength if $a/h < 3.0$.

2) The tension field method, which may be used for webs with transverse stiffeners at the supports plus intermediate transverse stiffeners, provided that adjacent panels or end posts provide anchorage for the tension fields, but only for webs for which $a/h < 3.0$. It has been found that the tension field method tends to underestimate the strength if $a/h \geq 3.0$.

For both of these methods, the transverse stiffeners are considered to be sufficiently stiff that they will remain straight until the web buckles. In this regard, ENV 1993-1-1 (1992) of Eurocode 3 suggests that the following stiffness criterion for the transverse stiffeners must be satisfied:

$$I_s \geq \begin{cases} 1.5h^3t_w^3/a^2 & \text{for } a/h < \sqrt{2} \\ 0.75ht_w^3 & \text{for } a/h \geq \sqrt{2} \end{cases} \tag{7.7}$$

where I_s is the moment of inertia of the transverse stiffener.

7.3.1.1 Simple Post-Critical Buckling Method

In this method, the ultimate shear load of a web is calculated as follows:

$$V_u = ht_w\tau_u \tag{7.8}$$

where τ_u is the simple post-critical shear strength, which is taken as Equation (4.81) or the formula of ENV 1993-1-1 (1992), namely,

$$\tau_u = \begin{cases} \tau_Y & \text{for } \lambda \leq 0.8 \\ [1 - 0.625(\lambda - 0.8)]\tau_Y & \text{for } 0.8 < \lambda < 1.2 \\ 0.9/\lambda\tau_Y & \text{for } \lambda \geq 1.2 \end{cases} \tag{7.9}$$

where

$$\lambda = \sqrt{\frac{\tau_Y}{\tau_E}} = \frac{h}{t_w}\frac{1}{37.4\varepsilon\sqrt{k_s}},$$

τ_E is the elastic web shear buckling stress and k_s and ε are defined in Equations (7.6).

7.3.1.2 Tension Field Method

The plate girder with intermediate transverse stiffeners normally has significant reserve strength after the web buckles in shear because the so-called tension field effect appears in the web, as shown in Figure 7.6a. As the applied loads increase, the stress inside the web is redistributed so that the diagonal tensile stresses continue to increase with the applied shear, whereas the diagonal compressive stresses remain substantially unchanged. In this regard, the web ultimate shear load, V_u, is typically given as a sum of the three contributions, namely,

$$V_u = V_{cr} + V_t + V_f \tag{7.10}$$

where V_{cr} is the beam-action strength, V_t is the tension field strength, and V_f is the frame-action strength.

(a)

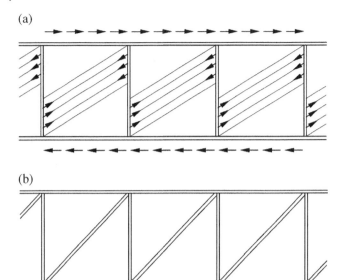

(b)

Figure 7.6 Tension field action in the web of a plate girder under shearing force: (a) tension field action in the web; (b) truss-like structure.

For practical purposes, the frame-action strength is often neglected, that is, $V_f = 0$. The beam-action strength, V_{cr} is obtained by

$$V_{cr} = ht_w \tau_{cr} \tag{7.11}$$

where τ_{cr} is the critical shear buckling stress, which can be obtained by the plasticity correction of the elastic shear buckling stress, τ_E, using the Johnson–Ostenfeld formulation method as described in Equation (2.93).

τ_E is often estimated with the assumption that all four web edges are simply supported, given by

$$\tau_E = k_s \frac{\pi^2 E}{12(1 - v^2)} \left(\frac{t_w}{h}\right)^2 \tag{7.12a}$$

where E is Young's modulus, v is Poisson's ratio, and k_s is the elastic shear buckling coefficient for the web with intermediate transverse stiffeners as defined in Equation (7.6) or

$$k_s = \begin{cases} 5.34 + 4(h/a)^2 & \text{for } a/h \geq 1 \\ 4.0 + 5.34(h/a)^2 & \text{for } a/h < 1 \end{cases} \tag{7.12b}$$

Two models are typically adopted for strength predictions related to the tension field action: the so-called Basler model (Basler 1961) and the so-called Cardiff model (Porter et al. 1975). In the Basler model, the flanges are assumed to be too flexible to support any lateral loading induced by the tension field. The yield band, which determines the tension field strength, is resisted by the transverse stiffeners alone. The width of the tensile band depends on the slope of the band that is chosen to maximize the shear strength. The Basler model may be considered to provide a lower bound to the web ultimate shear strength.

The plate girder subject to a tension field is considered to behave as a truss structure together with the flanges and vertical stiffeners to transfer additional shearing force, as shown in Figure 7.6b. However, the plate girder may not sustain further increase of shear if the tension field material yields. The contribution due to the tension field action to the load-carrying capacity is then given by

$$V_t = ht_w\tau_{tf} \tag{7.13}$$

where τ_{tf} is the shear strength due to the tension field contribution.

Using the Basler model, τ_{tf} can be approximately calculated based on the model of the truss-like structure, which neglects the contribution made by the bending resistance of the flanges, with results as follows:

$$\tau_{tf} = \frac{\sigma_{Yw}}{2}\frac{1 - \tau_{cr}/\tau_Y}{\sqrt{1 + (a/h)^2}} \tag{7.14}$$

where τ_{cr} is defined in Equation (7.11).

It is noted that the transverse stiffeners in such a case in which the tension field is used in design should be sufficiently strong that they can sustain and transmit the forces caused by the tension field action in the web. To achieve this, the following criteria need to be satisfied:

$$A_s \geq \frac{P_s}{\sigma_{Yw}}, \quad I_s \geq \frac{P_s h^2}{\pi^2 E} \tag{7.15a}$$

where A_s and I_s are the cross-sectional area and the moment of inertia of the transverse stiffener, which is similar to Equation (7.7), and P_s is the stiffener force due to the tension field action, given by (Trahair and Bradford 1988):

$$P_s = \frac{\sigma_{Yw}ht_w}{2}\left(1 - \frac{\tau_{cr}}{\tau_Y}\right)\left[\frac{a}{h} - \frac{(a/h)^2}{\sqrt{1 + (a/h)^2}}\right] \tag{7.15b}$$

Some valuable improvements to the Basler model were made by Rockey and his colleagues at University College, Cardiff (Porter et al. 1975), and the result is often called the Cardiff model. The Cardiff model accounts for the effects of the bending stiffness of the flanges on the width of the diagonal tension band. Although the diagonal tension band is composed of three parts, as shown in Figure 7.6a, the central part is anchored to the transverse stiffeners, while the other two parts are anchored to the lower and upper flanges. Therefore, the strength related to the tension field action is determined by the vertical component of the force in the band at collapse. If the flanges have an infinite flexural stiffness for bending in the plane of the web, a pure tension field will develop. In this case, the anchorage lengths on the flanges become equal to the spacing, a, of the intermediate transverse stiffeners (or the web length). For very flexible flanges, the tension field is anchored on the adjacent webs only. In practice, the anchorage lengths will span only part of the web length because the flanges have a finite flexural rigidity. For an elaborate description of the Basler and Cardiff models, interested readers may refer to Maquoi (1992). The descriptions earlier are only for internal webs. Maquoi (1992) may also be referred to for information on end webs.

7.3.2 Ultimate Strength Under Bending Moment

The collapse behavior of a plate girder under bending is governed by buckling of the web and the compression flange. The elastic buckling stress, σ_{bE}, of the plate girder web under longitudinal bending may be calculated from Equation (3.2), considering that all four web edges are simply supported as follows:

$$\sigma_{bE} = k_b \frac{\pi^2 E}{12(1-\nu^2)} \left(\frac{t_w}{h}\right)^2 \tag{7.16}$$

where

$$k_b = \begin{cases} 2.39 & \text{for } a/h \geq \dfrac{2}{3} \\ 15.9 + 1.87(h/a)^2 + 8.6(a/h)^2 & \text{for } a/h < \dfrac{2}{3} \end{cases}.$$

The inelastic (or critical) buckling strength, σ_{bcr}, of the web under bending is then estimated by the plasticity correction of the elastic buckling strength using the Johnson–Ostenfeld formulation method described in Equation (2.93). When the web buckles in bending, the corresponding critical bending moment, M_{wcr}, for the plate girder is given by

$$M_{wcr} = \frac{2I}{h}\sigma_{bcr} \tag{7.17}$$

where I is the moment of inertia of the plate girder cross section.

Under longitudinal bending, M, the compression flange is subjected to the axial compressive stress, given by

$$\sigma_f = -\frac{M}{2I}(h + t_f) \tag{7.18}$$

The compression flange may buckle before or after the web fails, whereas the failure pattern of the former type, that is, buckling of the compression flange before failure of the web, is quite undesirable. The buckling of the compression flange may be estimated from Equation (5.36), considering the half flange with regard to the web with boundary conditions simply supported at three edges and free at one edge, as shown in Figure 5.8. The critical buckling strength, σ_{fcr}, of the compression flange is then approximately computed by the plasticity correction of the corresponding elastic buckling strength using the Johnson–Ostenfeld formulation method described in Equation (2.93).

To prevent the possibility of the compression flange buckling in the plane of the plate girder web, the following criterion needs to be satisfied (ENV 1993-1-1 1992):

$$\frac{h}{t_w} \leq 0.55 \frac{E}{\sigma_{Yf}} \sqrt{\frac{A_w}{A_{fc}}} \tag{7.19}$$

where A_w is the area of the web, A_{fc} is the area of the compression flange, and σ_{Yf} is the yield stress of the compression flange.

The critical bending moment at buckling of the compression flange may be given by

$$M_{fcr} = \frac{2I\sigma_{fcr}}{h + t_f} \tag{7.20}$$

For a plate girder with unequal flanges, the buckling of the flanges may be checked for the weaker flange. The plastic bending moment of the plate girder without local buckling is given by

$$M_P = M_{Pw} + M_{Pf} \tag{7.21}$$

where $M_{Pf} = hb_f t_f \sigma_{Yf}$ is the plastic moment of the flanges and $M_{Pw} = (h^2 t_w/4)\sigma_{Yw}$ is the plastic moment of the web.

In practice, the flanges are normally designed so that they do not buckle until the plate girder reaches the ultimate strength. In this case, the ultimate strength behavior of the plate girder is primarily governed by buckling of the web. Therefore, two types of failure regimes should be considered for the calculation of the ultimate strength, depending on flange failure either before or after buckling of the web.

7.3.2.1 Mode I

If $M_{wcr} > M_Y$, the web may not buckle until the flange fails or yields. In this case, the plate girder is considered to reach the ultimate strength if the compression flange yields. This results in

$$M_u = M_Y \tag{7.22}$$

where M_u is the ultimate bending moment of the plate girder and $M_Y = (2I/h)\sigma_{Yf}$ is the critical bending moment at the yielding of the compression flange.

7.3.2.2 Mode II

The axial stress distribution over the plate girder cross section immediately after the web buckles in bending may be idealized, as shown in Figure 7.7. The plate girder may sustain a further increase of bending, but only the flange and the web with the effective section will be available in the compression zone, whereas all sections in the tension side

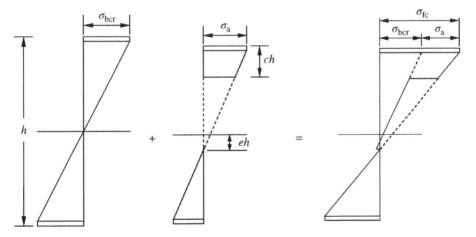

Figure 7.7 Idealized stress distribution over the plate girder cross section under bending after buckling of the web (c, effective section coefficient).

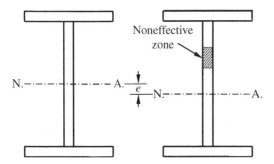

Figure 7.8 Effective cross section of a plate girder.

are still fully effective, as shown in Figure 7.8. The axial stress, σ_{fc}, of the compression flange is then computed by

$$\sigma_{fc} = -\frac{h+t_f}{2}\left[\frac{M_{wcr}}{I} + \frac{M-M_{wcr}}{I_e}(1+2e)\right] \tag{7.23}$$

where

$$I_e = \left(\frac{1}{2}+2e^2\right)h^2 b_f t_f + \frac{1}{3}h^3 t_w\left[\frac{1}{4}+3e^2-\left(\frac{1}{2}+e-c\right)^3\right]$$

is the effective moment of inertia of the plate girder cross section, with

$$e = \left(\frac{1}{2}+c+2\frac{b_f t_f}{h t_w}\right)-\sqrt{2\left(1+2\frac{b_f t_f}{h t_w}\right)\left(c+\frac{b_f t_f}{h t_w}\right)}$$

The plate girder is considered to collapse if the axial stress of the compression flange reaches the yield stress, namely, $\sigma_{fc} = -\sigma_{Yf}$. This results in

$$M_u = M_{wcr} + (M_Y - M_{wcr})\frac{I_e}{I}\frac{1}{1+2e}. \tag{7.24}$$

Predictions of the ultimate bending strength are often made by applying the concept of the effective cross section associated with the effective width of the compression elements. The plastic bending capacity is then calculated as for the procedure described in Chapter 2 but with the effective cross sections, as shown in Figure 7.8. It is noted that even for uniaxial compressive loading, the additional moment induced by the shift of the neutral axis can develop in the effective cross section and must be considered in the strength calculations.

The shear lag effects in flanges may be neglected if the flange breadth is less than 10% of the length between the points of zero moment for the internal elements or if the outstand is less than 5% of the length between the points of zero moment for the outstand elements. When these limits are exceeded, the shear lag effects cannot be neglected and an effective breadth of the flanges should be used. More elaborate descriptions of the possible interaction between shear lag and plate buckling are described in Chapter 2.

For other types of ultimate bending strength design formulations for plate girders with stiffened webs, interested readers may refer to Kitada and Dogaki (1997).

7.3.3 Ultimate Strength Under Combined Shearing Force and Bending Moment

Plate girders are likely to be subjected to combined bending and shearing forces. For plate girders with slender webs, the ultimate strength may be reached after the webs buckle. The ultimate strength interactive relationship of such plate girders under combined bending and shearing forces is often represented by a piecewise linear curve, as shown in Figure 7.9 (ENV 1993-1-1 1992).

(a)

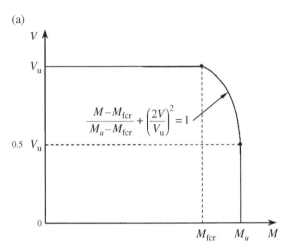

$$\frac{M - M_{\text{fcr}}}{M_u - M_{\text{fcr}}} + \left(\frac{2V}{V_u}\right)^2 = 1$$

(b)

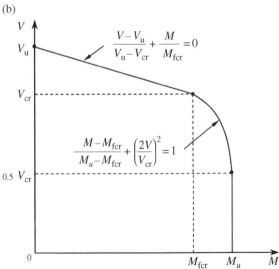

$$\frac{V - V_u}{V_u - V_{\text{cr}}} + \frac{M}{M_{\text{fcr}}} = 0$$

$$\frac{M - M_{\text{fcr}}}{M_u - M_{\text{fcr}}} + \left(\frac{2V}{V_{\text{cr}}}\right)^2 = 1$$

Figure 7.9 A schematic of the ultimate strength relationship of plate girders under combined bending and shearing forces: (a) simple post-critical buckling method basis; (b) tension field method basis.

It is more convenient to have a closed-form expression of the ultimate strength inter-active relationship for plate girders under combined bending and shearing force. For this purpose, the following formula may be used:

$$\left(\frac{M}{M_u}\right)^4 + \left(\frac{V}{V_u}\right)^4 = 1 \tag{7.25}$$

where V_u is the ultimate shear force, as defined in Section 7.3.1, and M_u is the ultimate bending moment as defined in Section 7.3.2. The validity of Equation (7.25) has been confirmed by comparison to test results with plate girders under combined bending and shearing force (Fukumoto et al. 1985, Mikami et al. 1991). Equation (7.25) was found to agree well with the lower limit of the test results, with related modeling error given by a mean of 0.856 and a coefficient of variation of 0.08 for $\sigma_{uf} \le \sigma_{uw}$ and a mean of 0.926 and a coefficient of variation of 0.05 for $\sigma_{uf} > \sigma_{uw}$, where σ_{uf} is the ultimate strength of the compression flange and σ_{uw} is the ultimate strength of the web.

7.3.4 Ultimate Strength Under Patch Load

Plate girders used in the field of civil engineering are sometimes subjected to patch loads, as shown in Figure 7.10. The collapse strength of an unstiffened web subject to patch (transverse) loads applied through a plate girder flange is governed by one of the following three failure modes (ENV 1993-1-1 1992):

- Crushing of the web close to the flange, accompanied by plastic deformation of the flange
- Crippling of the web in the form of localized buckling and crushing of the web close to the flange, accompanied by plastic deformation of the flange
- Buckling of the web over most of the depth of the plate girder

Two types of load application are normally considered: (i) forces applied through one flange and resisted by shear forces in the web and (ii) forces applied to one flange and transferred through the web directly to the other flange. For the first load type, the capacity of the web to lateral forces may be determined as the lesser value of the two strengths due to crushing and crippling. For the other load type, the web capacity may be taken as the lesser value of the two strengths due to crushing and buckling. The crippling strength

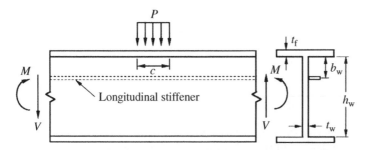

Figure 7.10 A plate girder with the unstiffened web under patch load.

of a web with intermediate transverse stiffeners is similar to that of an unstiffened web, with the increase due to the stiffeners.

Dogaki et al. (1992a) studied the ultimate strength of longitudinally stiffened plate girders under patch loading and concluded that the optimal location of the longitudinal stiffener close to the plate girder flange under patch loading is about $b_w = 0.15h_w$ (see Figure 7.10). Dogaki et al. (1992b) then proposed an empirical expression of the ultimate strength, P_u, of plate girders under (concentrated) patch loading by curve fitting based on their own test results:

$$\frac{P_u}{2V_P} = \frac{0.594}{\lambda} + 0.069 \tag{7.26}$$

where V_P is defined in Equation (7.5), $\lambda = \sqrt{(2V_P/P_E)}$ is the buckling parameter, and P_E is the elastic buckling strength of the plate girder web under patch loading, considering the effects of the flexural and torsional rigidities of the flange.

For plate girders without longitudinal stiffeners under patch loading, Takimoto (1994) proposed a closed-form expression of the ultimate strength, P_u:

$$P_u = \left(25t_w^2\sigma_{Yw} + 4t_w t_f \sigma_{Yf}\right)\left(1 + \frac{c + 2t_f}{2h_w}\right) \tag{7.27}$$

The mean and coefficient of variation of the accuracy of Equation (7.27) when compared to the test results for 143 specimens on plate girders under patch loading were 0.984 and 0.15, respectively. For more details on the ULS design of plate girder webs to patch loading, Granath et al. (2000) or ENV 1993-1-1 (1992) may be referred to.

7.3.5 Ultimate Strength Under Combined Patch Load, Shearing Force, and Bending Moment

An ultimate strength interactive relationship for the plate girders under combined patch load, bending moment, and shearing force was proposed by Takimoto (1994) as follows:

$$\left(\frac{P}{P_u}\right)^2 + \left(\frac{M}{M_u}\right)^4 + \left(\frac{V}{V_u}\right)^4 = 1 \tag{7.28}$$

where V_u is defined in Section 7.3.1, M_u is defined in Section 7.3.2, and P_u is defined in Section 7.3.4.

7.4 Ultimate Strength of Box Columns

Plated structures with a box-type cross section are often used in marine and land-based applications. Such structures are designed to sustain axial compression, bending moment, shearing force, or their combination, as shown in Figure 7.11. When subjected to predominantly axial compression, the structure is called a box column. This section describes the ultimate strength formulations of box columns with or without diaphragms (or transverse bulkheads) under axial compression.

The ultimate strength of a box column may be obtained as the sum of the strengths of the individual flanges or webs. In this case, the interactive effect between the webs and

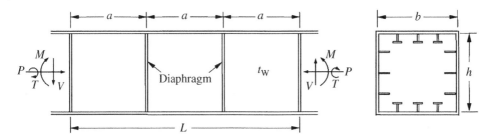

Figure 7.11 Plated structures with a box-type cross section under axial compression, shearing force, bending, torsion, or their combination.

flanges may be negligible, and their edges are considered to be simply supported. For short box columns, the ultimate strength is governed by local buckling of flanges or webs, whereas, for slender box columns, it is affected by both the global buckling of the box column and the local buckling of its components. The ultimate strength, σ_{uL}, of a short box column under axial compression, considering the local buckling and collapse of component elements, may therefore be obtained by

$$\sigma_{uL} = \frac{1}{A_t} \sum_{i=1}^{4} A_i \sigma_{upi} \tag{7.29}$$

where A_i and σ_{upi} are the cross-sectional area and the ultimate compressive strength of the ith panel (i.e., flange or web panels), respectively, and A_t is the total cross-sectional area.

The ultimate strength, σ_{uG}, of a long box column (without considering local buckling) in the global collapse mode may be obtained by the plasticity correction of the corresponding Euler buckling strength, σ_{EG}, using the Johnson–Ostenfeld formulation method described in Equation (2.93). If the effect of initial imperfections is not considered, the following Euler global buckling stress, σ_{EG} may be used for a box column simply supported at both ends:

$$\sigma_{EG} = \frac{\pi^2 EI}{A_t a^2} \tag{7.30}$$

where I is the moment of inertia of the cross section with regard to the weaker axis, L is the length of the box column, and E is Young's modulus.

For a box column with a "medium" length, the interactive effect between local and overall buckling may play a significant role. In this case, the ultimate strength, σ_u, of the box column may be estimated from the Johnson–Ostenfeld formulation method, but by correcting the yield stress, σ_Y, as follows (AISC 1969):

$$\sigma_u = \begin{cases} \sigma_{EG} & \text{for } \sigma_{EG} \le 0.5 c \sigma_Y \\ c \sigma_Y (1 - c \sigma_Y / 4 \sigma_{EG}) & \text{for } \sigma_{EG} > 0.5 c \sigma_Y \end{cases} \tag{7.31a}$$

where c is the knockdown factor applied to the yield stress, defined as follows:

$$c = \frac{\sigma_{uL}}{\sigma_Y} \tag{7.31b}$$

Alternatively, the ultimate strength of a box column considering the effect of local buckling may be calculated using the reduced cross section associated with the effective width of the plate elements (ENV 1993-1-1 1992). It is of interest to note that a "uniform" box column may generally buckle in the global mode when the elastic buckling stress, σ_{EL}, of the individual plate elements is greater than the global elastic buckling stress, σ_{EG}, of the box column, and vice versa for the local buckling mode. The transition from the local buckling mode to the global buckling mode may take place when σ_{EL} equals σ_{EG}, namely,

$$\sigma_{EL} = \sigma_{EG} \tag{7.32}$$

For a square cross-sectional uniform box column without stiffeners simply supported at both ends, for instance, we have

$$\sigma_{EL} = \frac{4\pi^2 E}{12(1-\nu^2)} \left(\frac{t}{b}\right)^2, \quad \sigma_{EG} = \frac{\pi^2 EI}{A_t L^2}, \quad A_t = 4bt, \quad I = \frac{2}{3}b^3 t \tag{7.33}$$

where ν is Poisson's ratio.

Substitution of Equation (7.33) into Equation (7.32) yields

$$\frac{L}{b} = \sqrt{\frac{1-\nu^2}{2} \frac{b}{t}} \tag{7.34a}$$

From Equation (7.34a), it is considered that the global buckling mode may take place in a square-section uniform box column (without stiffeners) that is simply supported at both ends if the following condition is satisfied:

$$\frac{L}{b} > \sqrt{\frac{1-\nu^2}{2} \frac{b}{t}} \tag{7.34b}$$

The effects of initial imperfections on the ultimate strength of box columns can be accounted for as described in Chapters 1 and 4. For the ultimate strength of welded box columns accounting for the effects of initial imperfections, interested readers may refer to Schafer & Pekoz (1998), AASHTO (2010), and Susanti et al. (2014), among others.

7.5 Ultimate Strength of Box Girders

When a plated structure with a box-type cross section as shown in Figure 7.11 is subjected to predominantly bending moment, it is termed a box girder. This section describes the ultimate strength formulations of box girders with diaphragms (or transverse bulkheads) under bending moment, shearing force, torsional moment, or their combination.

7.5.1 Simple-Beam Theory Method

The simple-beam theory method is useful to analyze the strength of the beams because a box girder under bending moments can be dealt with as a beam. In the simple-beam

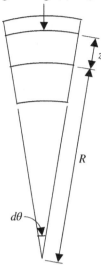

Elongated length, $(R+z)\,d\theta$

Figure 7.12 Schematic of longitudinal strain in an infinitesimal element of a deflected beam.

theory, the following assumptions (called the Bernoulli–Euler hypothesis) are adopted (Hughes & Paik 2013):

- Plane cross section remains plane.
- The beam is essentially prismatic with no opening or discontinuities.
- Other types of load effects, for example, transverse and longitudinal deflections and distortions caused by shear and/or torsion, do not affect the bending response, and they may be dealt with separately.
- The material is homogeneous and elastic.

With the abovementioned assumptions, the longitudinal strain ε_x of a beam deflected by the action of a bending moment as illustrated in Figure 7.12 can be determined as follows:

$$\varepsilon_x = \frac{(R+z)d\theta - Rd\theta}{Rd\theta} = \frac{z}{R} \tag{7.35}$$

where R and $d\theta$ are the radius and infinitesimal angle of a deflected beam element, respectively, as defined in Figure 7.12. As the coordinate z is set to be zero at the horizontal neutral axis of the beam cross section, it is obvious that ε_x varies linearly in the vertical direction with regard to z.

The longitudinal (bending) stress σ_x with respect to the horizontal neutral axis of the beam cross section can be determined in the linear elastic regime as follows:

$$\sigma_x = E\varepsilon_x = E\frac{z}{R} \tag{7.36}$$

Since a pure bending moment is applied without axial force, the following equilibrium condition should be satisfied over the beam cross section:

$$\int \sigma_x dA = 0 \quad \text{or} \quad \int zdA = 0 \tag{7.37}$$

The vertical bending moment is calculated by an integration of the first moment associated with the longitudinal stress over the beam cross section as follows:

$$M_y = \int z\sigma_x dA = \frac{EI}{R} \tag{7.38}$$

where M_y is the vertical bending moment about the horizontal neutral axis and $I = \int z^2 dA$ is the moment of inertia of the beam cross section.

In Equation (7.38), the radius R is eliminated using Equation (7.36) to compute the bending stress of the beam at a height z from the horizontal neutral axis as follows:

$$\sigma_x = \frac{M_y}{I}z \tag{7.39}$$

7.5.1.1 Maximum Bending Stress

According to the simple-beam theory, the bending stress at the cross section of a box girder under a bending moment is calculated from Equation (7.39) as follows:

$$\sigma = \frac{M}{I} z \qquad (7.40)$$

where σ is the bending stress, M is the applied bending moment, I is the moment of inertia, and z is the distance from the neutral axis position of the beam cross section to the location of the bending stress calculation in the direction of the depth of the box girder.

The maximum bending stress will develop at the outmost fiber of the cross section of the box girder as shown in Figure 7.13a and can thus be obtained from Equation (7.40) as follows:

$$\sigma_D = \frac{M}{Z_D} \text{ at deck (upper flange)} \qquad (7.41a)$$

$$\sigma_B = \frac{M}{Z_B} \text{ at bottom (lower flange)} \qquad (7.41b)$$

where σ_D and σ_B are the bending stresses at the deck (upper flange) and bottom (lower flange), respectively, and Z_D and Z_B are the section moduli at the deck and bottom, respectively.

7.5.1.2 Section Modulus

In Equation (7.41), the two components of the section modulus are defined as follows:

$$Z_D = \frac{I}{z_D} \qquad (7.42a)$$

$$Z_B = \frac{I}{z_B} \qquad (7.42b)$$

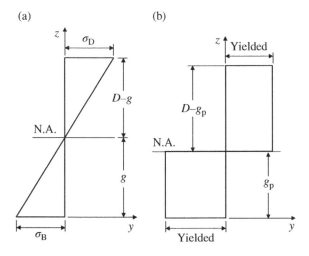

(a)   (b)

Figure 7.13 Bending stress distribution in a box girder section in (a) linear elastic state and (b) full plastic state (N.A., neutral axis).

where z_D and z_B are the distances from the neutral axis position of the cross section to the deck and bottom, respectively.

In Equation (7.42), z_D and z_B can be obtained as follows:

$$z_D = D - g - \frac{t_D}{2} \tag{7.43a}$$

$$z_B = g - \frac{t_B}{2} \tag{7.43b}$$

where D is the depth of the box girder ($= h$ in Figure 7.11), t_D is the representative (equivalent) thickness of the deck plate, t_B is the representative (equivalent) thickness of the bottom plate, and g is the distance from the baseline of the box girder to the neutral axis position. The term $t_D/2$ or $t_B/2$ represents that the bending stress σ_D or σ_B is evaluated at the mid-thickness of the deck or bottom plate.

In Equation (7.43), g can be calculated as follows:

$$g = \frac{\sum_{i=1}^{n} a_i z_i}{\sum_{i=1}^{n} a_i} \tag{7.44}$$

where a_i is the cross-sectional area of the ith structural element (portion), z_i is the distance from the baseline to the neutral axis of the ith structural element (portion), and n is the total number of members to be included in the cross-sectional property calculation. The baseline is usually presumed to be located at the outer fiber of the bottom (lower flange) plate. To automate the computation of Equation (7.44), the cross section of a box girder can be idealized as an assembly of pure plate element (segment) models as shown in Figure 2.2d but with one plate element for the stiffener flange.

The moment of inertia I for the box girder cross section in Equation (7.42) can be calculated once g is determined from Equation (7.44) as follows:

$$I = \sum_{i=1}^{n} \left(a_i z_i^2 + i_i \right) - A g^2 \tag{7.45}$$

where $A = \sum_{i=1}^{n} a_i$ is the total area of the box girder cross section, a_i, z_i, g, and n are defined in Equation (7.44), and i_i is the moment of inertia for the ith structural element (portion) about its own neutral axis.

Inclined or curved plates as shown in Figure 7.14 are sometimes used, for example, at the corner of box girders. In this case, the neutral axis position and moment of inertia about its own neutral axis are approximately determined by

$$i = \frac{1}{12} a d^2, \quad z_0 = \frac{d}{2} \quad \text{for an inclined plating} \tag{7.46a}$$

$$i = \left(\frac{1}{2} - \frac{4}{\pi^2} \right) a r^2, \quad z_0 = \frac{(\pi - 2) r}{\pi} \quad \text{for a curved plating} \tag{7.46b}$$

where z_0 is the distance from the bottom to the neutral axis position, i is the moment of inertia about its own neutral axis, a is the cross-sectional area of the inclined or curved plating, d is the projected depth of the inclined plating, and r is the radius of the curvature of the curved plating, as defined in Figure 7.14.

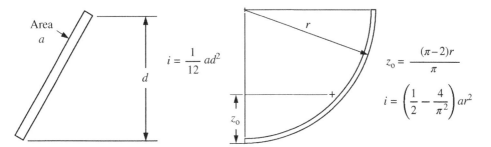

Figure 7.14 Inclined or curved plating in the box girder.

7.5.1.3 First-Yield Bending Moment

The first-yield bending moment of a box girder can be a good indication of the structural capacity (strength) where local buckling is not considered. The first-yield bending moment is determined when the maximum bending stresses in Equation (7.41) reach the yield stress of the material for the first time as follows:

$$M_{YD} = Z_D \sigma_{YeqD} \text{ at deck (upper flange)} \tag{7.47a}$$

$$M_{YB} = Z_B \sigma_{YeqB} \text{ at bottom (lower flange)} \tag{7.47b}$$

where M_{YD} and M_{YB} are the first-yield bending moments at the deck and bottom, respectively, and σ_{YeqD} and σ_{YeqB} are the representative (equivalent) yield stresses of the deck and bottom plate panels, respectively.

7.5.1.4 First-Collapse Bending Moment

Another indication of the structural capacity is the first-failure (collapse) collapse bending moment when the compressed flange, that is, the deck plate panels in sagging moment or the bottom plate panels in hogging moment, reaches the ultimate strength in axial compression:

$$M_{fD} = Z_D \sigma_{uD} \tag{7.48a}$$

$$M_{fB} = Z_B \sigma_{uB} \tag{7.48b}$$

where M_{fD} and M_{fB} are the first-collapse bending moments at the deck or bottom plate panels, respectively, and σ_{uD} and σ_{uB} are the ultimate strength of the deck or bottom plate panels in axial compression. It is noted that yielding or collapse of structural members except for the compressed flange is not considered in Equation (7.48).

Equation (7.48) can be used to predict the first-collapse bending moments of box girders in sagging or hogging conditions as follows:

$$M_{fs} = M_{fD} = Z_D \sigma_{uD} \tag{7.49a}$$

$$M_{fh} = M_{fB} = Z_B \sigma_{uB} \tag{7.49b}$$

where M_{fs} and M_{fh} are the first-failure (collapse) bending moments for sagging or hogging, respectively.

It is noted that box girders can usually sustain further loading even after reaching the first-yield or first-failure status because the structural failures can grow into the vertically positioned structures, such as side shell structures and/or longitudinal bulkheads, until the box girders reach their ULS.

7.5.1.5 Full Plastic Bending Moment

It is sometimes of interest to know the full plastic bending capacity of a box girder cross section, which can be used as an upper limit of the ultimate strength, as described in Chapter 2. The neutral axis of the full plastic cross section above the baseline can be determined, as shown in Figure 7.13b, as follows:

$$g_p = \frac{\sum_{i=1}^{n} a_i \sigma_{Yi} z_i}{\sum_{i=1}^{n} a_i \sigma_{Yi}} \tag{7.50}$$

where g_p is the distance from the baseline to the neutral axis position of the full plastic cross section and σ_{Yi} is the material yield stress of the ith structural element.

The full plastic bending moment M_P is then determined as follows:

$$M_P = \sum_{i=1}^{n} a_i \sigma_{Yi} |z_i - g_p| \tag{7.51}$$

7.5.1.6 Exercise for Cross-Sectional Property Calculations

An exercise for the calculations of the cross-sectional property and the first-yield or first-collapse bending moment using the simple-beam theory method is now considered. Figure 7.15 shows an example of a box girder with 4.5 m deep and 7.5 m wide. The spacing between the transverse bulkheads (or diaphragms) is 8 m. The structure is made of

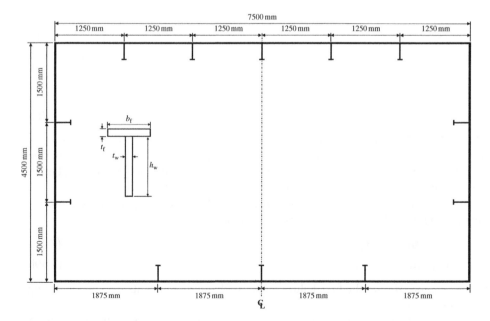

Figure 7.15 An example of a box girder.

mild steel with a yield stress of $\sigma_Y = 235$ MPa and an elastic modulus of $E = 205.8$ GPa. Figure 7.16 shows a modeling technique for the box girder cross section in which the plating between the stiffeners and the webs or flanges of the stiffeners are considered as a rectangular plate element (segment). Tables 7.1 and 7.2 summarize the calculations of the cross-sectional properties of the box girder. In this case, the neutral axis, the moment of inertia, the section modulus, and the full plastic bending moment of the box girder cross section can be calculated as follows:

Height of neutral axis from the baseline: $g = \dfrac{\sum a_i z_i}{\sum a_i} = \dfrac{1.8059}{0.7676} = 2.353\,\text{m}$

Total cross-sectional area: $A = 0.768\,\text{m}^2$

Moment of inertia: $I = \displaystyle\sum_{i=1}^{n}\left(a_i z_i^2 + i_i\right) - Ag^2 = 6.3258 + 0.3039 - 4.2489 = 2.381\,\text{m}^4$

Distance from the neutral axis to the deck: $z_D = D - g - \dfrac{t_D}{2} = 4.5 - 2.353 - 0.01 = 2.137\,\text{m}$

Distance from the neutral axis to the bottom: $z_B = g - \dfrac{t_B}{2} = 2.353 - 0.01 = 2.343\,\text{m}$

Section modulus at deck: $Z_D = \dfrac{I}{z_D} = \dfrac{2.3809}{2.1372} = 1.114\,\text{m}^3$

Section modulus at bottom: $Z_B = \dfrac{I}{z_B} = \dfrac{2.3809}{2.3428} = 1.016\,\text{m}^3$

Distance from the baseline to the neutral axis of the full plastic cross section:

$$g_p = \frac{\sum_{i=1}^{n} a_i \sigma_{Yi} z_i}{\sum_{i=1}^{n} a_i \sigma_{Yi}} = \frac{\sum_{i=1}^{n} a_i \sigma_{Yi} z_i}{\sum_{i=1}^{n} a_i \sigma_{Yi}} = \frac{424.3838}{180.3747} = 2.353\,\text{m}$$

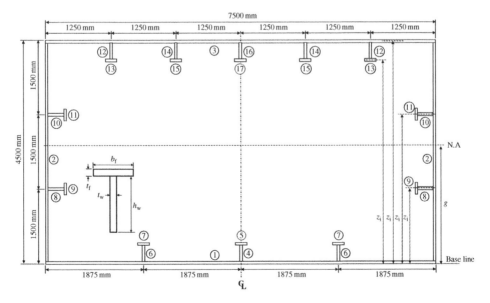

Figure 7.16 The pure plate element model for calculating the cross-sectional properties of a box girder.

Table 7.1 Summary of the cross-sectional property calculations of a box girder for the section modulus.

No.	Segment type	Number of segment	Breadth/ height	Thickness	a_i (m²)	z_i (m)	a_iz_i (m³)	$a_iz_i^2$ (m⁴)	i_i (m⁴)
			Scantlings (mm)						
1	Plate	1	7500	20	0.150	0.010	0.002	0.000	0.000
2	Plate	2	4460	20	0.089	2.250	0.201	0.452	0.148
3	Plate	1	7500	20	0.150	4.490	0.674	3.024	0.000
4	Web	1	872	18	0.016	0.456	0.007	0.003	0.001
5	Flange	1	300	28	0.008	0.906	0.008	0.007	0.000
6	Web	2	872	18	0.016	0.456	0.007	0.003	0.001
7	Flange	2	300	28	0.008	0.906	0.008	0.007	0.000
8	Web	2	872	18	0.016	1.500	0.024	0.035	0.000
9	Flange	2	300	28	0.008	1.500	0.013	0.019	0.000
10	Web	2	872	18	0.016	3.000	0.047	0.141	0.000
11	Flange	2	300	28	0.008	3.000	0.025	0.076	0.000
12	Web	2	872	18	0.016	4.044	0.063	0.257	0.001
13	Flange	2	300	28	0.008	3.594	0.030	0.109	0.000
14	Web	2	872	18	0.016	4.044	0.063	0.257	0.001
15	Flange	2	300	28	0.008	3.594	0.030	0.109	0.000
16	Web	1	872	18	0.016	4.044	0.063	0.257	0.001
17	Flange	1	300	28	0.008	3.594	0.030	0.109	0.000
Total					0.768		1.806	6.326	0.304

Full plastic bending moment: $M_P = \sum_{i=1}^{n} a_i \sigma_{Yi} |z_i - g_p| = 294.758$ MNm, where it is cautioned that any element (segments) having the full plastic neutral axis inside it should further be separated into two segments. Element number 2 indicated in Table 7.2 is an example of such a case.

The first-yield bending moment of the box girder is then calculated as follows:

At deck plate: $M_{YD} = Z_D \sigma_{YeqD} = 1.114 \text{m}^3 \times 235 \text{N/mm}^2 = 261\ 790 \times 10^6$ Nmm
At bottom plate: $M_{YB} = Z_B \sigma_{YeqB} = 1.016 \text{m}^3 \times 235 \text{N/mm}^2 = 238\ 818 \times 10^6$ Nmm

For calculation of the first-collapse bending moment of the box girder, the ultimate compressive stress of the compressed flanges should be calculated. The ultimate strength formulations described in Chapter 6 will be used for this purpose. Alternatively, a simpler approach is to use the Paik–Thayamballi formulation, which is given as Equation (2.99). Figure 7.17 shows the representative plate-stiffener combination models at the deck and at the bottom of the box girder with a fully effective cross section. The ultimate compressive stresses at the deck or bottom are then obtained as a function of the plate slenderness

Table 7.2 Summary of the cross-sectional property calculations of a box girder for the full plastic bending capacity.

No.	Segment type	Number of segment	a_i (m²)	z_i (m)	$a_i z_i$ (m³)	$a_i \sigma_{Yi}$ (MN)	$a_i \sigma_{Yi} z_i$ (MNm)	$a_i \sigma_{Yi}\|z_i - g_p\|$ (MNm)
1	Plate	1	0.150	0.010	0.002	35.250	0.353	82.583
2	Plate	2	0.047	1.197	0.056	11.059	13.232	12.788
			0.042	3.427	0.144	9.903	33.932	10.632
3	Plate	1	0.150	4.490	0.674	35.250	158.273	75.337
4	Web	1	0.016	0.456	0.007	3.689	1.682	6.996
5	Flange	1	0.008	0.906	0.008	1.974	1.788	2.856
6	Web	2	0.016	0.456	0.007	3.689	1.682	6.996
7	Flange	2	0.008	0.906	0.008	1.974	1.788	2.856
8	Web	2	0.016	1.500	0.024	3.689	5.533	3.146
9	Flange	2	0.008	1.500	0.013	1.974	2.961	1.683
10	Web	2	0.016	3.000	0.047	3.689	11.066	2.387
11	Flange	2	0.008	3.000	0.025	1.974	5.922	1.278
12	Web	2	0.016	4.044	0.063	3.689	14.917	6.238
13	Flange	2	0.008	3.594	0.030	1.974	7.095	2.450
14	Web	2	0.016	4.044	0.063	3.689	14.917	6.238
15	Flange	2	0.008	3.594	0.030	1.974	7.095	2.450
16	Web	1	0.016	4.044	0.063	3.689	14.917	6.238
17	Flange	1	0.008	3.594	0.030	1.974	7.095	2.450
Total			0.768		1.806	180.375	424.384	294.758

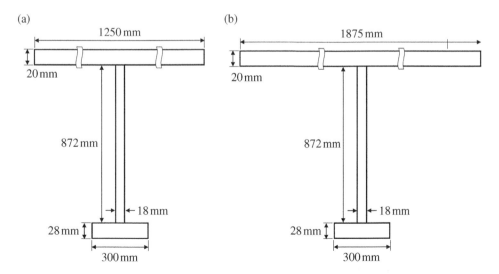

Figure 7.17 The representative plate-stiffener combination models: (a) at the deck; (b) at the bottom.

ratio β and the column slenderness ratio λ for a fully effective box girder cross section, defined in Table 2.1, as follows:

At deck panels: $\beta = 2.112$, $\lambda = 0.235$, $\sigma_{uD} = 172.712\,\text{MPa}$
At bottom panels: $\beta = 3.168$, $\lambda = 0.248$, $\sigma_{uB} = 138.619\,\text{MPa}$

The first-failure bending moments in a sagging or hogging condition are then determined as follows:

In sagging condition: $M_{fs} = Z_D\sigma_{uD} = 1.114\,\text{m}^3 \times 172.712\,\text{N/mm}^2 = 175\,518\,\text{MNmm}$
In hogging condition: $M_{fh} = Z_B\sigma_{uB} = 1.016\,\text{m}^3 \times 138.619\,\text{N/mm}^2 = 140\,871\,\text{MNmm}$

7.5.2 The Caldwell Method

The simple-beam theory method cannot accommodate the local failure of structural elements except for the flanges of the box girder at the outer fibers. Instead, if the bending stress distribution over the cross section of the box girder can be recognized at the ULS, the presumed stresses are integrated across the cross section to calculate the corresponding ultimate bending moment. This method, called the presumed stress-based method, accounts for the effect of local structural failure more precisely than the simple-beam theory method.

The pioneer of the presumed stress distribution-based method for calculating the ultimate bending moments of a box girder and then a ship's hull girder was Caldwell (1965). He presumed a bending stress distribution over the cross section at the ULS under vertical bending moments, as shown in Figure 7.18, in which all of the materials in compression have reached their ultimate strength with buckling, and all of the materials in tension have yielded. He then calculated the ultimate bending moments by integrating the presumed bending stresses over the hull's cross section. However, the stress distribution used in the Caldwell method is too optimistic, resulting in overestimated calculations of the ultimate bending moments.

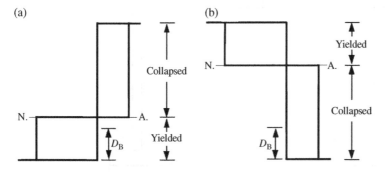

Figure 7.18 Caldwell's presumption of the bending stress distribution at the ultimate limit state under a vertical bending moment for a simplified cross section of a ship's hull under sagging or hogging (N.A., neutral axis): (a) sagging; (b) hogging (Caldwell 1965).

7.5.3 The Original Paik–Mansour Method

Experimental studies of large-scale ship's hull girder models (e.g., Dow 1991) and numerical studies of full-scale ships (e.g., Rutherford & Caldwell 1990, Paik et al. 1996) have demonstrated that the overall collapse of a ship's hull girder under a vertical bending moment is governed by the collapse of the compressed flange, although some degree of reserve strength remains after the compressed flange has collapsed.

This is the case because, after the buckling of the compressed flange, the neutral axis of the ship's hull cross section moves toward the tensioned flange, and a further increase in the applied bending moment is sustained until this flange yields. At later stages of this process, the vertical structures around the compressed and tensioned flanges (e.g., the longitudinal bulkheads or side shell structures) may also fail. In the vicinity of the final neutral axis position, however, the vertical structures usually remain in a linear elastic state until the overall collapse of the hull girder. Depending on the geometrical and material properties of the hull's cross section, these parts may of course fail, which corresponds with Caldwell's (1965) presumption.

Figure 7.19 shows an example of typical bending stresses across the hull girder cross section of a single-hulled oil tanker at the ULS under a vertical hogging bending moment, as obtained through numerical investigations (Paik et al. 1996). It is evident from this figure that the compressed flange (the bottom panel) collapses, and the tensioned flange (the deck panel) yields, until the ultimate strength has been reached, whereas the vertical structures in the vicinity of the neutral axis position remain intact (linear elastic). Hence, the approach based on Caldwell's presumed bending stress distribution can result in the strength of a ship's hull against collapse being greatly overestimated.

Paik and Mansour (1995) subsequently suggested the bending stress distribution over the hull cross section at the ULS that is shown in Figure 7.20. In the sagging condition, regions 1 and 2 are under tension and regions 3 and 4 are under compression. Region 1

Figure 7.19 Example of typical bending stress distribution across the cross section of a ship's hull girder at the ultimate limit state under a hogging bending moment (+, tension; –, compression), obtained through numerical investigations (Paik et al. 1996).

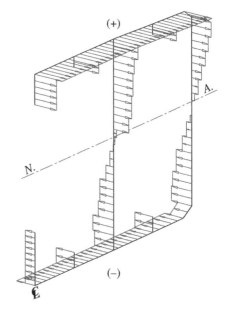

(a)

(b)

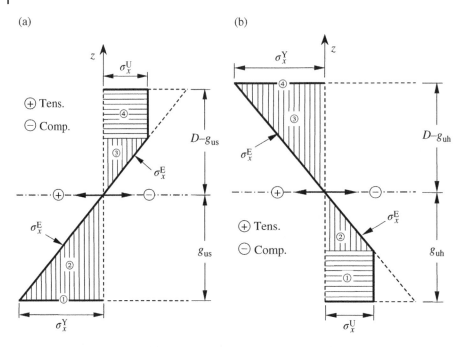

Figure 7.20 Paik and Mansour's original presumption of the bending stress distribution across the cross section of a ship's hull at the ultimate limit state under sagging or hogging conditions (+, tension; −, compression): (a) sagging condition; (b) hogging condition (the superscripts *U*, *Y*, and *E* denote the ultimate strength, yielding, and elastic region, respectively) (Paik & Mansour 1995).

represents the outer bottom panels, which have yielded to reach yield stress σ_x^Y, and region 4 the upper deck panels and upper part of the vertical structures, which have buckled and collapsed to reach ultimate stress σ_x^U. Regions 2 and 3, however, remain in a linear elastic or unfailed state, reaching an elastic stress of σ_x^E.

In the hogging condition, regions 1 and 2 are under compression, and regions 3 and 4 are under tension. Region 1, which represents the outer bottom panels and the lower part of the vertical structures, has buckled and collapsed to reach ultimate stress σ_x^U, and region 4, which represents the upper deck panels, has yielded to reach yield stress σ_x^Y. Regions 2 and 3 remain in the linear elastic regime, reaching elastic stress σ_x^E.

The height of region 4 (the upper part of the vertical structures) in the sagging condition, or that of region 1 (the lower part of the vertical structures) in the hogging condition, after buckling and collapse is assigned on the basis of the geometrical and material properties of the ship's hull structure. Under a vertical bending moment, the summation of axial forces over the entire cross section of the hull becomes zero:

$$\int \sigma_x dA = 0 \tag{7.52}$$

where $\int ( )dA$ is the integration across the entire cross section of the hull.

The height of region 4 in the sagging condition or that of region 1 in the hogging condition can be defined by solving Equation (7.52). The distance g_u from the ship's baseline (reference position) to the horizontal neutral axis of the cross section of the ship's hull at the ULS can then be obtained as follows:

$$g_u = \frac{\sum_{i=1}^{n} |\sigma_{xi}| a_i z_i}{\sum_{i=1}^{n} |\sigma_{xi}| a_i} \tag{7.53}$$

where z_i is the distance from the baseline (reference position) to the horizontal neutral axis of the ith structural component, σ_{xi} is the longitudinal stress of the ith structural component following the presumed stress distribution, a_i is the cross-sectional area of the ith structural component, and n is the total number of structural components. g_u is denoted by g_{us} in a sagging condition and by g_{uh} in a hogging condition.

The ultimate bending moment is then calculated as the first moment of the bending stresses about the neutral axis position:

$$M_{us} = \sum_{i=1}^{n} \sigma_{xi} a_i (z_i - g_{us}) \tag{7.54a}$$

$$M_{uh} = \sum_{i=1}^{n} \sigma_{xi} a_i (z_i - g_{uh}) \tag{7.54b}$$

where n is the total number of structural components and M_{us} (negative value) and M_{uh} (positive value) are the ultimate vertical bending moment for sagging or hogging, respectively.

7.5.4   The Modified Paik–Mansour Method

The original Paik–Mansour formulation method described in Section 7.5.3 does not allow the expansion of the yielded area to the vertical members under tensile loads, although it presumes that the tension flange, that is, the deck panels in the hogging condition and the outer bottom panels in the sagging condition, has yielded at the ULS of the hull girders subject to vertical bending moments.

However, depending on the geometric and/or material properties of the ship's hull cross sections, the vertical members close to the tension flange may also have yielded until the hull girders reach the ULS. Therefore, the bending stress distribution at the ULS presumed in the original Paik–Mansour method is now modified to that shown in Figure 7.21, where h_Y is the height of the yielded area under axial tension and h_C is the height of the collapsed area under axial compression.

To determine the heights of yielded and collapsed regions, that is, h_Y and h_C, Equation (7.52) is insufficient because two unknowns are available. In this regard, the following iteration process is required to determine h_Y and h_C:

1) Develop the structural model with nodal points for the cross section using the plate-stiffener combination elements and/or plate elements.
2) Calculate the ultimate axial compressive stresses of the individual elements.
3) Divide the depth into a number of segments (parts).
4) Keeping h_Y at a constant value starting from $h_Y = 0$, increase h_C starting from $h_C = 0$.

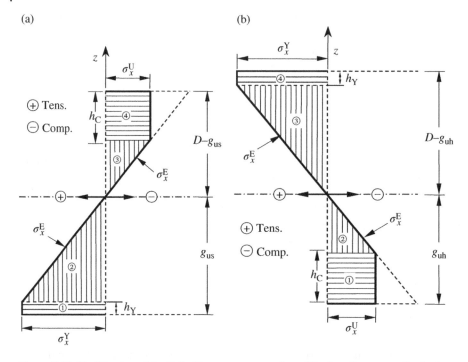

Figure 7.21 Modification of the Paik–Mansour presumption of the bending stress distribution across the cross section of a box girder at the ultimate limit state under sagging or hogging conditions (+, tension; –, compression) (superscripts *U*, *Y*, and *E* denote the ultimate strength, yielding, and elastic region, respectively): (a) sagging; (b) hogging (Paik et al. 2013).

5) Assign the linear elastic stresses of the individual elements in regions 2 and 3 linearly between the average values of the ultimate stresses in the collapsed region (i.e., region 4 under sagging or region 1 under hogging) and the yield stresses in the yielded region (i.e., region 1 under sagging or region 4 under hogging).
6) Calculate the total axial forces (positive sign) in tension and the total axial forces (negative sign) in compression across the entire cross section.
7) Repeat steps (4) to (6) varying h_C together with h_Y until the difference between the numerical values of these axial forces is acceptably small.

Because the stress distribution has been presumed and the heights of the yielded and collapsed regions have been determined, the distance g_u from the baseline (reference position) to the horizontal neutral axis of the cross section of the box girder at the ULS can be obtained from Equation (7.53). Also, the ultimate strength of the box girder under vertical bending moment can be determined from Equation (7.54).

7.5.5 Interactive Relationship Between Vertical and Horizontal Bending

An ultimate interactive relationship between the vertical and horizontal bending moments is considered as follows (Hughes & Paik 2013):

$$\left(\frac{M_V}{M_{Vu}}\right)^{c_1} + \left(\frac{M_H}{M_{Hu}}\right)^{c_2} = 1 \tag{7.55}$$

where M_V and M_H are the applied vertical and horizontal bending moments, respectively; M_{Vu} and M_{Hu} are the ultimate vertical and horizontal bending moments, respectively; and c_1 and c_2 are coefficients, which may be taken as $c_1 = 1.85$ and $c_2 = 1.0$. The ultimate horizontal bending moment can be determined in a manner similar to that of the ultimate vertical bending moment, as described in Section 7.5.3 when the bending moment is applied in the horizontal direction.

7.5.6 Interactive Relationship Between Combined Vertical or Horizontal Bending and Shearing Force

The ultimate interactive relationships between the vertical bending moment and the shearing force or between the horizontal bending moment and the shearing force are considered as follows (Hughes & Paik 2013):

$$\left(\frac{M_V}{M_{Vu}}\right)^{c_3} + \left(\frac{F}{F_u}\right)^{c_4} = 1 \tag{7.56a}$$

$$\left(\frac{M_H}{M_{Hu}}\right)^{c_5} + \left(\frac{F}{F_u}\right)^{c_6} = 1 \tag{7.56b}$$

where M_V and F are the applied vertical bending moment and shearing force, respectively; M_{Vu} and F_u are the ultimate vertical bending moment and ultimate shearing force, respectively; and c_3, c_4, c_5, and c_6 are coefficients, which may be taken as $c_3 = 2.0$, $c_4 = 5.0$, $c_5 = 2.5$, and $c_6 = 5.5$. The ultimate shearing force can be calculated as follows:

$$F_u = \sum_{i=1}^{n} a_i \tau_{ui} \tag{7.56c}$$

where a_i is the cross-sectional area of the ith structural element, τ_{ui} is the ultimate shear stress of the ith structural element, and n is the total number of structural elements.

7.5.7 Interactive Relationship Between Combined Vertical Bending, Horizontal Bending, and Shearing Force

An approach similar to that presented in Section 3.8 can be applied to derive the ultimate interactive relationship among vertical bending M_V, horizontal bending M_H, and shearing force F together with Equations (7.55) and (7.56) as follows (Hughes & Paik 2013):

$$\left(\frac{M_V}{M_{Vu}F_1}\right)^{c_1} + \left(\frac{M_H}{M_{Hu}F_2}\right)^{c_2} = 1 \tag{7.57}$$

where

$$F_1 = \left[1 - \left(\frac{F}{F_u}\right)^{c_4}\right]^{1/c_3}, \quad F_2 = \left[1 - \left(\frac{F}{F_u}\right)^{c_6}\right]^{1/c_5}.$$

7.5.8 Effect of Torsional Moment

For a box girder with large deck openings, the analysis of warping stresses and hatch opening deformations is an essential part of the structural response analysis.

For a thin-walled beam with an open cross section, the torsional stiffness is much less than that with a closed section. This implies that for a given level of torsion, the open section may twist much more due to its low torsional rigidity.

In contrast to the uniform (i.e., St. Venant) torsion of solid beams and a rather special class of warping-free thin-walled beams, nonuniform axial deformation (i.e., warping) usually occurs in the case of thin-walled beams with open sections such that an initially plane cross section will no longer remain plane. This would mean that torsion will develop axial (warping) stresses as well as shear stresses when the warping displacements are restrained as shown in Figure 7.22.

In actual structures, warping displacements are normally only partly restrained and thus the analysis of warping stresses as well as hatch opening deformations is in principle an essential part of the structural response analysis. When the cross sections are free to warp, warping stresses normal to the cross section will not be introduced. However, at cross-sectional discontinuities such as at the transition between the two neighboring areas with respect to the diaphragm (or transverse bulkhead), and at heavy cross deck beams, the warping deformations will be restrained to varying degrees. The restraint at these locations induces warping stresses, which for container ships with large deck openings are significant and so the warping stresses and the associated deformations (e.g., hatch opening distortions) must be accounted for in design.

Related to the effect of torsion on vertical bending capacity, however, it has been shown that torsion is not a very sensitive load component affecting the ultimate vertical bending moment of a box girder as long as the magnitude of torsion is not predominant (Paik et al. 2001). However, it should also be noted that the ultimate bending strength of a box girder with low torsional rigidity can be reduced significantly when torsion loads are large. Paik et al. (2001) suggested the following formulations that may fit the ultimate hull strength interaction relationship of a box girder with deck opening under combined torsion M_T and vertical bending M_V:

$$\left(\frac{M_V}{M_{Vu}}\right)^{c_7} + \left(\frac{M_T}{M_{Tu}}\right)^{c_8} = 1 \tag{7.58}$$

where M_{Vu} is the ultimate vertical bending moment and M_{Tu} is the ultimate torsional moment. c_7 and c_8 are coefficients, which may be taken as $c_7 = c_8 = 3.1$ for sagging and $c_7 = c_8 = 3.7$ for hogging.

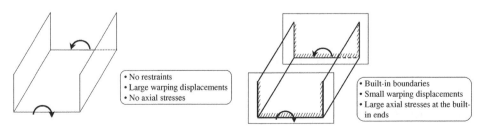

Figure 7.22 Warping displacements and stresses for an open cross-section thin-walled beam under torsion due to end restraints.

7.6   Effect of Age Related Structural Degradation

Aging plate assemblies may have suffered structural degradation such as corrosion and fatigue cracks over time. For general corrosion that uniformly reduces the wall thickness of structural members, the ultimate strength or effectiveness of primary strength members can be evaluated by excluding the corrosion diminution (reduction in thickness). For structural members with fatigue cracks, the cross-sectional area associated with the cracking damage may in turn be reduced in strength calculations, as described in Chapters 4 and 9. For applied examples in box girders with age related damage, interested readers may refer to Sharifi and Paik (2009, 2011, 2014), among others.

7.7   Effect of Accident Induced Structural Damage

The effect of accident-related structural damage such as local denting on the ultimate strength of plate assemblies can be accounted for where the buckling and ultimate strength of damaged plate elements can be evaluated as described in Chapters 4 and 10. When the damage is significant enough in size or extent, the damaged parts may be excluded from the calculations of effective plate width and the ultimate strength of individual members involving the damage. Similar exclusion may be applied even for structural members subject to tension as long as they have suffered serious structural damage.

To facilitate the rapid planning of salvage and rescue operations of structures after accidents, the residual ultimate strength of damaged structures must be assessed quickly

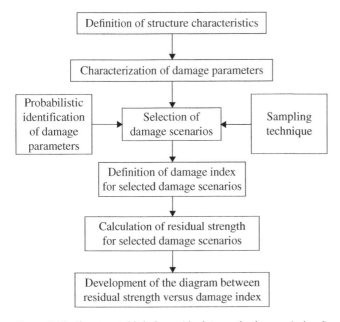

Figure 7.23 Flow to establish the residual strength–damage index diagram (R–D diagram) proposed by Paik et al. (2012).

and accurately, together with the location and extent of the damage. Paik et al. (2012) proposed the damage index-based method for the safety assessment of structures that have suffered structural damage due to accidents, where a diagram relating the residual ultimate strength performance to the damage index is established. This diagram is useful as a first-cut assessment of a structure's safety immediately after suffering structural damage. Figure 7.23 shows the flow of the method for the development of the residual strength versus damage index diagram (R-D diagram) proposed by Paik et al. (2012).

References

AASHTO (2010). *LRFD bridge design specifications*. American Association of State and Highway Transportation Officials, Washington DC.

AISC (1969). *Specification for the design, fabrication and erection of structural steel for buildings*. American Institute of Steel Construction, Chicago.

Basler, K. (1961). Strength of plate girders in shear. *ASCE Journal of the Structural Division*, **87**(ST7): 151–180.

Caldwell, J.B. (1955). The strength of corrugated plating for ships' bulkheads. *Transactions of the Royal Institution of Naval Architects*, **97**: 495–522.

Caldwell, J.B. (1965). Ultimate longitudinal strength. *Transactions of the Royal Institution of Naval Architects*, **107**: 411–430.

Dogaki, M., Nishijima, Y. & Yonezawa, H. (1992a). Nonlinear behaviour of longitudinally stiffened webs in combined patch loading and bending. In *Constructional steel design: world developments*, Edited by Dowling, P.J., Harding, J.E., Bjorhovde, R. & Martinez-Romero, E., Elsevier Applied Science, London, 141–150.

Dogaki, M., Yonezawa, H. & Tanabe, T. (1992b). Ultimate strength of plate girders with longitudinal stiffeners under patch loading. *Proceedings of the 3rd Pacific Structural Steel Conference*, The Japan Society of Steel Construction, Tokyo, October 26–28: 507–514.

Dow, R.S. (1991). Testing and analysis of 1/3-scale welded steel frigate model. *Proceedings of the International Conference on Advances in Marine Structures*, Dunfermline, Scotland: 749–773.

ENV 1993-1-1 (1992). *Eurocode 3: design of steel structures, part 1.1 general rules and rules for buildings*. British Standards Institution, London.

Fukumoto, Y., Maegawa, K., Itoh, Y. & Asari, Y. (1985). Lateral-torsional buckling tests of welded I-girders under moment gradient. *Proceedings of the Japan Society of Civil Engineers*, **362**: 323–332 (in Japanese).

Granath, P., Thorsson, A. & Edlund, B. (2000). I-shaped steel girders subjected to bending moment and travelling patch loading. *Journal of Constructional Steel Research*, **54**: 409–421.

Hughes, O.F. & Paik, J.K. (2013). *Ship structural analysis and design*. The Society of Naval Architects and Marine Engineers, Alexandria, VA.

Ji, H.D., Cui, W.C. & Zhang, S.K. (2001). Ultimate strength analysis of corrugated bulkheads considering influence of shear force and adjoining structures. *Journal of Constructional Steel Research*, **57**: 525–545.

Kitada, T. & Dogaki, M. (1997). Plate and box girders. Chapter 6 in *Structural stability design: steel and composite structures*, Edited by Fukumoto, Y., Pergamon Press, Oxford, 185–228.

Maquoi, R. (1992). Plate girders. Chapter 2.6 in *Constructional steel design: an international guide*, Edited by Dowling, P.J., Harding, J.E., Bjorhovde, R. & Martinez-Romero, E., Elsevier Applied Science, London, 133–173.

Mikami, I., Harimoto, S., Yamasato, Y. & Yoshimura, F. (1991). Ultimate strength tests of steel plate girders under repetitive shear. *Technical Report of the Kansai University, Japan*, **33**: 145–164.

Paik, J.K., Kim, D.H., Park, D.H. & Kim, M.S. (2012). A new method for assessing the safety of ships damaged by grounding. *Transactions of the Royal Institution of Naval Architects*, **154** (A1): 1–20.

Paik, J.K., Kim, D.K., Park, D.H., Kim, H.B. Mansour, A.E. & Caldwell, J.B. (2013). Modified Paik-Mansour formula for ultimate strength calculations of ship hulls. *Ships and Offshore Structures*, **8**(3–4): 245–260.

Paik, J.K. & Mansour, A.E. (1995). A simple formulation for predicting the ultimate strength of ships. *Journal of Marine Science and Technology*, **1**(1): 52–62.

Paik, J.K., Thayamballi, A.K. & Che, J.S. (1996). Ultimate strength of ship hulls under combined vertical bending, horizontal bending, and shearing forces. *Transactions of the Society of Naval Architects and Marine Engineers*, **104**: 31–59.

Paik, J.K., Thayamballi, A.K. & Chun, M.S. (1997). Theoretical and experimental study on the ultimate strength of corrugated bulkheads. *Journal of Ship Research*, **41**(4): 301–317.

Paik, J.K., Thayamballi, A.K., Pedersen, P.T. & Park, Y.I. (2001). Ultimate strength of ship hulls under torsion. *Ocean Engineering*, **28**: 1097–1133.

Porter, D.M., Evans, H.R. & Rockey, K.C. (1975). The collapse behavior of plate girders loaded in shear. *The Structural Engineer*, **53**: 313–325.

Rutherford, S.E. & Caldwell, J.B. 1990. Ultimate longitudinal strength of ships: a case study. *Transactions of the Society of Naval Architects and Marine Engineers*, **98**: 441–471.

Schafer, B.W. & Pekoz, T. (1998). Computational modeling of cold-formed steel: characterizing geometric imperfections and residual stresses. *Journal of Constructional Steel Research*, **47**: 193–210.

Sharifi, Y. & Paik, J.K. (2009). Environmental effects on ultimate strength reliability of corroded steel box girder bridges. *Structural Longevity*, **2**(2): 81–101.

Sharifi, Y. & Paik, J.K. (2011). Ultimate strength reliability analysis of corroded steel-box girder bridges. *Thin-Walled Structures*, **49**(1): 157–166.

Sharifi, Y. & Paik, J.K. (2014). Maintenance and repair scheme for corroded stiffened steel box girder bridges based on ultimate strength reliability and risk assessments. *Journal of Engineering Structures and Technologies*, **6**(3): 95–105.

Susanti, L., Kasai, A. & Miyamoto, Y. (2014). Ultimate strength of box section steel bridge compression members in comparison with specifications. *Case Studies in Structural Engineering*, **2**: 16–23.

Takimoto, T. (1994). Plate girders under patch loading. In *Ultimate strength and design of steel structures*. The Japan Society of Civil Engineers, Tokyo, 122–127 (in Japanese).

Trahair, N.S. & Bradford, M.A. (1988). *The behaviour and design of steel structures*. Chapman and Hall, London and New York.

Zhao, Y.J., Yan, R.J. & Wang, H.X. (2015). Experimental and numerical investigations on plate girder with perforated web under axial compression and bending moment. *Thin-Walled Structures*, **97**: 199–206.

8

Ultimate Strength of Ship Hull Structures

8.1 Introduction

Ship's hull structure is much more complex in geometry and larger in size than other types of plate assemblies described in Chapter 7. The ultimate strength behavior of a ship's hull girder is very unique, and thus this chapter describes not only the ultimate strength but also the characteristics of hull girder loads. It is noted that the theories and methodologies described in this chapter can be commonly applied to steel- and aluminum-ship hull structures.

8.2 Characteristics of Ship's Hull Structures

Figure 8.1 shows mid-ship cross sections of typical merchant ships or offshore installations. Table 8.1 indicates the cross-sectional properties of the vessels. It is found that the vessel's structural characteristics vary significantly depending on the cargo types or missions, among other factors.

To understand the structural characteristics of merchant ship hull structures, some important properties of existing bulk carrier structures are further surveyed in this section as an illustrative example (Paik & Thayamballi 1998). These structural parameter studies would be of value in judging, describing, and generalizing the nature of structural failure behavior to be expected for given operational, extreme, and accidental load levels.

Figure 8.2a shows the relationship between ship length and deadweight for bulk carriers with a large proportion of which are conventional. It is evident that the said relationship is present and definite, with moderate degree of scatter. Typically the number of holds is seven for the Panamax class and nine for the Capesize class, with some variability. There are essentially three types of holds in such vessels insofar as the structure is concerned, that is, ore holds, light holds, and ballast holds. When carrying dense cargo such as iron ore, an alternate hold loading condition is common, with the light holds empty.

Figure 8.2b shows the relationship between ship length and maximum hold length. It is apparent from this figure that the maximum hold length tends to decrease with increase in the ship length. It is of interest that the hold lengths in the Panamax are comparable to the Capesize, but the Capesize hold areas and volumes are normally greater than those of a Panamax (e.g., areas ~1.5 times for hold No.1, most forward, and 1.25 times in the case of the other holds, in one particular set of vessels). Hence normally any postulated

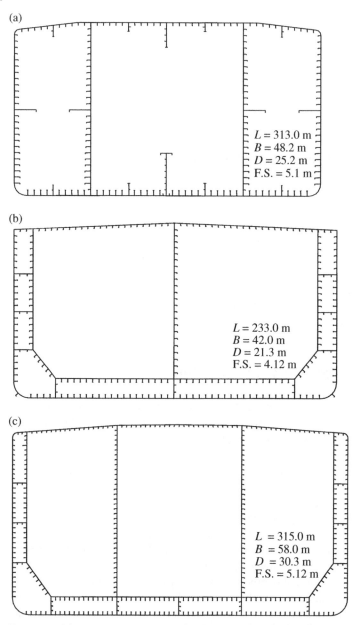

(a)

$L = 313.0$ m
$B = 48.2$ m
$D = 25.2$ m
F.S. $= 5.1$ m

(b)

$L = 233.0$ m
$B = 42.0$ m
$D = 21.3$ m
F.S. $= 4.12$ m

(c)

$L\ = 315.0$ m
$B\ = 58.0$ m
$D\ = 30.3$ m
F.S. $= 5.12$ m

Figure 8.1 Schematic representation of mid-section: (a) a 254 000 DWT single hull tanker; (b) a 105 000 DWT double hull tanker with one center-longitudinal bulkhead; (c) a 313 000 DWT double hull tanker with two side-longitudinal bulkheads; (d) a 170 000 DWT single-sided bulk carrier; (e) a 169 000 DWT double-sided bulk carrier; (f) a 3 500 TEU container vessel; (g) a 5 500 TEU container vessel; (h) a 9 000 TEU container vessel; (i) a 113 000 DWT FPSO (floating production storage offloading unit); (j) a 165 000 DWT shuttle tanker (*B*, ship breadth; *D*, ship depth; DWT, deadweight; F.S., frame spacing; *L*, ship length, TEU, twenty equivalent unit).

(d)

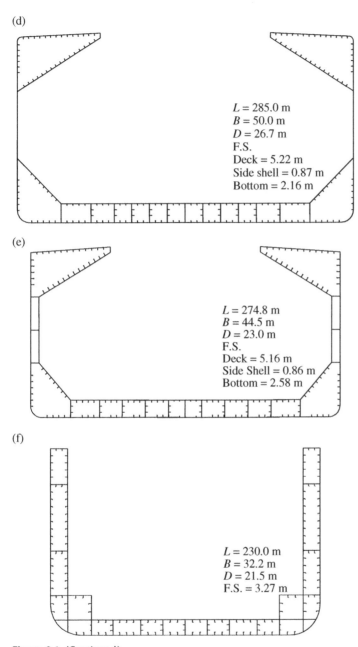

$L = 285.0$ m
$B = 50.0$ m
$D = 26.7$ m
F.S.
Deck = 5.22 m
Side shell = 0.87 m
Bottom = 2.16 m

(e)

$L = 274.8$ m
$B = 44.5$ m
$D = 23.0$ m
F.S.
Deck = 5.16 m
Side Shell = 0.86 m
Bottom = 2.58 m

(f)

$L = 230.0$ m
$B = 32.2$ m
$D = 21.5$ m
F.S. = 3.27 m

Figure 8.1 (Continued)

(g)

(h)

(i)

(j)

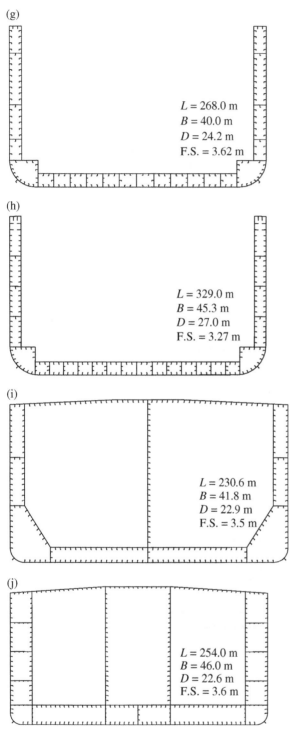

$L = 268.0$ m
$B = 40.0$ m
$D = 24.2$ m
F.S. $= 3.62$ m

$L = 329.0$ m
$B = 45.3$ m
$D = 27.0$ m
F.S. $= 3.27$ m

$L = 230.6$ m
$B = 41.8$ m
$D = 22.9$ m
F.S. $= 3.5$ m

$L = 254.0$ m
$B = 46.0$ m
$D = 22.6$ m
F.S. $= 3.6$ m

Figure 8.1 (Continued)

Table 8.1 Hull cross-sectional properties of the 10 typical merchant ships or offshore installations.

Item		SHT	DHT#1	DHT#2	Bulk#1	Bulk#2	Cont#1	Cont#2	Cont#3	FPSO	Shuttle
LBP (L) (m)		313.0	233.0	315.0	282.0	273.0	230.0	258.0	305.0	230.6	254.0
Breadth (B) (m)		48.2	42.0	58.0	50.0	44.5	32.2	40.0	45.3	41.8	46.0
Depth (D) (m)		25.2	21.3	30.3	26.7	23.0	21.5	24.2	27.0	22.9	22.6
Draught (d) (m)		19.0	12.2	22.0	19.3	15.0	12.5	12.7	13.5	14.15	15.0
Block coeff. (C_b)		0.833	0.833	0.823	0.826	0.8374	0.6839	0.6107	0.6503	0.8305	0.831
Design speed (knots)		15.0	16.25	15.5	15.15	15.9	24.9	26.3	26.6	15.4	15.7
DWT or TEU		254 000 DWT	105 000 DWT	313 000 DWT	170 000 DWT	169 000 DWT	3 500 TEU	5 500 TEU	9 000 TEU	113 000 DWT	165 000 DWT
Cross-sectional area (m²)		7.858	5.318	9.637	5.652	5.786	3.844	4.933	6.190	4.884	6.832
Height to neutral axis from baseline (m)		12.173	9.188	12.972	11.188	10.057	8.724	9.270	11.614	10.219	10.568
I	Vertical (m⁴)	863.693	359.480	1346.097	694.307	508.317	237.539	397.647	682.756	393.625	519.674
	Horizontal (m⁴)	2050.443	1152.515	3855.641	1787.590	1530.954	648.522	1274.602	2120.311	1038.705	1651.479
Z	Deck (m³)	66.301	29.679	77.236	44.354	39.274	18.334	26.635	44.376	31.040	43.191
	Bottom (m³)	70.950	39.126	103.773	62.058	50.544	27.228	42.894	58.785	38.520	49.175
σ_Y	Deck	HT32	HT32	HT32	HT40	HT36	HT36	HT36	HT36	HT32	HT32
	Bottom	HT32	HT32	HT32	HT32	HT32	HT32	HT32	HT32	HT32	HT32
M_P	Vertical moment (GNm)	22.615	11.930	32.481	20.650	15.857	8.881	12.179	18.976	12.451	15.669
	Horizontal moment (GNm)	31.202	19.138	54.465	31.867	26.714	14.967	21.763	33.229	19.030	25.105

Notes: σ_Y, yield stress; Bulk#1, single-sided bulk carrier; Bulk#2, double-sided bulk carrier; Cont#1, 3500 TEU container vessel; Cont#2, 5500 TEU container vessel; Cont#3, 9000 TEU container vessel; DHT#1, double hull tanker with one center-longitudinal bulkhead; DHT#2, double hull tanker with two longitudinal bulkheads; FPSO, floating, production, storage, and offloading system; HT32, high tensile steel with a yield stress of 315 MPa; HT36, high tensile steel with a yield stress of 355 MPa; I, moment of inertia; M_P, full plastic bending moment; SHT, single hull tanker; Shuttle, shuttle tanker; Z, section modulus.

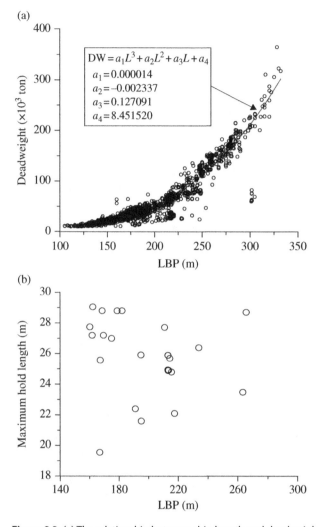

Figure 8.2 (a) The relationship between ship length and deadweight; (b) the relationship between ship length and maximum hold length, for conventional bulk carriers (DW, deadweight; LBP (*L*), length between perpendiculars; symbols, real ship data).

flooding of the Capesize hold No.1 is of more load consequence than flooding of the Panamax hold No.1.

Figure 8.3 shows some important properties of conventional bulk carrier hull structures of Handymax class or larger. The bulk carrier double bottom height (and also width of flat part of inner bottom) increases remarkably as the vessel becomes larger; see Figure 8.3a. As shown in Figure 8.3b, the slenderness ratio of outer bottom plating decreases (e.g., the bottom plate thickness increases) with increase in the vessel size. The ratio of actual hull section modulus to class rule required section modulus decreases as the vessel becomes larger (see Figure 8.3c), which implies that some vessels are built to satisfy the rule requirements with little additional margin. Figure 8.3d indicates the usage

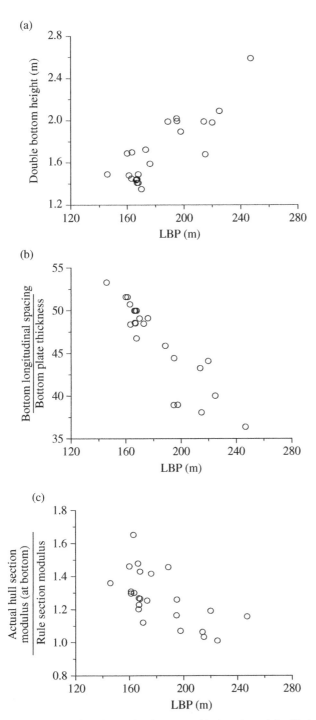

Figure 8.3 (a) The relationship between ship length and double bottom height; (b) the relationship between ship length and bottom plate slenderness ratio; (c) the relationship between ship length and hull section modulus at bottom; (d) the relationship between ship length and HTS usage, for conventional bulk carriers (LBP, length between perpendiculars; symbols, real ship data).

(d)

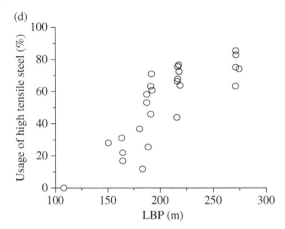

Figure 8.3 (Continued)

Table 8.2 Structural characteristics of the plating between stiffeners in selected existing bulk carriers.

	a/b		$(b/t)\sqrt{\sigma_Y/E}$	
Structure	**Range**	**Average**	**Range**	**Average**
Outer bottom plate	2.9–3.4	3.2	1.6–2.1	1.9
Inner bottom plate	2.9–3.4	3.2	1.3–1.8	1.6
Bottom floor	2.0–2.9	2.6	2.3–3.0	2.4
Bottom girder	3.2–4.0	3.6	1.8–2.8	2.3
Side shell	3.2–3.3	3.3	2.0–2.2	1.6
Deck plate	4.7–6.7	5.7	1.0–2.0	1.6
Longitudinal bulkhead plate	3.2–3.3	3.3	2.2–2.4	2.3
Topside wing tank bottom plate	4.9–7.5	6.3	1.9–2.7	2.3
Topside wing tank web	1.0–1.6	1.3	2.2–2.9	2.5
Hopper bottom plate	1.9–3.7	2.8	1.6–2.5	1.9
Hopper web	1.0–1.6	1.3	2.0–2.7	2.5

Note: σ_Y, yield strength; a, plate length; b, plate breadth; E, Young's modulus; t, plate thickness.

of high tensile steel in conventional bulk carrier hull structures. More than 70% of the hull structures in Panamax and Capesize bulk carriers can in some cases be of high tensile steel. All statistics indicated are for a selected cross section of vessels.

Table 8.2 indicates typical ranges and average values of the aspect ratio and slenderness ratio for bulk carrier plating between stiffeners. The plate aspect ratio in the bottom and side shell is about 3.0 in way of longitudinal framing (and aspect ratios 30–40 in way of transverse framing), while it is about 6.0 in the deck. It is also seen from Table 8.2 that the slenderness ratio of longitudinal plating is less than 2.0, indicating that such plating of bulk carriers is usually stocky in a buckling sense. It is noted that for long plates under compression, if the plate slenderness ratio is less than about 1.9, buckling normally occurs in the elastic–plastic or plastic regime, in which the buckled plate will have little residual strength margin beyond first buckling, while thin plate elements with the plate slenderness ratios greater than about 2.5 will normally first buckle in the elastic regime, with a fair amount of post-buckling strength reserve until the ultimate limit state is reached.

Bending and torsional rigidities of stiffeners and support members in bulk carriers are also important aspects. The rigidities of transverse members are usually larger than those of longitudinal members. The bending rigidity of support members is large enough compared with that of plate elements so that the relative lateral deformation of plate elements along the edges could be ignored in calculating the plate buckling strength. However, the normalized torsional rigidity of support members, that is, the ratio of torsional rigidity of support members to the bending rigidity of plating between supporting members, is less than about 1.0 for longitudinal stiffeners and less than about 2.0 for transverse frames. This could indicate that although the ship plating has a finite amount of rotational restraints along the edges implying that idealized boundary conditions, that is, simply supported or clamped, never occur, there is still the possibility of flexural-torsional failure of stiffeners, particularly in the longitudinal direction that needs to be controlled by design.

The various obvious generalizations of this section relate of course to strength and thus are only indirectly indicative of the structural performance considering loads as well. Also, it is noted that part of the conventional bulk carrier side structure is transversely framed, for which the limiting rigidities and slenderness would be different from those noted previously. In conventional bulk carriers, each side shell frame may not form part of a continuous ring, that is, the transverse frame spacing in the topside tanks can be different from that in the cargo hold, which in turn can be different from the frame spacing in the double bottom, although the recently suggested double skin bulk carrier design is a clear exception.

8.3 Lessons Learned from Accidents

In the past, there have been several vessel casualties including total losses including overall collapse of a ship's hull. Figure 8.4a shows an example of such an accident to a ship. In this case, a Capesize bulk carrier collapsed due to human error during discharge in port of its 126 000 tons of iron ore cargo. While this 23-year-old 139 800 DWT ship did not separate into two, the bottom of its mid-body reportedly touched the sea bed, and the hull girder, in fact, collapsed. After having emptied the fore and aft holds among the five cargo holds, buckling collapse took place in the vessel's deck, while the central hold was still full. It is clear that this incident was primarily a result of improper unloading of the cargo from the ship. But it does serve to indicate that a ship, like any other structure, has a finite strength, and whether for routine design purposes, damage investigations, or to determine the ongoing effects of age related structural degradation, relevant procedures to compute that strength accurately are a necessity.

Total losses by ship's hull collapse have also occurred at sea as shown in Figure 8.4b. Those include the sinking of the M.V. Derbyshire, which was a double side-hulled Capesize bulk carrier of 281.94 m in length between the perpendiculars, 44.2 m in the beam, and 25 m in depth (Faulkner 1998, Paik & Faulkner 2003, Paik et al. 2008, Hughes & Paik 2013). Her maximum deadweight was 173 218 tons. She was 5 years old at the time of the accident and was believed to have suffered almost no age related degradation such as corrosion wastage. Another distinct characteristic of the ship is

(a)

(b)

Figure 8.4 Collapse of ship's hull girder: (a) during unloading of cargo at port; (b) at sea.

that she had a double-sided hull arrangement that aimed to prevent unintended water ingress into the cargo holds from the failure of the side shell structures. On September 9, 1980, she sank in the northwest Pacific, some 400 miles south of Shikoku Island, Japan, during typhoon Orchid while on a voyage from Canada to Japan carrying fine iron ore concentrates. On her last voyage from the Sept Isles, Canada, to Yokohama, Japan, she was carrying about 158 000 tons of fine ore concentrates distributed across seven of her nine holds. Her estimated displacement as she approached Japan was about 194 000 tons, indicating a mean draught of approximately 17 m. Just before sinking, she was within the most dangerous ambit of typhoon Orchid, and the significant wave height soon before her sinking was reportedly 14 m. There was no distress signal, and only two sightings of oil upwelling were seen some days later to indicate the position of the sunken craft. A damaged lifeboat from the ship was sighted, but this was not recovered and subsequently sank. This and the absence of a distress signal were taken to imply that she sank very quickly.

Based on the lessons learned from the accidents, the possible causes of the ship's total losses can be categorized into three groups, namely, (i) loss of reserve buoyancy (or floating capability), (ii) hull girder breakage, and (iii) loss of stability, initiated perhaps totally or partly by unintended water ingress into cargo holds. Figure 8.5 shows total loss scenarios of a ship developed by Paik and Thayamballi (1998).

As an illustrative example with bulk carriers, a large proportion of the vessel incidents reported were carrying iron ore or coal, the former being one of the denser types of cargo and the latter being one of the more corrosive. It has been indicated that the age of most vessels concerned was over 15 years, and so significant defects related to corrosion and fatigue may have been present. Decrease of the residual strength and/or increase of the applied hull girder loads in a flooding event may possibly lead to hull girder breakage (Paik 1994, Faulkner 1998, Paik & Faulkner 2003, Paik et al. 2008). Related to this, there is the possibility that part of the side shell forward could have been lost due to a combination of circumstances, for example, excessive corrosion and cracking damages together with perhaps the shifting of solid cargo due to roll in rough weather. This would cause ingress of sea water into the cargo holds. Water ingress into the forward hold could occur also through failed hatch covers.

In such postulated accident scenarios as indicated in Figure 8.5, even if the vessel could initially survive with one of the compartments flooded, the ingress of sea water into a cargo hold could amplify the applied loads and also otherwise lead to vessel loss by progressive flooding (particularly when the watertight transverse bulkheads are insufficient to withstand the increased static and dynamic pressures). The relevant flooded load components may be increased by pitch motion, whose severity can depend on the cargo density and other characteristics of the vessel. Progressive flooding after the collapse of corrugated bulkheads in a flooded condition is thought to be implicated in some of the

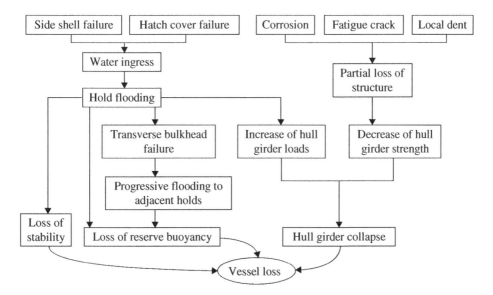

Figure 8.5 Total loss scenarios of bulk carriers developed by Paik and Thayamballi (1998).

bulk carrier losses. Also, with progressive hold flooding, the vessel in some cases could lose stability in rough seas, potentially leading to capsizing (Turan & Vassalos 1994).

From the point of view on a ship structural design, important lessons include, first, abnormal waves not expected in the structural design can occur and amplify the maximum hull girder loads, which may reach or even exceed the corresponding design values. Second, unintended water ingress into cargo holds, which may occur due to hatch cover failure, can further amplify the hull girder loads. Third, the allowable working stress design approach that was applied in the structural design of vessel hulls cannot deal with this issue, and thus the ultimate limit state design method should be employed to prevent hull girder collapse accidents.

8.4 Fundamentals of Vessel's Hull Girder Collapse

A ship's hull in the intact condition will sustain applied loads smaller than the design loads, and, in normal seagoing and approved cargo loading conditions, it will not suffer any structural damages such as buckling and collapse. However, the loads acting on the ship's hull are uncertain both due to the nature of rough seas and because of possibly unusual loading/unloading of cargo, the latter due to human error. In rare cases, applied loads may hence exceed design loads and the ship's hull may collapse globally. Since aging ships may have suffered structural deterioration due to corrosion and fatigue, related weakening in their structural resistance may play a part as well.

Figure 8.6 shows illustrative examples of the progressive collapse behavior of the typical merchant ships or offshore installations, indicated in Figure 8.1 and Table 8.1, under vertical bending moment until and after the ultimate limit state is reached, where the level of initial imperfections is varied (Paik et al. 2002).

As applied loads increase beyond the design loads, structural members of the vessel's hull girder buckles in compression and yields in tension. The vessel's hull girder can normally carry further loading beyond the onset of limited member buckling or yielding, but the structural effectiveness of any such failed member clearly decreases, and its individual stiffness can even become "negative," with their internal stress being redistributed to adjacent intact members. The most highly compressed member will, deterministically speaking, collapse earlier, and the stiffness of the overall hull girder decreases gradually. As loads continue to increase, buckling and collapse of more structural members occur progressively until the ultimate limit state is reached for the hull girder as a whole.

Figure 8.7 shows the change of the neutral axis position for a single hull tanker hull girder as the bending moment increases. It is also apparent from this figure that there is still some residual strength even after buckling collapse of the compression flange. This is due to a shift of the neutral axis toward the tension flange, resulting from loss of effectiveness of the collapsed compression flange. Of interest, as the bending moment increases, the neutral axial position changes quickly and becomes stable, as shown in Figure 8.7, because the neutral axis is calculated for partially effective hull cross section after the bending moment is applied, while it is estimated for fully effective cross section before loading. This implies that the section moduli calculated for fully effective hull cross section may not always be a real indication of the ship hull sectional load resistive properties. The ultimate hogging moment of the tanker hull is larger than the ultimate sagging moment as usual.

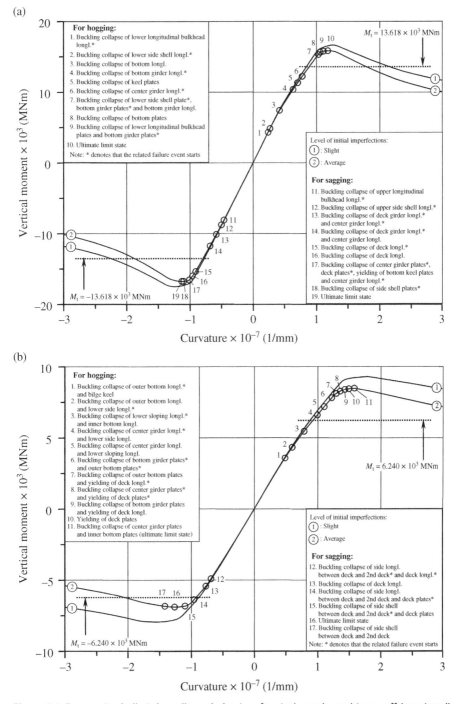

Figure 8.6 Progressive hull girder collapse behavior of typical merchant ships or offshore installations under vertical bending moment, varying the level of initial imperfections: (a) a 254 000 DWT single hull tanker; (b) a 105 000 DWT double hull tanker; (c) a 313 000 DWT double hull tanker; (d) a 170 000 DWT single-sided bulk carrier; (e) a 169 000 DWT double-sided bulk carrier; (f) a 3 500 TEU container vessel; (g) a 5 500 TEU container vessel; (h) a 9 000 TEU container vessel; (i) a 113 000 DWT FPSO under vertical moment varying the level of initial imperfections, as obtained by ALPS/HULL (2017); (j) a 165 000 DWT shuttle tanker (Paik et al. 2002).

(c)

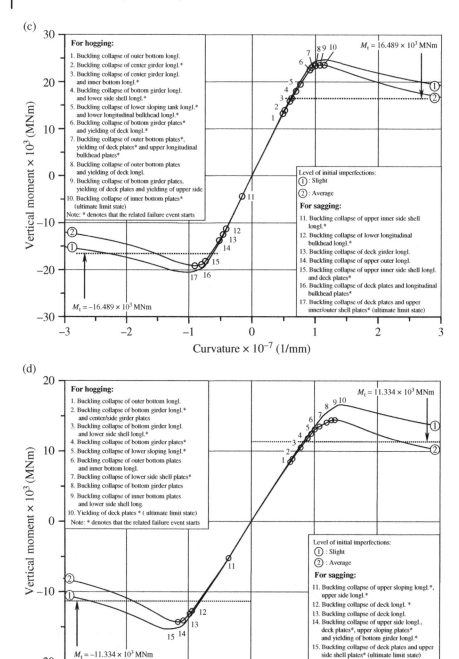

(d)

Figure 8.6 (Continued)

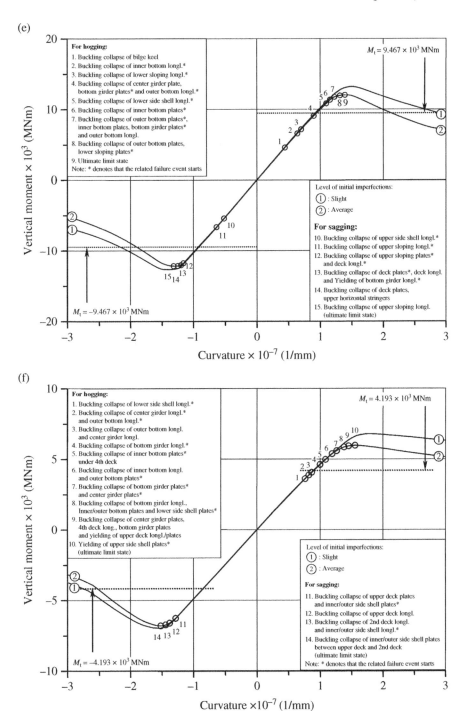

Figure 8.6 (Continued)

(g)

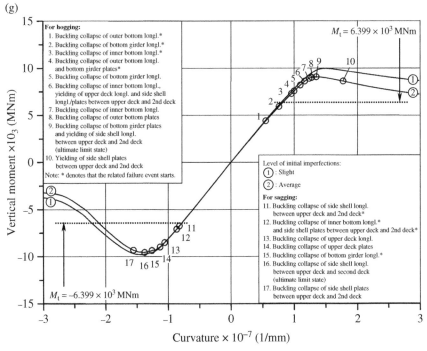

(h)

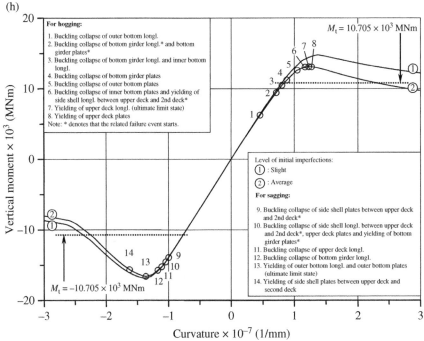

Figure 8.6 (Continued)

(i)

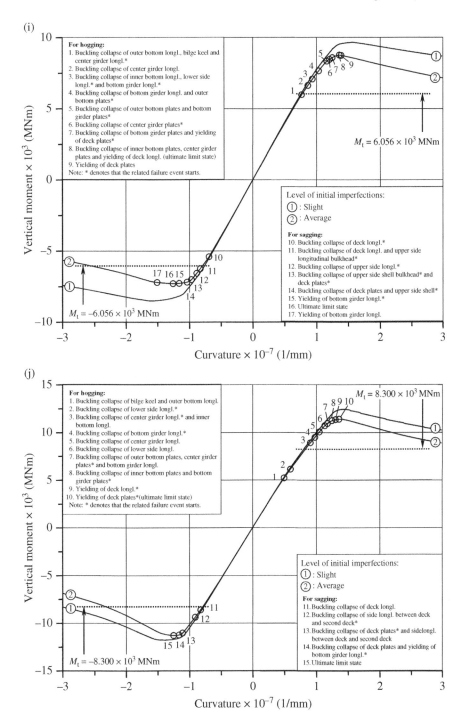

(j)

Figure 8.6 (Continued)

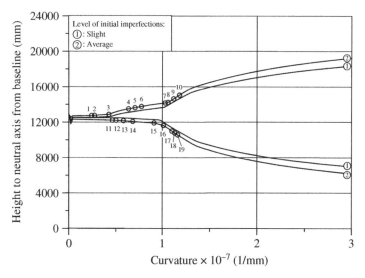

Figure 8.7 The change of the neutral axis due to structural failure for a 254 000 DWT single hull tanker.

When the structural safety of a ship's hull is considered, the ultimate hull girder strength must then be accurately evaluated. It is also helpful in this regard if one can derive simple expressions for calculation of the hull ultimate strength so that these can be used for the ready formulation of failure functions to be used in reliability analysis and for use in the early stages of structural design.

Most classification society criteria and procedures for ship structural design have been based on the first yield of hull structures together with buckling checks for structural components (i.e., not for the whole hull structure). These methods have proven themselves to be effective for intact vessels in normal seas and under loading conditions. However, their applicability to assess the survivability of vessels in damaged or accidental situations, that is, deterioration due to corrosion, fatigue, collisions, grounding, or overloading, is somewhat less certain. In these cases, it is necessary to account more precisely for the interacting effects between yielding, buckling, and, sometimes, crushing and fracture of local components and the related effects on the global behavior of the structural system.

While service proven, the traditional design criteria and associated linear elastic stress calculations do not necessarily define the true ultimate limit state that is the limiting condition beyond which a ship's hull will fail to perform its function. Neither do such procedures help understand the likely sequence of local failure prior to reaching the ultimate limit state. It is of course important to determine the true ultimate strength if one is to obtain consistent measures of safety that can form a fairer basis for comparisons of vessels of different sizes and types. An ability to better assess the true margin of safety should also inevitably lead to improvements in regulations and design requirements.

A consequence of present day design procedures is that in some vessels the ultimate hogging moment is not always greater than the ultimate sagging moment even if the section modulus at bottom is larger than that at deck. This is the case in some bulk carriers where the deck panels are sturdier than the bottom panels. One might of course have (incorrectly) presumed that, as long as the section modulus at bottom is greater than that at deck, the ultimate hull girder strength in hogging will be greater than that in sagging. But this is not always true. Such differences will of course be better detected and corrected if one undertakes ultimate strength-based design.

This indicates the disadvantage of the conventional structural design procedures for vessels based on the allowable stress. The ultimate limit state design procedure can avoid such difficulties and facilitate the determination of the real safety margin of the structure. In the ultimate limit state design of ship hulls using Equation (1.17), the capacity is taken to be the relevant ultimate strength, while the demand is given in terms of the hull girder loads that may be calculated by the design rules of classification societies or by direct methods. This chapter presents relevant approaches to calculate the ship hull girder loads and the related ultimate strength.

8.5 Characteristics of Ship Structural Loads

Ship structures are subjected to various types of loads, which may be grouped according to their characteristics in time: static loads, low-frequency dynamic loads, high-frequency dynamic loads, and impact loads (Hughes & Paik 2013).

Static loads are those that arise from the ship's weight and buoyancy. Low-frequency dynamic loads occur at frequencies that are sufficiently low compared to the frequencies of the vibratory response of the ship's hull (and its parts, as the case may be) that the resulting dynamic effects on the structural response are relatively small. Such loads include hull pressure variations induced by waves or oscillatory ship motions and inertial reaction forces from the acceleration of the mass of the ship and its cargo or ballast water. High-frequency dynamic loads have frequencies that approach or exceed the lowest natural frequency of the hull girder. A typical example is wave-induced "springing" (flexural vibration of the hull girder), which may occur when the natural period of the hull girder is close to the period of the shorter components of the encountered waves. Because steady-state springing occurs at a higher frequency than that of ordinary wave-induced bending, it increases the number of stress cycles during the ship's lifetime, thus possibly increasing the fatigue damage to the hull girder. Impact loads are those with a duration even shorter than the period of high-frequency dynamic loads. Examples of impact loads are slamming and green water impact on deck. Slamming causes a sudden upward acceleration and deflection of the bow and excites hull girder flexural vibration in the first two or three modes, typically with a period in the range of 0.5–2 s. This transient slam-induced vibration is termed "whipping."

In ship structural analysis and design, the most common loads are the static and low-frequency dynamic loads; the latter are usually treated as static or quasi-static loads. High-frequency dynamic loads can be important in specific design cases, such as the long and slender Great Lakes vessels. The local effect of impact loads, particularly pressure, must usually be considered in some manner.

Because the characteristics of ship structural loads vary significantly depending on loading, operating conditions, and sea states, all potential conditions during the ship's lifetime must be taken into account in the analysis and design of ship structures. Therefore, flooding and damaged conditions should also be considered.

8.6 Calculations of Ship's Hull Girder Loads

The important components of hull girder loads are vertical bending, horizontal bending, sectional shear, and torsional moment, as shown in Figure 8.8. These arise from the distribution of local pressures, including sea and cargo loads. The basic theory to calculate hull girder loads may be found in textbooks such as that by Hughes and Paik (2013). For calculation of the design hull girder loads of merchant ships, the classification societies provide simplified formulas or guidelines, with a direct calculation of hull girder loads from first principles usually recommended in cases that involve unusual structures, patterns of loading, or operational conditions.

As the most important hull girder load component, the total vertical bending moment M_t for a ship's hull is defined as the extreme algebraic sum of the still-water moment M_{sw} and the wave-induced moment M_w as follows:

$$M_t = M_{sw} + M_w \tag{8.1}$$

where M_{sw} is taken as the maximum value of the still-water bending moment from the worst load condition on the ship, considering both hogging and sagging. For merchant ships, the design value of M_{sw} may be taken as the maximum allowable still-water bending moment approved by a ship classification society. M_w is taken as the extreme wave-induced bending moment that the ship is likely to encounter during its lifetime.

To assess the safety and reliability of damaged ship structures, the hydrodynamic "strip" theory (so-called because the hull is idealized as a series of short prismatic sections, or girthwise "strips") may be used to determine M_w. For long-term prediction of wave-induced loads, loads that are likely to be exceeded only once during the vessel's design lifetime are considered, including all sea states that may be encountered, whereas short-term prediction is carried out based on a ship encountering a storm of a specific persistence or duration (e.g., 3 h).

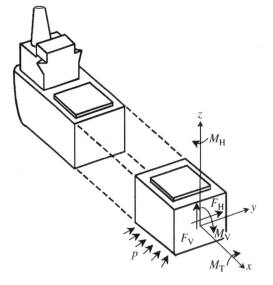

Figure 8.8 Hull girder sectional load components.

Conventionally, a long-term analysis has normally been used to determine M_w for the design of newly built ships, whereas a short-term analysis is typically necessary to predict M_w for ships in specific sea conditions, such as for damaged ships. In calculating M_w, a second-order strip theory may be used in which it is necessary to distinguish between sagging and hogging wave-induced bending moments.

To approximately account for the correlation between still-water and wave-induced bending moments, the following type of equation can be used to calculate the total bending moment:

$$M_t = k_{sw}M_{sw} + k_w M_w \tag{8.2}$$

where k_{sw} and k_w are the load combination factors for still-water and wave-induced bending moments, respectively, which account for the nonsimultaneous occurrence of extreme still-water and wave-induced loads.

To consider dynamic load effects, the total bending moment may be given by

$$M_t = k_{sw}M_{sw} + k_w(M_w + k_d M_d) \tag{8.3}$$

where k_d is the load combination factor related to the dynamic bending moment, M_d, that arises from either slamming or whipping. M_d is taken as the extreme dynamic bending moment in the same wave condition (e.g., sea state) as the wave-induced bending moment, whereas the effects of ship hull flexibility are accounted for in the computation of M_d. In very high sea states, M_d is normally ignored because the possibility of whipping is usually low. To consider the hull girder effects of slamming in oceangoing merchant ships, it has been suggested that $M_d = 0.15\,M_w$ be used for tankers in sagging, but $M_d = 0$ for those in hogging (Mansour & Thayamballi 1994).

Although the external pressure loads imposed on the ship's hull in seaways can be calculated in terms of sea water heads, the internal pressure loads must be determined for each fully loaded cargo hold and ballast tank, as caused by the dominant ship motions (pitch and roll) and the resulting accelerations. The relative phasing of all such dynamic loads is important in defining the total load.

8.6.1 Still-Water Loads

A detailed distribution of the still-water moment along the ship's length can be calculated by a double integration of the difference between the weight force and the buoyancy force, using simple beam theory.

The sectional shear force $F(x_1)$ at location x_1 is estimated by the integral of the load curve, which represents the difference between the weight and buoyancy curves, namely,

$$F(x_1) = \int_0^{x_1} f(x)dx \tag{8.4}$$

where $f(x) = b(x) - w(x)$ is the net load per unit length in still water, $b(x)$ is the buoyancy per unit length, and $w(x)$ is the weight per unit length.

The bending moment $M(x_1)$ at location x_1 is estimated as the integral of the shear curve indicated in Equation (8.4) as follows:

$$M(x_1) = \int_0^{x_1} F(x)dx \tag{8.5}$$

8.6.2 Long-Term Still-Water and Wave-Induced Loads: IACS Unified Formulas

Ship classification societies have over time established their own individual design guidelines based on their own studies and experience. This resulted in a variety of different requirements for ship hull girder structural design. The International Association of Classification Societies (IACS) has now unified these requirements in the case of the hull girder longitudinal strength.

This unification is important because the longitudinal strength of a ship governs basic scantlings of the primary strength members, such as the strength decks, side shells, bottom structures, and longitudinal bulkheads, thereby having a great effect on the hull weight, cargo deadweight, and ship price. The unified standard for a ship's longitudinal strength was approved by the IACS in May 1989, and it is now implemented as the longitudinal strength requirement of most classification societies.

In the IACS unified requirements (IACS 2012), the design still-water vertical bending moment M_{sw} is calculated as usual, by considering all appropriate loading conditions, while the following formulation is given as a guide:

$$M_{sw} = \begin{cases} +0.015C_1L^2B(8.167 - C_b) \text{ kNm for hogging} \\ -0.065C_1L^2B(C_b + 0.7) \text{ kNm for sagging} \end{cases} \tag{8.6}$$

where L is the ship length (m), B is the ship breadth (m), and C_b is the block coefficient.

The design wave-induced vertical bending moment, M_w, is given by unified formulas to represent a once-in-25-years occurrence under north Atlantic wave conditions. With the hogging moment taken as positive and the sagging moment taken as negative, the applicable formulas are as follows:

$$M_w = \begin{cases} +0.19C_1C_2L^2BC_b \text{ kNm for hogging} \\ -0.11C_1C_2L^2B(C_b + 0.7) \text{ kNm for sagging} \end{cases} \tag{8.7}$$

The coefficient C_1 in Equations (8.6) and (8.7) is determined as a function of vessel length (m) as follows:

$$C_1 = \begin{cases} 0.0792L & \text{for } L \leq 90 \\ 10.75 - [(300 - L)/100]^{1.5} & \text{for } 90 \leq L \leq 300 \\ 10.75 & \text{for } 300 < L \leq 350 \\ 10.75 - [(L - 350)/150]^{1.5} & \text{for } 350 < L \leq 500 \end{cases} \tag{8.8}$$

The coefficient C_2 in Equation (8.7) is taken as 1.0 around the mid-ship location, that is, between $0.4L$ and $0.65L$ measured from the after-perpendicular of the ship; a value less than 1.0 is used to represent the distribution at the other locations.

8.6.3 Long-Term Wave-Induced Loads: Direct Calculations

Using a direct method such as strip theory or panel theory, detailed information on the values and distribution of long-term wave-induced loads along the ship's length can be estimated for the anticipated environmental conditions.

The calculation of the hull girder loads in waves requires information related to the time-variant distribution of fluid forces over the wetted surface of the ship's hull together with the distribution of the inertia forces. The time-variant fluid forces depend on the wave-induced motions of the water and the corresponding motions of the ship. The distribution of inertial forces is estimated by multiplying the local mass of the ship with the local absolute value of acceleration.

The shear force and bending moment are then obtained at any instant by computing the first and second integrals of the distribution of the sectional force per unit length along the ship's length, respectively. The wave-induced sectional shear force $F_w(x_1)$ at location x_1 is then given by

$$F_w(x_1) = \int_0^{x_1} f_w(x)dx \qquad (8.9)$$

where $f_w(x) = d_f(x) - d_i(x)$ is the net load per unit length in waves, $d_f(x)$ is the time-variant fluid force per unit length, and $d_i(x)$ is the inertial force per unit length.

The wave-induced bending moment is then obtained by the integral of the wave-induced shear force as follows:

$$M_w(x_1) = \int_0^{x_1} F_w(x)dx \qquad (8.10)$$

To obtain a part of the time-variant fluid forces and the inertial forces, the wave-induced motions of the ship should be analyzed in advance of the loads themselves. The solutions for these ship motions and the resulting forces are normally obtained with the so-called strip theory. Related procedures may be found in most classification society guidance and in standard textbooks (Jensen 2001, Hughes & Paik 2013). The total hull girder loads are then obtained by a relevant sum of the still-water and wave-induced loads, as previously noted.

8.6.4 Short-Term Wave-Induced Loads: Simplified Direct Calculations Using Parametric Seakeeping Tables

The calculations of the extreme wave-induced loads of a ship during a short-term sea state may also be calculated using direct methods. In many cases and within predefined ranges of validity, time savings in this regard can be achieved by use of parametric seakeeping tables considering variations in ship size, significant wave height, and ship speed, such as those developed by Loukakis and Chryssostomidis (1975). The Loukakis–Chryssostomidis seakeeping tables are designed to efficiently determine the root-mean-square (rms) value of the wave-induced bending moment given the values of significant wave height (H_s), B/T ratio (where B is the ship's beam and T is its draught), L/B ratio (where L is the ship's length), ship operating speed (V), block coefficient (C_b), and sea-state persistence time.

The most probable extreme value of the wave-induced loads, M_w, that is, mode, which we may refer to as a mean for convenience, and its standard deviation, σ_w, can then be computed based on upcrossing analysis as follows:

$$M_w = \left[\sqrt{2\lambda_0 \ln N} + \frac{0.5772}{\sqrt{2\lambda_0 \ln N}} \right] \rho g L^4 \times 10^{-16} \text{(GNm)} \qquad (8.11a)$$

$$\sigma_w = \frac{\pi}{\sqrt{6}} \sqrt{\frac{\lambda_0}{2 \ln N}} \rho g L^4 \times 10^{-16} (\text{GNm}) \qquad (8.11b)$$

where $\sqrt{\lambda_0}$ is the rms value of the short-term wave-induced bending moment process, ρ is the density, g is the acceleration of gravity, and N is the expected number of wave-bending peaks, which is usually estimated as $N = S/\sqrt{13H_s} \times 3600$, where H_s is the significant wave height (m) and S is the storm persistence time (h). Alternatively, if a wave peak normally occurs every 6–10 s and in a 3-hour storm, for example, $N = 3 \times 60 \times 60/10 \approx 1000$.

Figures 8.9 and 8.10 show the variation of the wave-induced bending moments for selected vessels in a storm during 3 h (short-term) persistence time as obtained by the simplified direct method with various effective wave heights or ship speeds. The IACS design wave-induced moment values are also compared. It is apparent from Figure 8.9 that the wave-induced bending moment can theoretically be forced to exceed the IACS design value if the significant wave height is continually increased and the vessel's speed is kept high, which of course are unrealistic scenarios. Figure 8.10 shows that the wave-induced bending moments increase nearly linearly with the increase in the ship's speed. Furthermore, the figures represent that for the present hypothetical ships, the wave-induced bending moment should never exceed the IACS design values for reasonable ship speeds and significant wave heights.

8.7 Minimum Section Modulus Requirement

Since the section modulus of a ship is indicative of the ship's longitudinal strength, all classification societies have established relevant requirements so that the section modulus should be greater than a prescribed value.

The simple beam theory method together with the allowable working stress method indicated in Equation (1.16) is useful to establish the requirement for the minimum section modulus:

$$\sigma = \frac{M}{Z} < \sigma_a \qquad (8.12)$$

where M is the applied bending moment, Z is the section modulus, and σ_a is the allowable stress.

Because the design bending moments have been defined as the sum of the still-water and wave-induced bending moments and, because the allowable stress is assumed, Equation (8.12) gives the following requirement to satisfy for steel ship structures:

$$Z_{min} > \frac{k}{\sigma_a \times 10^6} |M_{sw} + M_w| (\text{m}^3) \qquad (8.13)$$

where Z_{min} is the minimum section modulus and k is the high tensile steel factor, which is defined as indicated in Table 8.3. In Equation (8.12) or (8.13), the allowable stress σ_a may be 190 MPa for the net thickness without a corrosion margin or 175 MPa for the gross thickness with a corrosion margin (Hughes & Paik 2013).

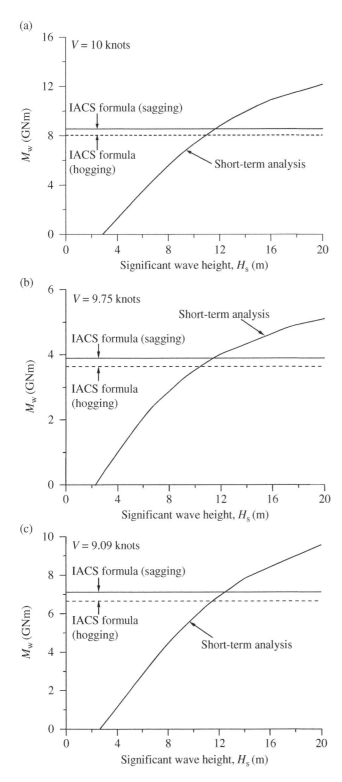

Figure 8.9 Variation of wave-induced bending moments plotted versus the significant wave height for (a) a 254 000 DWT oil tanker; (b) a 105 000 DWT oil tanker and (c) a 170 000 DWT bulk carrier, as obtained by the Loukakis–Chryssostomidis seakeeping table method.

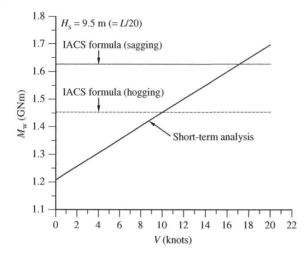

Figure 8.10 Variation of wave-induced bending moments plotted versus the ship speed for a 254 000 DWT oil tanker, as obtained by the Loukakis–Chryssostomidis seakeeping table method.

Table 8.3 High tensile steel factor for gross scantlings with a corrosion margin.

Steel type	Yield stress, σ_Y (MPa)	k	Allowable stress, $\sigma_a = 175/k$ (MPa)	σ_a/σ_Y
AH24	235	1.00	175.0	0.745
AH27	265	0.93	188.17	0.710
AH32	315	0.78	224.36	0.713
AH36	355	0.74	236.49	0.666
AH40	390	0.72	243.06	0.623

8.8 Determination of Ultimate Hull Girder Strength

The same methods used for box girders described in Section 7.5 can be used to calculate the ultimate strength of ship's hull girders. To compute the cross-sectional property calculations such as the section modulus, a ship's hull girder may be idealized as an assembly of pure plate element (segment) models as illustrated in Figure 2.2d although the stiffener flange can be modeled using one plate element. The modified Paik–Mansour method described in Section 7.5.4 can be used to calculate the ultimate hull girder strength where the structure may be idealized as an assembly of the plate–stiffener combination models as illustrated in Figure 2.2a. The ultimate compressive strength of the plate–stiffener combination models can be predicted by the Euler column buckling strength with the plasticity correction using the Johnson–Ostenfeld formulation method described in Section 2.9.5.1, when initial imperfections are insignificant or by the Perry–Robertson formulation method described in Section 2.9.5.2, or the Paik–Thayamballi empirical formulation method described in Section 2.9.5.3, when initial imperfections are significant.

Figure 8.11a shows an illustrative example of a bulk carrier's hull girder cross-sectional model as an assembly of plate–stiffener combination models. Figure 8.11b and c shows

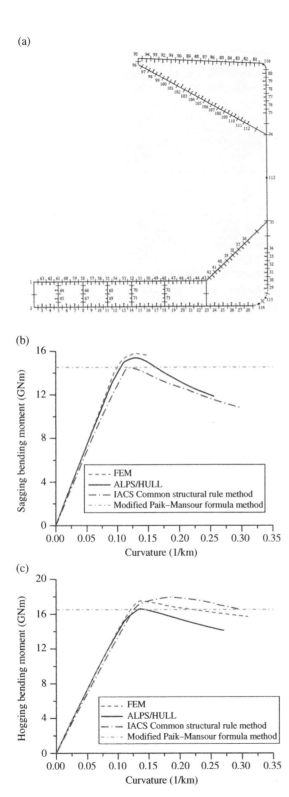

Figure 8.11 (a) Structural modeling for a bulk carrier's hull girder using both plate–stiffener combination models; (b) ultimate strength behavior under a sagging moment; (c) ultimate strength behavior under a hogging moment (Paik et al. 2013).

Table 8.4 Heights of the collapsed and yielded parts of the bulk carrier's hull girder under vertical bending moment, obtained by the original or modified Paik–Mansour method described in Section 7.5.3 or 7.5.4.

Method	Hogging (mm)		Sagging (mm)	
	h_C	h_Y	h_C	h_Y
Original P-M	–	–	17 935.0	0.0
Modified P-M	1 654.1	13.7	17 935.0	0.0

the results of the ultimate strength behavior comparison for the bulk carrier's hull girder subject to a vertical sagging or hogging bending moment by comparison to various computations including the nonlinear finite element method described in Chapter 12, ALPS/HULL (2017) intelligent supersize finite element method described in Chapter 13, and IACS common structural rules method (IACS 2012). Table 8.4 presents the heights of the collapsed and yielded parts, as determined by the original and modified Paik–Mansour methods described in Sections 7.5.3 and 7.5.4.

It is apparent from these results that the pure hogging bending moment condition cannot be achieved for this case with the original Paik–Mansour method, as it does not permit the expansion of the yielded part except for the tension flange. However, the modified Paik–Mansour method is able to achieve this condition because the tension flange is allowed to expand, with $h_Y = 13.7$ mm. It is however considered that the original Paik–Mansour method may also be useful for those structures in that the expansion of the yielded region in the tension side is insignificant, considering that the original Paik–Mansour method gives the ultimate strength formulations in a closed form and does not need an iteration process as described in Section 7.5.3.

8.9 Safety Assessment of Ships

Table 8.5 indicates safety measure calculations for the 10 typical merchant ships or offshore installations, described in Section 8.2 and Table 8.1, under vertical bending moments alone with average level of initial imperfections but without structural damage. The total bending moment is calculated from Equation (8.2) when $k_{sw} = k_w = 1.0$, while M_{sw} and M_w are determined from the IACS unified formula. Z_{min} is also computed from Equation (8.13). M_u is the ultimate vertical moment of the ship with an average level of initial imperfections, which was computed by the ALPS/HULL intelligent supersize finite element method described in Chapter 13.

For the safety measure calculations based on the ultimate strength, the partial safety factor of 1.0 is adopted in this regard. As evident from Table 8.5, the safety measure based on section modulus (allowable stress design approach) has greater margin at bottom than at deck for all of the vessels considered. In some vessels such as bulk carriers and container vessels, however, the safety measure based on ultimate hull strength (ultimate limit state design approach) has lesser margin in hogging than in sagging.

Table 8.5 Safety measure calculations for the 10 typical merchant ships or offshore installations.

Item		SHT	DHT#1	DHT#2	Bulk#1	Bulk#2	Cont#1	Cont#2	Cont#3	FPSO	Shuttle
Z (m^3)	Deck	66.301	29.679	77.236	44.354	39.274	18.334	26.635	44.376	31.040	43.191
	Bottom	70.950	39.126	103.773	62.058	50.544	27.228	42.894	58.785	38.520	49.175
Z_{min} (m^3)	Deck	60.699	27.814	73.494	44.040	38.950	17.252	26.327	44.042	26.991	36.992
	Bottom	60.699	27.814	73.494	50.516	42.196	18.689	28.521	47.712	26.991	36.992
$\dfrac{Z}{Z_{min}}$	Deck	1.092	1.067	1.051	1.007	1.008	1.063	1.012	1.008	1.150	1.168
	Bottom	1.169	1.407	1.412	1.228	1.198	1.457	1.504	1.232	1.427	1.329
M_{sw} (GNm)	Sag	-5.058	-2.318	-6.125	-4.210	-3.516	-1.557	-2.377	-3.976	-2.249	-3.083
	Hog	5.584	2.559	6.815	4.673	3.868	1.943	3.162	5.107	2.488	3.409
M_w (GNm)	Sag	-8.560	-3.923	-10.365	-7.124	-5.951	-2.636	-4.022	-6.729	-3.806	-5.217
	Hog	8.034	3.682	9.674	6.661	5.599	2.250	3.237	5.597	3.568	4.891
M_t (GNm)	Sag	-13.618	-6.240	-16.489	-11.334	-9.467	-4.193	-6.399	-10.705	-6.056	-8.300
	Hog	13.618	6.240	16.489	11.334	9.467	4.193	6.399	10.705	6.056	8.300
M_u (GNm)	Sag	-16.767	-6.899	-19.136	-14.281	-12.165	-6.800	-9.571	-16.599	-7.282	-11.280
	Hog	15.826	8.485	23.566	14.434	12.027	5.953	9.049	13.075	8.760	11.404
$\dfrac{M_u}{M_t}$	Sag	1.231	1.106	1.161	1.260	1.285	1.622	1.496	1.551	1.202	1.359
	Hog	1.162	1.360	1.429	1.274	1.270	1.420	1.414	1.221	1.446	1.374

Note: Z_{min} = minimum required section modulus, Equation (8.13), $M_t = M_{sw} + M_w$; M_u = ultimate vertical moment of vessel hulls with average level of initial imperfections, but without structural damage, as obtained by ALPS/HULL intelligent supersize finite element method.

8.10   Effect of Lateral Pressure Loads

It is essential to account for the effect of lateral pressure loads on the ultimate hull girder strength. Kim et al. (2013b) investigated the progressive hull collapse behavior of a Suezmax-class double-hull oil tanker, with a length of $L = 264$m, a breadth of $B = 48$m, a depth of $D = 23.2$m, a draught of $d = 16$m, and a block coefficient of $C_b = 0.843$, subject to vertical bending moments, with a focus on the effects of lateral pressure loads. Two typical loading conditions are considered, namely, the full-load and ballasted conditions, together with two types of vessel conditions such as anchored (static lateral pressure) and operating conditions (static + dynamic lateral pressure) on the basis of CSR (IACS 2012), as shown in Figures 8.12 and 8.13.

Figure 8.14 shows the progressive hull collapse behavior of the ship in the full-load condition and in the ballasted condition, respectively, obtained by the nonlinear finite element method, as described in Chapter 12. It is apparent from Figure 8.14 that dynamic lateral pressure loads give rise to the greater effect on the ultimate hull girder strength behavior than static effect. The reduction of ultimate hull girder strength arising from these loads in either the full-load condition or ballasted condition together with lateral pressure was less than 10% in this specific case.

(a)

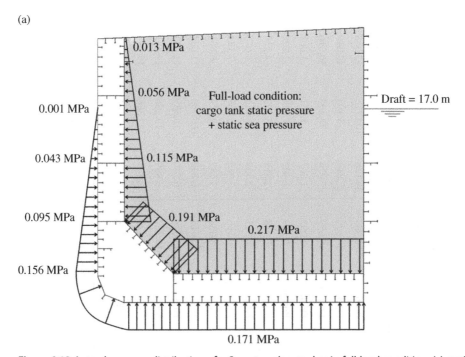

Figure 8.12 Lateral pressure distribution of a Suezmax-class tanker in full-load condition: (a) static pressure; (b) dynamic pressure; (c) combined (static + dynamic) pressure (IACS 2012).

(b)

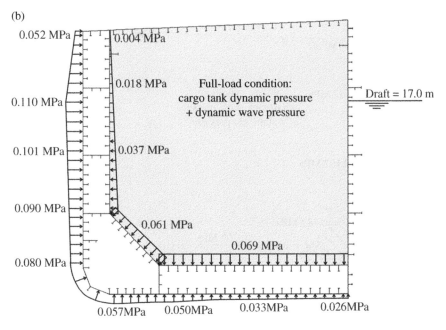

0.052 MPa

0.004 MPa

0.018 MPa

Full-load condition:
cargo tank dynamic pressure
+ dynamic wave pressure

Draft = 17.0 m

0.110 MPa

0.101 MPa

0.037 MPa

0.090 MPa

0.061 MPa

0.069 MPa

0.080 MPa

0.057MPa   0.050MPa   0.033MPa   0.026MPa

(c)

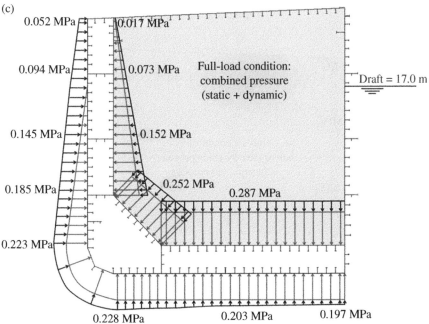

0.052 MPa

0.017 MPa

0.094 MPa

0.073 MPa

Full-load condition:
combined pressure
(static + dynamic)

Draft = 17.0 m

0.145 MPa

0.152 MPa

0.185 MPa

0.252 MPa

0.287 MPa

0.223 MPa

0.228 MPa   0.203 MPa   0.197 MPa

Figure 8.12 (Continued)

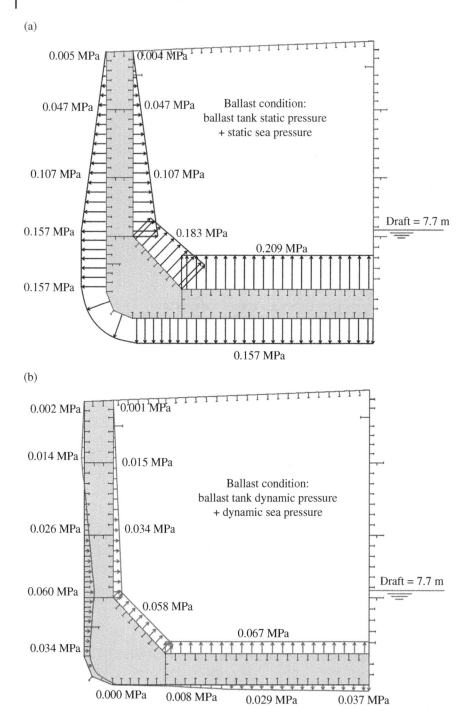

Figure 8.13 Lateral pressure distribution of a Suezmax-class tanker in ballast condition: (a) static pressure; (b) dynamic pressure; (c) combined (static + dynamic) pressure (IACS 2012).

(c)

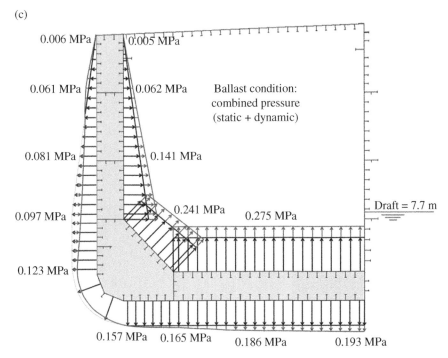

0.006 MPa 0.005 MPa

0.061 MPa 0.062 MPa Ballast condition:
combined pressure
(static + dynamic)

0.081 MPa 0.141 MPa

0.097 MPa 0.241 MPa 0.275 MPa Draft = 7.7 m

0.123 MPa

0.157 MPa 0.165 MPa 0.186 MPa 0.193 MPa

Figure 8.13 (Continued)

(a)

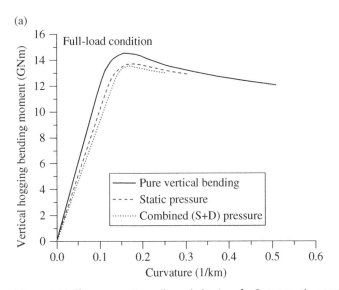

Full-load condition

— Pure vertical bending
---- Static pressure
········ Combined (S+D) pressure

Figure 8.14 The progressive collapse behavior of a Suezmax-class tanker hull under vertical bending moment accounting for the effect of lateral pressure loads: (a) hogging in the full-load condition; (b) sagging in the full-load condition; (c) hogging in the ballasted condition; (d) sagging in the ballast condition.

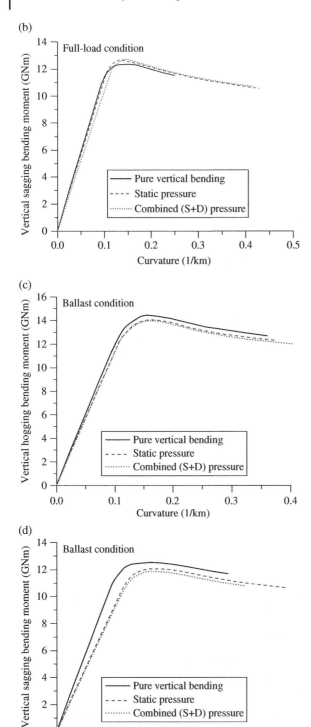

Figure 8.14 (Continued)

8.11 Ultimate Strength Interactive Relationships Between Combined Hull Girder Loads

8.11.1 Combined Vertical and Horizontal Bending

In some vessels subject to predominantly vertical bending moments at sea, the effect of horizontal moment is significant, and thus it is important to evaluate the effect of horizontal bending moment on the ultimate vertical bending moment of a ship's hull girder.

Figure 8.15 shows the progressive collapse behavior of some selected hull girders such as a 313 000 DWT double hull tanker with two longitudinal bulkheads, a 170 000 DWT

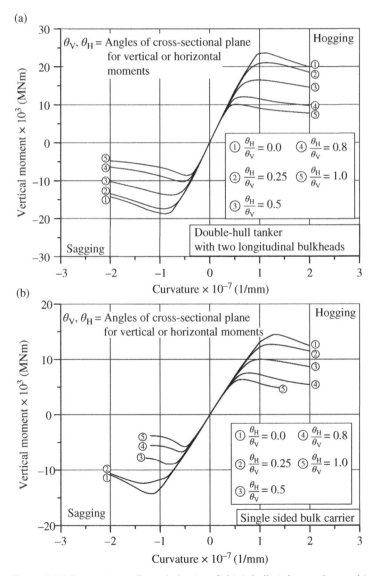

Figure 8.15 Progressive collapse behavior of ship's hull girders under combined vertical and horizontal bending moments obtained by ALPS/HULL intelligent supersize finite element method: (a) a 313 000 DWT double hull tanker with two longitudinal bulkheads; (b) a 170 000 DWT single-sided bulk carrier; (c) a 9 000 TEU container vessel.

(c)

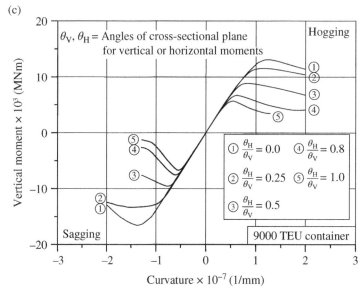

Figure 8.15 (Continued)

single-sided bulk carrier, and a 9 000 TEU container vessel, described in Section 8.2. Figure 8.16 represents the ultimate hull girder interaction relationship of typical merchant ships or offshore installations, described in Section 8.2, between vertical and horizontal bending moments, obtained by the ALPS/HULL intelligent supersize finite element method described in Chapter 13, accounting for the effects of initial

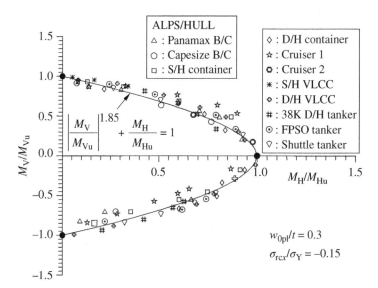

Figure 8.16 Ultimate hull girder strength interactive relationship between vertical and horizontal bending moments obtained by the ALPS/HULL intelligent supersize finite element method.

imperfections. It is evident that the effect of horizontal moment on the ultimate hull girder strength is of significance, and Equation (7.55) can be used to predict the ultimate hull girder strength under combined vertical and horizontal bending moments. It is noted that the horizontal bending moment is typically not the maximum when the vertical moment is the maximum, and thus a relevant consideration for load combination is necessary in performing a design check using the results of Figure 8.16.

8.11.2 Combined Vertical Bending and Shearing Force

The hull girder collapse under combined vertical bending moment and shearing forces can be evaluated using Equation (7.56a). Figure 8.17 shows the ultimate hull girder strength interaction relation of typical merchant ships or offshore installations, described in Section 8.2, under combined vertical bending moment and shear forces. The ultimate hull girder strengths predicted from Equation (7.56a) are compared with ALPS/HULL intelligent supersize finite element method solutions as described in Chapter 13. It is considered that Equation (7.56a) is in reasonably good agreement with more refined method solutions.

8.11.3 Combined Horizontal Bending and Shearing Force

The hull girder collapse under combined horizontal bending moment and shearing forces can be evaluated using Equation (7.56b). Figure 8.18 shows the ultimate hull girder strength interactive relationship of typical merchant ships or offshore installations, described in Section 8.2, under combined horizontal bending moment and shear forces. The ultimate hull girder strengths predicted from Equation (7.56b) are compared with

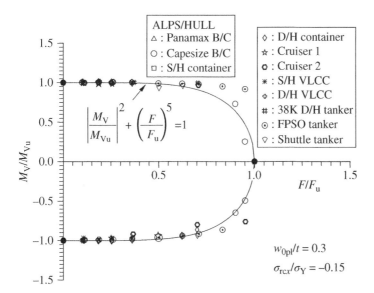

Figure 8.17 Ultimate strength interactive relationship between vertical bending moment and shear forces obtained by the ALPS/HULL intelligent supersize finite element method.

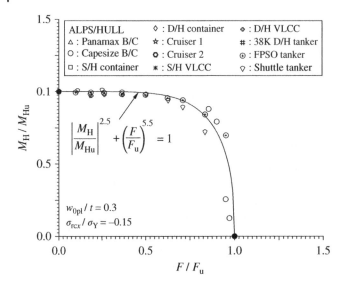

Figure 8.18 Ultimate hull girder strength interactive relationship between horizontal bending moment and shear forces obtained by the ALPS/HULL intelligent supersize finite element method.

ALPS/HULL intelligent supersize finite element method solutions, taking into account the effects of initial imperfections, as described in Chapter 13. It is considered that Equation (7.56b) is in reasonably good agreement with more refined method solutions.

8.11.4 Combined Vertical Bending, Horizontal Bending, and Shearing Force

Applying an approach similar to that presented in Section 3.8, an interactive relationship involving the three load components, namely, vertical bending, horizontal bending, and shearing forces can be derived based on three sets of the interactive relationships between two load components each, that is, $M_V - M_H$ relationship, $M_V - F$ relationship, and $M_H - F$ relationship. Figure 8.19 shows a schematic representation of the process for

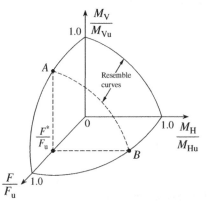

Figure 8.19 A schematic representation of the derivation of the interactive relationship involving three load components.

deriving the interaction relationship between three load components. The resulting interaction equation is then given by

$$\Gamma_u = \left(\frac{M_V}{M_{Vu}F_{VR}}\right)^{c_1} + \left(\frac{M_H}{M_{Hu}F_{HR}}\right)^{c_2} - 1 = 0 \tag{8.14}$$

where $F_{VR} = \{1-(F/F_u)^{c_4}\}^{1/c_3}$, $F_{HR} = \{1-(F/F_u)^{c_6}\}^{1/c_5}$, c_1 and c_2 are the coefficients defined in Equation (7.55), and c_3-c_6 are the coefficients defined in Equation (7.56). Γ_u thus represents the ship hull collapse function, where F_{VR} and F_{HR} indicate reduction factors due to shearing force. If the value of Γ_u is less than zero, that is, $\Gamma_u < 0$, then the ship hull is considered to be in a collapse-free condition, but it will possibly collapse if $\Gamma_u \geq 0$.

8.11.5 Effect of Torsional Moment

In vessel hull structures, warping displacements are normally only partly restrained, and thus the analysis of warping stresses as well as hatch opening deformations is in principle an essential part of the structural response analyses of vessels. Related to the effect of torsion on the bending moment capacity, however, it has been realized that torsion is not a very sensitive load component affecting the ultimate vertical bending moment of ship hulls as long as the magnitude of torsion is not predominant (Paik et al. 2001). However, it should also be noted that the ultimate bending strength of ship hulls with low torsional rigidity can be reduced significantly when torsional moments are large. Figure 8.20 shows the ultimate hull girder strength interaction relation of container

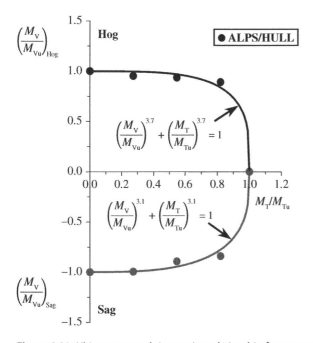

Figure 8.20 Ultimate strength interactive relationship for a container vessel under combined vertical bending and torsional moments obtained by the ALPS/HULL intelligent supersize finite element method.

vessels under combined vertical bending and torsional moments obtained by the ALPS/HULL intelligent supersize finite element method described in Chapter 13. It is apparent that Equation (7.58) can be used to predict the ultimate hull girder strength in this case.

8.12   Shakedown Limit State Associated with Hull Girder Collapse

One of the hull girder collapse accidents is the MOL COMFORT accident of an 8000 TEU containership occurred on June 17, 2013. Extensive investigations of the causes associated with the accident have been undertaken (ClassNK 2014), but some uncertainty still exists (Koh & Paik 2016). Exiting methodologies for calculating the ultimate hull girder strength presume that the hull girder loads are applied monotonically. However, hull girder loads are likely to be cyclic in accordance with wave actions, and furthermore several hull girder loads could be extreme even though they may not lead to hull girder collapse. In this case, load effects in local regions may well exceed yield stress resulting in plastic behavior.

Shakedown occurs when a structure responds steadily under cyclic loadings within appropriate limits after one or a few cycles of loading. Shakedown theorems have been developed to examine the plastic behavior of continua and structures that are subjected to cyclic loadings of relatively large ranges. Elastic shakedown means that a structure always responds elastically after initial plastic flow during earlier loading cycles. Elastic shakedown limits can be used to assess the safe range of the cyclic loading of a structure. The term "shakedown" was coined by Gruning in 1929. Melan (1938) described a more general theorem referred to as static shakedown, and Koiter (1956) subsequently presented a kinematic shakedown theorem. These theorems were summarized by Williams (2005) as follows:

- Melan's static shakedown theorem (lower bound) states that

> if any system of self-equilibrating residual stresses can be found which, in combination with the stresses due to the cyclic load, do not exceed yield at any time, then elastic shakedown will take place.

- Koiter's kinematical shakedown theorem (upper bound) states that

> if any kinematically acceptable mechanism of incremental plastic collapse can be found in which the rate of work done by the elastic stresses due to the load exceeds the rate of plastic dissipation, then incremental collapse will take place.

Jones (1975) noted that "these theorems can be used to predict the magnitude of the shakedown load but they do not provide any information on the number of load cycles required to reach a shakedown state." In general, the shakedown theorem can be used directly to determine whether any shakedown state exists in a structure and, if it does exist, to define the shakedown load range.

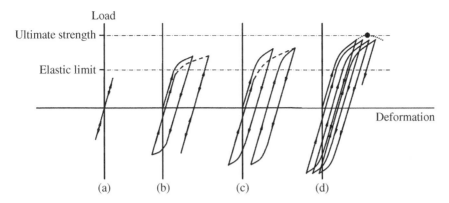

Figure 8.21 Structural responses under cyclic loadings with different ranges: (a) elastic; (b) elastic shakedown; (c) plastic shakedown (cyclic collapse); (d) ratchetting (incremental collapse).

Figure 8.21 presents the different forms of structural responses under cyclic loadings with different load ranges. If the load is lower than the first yield load of the structure, then the response will be wholly elastic as shown in Figure 8.21a. In this situation, the structure may eventually fail after a large number of load cycles due to high-cycle fatigue, which is of interest for the design service life of the structure. When the cyclic load falls between the first yield and the static ultimate load, three types of elastic–plastic behavior may occur as shown in Figure 8.21b elastic shakedown, (c) plastic shakedown, and (d) ratchetting. In the elastic shakedown case (b), plastic flow occurs in the first or initial few cycles, which produces a self-equilibrating residual stress or residual strain in the structure, such that subsequently the structure exhibits only an elastic response. At higher load case (c), each loading cycle leads to both the elastic and plastic deformations of the structure. The display of a steady state of noncumulative cyclic plasticity can be called plastic shakedown. Structures exhibiting plastic shakedown will fail after a finite number of loading cycles due to low-cycle fatigue (Abdel-Karim 2005). Under larger load case (d), each loading cycle generates net increments of plastic deformation, which accumulates until the incremental collapse of the structure occurs.

The unloading and reloading processes, together with any residual stress distribution for the elastic shakedown state of a beam, are described in the literature (Hodge 1959, König 1971, Sawczuk 1974, Jones 1975, Borkowske & Kleiber 1980, Williams 2005). It is apparent that under cyclic loading, in addition to high-cycle fatigue due to exhaustion of the service life in the case corresponding to Figure 8.21a, a structure may fail under a loading that is lower than the ultimate monotonic loading, as marked by the black point in Figure 8.21d. In general practice, therefore, a structure is theoretically safe up to the elastic shakedown limit for its entire service life. In other words, the corresponding elastic shakedown limit can be treated as a kind of cyclic load-carrying capacity of the structure.

Shakedown theorems were first applied to ships' hull girders by Jones (1975), who modeled a ship as a beam without taking the limitation of buckling into account. His results showed that the vertical bending moments associated with an elastic shakedown limit are always lower or at most equal to the static ultimate vertical bending moments. He suggested that the shakedown limit should be used as a basis for failure evaluation rather than the ultimate strength as described in Section 8.8.

The shakedown phenomenon was developed for infinitesimal displacements, as in classical beam theory, and is discussed further by Jones (1976). However, the phenomenon of pseudo-shakedown was introduced in Jones (1973) for a rigid, perfectly plastic rectangular plate subjected to repeated dynamic pressure pulses. Pseudo-shakedown may develop only in a rigid-plastic structure that is subjected to identical repeated loads producing stable finite deflections (e.g., axially restrained beams, circular, rectangular, and arbitrarily shaped plates, and axially restrained cylindrical shells). This phenomenon has been observed for the repeated wave impact of ship bows (Yuhara 1975) and analyzed successfully with a pseudo-shakedown analysis (Jones 1977). Some more details about the pseudo-shakedown phenomenon can be found in Jones (1997).

Jones (1975) applied these theorems to calculate the elastic shakedown limit of a ship's hull girder without considering any residual bending moment by idealizing the ship as a beam with free ends. Zhang et al. (2016) extended these theorems to calculate the elastic shakedown of a ship's hull girder that takes local buckling effects into account.

8.13 Effect of Age Related Structural Degradation

Aging ship hull structures may have suffered structural degradation such as corrosion and fatigue cracks over time. As general corrosion uniformly reduces the wall thickness of structural members, the ultimate strength or effectiveness of primary strength members can be evaluated by excluding the corrosion diminution (reduction in thickness). For structural members with fatigue cracks, the cross-sectional area associated with the cracking damage may in turn be reduced in strength calculations, as described in Chapters 4 and 9.

Age related damage is essentially time dependent, and subsequently the ultimate hull girder strength with age related damage is also time dependent. Paik et al. (2003) investigated the time-dependent ultimate hull girder strengths of selected vessels. Figure 8.22 shows the effects of age related damages on the ultimate hull girder strength of a 105 000 DWT double hull oil tanker over time. It is apparent from Figure 8.22 that the ultimate hull girder strength decreases as the vessel gets older because the corrosion depth and cracking size (length) grow over time.

To efficiently keep the ship's safety and reliability higher than a critical level, a proper and cost-effective scheme of repair and maintenance must be established, and repair strategies of heavily damaged members should be considered. The classification society rules usually require keeping the longitudinal strength of an aging ship at the level of higher than 90% of the initial state of new building. While the rule requirement is in fact based on the ship's section modulus, it may also be applied for establishing repair schemes so that the ultimate hull girder strength of an aging ship must be greater than 90% of the original ship even later in life.

The renewal criterion of any structural member is based on the member's ultimate strength rather than the plate thickness. This is because the latter approach, which is based on the percentage of member thickness loss, cannot reveal the effects of fatigue cracking or local dent damage and even pit corrosion, while it may properly handle the thickness reduction due to uniform corrosion. On the other hand, the former approach, which is based on member's ultimate strength, is adequate to measure the

(a)

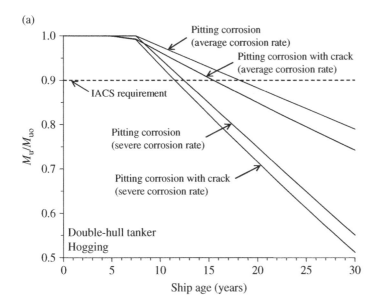

(b)

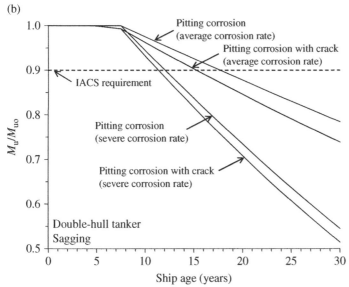

Figure 8.22 Time-dependent ultimate hull girder strength of a 105 000 DWT double hull oil tanker with age related deterioration in (a) hogging and (b) sagging.

reduced strength of damaged structures. In the former approach, therefore, any structural member category will be repaired if the percentage of its ultimate strength loss due to age related degradation and/or mechanical denting exceeds a critical value.

Figure 8.23 shows the time-dependent ultimate hull girder strength of a 105 000 DWT double hull oil tanker after repair of heavily damaged structural members so that the ultimate hull girder strength must always be greater than 90% of the original state.

(a)

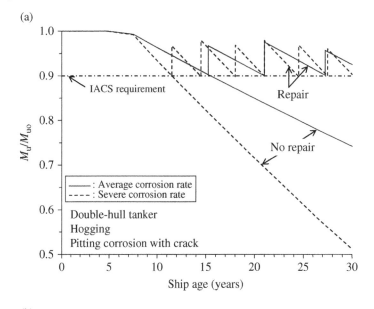

(b)

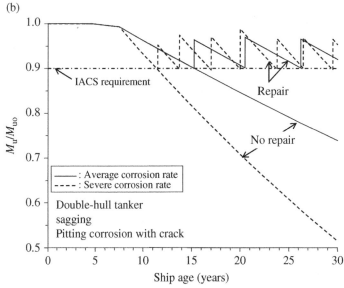

Figure 8.23 Repair and the resulting time-dependent ultimate hull girder strength of a 105 000 DWT double oil tanker with age related deterioration in (a) hogging and (b) sagging.

It is apparent from Figure 8.23 that the structural safety of aging vessels can be controlled by proper repair and maintenance strategies and repair criteria based on the member's ultimate strength can be better controlled.

For more applied examples in ship hulls with age related damage, interested readers may refer to Paik (1994), Paik et al. (1998a, 2003), and Kim et al. (2012, 2015a), among others.

8.14   Effect of Accident Induced Structural Damage

The various approaches to plate assemblies described in Chapter 7 may also be applied to account for the effect of accident-related structural damage such as local denting on the ultimate strength of ship hull structures, where the buckling and ultimate strength of damaged plate elements can be evaluated as described in Chapters 4 and 10. The damaged parts may be excluded from the calculations of effective plate width and the ultimate strength of individual members involving the damage when the damage is significant enough in size or extent. Even for structural members subject to tension, similar exclusion may be applied as long as they have suffered serious structural damage.

Figure 8.24 shows an illustrative example of a 307 000 DWT double hull tanker hull cross section with collision or grounding-induced damages. Figure 8.25 shows a

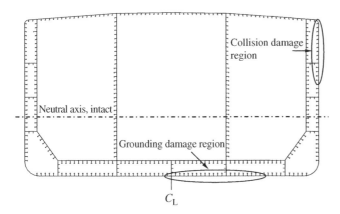

Figure 8.24 A 307 000 DWT double hull oil tanker's hull cross section with collision or grounding damage.

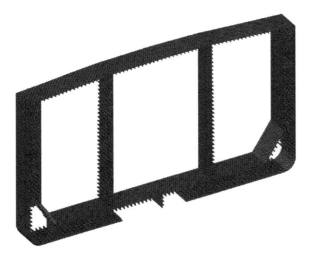

Figure 8.25 A nonlinear finite element method model for the progressive hull collapse analysis of a double hull oil tanker with grounding damage.

(a)

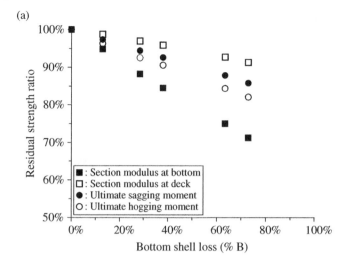

(b)

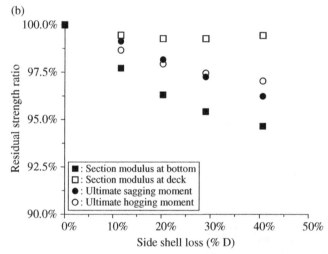

Figure 8.26 Effect of accident induced damage on the residual strength ratio for a 307 000 DWT double hull oil tanker with (a) grounding damage and (b) collision damage.

nonlinear finite element method model for the ship with grounding damage. While the grounding damage is present in bottom structures, the collision damage exists from the upper side shell to a known or predefined extent. In the analysis, the damaged members in both tension and compression are excluded. The residual strength ratio can be defined as the ratio of the strength measure of the damaged ship to that of the intact ship based on either the section modulus or the ultimate hull girder strength. Figure 8.26 shows the variation of the residual strength ratio as a function of such grounding or collision damage amount. The ultimate hull girder strength with grounding or collision damages is computed by the ALPS/HULL intelligent supersize finite element method described in Chapter 13. These results are obtained subject to the assumption that the damaged neutral axis is parallel to the original and should thus be viewed as notional, especially for any significant amount of damage. It is apparent from Figure 8.26 that the accident induced damages can significantly reduce the safety measure of ship's hulls.

As described in Section 7.7, it is highly desirable to assess the residual ultimate strength of damaged ship structures quickly and accurately to facilitate the rapid planning of salvage and rescue operations immediately after accidents where the location and extent of the damage is known. Paik et al. (2012) proposed a method using the residual strength–damage index diagram (R-D diagram), which is established in advance. Figure 8.27 shows

(a)

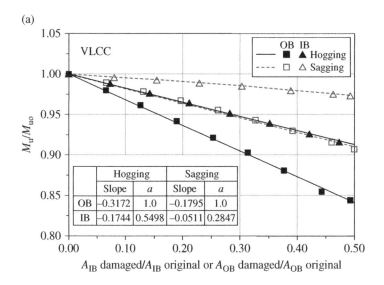

(b)

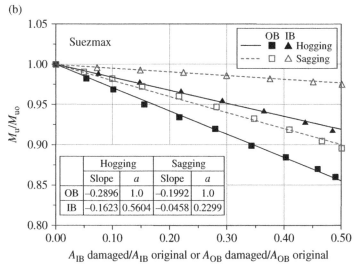

Figure 8.27 The residual ultimate strength-grounding damage index diagrams for (a) a VLCC class double hull oil tanker; (b) a Suezmax-class double-hull oil tanker; (c) an Aframax-class double-hull oil tanker and (d) a Panamax-class double-hull oil tanker.

(c)

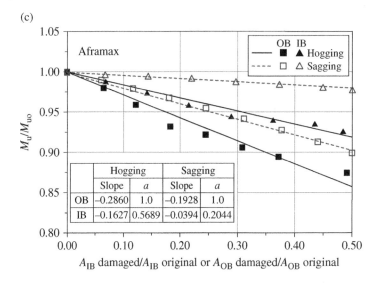

(d)

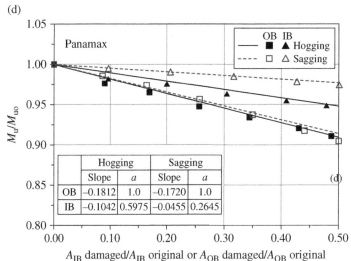

Figure 8.27 (Continued)

an example of such R-D diagrams associated with the residual ultimate hull girder strength with grounding damage for various sizes of double hull oil tankers, established by Paik et al. (2012). Youssef et al. (2017) established the R-D diagram for a ship damaged by collisions.

For more applied examples in ship hulls with accident-related damage, interested readers may refer to Paik et al. (1998b, 2012), Kim et al. (2013a, 2013c, 2014a, 2014b, 2015b), and Youssef et al. (2014, 2016, 2017), among others.

References

Abdel-Karim, M.D. (2005). Shakedown of complex structures according to various hardening rules. *International Journal of Pressure Vessels and Piping*, **82**(6): 427–458.

ALPS/HULL (2017). *A computer program for the progressive collapse analysis of ship's hull structures*. MAESTRO Marine LLC, Stevensville, MD.

Borkowske, A. & Kleiber, M. (1980). On a numerical approach to shakedown analysis of structures. *Computer Methods in Applied Mechanics and Engineering*, **22**(1): 101–119.

ClassNK (2014). *Investigation report on structural safety of large container ships: the Investigative Panel on Large Container Ship Safety*. ClassNK, Tokyo.

Faulkner, D. (1998). An independent assessment of the sinking of the M.V. Derbyshire. *Transactions of the Society of Naval Architects and Marine Engineers*, **106**: 59–103.

Hodge, P.G. (1959). *Plastic analysis of structures*. McGraw-Hill Series in Engineering Sciences, New York.

Hughes, O.F. & Paik, J.K. (2013). *Ship structural analysis and design*. The Society of Naval Architects and Marine Engineers, Alexandria, VA.

IACS (2012). *Common structural rules*. International Association of Classification Societies, London.

Jensen, J.J. (2001). *Load and global response of ships*. Elsevier, London.

Jones, N. (1973). Slamming damage. *Journal of Ship Research*, **17**(2): 80–86.

Jones, N. (1975). On the shakedown limit of a ship's hull girder. *Journal of Ship Research*, **19**(2): 118–121.

Jones, N. (1976). Plastic behavior of ship structures. *Transactions of the Society of Naval Architects and Marine Engineers*, **84**: 115–145.

Jones, N. (1977). Damage estimates for plating of ships and marine vehicles. *Proceedings of International Symposium on Practical Design in Shipbuilding*, Tokyo: 121–128.

Jones, N. (1997). Dynamic plastic behaviour of ship and ocean structures. *Transactions of the Royal Institution of Naval Architects*, **139**(Part A): 65–97.

Kim, D.K., Kim, H.B., Hairil Mohd, M. & Paik, J.K. (2013a). Comparison of residual strength-grounding damage index diagrams for tankers produced by the ALPS/HULL ISFEM. *International Journal of Naval Architect and Ocean Engineering*, **5**(1): 47–61.

Kim, D.K., Kim, S.J., Kim, H.B., Zhang, X.M., Li, C.G. & Paik, J.K. (2015a). Ultimate strength performance of bulk carriers with various corrosion additions. *Ships and Offshore Structures*, **10**(1): 59–78.

Kim, D.K., Kim, B.J., Seo, J.K., Kim, H.B., Zhang, X.M. & Paik, J.K. (2014a). Time-dependent residual ultimate longitudinal strength-grounding damage index (R-D) diagram. *Ocean Engineering*, **76**: 163–171.

Kim, D.K., Liew, M.S., Youssef, S.A.M., Hairil Mohd, M., Kim, H.B. & Paik, J.K. (2014b). Time-dependent ultimate strength performance of corroded FPSOs. *Arabian Journal of Science and Engineering*, **39**(11): 7673–7690.

Kim, D.K., Park, D.K., Kim, J.H., Kim, S.J., Seo, J.K. & Paik, J.K. (2012). Effect of corrosion on the ultimate strength of double hull oil tankers – part II: hull girders. *Structural Engineering and Mechanics*, **42**(4): 531–549.

Kim, D.K., Park, D.H., Kim, H.B., Kim, B.J., Seo, J.K. & Paik, J.K. (2013b). Lateral pressure effects on the progressive hull collapse behavior of a Suezmax-class tanker under vertical bending moments. *Ocean Engineering*, **63**: 112–121.

Kim, D.K., Pedersen, P.T., Paik, J.K., Kim, H.B., Zhang, X.M. & Kim, M.S. (2013c). Safety guidelines of ultimate hull girder strength for grounded container ships. *Safety Science*, **59**: 46–54.

Kim, Y.S., Youssef, S.A.M., Ince, S.T., Kim, S.J., Seo, J.K., Kim, B.J., Ha, Y.C. & Paik, J.K. (2015b). Environmental consequences associated with collisions involving double hull oil tankers. *Ships and Offshore Structures*, **10**(5): 479–487.

Koiter, W.T. (1956). A new general theorem on shakedown of elastic-plastic structures. *Proceedings of the Koninklijke Nederlandse Akademie Van Wetenschappen*, **B59**: 24–32.

Koh, T.J. & Paik, J.K. (2016). Structural failure assessment of a post-Panamax class containership: lessons learned from the MSC Napoli accident. *Ships and Offshore Structures*, **11**(8): 847–859.

Konig, J.A. (1971). A method of shakedown analysis of frames and arches. *International Journal of Solids Structures*, **7**(4): 327–344.

Loukakis, T.A. & Chryssostomidis, C. (1975). Seakeeping standard series for cruiser-stern ships. *Transactions of the Society of Naval Architects and Marine Engineers*, **83**: 67–127.

Mansour, A.E. & Thayamballi, A.K. (1994). *Probability based ship design; loads and load combination*. **SSC-373**, Ship Structure Committee, Washington, DC.

Melan, E. (1938). Der spannungsgudstand eines Henky-Mises schen kontinuums bei verlandicher bealstung. *Wissenschaften Wien, Series 2A*, **147**: 73.

Paik, J.K. (1994). Hull collapse of an aging bulk carrier under combined longitudinal bending and shearing force. *Transactions of the Royal Institution of Naval Architects*, **136**: 217–228.

Paik, J.K. & Faulkner, D. (2003). Reassessment of the M.V. Derbyshire sinking with the focus on hull-girder collapse. *Marine Technology*, **40**(4): 258–269.

Paik, J.K., Kim, D.H., Park, D.H. & Kim, M.S. (2012). A new method for assessing the safety of ships damaged by grounding. *Transactions of the Royal Institution of Naval Architects*, **154** (A1): 1–20.

Paik, J.K., Kim, D.K., Park, D.H., Kim, H.B. Mansour, A.E. & Caldwell, J.B. (2013). Modified Paik-Mansour formula for ultimate strength calculations of ship hulls. *Ships and Offshore Structures*, **8**(3–4): 245–260.

Paik, J.K., Kim, S.K., Yang, S.H. & Thayamballi, A.K. (1998a). Ultimate strength reliability of corroded ship hulls. *Transactions of the Royal Institution of Naval Architects*, **140**: 1–18.

Paik, J.K., Seo, J.K. & Kim, B.J. (2008). Ultimate limit state assessment of the M.V. Derbyshire hull structure. *Journal of Offshore Mechanics and Arctic Engineering*, **130**(2): 021002-1–021002-9.

Paik, J.K. & Thayamballi, A.K. (1998). The strength and reliability of bulk carrier structures subject to age and accidental flooding. *Transactions of the Society of Naval Architects and Marine Engineers*, **106**: 1–40.

Paik, J.K., Thayamballi, A.K., Pedersen, P.T. & Park, Y.I. (2001). Ultimate strength of ship hulls under torsion. *Ocean Engineering*, **28**: 1097–1133.

Paik, J.K., Thayamballi, A.K. & Yang, S.H. (1998b). Residual strength assessment of ships after collision and grounding. *Marine Technology*, **35**(1): 38–54.

Paik, J.K., Wang, G., Kim, B.J. & Thayamballi, A.K. (2002). Ultimate limit state design of ship hulls. *Transactions of the Society of Naval Architects and Marine Engineers*, **110**: 173–198.

Paik, J.K., Wang, G., Thayamballi, A.K., Lee, J.M. & Park, Y.I. (2003). Time-dependent risk assessment of aging ships accounting for general/pit corrosion, fatigue cracking and local denting damage. *Transactions of the Society of Naval Architects and Marine Engineers*, **111**: 159–197.

Sawczuk, A. (1974). Shakedown analysis of elastic-plastic structures. *Nuclear Engineering and Design*, **28**(1): 121–136.

Turan, O. and Vassalos, D. (1994). Dynamic stability assessment of damaged passenger ships. *Transactions of the Royal Institution of Naval Architects*, **136**: 79–104.

Williams, J.A. (2005). The influence of repeated loading, residual stresses and shakedown on the behaviour of tribological contacts. *Tribology International*, **38**(9): 786–797.

Youssef, S.A.M., Faisal, M., Seo, J.K., Kim, B.J., Ha, Y.C., Paik, J.K., Cheng, F. & Kim, M.S. (2016). Assessing the risk of ship hull collapse due to collision. *Ships and Offshore Structures*, **11**(4): 335–350.

Youssef, S.A.M., Ince, S.T., Kim, Y.S., Paik, J.K., Cheng, F. & Kim, M.S. (2014). Quantitative risk assessment for collisions involving double hull oil tankers. *Transactions of the Royal Institution of Naval Architects*, **156**(A2): 157–174.

Youssef, S.A.M., Noh, S.H. & Paik, J.K. (2017). A new method for assessing the safety of ships damaged by collisions. *Ships and Offshore Structures*, **12**(6): 862–872.

Yuhara, T. (1975). Fundamental study of wave impact loads on ship bow. *Journal Society Naval Architects of Japan*, **137**: 240–245.

Zhang, X., Paik, J.K. & Jones, N. (2016). A new method for assessing the shakedown limit state associated with the breakage of a ship's hull girder. *Ships and Offshore Structures*, **11**(1): 92–104.

9

Structural Fracture Mechanics

9.1 Fundamentals of Structural Fracture Mechanics

Under the action of repeated loading, fatigue cracks may form in the stress concentration areas of a structure. Initial defects may form in the structure during the fabrication procedure and may conceivably remain undetected over time. Cracks may initiate and propagate from such defects. In addition to propagation under repeated cyclic loading, cracks may also grow in an unstable manner under monotonically increasing extreme loads, a circumstance that can conceivably lead to catastrophic failure of the structure. This possibility is of course tempered by the ductility of the material involved and the presence of reduced stress intensity regions in a complex structure that may serve to arrest cracks, even in an otherwise monolithic structure.

For residual strength assessment of aging structures under extreme loads, it is thus often necessary to account for a known (existing or premised) crack as a parameter of influence. This chapter is primarily concerned with the limit state assessment of plated structures with existing crack damage and under monotonic extreme loading. The propagation of a crack under cyclic loading is also briefly treated, because such technology may be necessary to predict the size of a crack subject to cyclic loading, at any given point in time, beginning with a known initial crack size.

Structural fracture modes associated with cracks may be classified into three groups: brittle fracture, ductile fracture, and rupture (Machida 1984). When the strain at which a material fractures is very low, it is called brittle fracture. In structures made of ductile material with adequate fracture toughness, however, the fracture strain can be comparatively large. When the material is broken by necking associated with large plastic flow, it is called rupture. Ductile fracture is an intermediate failure mode between brittle fracture and rupture.

The progress of ductile fracture from an existing sharp-tipped crack may be separated into four regimes: blunting of the initially sharp crack tip, initial crack growth, stable crack growth, and unstable crack propagation (Shih et al. 1977). The ductile fracture characteristics generally depend on the material's toughness, but it can also be affected by the loading rate and by environmental factors such as corrosion and temperature. For high-toughness materials, the crack tip may be significantly blunted and the stable crack growth regime can be substantial before fracture. For low-toughness materials, however, there is likely to be relatively little crack tip blunting, and unstable crack extension can occur even without a stable crack growth regime.

Ultimate Limit State Analysis and Design of Plated Structures, Second Edition. Jeom Kee Paik.
© 2018 John Wiley & Sons Ltd. Published 2018 by John Wiley & Sons Ltd.

In the rare situation in which the structure has been weakened by large cracks or large-scale plasticity associated with cracks, resulting in a decrease in structural stiffness, large deformations are likely to develop. Figure 9.1 shows a schematic representation of the nonlinear behavior of cracked structures under monotonic loading. It is noted that, for similar structures, the stiffness and ultimate strength of cracked structures are, as expected, less than those of structures without cracks.

Fracture behavior for ductile materials is quite different from that of brittle materials. Ductile materials generally exhibit slow stable crack growth accompanied by considerable plastic deformation. In other words, they resist crack growth during crack extension. The study of the fracture behavior of materials, components, and structures is now known as structural fracture mechanics. Structural fracture mechanics is hence the engineering discipline that can be used to quantify the conditions under which a load-bearing structure can fail due to the enlargement of a crack.

It is commonly agreed that the modern era of structural fracture mechanics originated with the work of Griffith (1920), who resolved the infinite crack tip stress dilemma inherent in the use of the theory of elasticity for cracked structures. However, for some time the study of fracture remained of scientific interest only. One reason for this was the apparent non-applicability of the Griffith theory to engineering materials (i.e., metals) whose fracture resistance values are typically orders of magnitude greater than that of brittle materials such as glass.

The next major contributions to the subject were made independently by Irwin (1948) and Orowan (1948), who extended Griffith's approach to metals by including the energy dissipated by local plastic flow. During this same period, Mott (1948) extended Griffith's theory to rapidly propagating cracks.

Irwin (1956) developed the concept of energy release rate and related it to Griffith's theory. Using the approach of Westergaard (1939), who developed a method to analyze

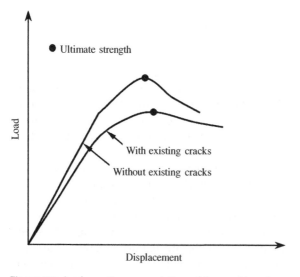

Figure 9.1 A schematic representation of the cracking damage effect on the ultimate strength behavior of plated structures.

stresses and displacements ahead of a sharp crack, Irwin (1957) showed that the stresses and displacements near the crack tip could be described with a single parameter that was related to the energy release rate. This crack tip-characterizing parameter is the stress intensity factor. During the same period, Williams (1957) also calculated the stress distribution at the crack tip but used a somewhat different technique from the Irwin approach. Both results were essentially identical.

Linear elastic fracture mechanics (LEFM) is generally found to be accurate for brittle materials. Direct application of LEFM to ductile materials has been found to yield overly conservative predictions. In the 1960s, it was realized that LEFM is not applicable when large-scale yielding at the crack tip precedes failure. To accommodate the effect of yielding at the crack tip, several researchers proposed approximate methods, mostly by correcting and expanding on LEFM (Dugdale 1960, Wells 1961, 1963, Barenblatt 1962). Although Dugdale (1960) proposed an idealized model based on a narrow strip of yielded material at the crack tip, Wells (1961, 1963) suggested displacement of the crack faces as an alternative fracture criterion when large-scale plasticity occurs at the crack tip. The Wells parameter is now known as the crack tip opening displacement (CTOD).

Rice (1968) introduced another parameter to characterize the nonlinear material behavior ahead of a crack tip. He generalized the energy release rate to nonlinear materials by idealizing plastic deformation as nonlinear elastic. The resulting parameter is the J-integral. During this same period, Hutchinson (1968) and Rice and Rosengren (1968) showed that the J-integral could be used to represent the characteristics of crack tip stress fields in the nonlinear elastic range of material behavior.

To apply fracture mechanics to structural design, a mathematical relationship between material toughness, stress, and flaw size must be established. Although these relationships for linear elastic problems had been available for some time, Shih and Hutchinson (1976) were perhaps the first to provide the theoretical framework to establish such a relationship for nonlinear problems. Shih (1981) also established a relationship between the J-integral and CTOD.

For an elaborate summary of early structural fracture mechanics research undertaken from 1913 to 1965, readers may refer to Barsom (1987). A comprehensive historical overview of structural fracture mechanics from 1960 to 1980 was provided by Anderson (1995). The details of structural fracture mechanics can be found in textbooks by Machida (1984), Kanninen and Popelar (1985), Broek (1986), Anderson (1995), Lotsberg (2016), and others. Many handbooks are also available for practical applications (e.g., Sih 1973, Tada et al. 1973, Rooke and Cartwright 1976, Murakami 1987).

This chapter describes the fundamentals of structural fracture mechanics, with the eventual aim of application to ductile structures. A simplified procedure for analysis of the limit state capacity of plated structures with premised cracks under monotonic extreme loads is presented. The ultimate limit state (ULS) criterion is still expressed by Equation (1.17). However, in this case, the "capacity" represents the ultimate strength of the cracked structure under monotonic extreme loads, and the "demand" indicates the extreme working stress or load. Any mathematical details omitted in this chapter may be found in the reference material noted earlier.

9.2 Basic Concepts for Structural Fracture Mechanics Analysis

Figure 9.2 represents a schematic of appropriate approaches for the fracture analysis of a cracked structure as a function of the material's fracture toughness. It is apparent from Figure 9.2 that for low-toughness materials, brittle fracture is predominant and LEFM is valid. For very high-toughness materials, however, rupture is dominant because of the large-scale plasticity that precedes structural collapse. In this case, limit-load analysis is more relevant. There is a transition between brittle fracture and rupture with an intermediate fracture toughness, which is termed the ductile fracture regime; in this case, nonlinear fracture mechanics concepts, now generally termed elastic–plastic fracture mechanics (EPFM), will be more relevant to assess the structure's failure characteristics.

In this section, these various basic concepts of structural fracture mechanics are described.

9.2.1 Energy-Based Concept

In the Griffith energy concept, fracture is considered to take place if the crack-growth-related energy exceeds the material's resistance to fracture. Mathematically, the following criterion must be satisfied for fracture to occur:

$$G \geq G_C \tag{9.1}$$

where G is the strain energy release rate or, alternatively, the crack driving force and G_C represents the material's resistance to crack growth.

For a cracked infinite plate under tensile stress σ, as shown in Figure 9.3, it can be shown that G and G_C are given by

$$G = \frac{\pi \sigma^2 a}{E} \tag{9.2a}$$

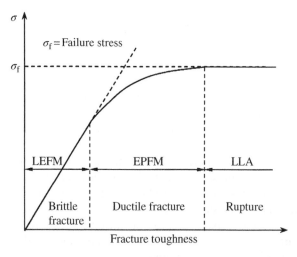

Figure 9.2 A schematic of appropriate approaches for fracture analysis as a function of material fracture toughness (LEFM, linear elastic fracture mechanics; EPFM, elastic–plastic fracture mechanics; LLA, limit-load analysis).

Figure 9.3 A cracked infinite plate under tensile loading.

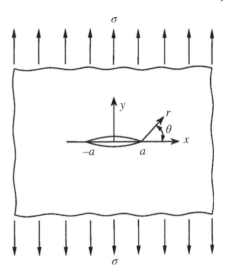

$$G_C = \frac{\pi \sigma_f^2 a}{E} \tag{9.2b}$$

where E is Young's modulus, a is the half crack length, and σ_f is the failure stress.

It can be surmised from Equation (9.2) that the failure stress, σ_f, is proportional to $1/\sqrt{a}$ for a constant value of G_C, implying that the failure stress decreases in that manner as the flaw grows.

9.2.2 Stress Intensity Factor Concept

For a cracked body with a linear elastic material as shown in Figure 9.4, stress components near the crack tip in the xy plane may be shown to be given by

$$\sigma_x = \frac{K_I}{\sqrt{2\pi r}} \cos\left(\frac{\theta}{2}\right) \left[1 - \sin\left(\frac{\theta}{2}\right) \sin\left(\frac{3\theta}{2}\right)\right] \tag{9.3a}$$

$$\sigma_y = \frac{K_I}{\sqrt{2\pi r}} \cos\left(\frac{\theta}{2}\right) \left[1 + \sin\left(\frac{\theta}{2}\right) \sin\left(\frac{3\theta}{2}\right)\right] \tag{9.3b}$$

$$\tau_{xy} = \frac{K_I}{\sqrt{2\pi r}} \cos\left(\frac{\theta}{2}\right) \sin\left(\frac{\theta}{2}\right) \cos\left(\frac{3\theta}{2}\right) \tag{9.3c}$$

where K_I is the mode I stress intensity factor. Mode I is the direct opening mode for a crack; crack modes are addressed later.

The dimensions of the stress intensity factor are given by [stress] × [length]$^{1/2}$ = [force] × [length]$^{-3/2}$ (e.g., kgf/mm$^{3/2}$, MN/m$^{3/2}$). It is evident from Equation (9.3) that each stress component is proportional to the stress intensity factor.

For the cracked plate shown in Figure 9.3, the mode I stress intensity factor is given by

$$K_I = \sigma\sqrt{\pi a} \tag{9.4}$$

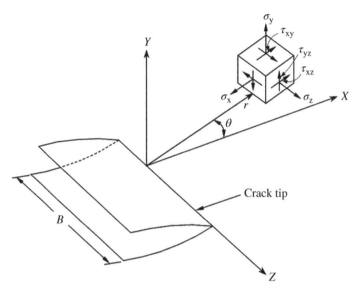

Figure 9.4 Local coordinate system and the resulting stress components for a cracked body (B = plate thickness).

From Equations (9.2a) and (9.4), one can obtain a relationship between K_I and G if necessary. In today's LEFM, it is considered that the fracture occurs when the following criterion is satisfied:

$$K \geq K_C \tag{9.5}$$

where K is the stress intensity factor and K_C is the critical stress intensity factor, which represents a measure of the material's resistance. It can be shown that the critical value of the stress intensity factor is related to the critical value of the crack driving force by the following equations:

$$K_C = \sqrt{EG_C} \text{ for plane stress} \tag{9.6a}$$

$$K_C = \sqrt{(1-v^2)EG_C} \text{ for plane strain} \tag{9.6b}$$

Standard test procedures exist for obtaining the mode I plane strain fracture toughness. Table 9.1 indicates a limited collection of such fracture toughness data for maraging steels according to Broek (1986). It can be seen from Table 9.1 that the K_{IC} values are between 50 and 350 kg/mm$^{3/2}$ for materials with a high yield strength, whereas those with a low yield strength have fracture toughness of the order of 500 kg/mm$^{3/2}$ or more. Depending on the materials' yield stress, the required specimen thickness may be on the order of 2–20 mm, but test specimens thinner than about 10 mm may not normally be useable because of buckling. In any event, because the combination of a high toughness and a low yield stress leads to extremely high values of $(K_{IC}/\sigma_Y)^2$, where σ_Y is the material yield stress, the required specimen thickness for a standard test may reach the order of 1 m, as indicated in Table 9.1.

Table 9.1 Illustrative collection of fracture toughness data for maraging steels (Broek 1986).

Material	Condition	σ_Y (kg/mm^2)	K_{IC} (kg/mm$^{3/2}$)	B_{min} (mm)
300	900°F 3 h	200	182	2.1
300	850°F 3 h	170	300	7.8
250	900°F 3 h	181	238	4.3
D6AC steel	Heat treated	152	210	4.8
	Heat treated	150	311	10.7
	Forging	150	178–280	–
4340 steel	Hardened	185	150	1.7
A533B	Reactor steel	35	$\cong$630	810
Carbon steel	Low strength	24	>700	2150

Note: B_{min} = minimum thickness required for the K_{IC} test specimen.

In today's practice, it is normally required that the following equation be satisfied for a through-thickness crack in a plate of thickness B, if plane strain conditions are to be assured at the crack tip in a static test:

$$B \geq 2.5 \left(\frac{K_{IC}}{\sigma_Y} \right)^2 \tag{9.7}$$

Apart from the fact that it is not practical to perform a valid K_{IC} test on most ductile materials, it is also not useful because materials with a thickness in the order of 1 m will never be used. This shows one of the limitations of LEFM, which is that it is readily applicable only to materials with a ratio of modulus to yield stress that is roughly lower than 200–250 at room temperature. Of course, LEFM may apply to low-strength steels at lower temperatures and/or higher loading rates, conditions under which the same material may behave in an appreciably more brittle manner. Such cases may or may not be of interest in specific applications that involve through-thickness cracks. In cases in which it can be used, once the crack situation (e.g., through-thickness crack in an infinite plate), fracture toughness, and applied loading details are known, the critical crack size at failure can be calculated with LEFM.

9.3 More on LEFM and the Modes of Crack Extension

In previous sections, we have based the discussion on mode I (direct tensile opening mode) through-thickness crack in an infinite plate. In this section, the concept and applications of the stress intensity factor as a representative parameter in LEFM are further described, also considering modes of crack extension other than mode I.

The stress intensity factor, K, is determined as a function of the crack size, geometric properties, and loading conditions. An investigation of crack tip stress and displacement fields and their relationship to K is important because these fields are typically those that govern the fracture process at the crack tip.

A cracked body as shown in Figure 9.4 is now considered. The crack plane lies in the xz plane, and the crack front is parallel to the y axis. In this case, three basic fracture modes are relevant, as depicted by Figure 9.5. Mode I is the opening or tensile mode in which the crack faces separate symmetrically with respect to the xy and xz planes. In mode II, the sliding or in-plane shearing mode, the crack faces slide relative to each other symmetrically about the xy plane but asymmetrically with respect to the xz plane. In the tearing or anti-plane or out-of-plane shearing mode, mode III, the crack faces also slide relative to each other but asymmetrically with respect to the xy and xz planes.

For plane problems with homogeneous, isotropic, and linear elastic materials, the stress intensity factors that correspond to the three modes are given as follows (see Kanninen & Popelar 1985, among others):

$$K_I = \lim_{r \to 0} \sigma_y|_{\theta=0} \sqrt{2\pi r} \tag{9.8a}$$

$$K_{II} = \lim_{r \to 0} \tau_{xy}|_{\theta=0} \sqrt{2\pi r} \tag{9.8b}$$

$$K_{III} = \lim_{r \to 0} \tau_{yz}|_{\theta=0} \sqrt{2\pi r} \tag{9.8c}$$

where r and θ are defined in Figure 9.4 and σ_y, τ_{xy}, and τ_{yz} are stress components defined in Figure 9.4.

An elastic body with a crack length of $2a$ and under uniform tensile stress, σ, is considered, although much of the treatment in Equations (9.9)–(9.12) is, in fact, more general and is applicable to arbitrary types of loading and crack geometry. The local

(a)   (b)

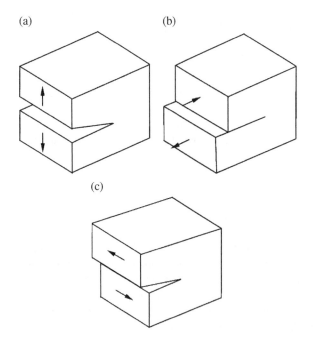

(c)

Figure 9.5 Three basic loading modes for a cracked body: (a) mode I, opening mode; (b) mode II, sliding mode (or in-plane shear mode); (c) mode III, tearing mode (or out-of-plane shear mode).

coordinate system of the body is defined as shown in Figure 9.4. The stresses and displacements at the crack tip may be given as follows (Machida 1984):

$$
\left\{ \begin{array}{c} \sigma_x \\ \sigma_y \\ \tau_{xy} \end{array} \right\} = \frac{K_\mathrm{I}}{\sqrt{2\pi r}} \cos\frac{\theta}{2} \left\{ \begin{array}{c} 1 - \sin(\theta/2)\sin(3\theta/2) \\ 1 + \sin(\theta/2)\sin(3\theta/2) \\ \sin(\theta/2)\cos(3\theta/2) \end{array} \right\}
\tag{9.9a}
$$

$$
\tau_{xz} = \tau_{yz} = 0
\tag{9.9b}
$$

$$
\sigma_z = \left\{ \begin{array}{ll} v(\sigma_x + \sigma_y) & \text{for plane strain state} \\ 0 & \text{for plane stress state} \end{array} \right.
\tag{9.9c}
$$

$$
\left\{ \begin{array}{c} u \\ v \\ w \end{array} \right\} = \frac{K_\mathrm{I}}{2\mu} \sqrt{\frac{r}{2\pi}} \left\{ \begin{array}{c} \cos(\theta/2)\left[\kappa - 1 + 2\sin^2(\theta/2)\right] \\ \sin(\theta/2)\left[\kappa + 1 - 2\cos^2(\theta/2)\right] \\ 0 \end{array} \right\}
\tag{9.9d}
$$

where $\kappa = 3 - 4v$ for the plane strain state and $\kappa = (3 - v)/(1 + v)$ for the plane stress state, $K_\mathrm{I} = \sigma\sqrt{(\pi a)}$, a is the crack length, $\mu = E/[2(1 + v)]$, E is the elastic modulus, v is Poisson's ratio, and u, v, and w are the translational displacements in the x, y, and z directions, respectively.

It is evident from Equation (9.9) that the stress or displacement components at the crack tip include a common parameter, K_I. The relative displacements used represent a distance between the two crack surfaces. This type of displacement is called mode I or the opening mode, as shown in Figure 9.5a.

A cracked body under shear stress τ is now considered. In this case, the stress and displacement components are given as follows:

$$
\left\{ \begin{array}{c} \sigma_x \\ \sigma_y \\ \tau_{xy} \end{array} \right\} = \frac{K_\mathrm{II}}{\sqrt{2\pi r}} \left\{ \begin{array}{c} -\sin(\theta/2)[2 + \cos(\theta/2)\cos(3\theta/2)] \\ \sin(\theta/2)\cos(\theta/2)\cos(3\theta/2) \\ \cos(\theta/2)[1 - \sin(\theta/2)\cos(3\theta/2)] \end{array} \right\}
\tag{9.10a}
$$

$$
\tau_{xz} = \tau_{yz} = 0
\tag{9.10b}
$$

$$
\sigma_z = \left\{ \begin{array}{ll} v(\sigma_x + \sigma_y) & \text{for plane strain state} \\ 0 & \text{for plane stress state} \end{array} \right.
\tag{9.10c}
$$

$$
\left\{ \begin{array}{c} u \\ v \\ w \end{array} \right\} = \frac{K_\mathrm{II}}{2\mu} \sqrt{\frac{r}{2\pi}} \left\{ \begin{array}{c} \sin(\theta/2)[(3-v)/(1+v) + 1 + 2\cos^2(\theta/2)] \\ -\cos(\theta/2)\left[(3-v)/(1+v) - 1 - 2\sin^2(\theta/2)\right] \\ 0 \end{array} \right\}
\tag{9.10d}
$$

where $K_\mathrm{II} = \tau\sqrt{(\pi a)}$. In this case, the displacements follow mode II, or the in-plane shear mode, as shown in Figure 9.5b.

When the body is subjected to uniform shear stress, s, in the direction normal to the xy plane, the stress and displacement components are given by

$$
\left\{ \begin{array}{c} \tau_{xz} \\ \tau_{yz} \end{array} \right\} = \frac{K_\mathrm{III}}{\sqrt{2\pi r}} \left\{ \begin{array}{c} -\sin(\theta/2) \\ \cos(\theta/2) \end{array} \right\}
\tag{9.11a}
$$

$$
\sigma_x = \sigma_y = \sigma_z = \tau_{xy} = 0
\tag{9.11b}
$$

$$w = \frac{2K_{III}}{\mu}\sqrt{\frac{r}{2\pi}}\sin\frac{\theta}{2} \tag{9.11c}$$

$$u = v = 0 \tag{9.11d}$$

where $K_{III} = s\sqrt{\pi a}$. In this case, the displacements follow mode III, or the anti-plane (or out-of-plane) shear mode, as shown in Figure 9.5c.

For an angled crack, similar expressions of stresses and displacements can be relevant (Anderson 1995). When the three modes noted earlier are combined, the stress or displacement components may be given as a sum of those for each mode as follows (Machida 1984):

$$\sigma_{ij}(r,\theta) = \frac{1}{\sqrt{2\pi r}}\left\{K_{I}f_{ij}^{I} + K_{II}f_{ij}^{II} + K_{III}f_{ij}^{III}\right\} \tag{9.12a}$$

$$u_i(r,\theta) = \frac{1}{2\mu}\sqrt{\frac{r}{2\pi}}\left\{K_{I}g_i^{I} + K_{II}g_i^{II} + 4K_{III}g_i^{III}\right\} \tag{9.12b}$$

where $f_{ij}^{I}, f_{ij}^{II}, f_{ij}^{III}$ are the stress functions of θ for modes I, II, and III, respectively, as defined in Equations (9.9)–(9.11), and $g_i^{I}, g_i^{II}, g_i^{III}$ are the displacement functions of θ for modes I, II, and III, respectively, as defined in Equations (9.9)–(9.11).

It is apparent that the K parameters are independent of the coordinate system but are affected by geometric properties and loading conditions (e.g., crack size, dimensions of the structure); they can hence be used as the measure of crack extension resistance at the crack tip as long as the structure remains in the linear elastic regime.

9.3.1 Useful *K* Solutions

In relation to LEFM, calculations of the stress intensity factor involve the largest part of the work. Generally speaking, analytical approaches and numerical approaches can be used to determine the stress intensity factors, and many essentially exploit the relationships between K and the crack tip stress field previously described. Some useful K solutions are now described below.

For plates with typical types of cracks under tensile stress as shown in Figure 9.6, the K value for mode I is approximately given as follows (Broek 1986):

1) Center crack, see Figure 9.6a:

$$K_I = F\sigma\sqrt{\pi a} \tag{9.13a}$$

where

$$F = \left(\sec\frac{\pi a}{b}\right)^{1/2}$$

2) Crack on one side, see Figure 9.6b:

$$K_I = F\sigma\sqrt{\pi a} \tag{9.13b}$$

where

$$F = 30.38\left(\frac{a}{b}\right)^4 - 21.71\left(\frac{a}{b}\right)^3 + 10.55\left(\frac{a}{b}\right)^2 - 0.23\left(\frac{a}{b}\right) + 1.12$$

(a)　　　　　　　(b)　　　　　　　(c)

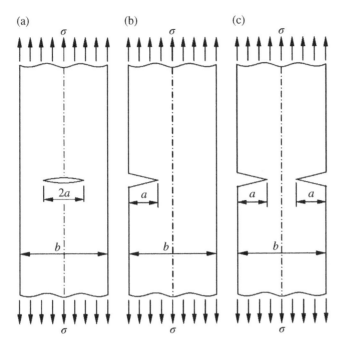

Figure 9.6 Typical crack locations in a plate under tensile stress: (a) center crack; (b) crack on one side; (c) crack on both sides.

3) Crack on both sides, see Figure 9.6c:

$$K_I = F\sigma\sqrt{\pi a} \tag{9.13c}$$

where

$$F = 15.44\left(\frac{a}{b}\right)^3 - 4.78\left(\frac{a}{b}\right)^2 + 0.43\left(\frac{a}{b}\right) + 1.12$$

If the plate width were infinite, the solution in the first of the aforementioned three cases would revert to the classical exact solution, namely, $K_I = \sigma\sqrt{(\pi a)}$ because $F = 1$ in this case.

9.3.2 Fracture Toughness Testing

Recall that fracture occurs if the K value of the structure reaches the critical K value, K_C, namely,

$$K \geq K_C \tag{9.14}$$

where K_C is sometimes called the fracture toughness, which is typically determined experimentally for a given material, crack, and loading situation. Under plane strain conditions, the notation K_{IC} is used. The fracture toughness parameter, K_C or K_{IC}, must be obtained by testing. Widely accepted test methods exist for K_{IC}, as noted in Section 9.2.2.

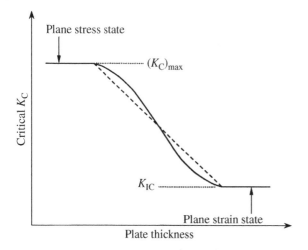

Figure 9.7 A schematic representation of the critical K_I value versus the plate thickness, B.

In such testing, K_C is determined once the ultimate fracture loads (or failure loads) and the crack sizes are obtained for a mechanical test specimen with the stress intensity factor known. In general, the fracture toughness, K_C, is affected by strain rate, temperature, and plate thickness. As the plate thickness decreases, the K_C value tends to significantly increase because the crack tip stress state approaches the plane stress case, and essentially mode II or III based on shear fracture, and mixtures of these with mode I fracture, is more likely to occur than pure mode I fracture.

For thicker plates, mode I fracture associated with the plane strain state is more likely to occur. In this case, the fracture toughness, K_C, is no longer a function of the plate thickness. Figure 9.7 shows a schematic representation of the critical K value at the crack tip versus the plate thickness. For a given plate thickness, in contrast to through-thickness cracks, surface cracks may sometimes exhibit plane strain behavior because of the related conditions at the crack tip.

It is noted that the critical stress intensity value given in this section, which is used as a material parameter to define the fracture toughness, should strictly speaking be referred to as the static fracture toughness because it relates to fracture under static loading. Other fracture toughness measures are useful in other situations, such as dynamic fracture toughness and crack arrest toughness, descriptions of which are beyond the scope of the present book; interested readers are referred, for example, to Broek (1986).

9.4 Elastic–Plastic Fracture Mechanics

As is apparent from Equation (9.13) in LEFM, the failure stress, $\sigma_f = K_{IC}/F\sqrt{(\pi a)}$, at the crack tip becomes infinite when the crack size, a, approaches zero. This is unrealistic because in real structures that behave in a ductile manner, the crack tip is likely to yield, and strictly speaking, LEFM may not be valid. For a body with relatively large flaws, LEFM may, of course, be dealt with approximately using the K values to an extent, as long as the plastic zone at the crack tip is small. The better alternative in this regard is to use the concepts of EPFM. As is now presented, the concepts of CTOD or the J-integral accommodate the effect of yielding at the crack tip in a more rigorous manner. These types of procedures are also variously called nonlinear fracture mechanics or post-yield fracture mechanics, in addition to EPFM.

9.4.1 Crack Tip Opening Displacement

Beyond the general yield condition, plastic deformation is likely to occur at the crack tip. The crack may propagate if the plastic strain at the crack tip exceeds a critical value. The change in the stress at the yielded crack tip may be small when the effect of strain hardening is neglected, and fracture will occur after large plastic deformation occurs at the crack tip. To account for limited amounts of crack tip yielding as an extension of LEFM, Dugdale (1960), Wells (1961, 1963), and Barenblatt (1962) independently introduced cohesive yield strip zones extending from the crack tip to account for the inelastic response of real materials in this region.

The CTOD concept as it is now used emanated from these early treatments. The plastic deformation at the crack tip can be measured in terms of the CTOD. Wells (1961, 1963) considered that fracture occurs if the CTOD exceeds a critical value. In LEFM, the crack opening displacement (COD) is given by (see Figure 9.8)

$$COD = 2v = \frac{4\sigma}{E}\sqrt{a^2 - x^2} \tag{9.15a}$$

The maximum COD occurs at the center of the crack, that is, at $x = 0$, as follows:

$$COD_{max} = \frac{4\sigma a}{E} \tag{9.15b}$$

9.4.1.1 The Irwin Approach

Equation (9.15) is the elastic solution of crack problems, whereas most engineering materials deform plastically. Strictly speaking, Equation (9.15) cannot be applied to the crack problems that involve plastic deformations at the crack tip. The size (distance) of the crack tip plastic zone can be approximately calculated by

$$\frac{K_I}{\sqrt{2\pi r_p^*}} = \sigma_Y \quad \text{or} \quad r_p^* = \frac{K_I^2}{2\pi \sigma_Y^2} = \frac{\sigma^2 a}{2\sigma_Y^2} \tag{9.16a}$$

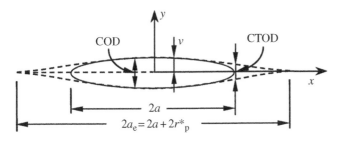

Figure 9.8 Crack opening displacement and CTOD.

where σ_Y is the material yield stress and r_p^* is the size of the crack tip plastic zone as shown in Figure 9.9.

Irwin (1956) assumed that due to the occurrence of plasticity, the equivalent crack tip size becomes longer than the physical size. In this regard, the COD is given by applying the plastic zone correction as follows:

$$\text{COD} = \frac{4\sigma}{E}\sqrt{\left(a+r_p^*\right)^2 - x^2} \tag{9.16b}$$

The CTOD is then found for $x = a$ as follows:

$$\text{CTOD} \equiv \delta = \frac{4\sigma}{E}\sqrt{\left(a+r_p^*\right)^2 - a^2} \approx \frac{4\sigma}{E}\sqrt{2ar_p^*} = \frac{4}{\pi}\frac{K_I^2}{E\sigma_Y} \tag{9.16c}$$

Measurement of CTOD is not straightforward, but it can be obtained from Equation (9.16c) by using the K value. By substituting Equation (9.16c) into Equation (9.16b), the following relationship between COD and CTOD is approximately obtained by regarding $\left(r_p^*\right)^2$ as infinitesimal, namely,

$$\text{COD} = \frac{4\sigma}{E}\sqrt{a^2 - x^2 + \left(\frac{E}{4\sigma}\right)^2 \delta^2} \tag{9.16d}$$

In testing, the COD can be measured easily so that the CTOD is determined from Equation (9.16d) in terms of the maximum COD, that is, at $x = 0$.

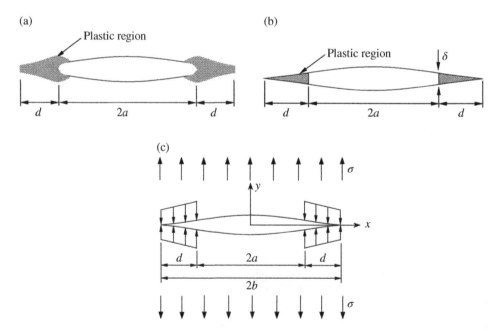

Figure 9.9 A schematic representation of the Dugdale approach (the shaded areas represent the plastic zones): (a) actual plastic zone; (b) idealized plastic zone; (c) stress equilibrium.

9.4.1.2 The Dugdale Approach

Dugdale (1960) treated yielding at the crack tip by replacing the yielded region with the equivalent elastic (unyielded) crack model.

As shown in Figure 9.9a, the crack tip is likely to yield, and the yielding may expand around the crack tip. In the Dugdale approach, however, yielding is assumed to be limited inside a region along the straight line of the crack, as shown in Figure 9.9b. This situation is considered equivalent to a virtual elastic structure with a crack length of $2b = 2a + 2d$, including the yielded region, and the yielded region is subjected to a "negative" internal pressure ("tensile" stress) equal to the yield stress, σ_Y, on the crack surfaces, which tends to "close" the virtual crack strip yielding zone opening caused by the external stress, as shown in Figure 9.9c. In this case, the K value at the tip of the virtual elastic crack must be zero, and hence the following applies:

$$K_\sigma + K_Y = 0 \tag{9.17a}$$

where K_σ is the K value due to applied stress σ and K_Y is the K value due to the closure yield stress, which is taken as $K_Y = -\sigma_Y$.

The extent of the plastic region, d, can be calculated using Equation (9.17a) as follows:

$$d = b - a = a\left[\sec\left(\frac{\pi\sigma}{2\sigma_Y}\right) - 1\right] \tag{9.17b}$$

where σ_Y is the material yield stress.

The CTOD value, δ, at $x = a$ may approximately be considered as the CTOD of the real structure, namely,

$$\delta = \frac{8a\sigma_Y}{\pi E}\ln\left[\sec\left(\frac{\pi\sigma}{2\sigma_Y}\right)\right] \tag{9.17c}$$

When $\sigma \ll \sigma_Y$, that is, representing small-scale yielding, Equation (9.17c) may be simplified to

$$\delta = \frac{\pi\sigma^2 a}{E\sigma_Y} = \frac{G_I}{\sigma_Y} = \frac{K_I^2}{E\sigma_Y} = \frac{J}{\sigma_Y} \tag{9.17d}$$

where G is as defined in Equation (9.2a) and J is the J-integral value as defined in Section 9.4.2.

When the finite element model is applied for EPFM analysis to obtain the CTOD for a structure, the COD value at the crack tip is often approximately taken as the extrapolated value, as shown in Figure 9.10 (Machida 1984).

9.4.1.3 CTOD Design Curve

The CTOD concept is now increasingly applied to fracture control of structural materials. It is also used as a quality control measure in the offshore industry. In applying the CTOD concept, it is considered that ductile fracture under a limited amount of crack tip plasticity occurs if the CTOD, δ, reaches the critical COD value, δ_C. The critical value involved may be determined by established test procedures as a function of factors such as the material properties, the plate thickness, and the crack type.

However, consistent application of the procedure to a real structure necessitates EPFM analysis to model crack tip behavior, for which special-purpose finite elements may be used.

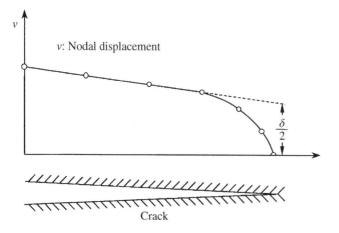

Figure 9.10 Extrapolation of the COD value at the crack tip in finite element analysis.

Because of this consideration, the CTOD has held greater attraction as a quality control aid; the larger the CTOD for a material, the greater its toughness, so to speak. CTOD values for some applications in this regard have been recommended based on successful experience.

The nondimensionalized critical CTOD value, Φ, is normally defined by

$$\Phi = \frac{\delta_C}{2\pi\varepsilon_Y a} \tag{9.18}$$

where δ_C is the critical CTOD value, ε_Y is the elastic yield strain, and a is the half crack length.

To facilitate the applicability of the CTOD to real steel structural design, Burdekin and Dawes (1971) and Dawes (1974) proposed the semiempirical expression of the normalized critical CTOD value as a function of the failure strain, ε_f, as follows:

$$\Phi = \begin{cases} (\varepsilon_f/\varepsilon_Y)^2 & \text{for } \varepsilon_f/\varepsilon_Y \leq 0.5 \\ (\varepsilon_f/\varepsilon_Y) - 0.25 & \text{for } \varepsilon_f/\varepsilon_Y > 0.5 \end{cases} \tag{9.19}$$

Although Equation (9.19) is primarily based on correlation test data for steel plates loaded in tension, it is generally called the CTOD design curve and was in fact fitted to the data to obtain "lower bound" or pessimistic predictions of fracture behavior. Figure 9.11 plots Equation (9.19). In the CTOD design curve approach, a relevant point associated with the applied strain and crack size in a structure along with the critical CTOD value for the material can be plotted in the figure. If the point lies above the design curve, it is considered that the structure is safe; otherwise, fracture is predicted to occur.

To apply the CTOD design curve approach to a complex structure, the British Standards document (BS 1980) suggests that the CTOD measure be based on the maximum total strain in the structural cross section, ε_{max}, which is estimated by

$$\varepsilon_{max} = \frac{1}{E}[K_t(S_m + S_b) + S_s] \tag{9.20}$$

Figure 9.11 The CTOD design curve.

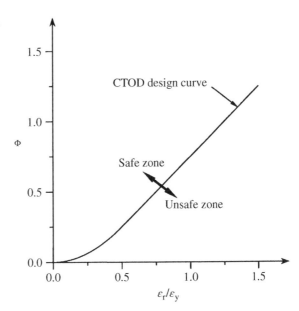

where K_t is the elastic stress concentration factor, S_m is the primary membrane stress, S_b is the primary bending stress, S_s is the secondary stress including thermal or residual stresses, and E is Young's modulus.

9.4.2   Other EPFM Measures: J-Integral and Crack Growth Resistance Curve

9.4.2.1   The J-Integral

The concept of the J-integral is useful in the analysis of ductile fracture mechanics involving small-scale plasticity at the crack tip in a somewhat rigorous way. The basic work on the J-integral was contributed from a theoretical point of view, primarily by Rice (1968) and by Hutchinson (1968). They envisaged a path-independent integral, called the J-integral, calculated along a contour around the tip of the crack, as a parameter that characterized the fracture behavior at the crack tip. The path independence of the integral followed from the principles of energy conservation, and the integral is in theory a nonlinear but elastic concept. Rice showed that the J-integral, when taken around a crack tip, was also equivalent to the change in potential energy for a virtual crack extension.

Without going into detail, within the context of LEFM, the following relationship between the stress intensity factor and the J-integral can be shown to exist, namely,

$$J = G \tag{9.21}$$

Hence, within the range of validity of LEFM, the four fracture parameters thus far introduced are all interrelated. For instance, for plane strain conditions in the "opening" mode (i.e., mode I), it can be shown that

$$G = J = \frac{1 - \nu^2}{E} K_I^2 = \delta \sigma_Y \tag{9.22}$$

In view of Equation (9.22), the issue of which of the four basic parameters involved in LEFM is the "most basic" is of little consequence. However, it turns out that it is considerably more important that *J* is potentially a better parameter to use when it becomes necessary to select the basis of nonlinear fracture mechanics for elastic–plastic conditions.

In seeking a fracture criterion that could predict fracture for both small- and large-scale plasticities, Begley and Landes (1972) recognized that the J-integral provides three distinct attractive features that make it useful for its intended purpose: (i) for linear elastic behavior, it is identical to *G*; (ii) for elastic–plastic behavior, it characterizes the crack tip region and hence would be expected to be equally valid under nonlinear conditions; and (iii) it can be evaluated experimentally in a convenient manner. This last feature follows from the path-independent property of the J-integral and its energy release rate interpretation.

The successful application of the J-integral to a case involving crack tip plasticity depends on two factors: (i) the ability to calculate the integral for a specific crack situation in a structure, which is evident in part from its path-independent property, although a specific detailed analysis may be needed and (ii) the ability to measure the critical value of the J-integral using an appropriate testing method, which will involve crack tip plasticity. Regarding the second factor, recall that we no longer have a convenient relationship between, say, *J* and *K*, as we did in LEFM. Without going into detail, the latter aspect is facilitated by the fact that *J* can be determined from the load versus displacement diagram for crack extension in a specific case, subject to the limitations of the deformation theory of plasticity. For a complete explanation, interested readers may consult Broek (1986).

As an illustration, consider the calculation of the J-integral in an example case in which crack tip plasticity is involved, and the needed experimental results are available for a plate with cracks under tensile loading, as shown in Figure 9.12a. The J-integral value in the elastic–plastic case may be shown to be approximated by

$$J = G + \frac{2}{bB}\left(\int_0^u P\,du - \frac{1}{2}Pu\right) \tag{9.23}$$

where *G* is the strain energy release rate, *B* is the plate thickness, and $\int_0^u P\,du - \frac{1}{2}Pu$ is represented by the shaded area of the force–displacement curve shown in Figure 9.12b.

In Equation (9.23), *G* is normally given as a function of the *K* parameter. For mode I fracture, the strain energy release rate, G_I, is given by

$$G_I = \frac{\kappa + 1}{8\mu}K_I^2 \tag{9.24}$$

where κ and μ are defined in Equation (9.9).

For combined fracture modes, *G* is given by

$$G = \frac{\kappa + 1}{8\mu}\left(K_I^2 + K_{II}^2\right) + \frac{1}{2\mu}K_{III}^2 = \frac{1}{E^*}\left(K_I^2 + K_{II}^2\right) + \frac{1}{2\mu}K_{III}^2 \tag{9.25}$$

where $E^* = E/(1 - v^2)$ for plane strain state, $E^* = E$ for plane stress state, and μ is defined in Equation (9.9).

(a)

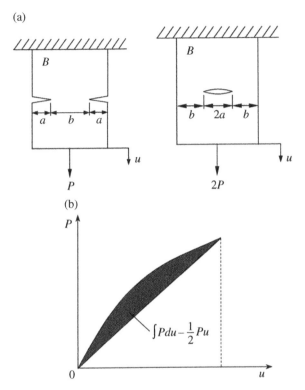

(b)

Figure 9.12 (a) Cracked plates under axial tensile loading; (b) the force-displacement curve of the plate with cracks under tensile loading.

In applying the J-integral criterion to EPFM, it is then considered that fracture occurs if the J-integral value of the structure reaches a critical J-integral value, J_C, namely,

$$J \geq J_C \tag{9.26}$$

9.4.2.2 The Crack Growth Resistance Curve

The concepts described in the previous sections lead to fracture indices that are applicable at the tip of a crack that is not propagating. For high-toughness materials, however, the structure may not reach the ULS immediately after satisfaction of the fracture criteria noted in the previous sections. This is to imply that even after initial crack extension, further substantial stable crack growth can occur until the ULS is reached as a result of unstable crack propagation. This might be true even under plane strain conditions at the crack tip. The concept of the crack growth resistance, termed R (resistance), associated with the potential energy required for crack extension is then useful to assess the characteristics of stable crack growth and to predict the circumstances of failure. The curve that represents the variation of R as a function of incremental crack extension is called the crack growth resistance curve (or R curve).

The use of the J-integral as the crack driving force parameter in a resistance curve approach was broached soon after the establishment of the J-integral as an elastic–plastic fracture parameter by Begley and Landes (1972), but it was the work of Paris and his cow-orkers (1979) that led to the acceptance of this concept. In essence, the crack growth resistance curve concept was simply reformulated as $J_R = J_R(\Delta a)$, where Δa denotes the extent of stable crack growth. Fracture instability then occurs when dJ/da exceeds dJ_R/da. Paris et al. formalized this concept by defining the parameters

$$T \equiv \frac{E}{\sigma_0^2} \frac{dJ}{da} \tag{9.27a}$$

$$T_R \equiv \frac{E}{\sigma_0^2} \frac{dJ_R}{da} \tag{9.27b}$$

where σ_0 is the material's flow stress. The dimensionless parameter, T, is known as the tearing modulus, with its critical value $T_R = T_R(\Delta a)$ taken to be a property of the material.

The Paris et al. concept is illustrated in Figure 9.13, and Figure 9.13a shows a typical J–resistance curve. It is important to recognize that all such relationships have a finite range of applicability. The limit is denoted by the value $(\Delta a)_{lim}$, which can be estimated from the ω parameter introduced by Hutchinson and Paris (1979), which is defined by

$$\omega = \frac{b}{J} \frac{dJ}{da} \tag{9.28}$$

where b denotes the smallest relevant dimension from the crack tip to the boundary of the cracked component.

Hutchinson and Paris showed that $\omega \gg 1$ for the theory to be valid, whereupon some value of $(\Delta a)_{lim}$ will designate the largest amount of crack growth (and J-integral value) for which the theory is valid. Assuming that the fracture instability point would occur before $(\Delta a)_{lim}$ is reached, its determination can be readily found via the J–T diagram shown in Figure 9.13b. Clearly, to use this approach, one requires the J–resistance curve

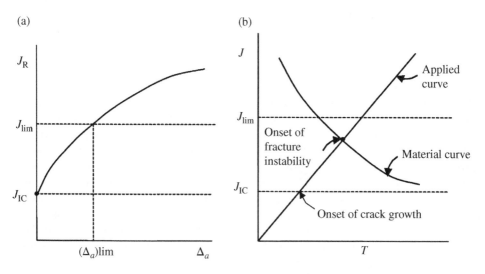

Figure 9.13 A schematic representation of the tearing modulus prediction of fracture instability: (a) the J–resistance curve; (b) the J–T diagram.

(and some means to accurately determine its slope) and estimates of J and T for the crack/structure/load conditions of interest.

The J–resistance curve will be unique only for limited amounts of stable crack growth; otherwise, the curves would exhibit geometry dependence. To overcome this deficiency, the J–resistance curve that properly reflects the degree of plastic constraint at the crack tip must be used. The triaxiality that is known to determine the degree of plastic constraint varies significantly as the primary loading on the remaining ligament changes from tension to bending. Thus, for extended amounts of crack extension, at least two fracture parameters would in principle be required to characterize the intensity of the deformation and the triaxiality.

For more elaborate descriptions, interested readers may refer to Machida (1984), Broek (1986), and Anderson (1995), among others.

9.5 Fatigue Crack Growth Rate and Its Relationship to the Stress Intensity Factor

Distinct from the stable crack extension or growth thus far described, the rate of cyclic growth of a fatigue crack in mode I (direct tensile opening mode) has also been correlated to fracture mechanics parameters such as the stress intensity factor or the energy release rate. This is useful for predicting the growth of a known initial crack when subject to cyclic loading.

Figure 9.14 illustrates the schematic variation in the crack size versus time in a cyclic loading (fatigue) situation, as obtained by the integration of Equation (9.29). The initial crack size, a_0, may be detected by nondestructive examination or close-up surveys with any certainty. The critical crack size, a_c, at failure, usually under the tensile maximum

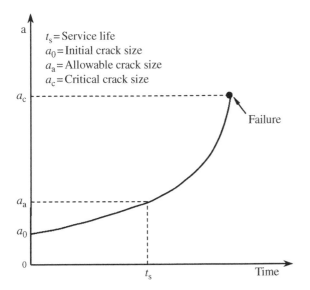

Figure 9.14 A schematic of the crack size variation versus time.

stress, can be computed from LEFM concepts where applicable, or estimated by other means, including the need to avoid severing important structural member ligaments. The maximum "allowable" crack size, a_a, may then be defined by dividing the critical crack size by a safety factor. In this regard, the service life of a structure can be consistently defined as the time during which the crack size grows from its initial size to the maximum allowable size.

In the LEFM approach, it has been recognized that the fatigue crack growth rate can be related to the cyclic elastic stress field at the tip of a long crack subjected to low to intermediate levels of cyclic stress (Paris et al. 1961). Later, investigators found that the crack growth rate curve is not necessarily linear for all the ranges of ΔK on a log–log scale. The general crack growth rate behavior for mode I cracks in metals is usually as shown in Figure 9.15.

The sigmoidal shape of the crack growth curve in Figure 9.15 suggests a subdivision into three regions. In region I, the crack growth rate goes asymptotically to zero as ΔK approaches a threshold value ΔK_{th}. This means that for stress intensities below ΔK_{th}, no crack growth occurs, that is, there is a fatigue limit.

A crack growth relationship for the threshold region was proposed by Donahue et al. (1972) as follows:

$$\frac{da}{dN} = C(\Delta K - \Delta K_{th})^m \tag{9.29}$$

The linear region of the log–log plot (i.e., region II) in Figure 9.15 follows a power law (Paris & Erdogan 1960) as given by

$$\frac{da}{dN} = C(\Delta K)^m \tag{9.30}$$

where da/dN represents the crack growth per cycle, ΔK is the stress intensity range at the crack tip being considered, and C and m are material constants, obtained on the basis of

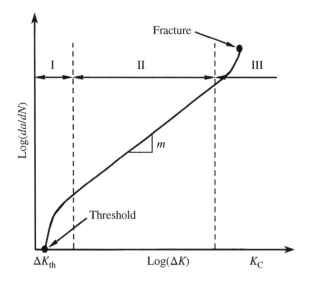

Figure 9.15 Crack growth rate curve shows the three regions.

Table 9.2 Crack growth relationships covering all regions.

Crack growth relation	Proposer
$\dfrac{da}{dN} = C\left(\dfrac{\Delta K - \Delta K_{th}}{K_C - K_{max}}\right)^m$, $\quad \Delta K_{th} = A(1-R)^\gamma$ $0.5 < \gamma < 1.0$	Priddle (1976), Schijve (1979)
$\dfrac{da}{dN} = C\left(\dfrac{\Delta K}{(1-R)^n}\right)^m$, $\quad m = 4, \quad n = 0.5$	Walker (1970)
$\dfrac{da}{dN} = \dfrac{A}{E\sigma_Y}(\Delta K - \Delta K_{th})^2\left(1 + \dfrac{\Delta K}{K_{IC} - K_{max}}\right)$ $\Delta K_{th} = \left(\dfrac{1-R}{1+R}\right)^2 (\Delta K)_0$	McEvily and Groeger (1977)
$\dfrac{da}{dN} = \dfrac{C(1+\beta)^m(\Delta K - \Delta K_{th})^n}{K_C - (1+\beta)\Delta K}$, $\quad \beta = \dfrac{K_{max} + K_{min}}{K_{max} - K_{min}}$	Erdogan (1963)

test data. According to Equation (9.30), the fatigue crack growth rate, *da/dN*, depends only on ΔK and is not sensitive to the *R* ratio as defined in Equation (9.31) in region II. Equation (9.30) is often called the Paris–Erdogan law (or the Paris law).

Region III crack growth exhibits a rapidly increasing growth rate toward "infinity" as the crack size increases, representing either ductile tearing and/or brittle fracture. This behavior led to the relationship proposed by Forman et al. (1967), namely,

$$\frac{da}{dN} = \frac{C(\Delta K)^m}{(1-R)K_C - \Delta K} \tag{9.31}$$

where K_C is the fracture toughness of the material and *R* is the *K* ratio defined by $R = K_{min}/K_{max}$.

Crack growth rate relationships that attempt to combine the behavior at high, intermediate, and low ΔK values also exist. Table 9.2 contains a sample collection of these relationships. Given an initial crack size, when the crack growth rate is known, the crack length at any instant of a structure's life can be calculated via integration if the fatigue loading history is also known. For further details, readers are referred to Broek (1986), among others. Then, given the crack size and other related details, one can study the effects of a fatigue crack on the ultimate strength, as described in the following section.

9.6 Buckling Strength of Cracked Plate Panels

9.6.1 Fundamentals

The buckling behavior of cracked plates has been studied in the literature (e.g., Stahl & Keer 1972, Roy et al. 1990, Shaw & Huang 1990, Riks et al. 1992, Lui 2001, Satish Kumar & Paik 2004, among others). With the elastic buckling stress known, the critical buckling stress may be determined by the plasticity correction using the Johnson–Ostenfeld formulation method, as described in Section 2.9.5.1.

Satish Kumar and Paik (2004) calculated the elastic buckling loads of cracked plates using the hierarchical trigonometric functions proposed by Beslin and Nicolas (1997), where the plates have edge cracks or central cracks with varying the crack size in $2c/b = 0.1$–0.5, in which $2c$ is the crack length, and b is the plate breadth, and they are subjected to uniaxial compressive loads, biaxial compressive loads, or in-plane shear loads.

The plate is discretized into several elements (each of dimension a_i, b_i, and h_i) based on crack location as shown in Figure 9.16a. Each element is represented by nondimensional (ξ–η) coordinate system as shown in Figure 9.16b, such that the location of the element corresponds to $-1 \leq \xi \leq 1$ and $-1 \leq \eta \leq 1$.

The equilibrium equation of the ith element in the x–y coordinate system is given as follows:

$$\frac{\partial^4 w^i}{\partial x^4} + 2\frac{\partial^4 w^i}{\partial x^2 \partial y^2} + \frac{\partial^4 w^i}{\partial y^4} + \frac{N_x \partial^2 w^i}{D\ \partial x^2} + \frac{N_y \partial^2 w^i}{D\ \partial y^2} + 2\frac{N_{xy} \partial^2 w^i}{D\ \partial x \partial y} = 0 \tag{9.32}$$

where w is the lateral deflection of the plate; D is the flexural rigidity of the plate that is defined as $Et^3/[12(1-v^2)]$ where E is the elastic modulus, t is the plate thickness, and v is the Poisson ratio; and N_x, N_y, and N_{xy} are axial loads in the x or y direction and shear loads, respectively.

Equation (9.32) can be transformed in the ξ–η coordinate system for the ith element as follows:

$$\frac{2Db}{a^3}\int_{-1}^{1}\int_{-1}^{1}\left[w,_{\xi\xi}^2 + \left(\frac{a}{b}\right)^4 w^i,_{\eta\eta}^2 + 2vw^i,_{\xi\xi}w^i,_{\eta\eta} + 2(1-v)w^i,_{\xi\eta}^2\right]d\xi d\eta$$

$$+ \left[\int_{-1}^{1}\int_{-1}^{1}\frac{b}{2a}N_x w^i,_{\xi}^2 + \frac{a}{2b}N_y w^i,_{\eta}^2 + N_{xy}w^i,_{\xi\eta}\right]d\xi d\eta = 0 \tag{9.33}$$

where a is the plate length and b is the plate breadth.

(a)   (b)

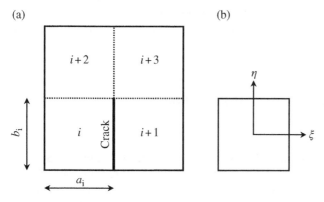

Figure 9.16 (a) Discretization of the plate with vertical edge crack into four elements; (b) ith plate element in (ξ–η) coordinate system.

The local displacement of the ith element is expressed using hierarchical trigonometric functions as follows:

$$w^i(\xi^i, \eta^i) = \sum_{m=1}^{M_i} \sum_{n=1}^{N_i} q_{mn}^i \varphi_m(\xi^i) \varphi_n(\eta^i) \tag{9.34}$$

where $\varphi_m(\xi^i)$ and $\varphi_n(\eta^i)$ are the trial functions, q_{mn} is the amplitude of displacement, M^i and N^i are the number of functions in x- and y-directions, respectively. The trigonometric set $\{\varphi_m(\xi^i)\}$ is defined as follows:

$$\varphi_m(\xi) = \sin(a_m\xi + b_m)\sin(c_m\xi + d_m) \tag{9.35}$$

where coefficients a_m, b_m, c_m, and d_m are given in Table 9.3.

The given functions enable the assembly of elements using displacement and slope compatibility across the boundaries of the connecting elements. Interested readers may consult Barrette et al. (2000). The various boundary conditions of the plate can be satisfied by selectively choosing the first four trigonometric functions (Beslin & Nicolas 1997).

Applying the Rayleigh–Ritz method, the element stiffness matrix of the ith element can be obtained as follows:

$$K_p^i = \left[K_{pmnrs}^i \right] = \frac{4D^i b^i}{a^3} \left[I_{mr}^{22} I_{ns}^{00} + \left(\frac{a^i}{b^i}\right)^4 I_{mr}^{00} I_{ns}^{22} + \nu^i \left(\frac{a^i}{b^i}\right)^2 \left(I_{mr}^{20} I_{ns}^{02} + I_{mr}^{02} I_{ns}^{20} \right) + 2\left(1 - \nu^i\right) \left(\frac{a^i}{b^i}\right) I_{mr}^{11} I_{ns}^{11} \right]$$

$$\tag{9.36}$$

The geometric stiffness matrix of the ith element is then given as follows:

$$K_{Gp}^i = \left[K_{Gpmnrs}^i \right] = \left[N_x \left(\frac{b^i}{a^i}\right) I_{mr}^{11} I_{ns}^{00} + N_y \left(\frac{a^i}{b^i}\right) I_{mr}^{00} I_{ns}^{11} + N_{xy} \left(I_{mr}^{10} I_{ns}^{01} + I_{mr}^{01} I_{ns}^{10} \right) \right]$$

$$\tag{9.37a}$$

where $I_{mr}^{\alpha\beta}$ is defined by the integrals as follows:

$$I_{mr}^{\alpha\beta} = \int_{-1}^{+1} \frac{-d^\alpha \varphi_m(\xi)}{d\xi^\alpha} \frac{-d^\beta \varphi_r(\xi)}{d\xi^\beta} d\xi \tag{9.37b}$$

Table 9.3 Coefficients a_m, b_m, c_m, and d_m relative to trigonometric set $\varphi_m(\xi) = \sin(a_m\xi + b_m)\sin(c_m\xi + d_m)$.

m	a_m	b_m	c_m	d_m
1	$\pi/4$	$3\pi/4$	$\pi/4$	$3\pi/4$
2	$\pi/4$	$3\pi/4$	$\pi/2$	$3\pi/2$
3	$\pi/4$	$-3\pi/4$	$\pi/4$	$-3\pi/4$
4	$\pi/4$	$-3\pi/4$	$\pi/2$	$-3\pi/2$
>4	$\pi/2(m-4)$	$\pi/2(m-4)$	$\pi/2$	$\pi/2(m-4)$

The assembly of the stiffness and geometric stiffness matrices is carried out using the displacement and slope compatibility conditions (Barrette et al. 2000). Using the variational principle to the entire cracked plate, the eigenvalue problem is given as follows:

$$[K] - \lambda_k [K_G] = \{0\} \tag{9.38}$$

where $[K]$ and $[K_G]$ are the global stiffness matrix and the global geometric stiffness matrix of the cracked plate, respectively, and λ_k is the buckling load.

Equation (9.38) can be used to determine the buckling loads of cracked plates. The finite element method solutions are also obtained using the discrete Kirchhoff–Mindlin triangular (DKMT) element (Satish Kumar & Mukhopadhyay 2000) to validate the analytical solutions. The DKMT element is capable to analyze both thin and thick plates due to its numerical flexibility by eliminating the effects of transverse shear in thin plates.

The dimensions of the plates to be analyzed are $a \times b \times t = 1000$ mm $\times$ 1000 mm $\times$ 10 mm in case of square plate and $a \times b \times t = 2000$ mm $\times$ 1000 mm $\times$ 10 mm in case of rectangular plate with aspect ratio of 2. The material properties are $E = 205.8$ GPa and $\nu = 0.3$. The plate is simply supported at all four edges. The buckling loads are presented in a nondimensionalized form in terms of buckling parameter, $\lambda = N_x b^2 / (\pi^2 D)$ for axial compression, or $\lambda = N_{xy} b^2 / (\pi^2 D)$ for edge shear.

9.6.2 A Plate with Edge Crack in Uniaxial Compression

Elastic buckling of a simply supported plate under uniaxial compressive loads is considered when the plate has vertical edge crack or horizontal edge crack. The plate is divided into 2 × 2 elements. Figure 9.17a and b show the variation of buckling loads of a square plate ($a/b = 1$) with vertical and horizontal edge cracks, respectively. Figure 9.17c shows the buckling behavior of a rectangular plate ($a/b = 2$) with varying crack size.

Twelve trigonometric functions (each in x- and y- directions) are used in the present computation. The results using the proposed method are found to compare well with those obtained using the finite element method. Table 9.4 shows the nondimensionalized buckling factors of the first five modes for a square plate with vertical edge crack. The analysis is carried out using eight trigonometric functions and also using 12 trigonometric functions. When the crack size is very small (i.e., $2c/b \le 0.2$), the buckling strength of the plate is the same as that of intact (uncracked) plate. However, when $2c/b > 0.3$, the plate witnessed reduction in buckling strengths. The decrease is found to be significant at higher modes.

It is observed that the loads are estimated accurately with eight trigonometric functions, and there is no significant improvement with increased number of functions. The present analysis is carried out with 2 × 2 mesh irrespective of plate size. In the conventional finite element method, the analysis requires variable number of elements based on plate size. Therefore, the present method provides an efficient solution with a few number of elements and equations, eliminating the rigors involved in modeling the cracked plates.

9.6.3 A Plate with Central Crack in Uniaxial Compression

Figure 9.18a and b show the variation of buckling parameter of a plate with horizontal and vertical central cracks, respectively, under uniaxial compressive loads. The plate is

(a)

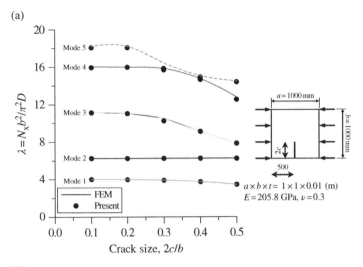

(b)

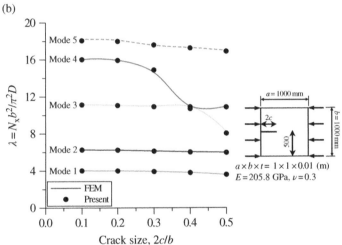

(c)

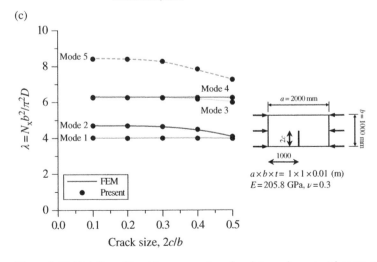

Figure 9.17 Variation of buckling parameter of a plate under uniaxial compressive loads: (a) $a/b = 1$ with vertical edge crack; (b) $a/b = 1$ with horizontal edge crack; (c) $a/b = 2$ with vertical edge crack.

Table 9.4 Nondimensionalized buckling loads, λ of the plate ($a/b = 1$) with vertical edge crack under axial compressive loads.

Crack size, 2c/b	Mode	FEM	Theory (8 × 8)	Theory (12 × 12)
0.1	1	4.01	4.04	4.01
	2	6.26	6.26	6.25
	3	11.15	11.11	11.11
	4	16.01	16.00	15.94
	5	18.21	18.09	18.06
0.2	1	4.00	4.01	3.99
	2	6.26	6.25	6.25
	3	11.07	11.02	11.03
	4	15.99	15.99	15.98
	5	18.20	18.09	18.07
0.3	1	3.94	3.95	3.91
	2	6.26	6.25	6.25
	3	10.53	10.57	10.24
	4	15.84	15.83	15.69
	5	16.42	16.55	15.84
0.4	1	3.76	3.79	3.75
	2	6.26	6.25	6.25
	3	9.14	9.27	9.11
	4	14.74	14.74	14.66
	5	15.06	14.98	14.92
0.5	1	3.44	3.50	3.45
	2	6.25	6.24	6.24
	3	7.82	7.97	7.84
	4	12.77	12.63	12.53
	5	14.49	14.40	14.39

divided into 2 × 3 elements due to the presence of a central crack. The buckling factors are computed using 12 trigonometric functions.

The elastic buckling load is found to reduce with increase in crack size. At higher modes, the reduction is noticed to be significant in case of plate with horizontal central crack. The results obtained by the proposed method compare excellently with those obtained using the finite element method. Table 9.5 shows the buckling loads of a plate with vertical central crack. The loads are computed using 8 and 12 trigonometric functions in each element of the plate. The results are found to converge with eight trigonometric functions.

(a)

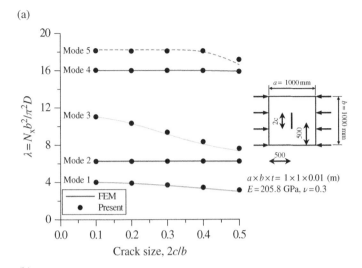

(b)

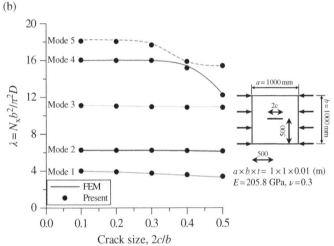

Figure 9.18 Variation of buckling parameter of a plate under uniaxial compressive loads: (a) $a/b = 1$ with vertical central crack; (b) $a/b = 1$ with horizontal central crack.

9.6.4 A Plate with Edge or Central Crack in Edge Shear

Figure 9.19a and b shows the variation of elastic shear buckling parameter of a square plate ($a/b = 1$) with vertical edge or central crack. The plate is divided into 2×2 elements to estimate the buckling load. The buckling parameters are computed using 12 trigonometric functions in each element in x- and y-directions. The results are found to compare excellently with those obtained by the finite element method. The increase in the number of trigonometric functions did not show any influence in convergence. Figure 9.19c shows the variation of buckling parameter in a rectangular plate ($a/b = 2$) with vertical edge crack under in-plane shear load.

Table 9.5 Nondimensionalized buckling loads, λ of the plate ($a/b = 1$) with vertical central crack under uniaxial compressive loads.

Crack size, 2c/b	Mode	FEM	Theory (8 × 8)	Theory (12 × 12)
0.1	1	3.99	4.01	3.99
	2	6.26	6.24	6.23
	3	11.02	11.04	11.00
	4	16.01	16.04	16.01
	5	18.21	18.13	18.11
0.3	1	3.64	3.71	3.63
	2	6.26	6.25	6.25
	3	9.14	9.37	9.07
	4	16.00	16.01	16.00
	5	18.20	18.09	18.06
0.5	1	2.98	3.15	3.03
	2	6.24	6.24	6.23
	3	7.35	7.62	7.45
	4	15.90	15.89	15.86
	5	16.60	17.16	16.66

9.6.5 A Plate with Vertical Edge Crack in Biaxial Compression

A simply supported square plate with vertical edge crack as shown in Figure 9.20 is analyzed under biaxial compressive loads, where the crack size is varied with $2c/b = 0.3$ and 0.5. The plate is divided into 2×2 elements, and the analysis is carried out using 8 trigonometric functions and also 12 trigonometric functions. Table 9.6 indicates the buckling parameter by comparison with the analytical solutions and the finite element method solutions. There is no significant variation in the results obtained using 8 trigonometric functions and those obtained using 12 functions. The buckling loads are found to decrease with increase in crack size as expected.

9.7 Ultimate Strength of Cracked Plate Panels

9.7.1 Fundamentals

It is well recognized that plated structures may have suffered age related deterioration such as fatigue cracking damage over time. The theories and methodologies described earlier have primarily focused on how to characterize the initiation and propagation of fatigue cracks due to repeated loading. As the cracking damage can reduce the ultimate strength of a structure under monotonic extreme loading as illustrated in Figure 9.1, it should be dealt with as a parameter of influence in the residual ultimate strength calculations of a cracked structure associated with Equation (1.17).

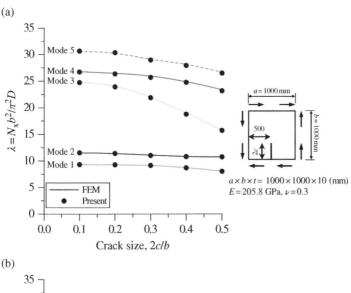

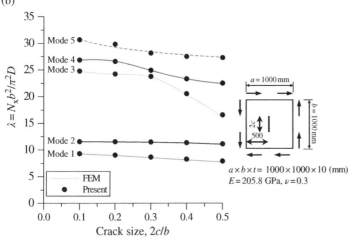

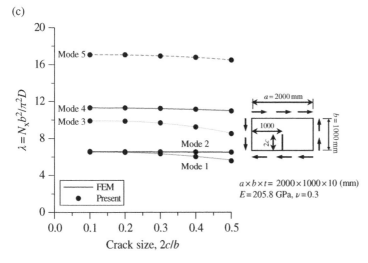

Figure 9.19 Variation of buckling parameter of a plate under edge shear: (a) $a/b = 1$ with vertical edge crack; (b) $a/b = 1$ with vertical central crack; (c) $a/b = 2$ with vertical edge crack.

Fatigue cracking is often observed in a stiffened plate structure along the weld intersection between the plating and stiffeners. The orientations of such cracking may be classified into three groups, namely, vertical, horizontal, and angular types, as shown in Figure 9.21a. The structure can of course be subjected to compressive "extreme" loads arising from axial compression or edge shear as shown in Figure 9.21a or tensile "extreme" loads as shown in Figure 9.21b.

In monotonic tensile loading, the crack size (length) may further increase (propagate) in a stable or unstable manner until the structure reaches the ULS, and therefore the ultimate tensile strength of a cracked structure must be less than that of an uncracked (intact) structure. On the other hand, a crack in a plate panel under monotonic compressive loading may close in the beginning, but it may open if buckling resulting in lateral deflection occurs. It is also possible that some lateral deflection may take place, because of either fabrication related initial deflections or additional local out-of-plane loading (lateral pressure). In cases with lateral deflection, the crack can open and reduce the plate panel collapse strength as out-of-plane deformation increases in compressive loading. It may therefore be pessimistically assumed that the effect of the cracking damage on the panel ultimate compressive strength is similar to that on the panel ultimate tensile strength.

In contrast to the methods described in the previous sections, therefore, one may postulate a simpler and intuitive model to predict the residual ultimate strength of a structure with premised cracks if one presupposes a very ductile material. For a plate panel

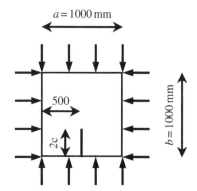

Figure 9.20 A square plate with vertical edge crack under biaxial compressive loads.

Table 9.6 Nondimensionalized buckling loads, λ of the plate ($a/b = 1$) with vertical central crack under biaxial compressive loads.

Crack size, $2c/b$	Mode	FEM	Theory (8×8)	Theory (12×12)
0.3	1	1.97	1.98	1.96
	2	4.89	4.89	4.83
	3	5.01	5.00	5.00
	4	7.99	7.97	7.97
	5	9.20	9.22	9.63
0.5	1	1.77	1.79	1.77
	2	4.56	4.56	4.54
	3	4.97	4.96	4.95
	4	6.35	6.28	6.22
	5	7.20	7.31	7.22

(a)

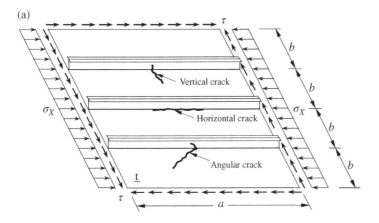

(b)

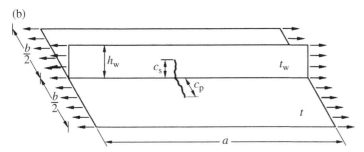

Figure 9.21 (a) Three types of fatigue cracking damage in compressive loads; (b) a stiffened panel with existing crack in tensile loads.

with a premised crack as shown in Figure 9.22, one may predict the ultimate strength on the basis of the reduced cross-sectional area, taking into account the loss of load-carrying material due to the crack damage. In this case, the ultimate strength of a plate panel with existing cracks and under monotonic extreme axial loading may be approximately obtained by (Paik et al. 2005, Paik & Satish Kumar 2006, Paik 2008, 2009)

$$\sigma_u = \frac{A_o - A_c}{A_o}\sigma_{uo} \tag{9.39}$$

Figure 9.22 A schematic of a plate with existing crack.

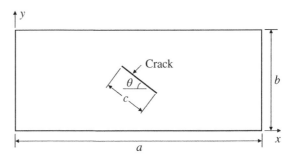

where σ_u and σ_{uo} are the ultimate strengths of the cracked and original (uncracked) plate panel, A_o is the original (uncracked) cross-sectional area, and A_c is the cross-sectional areas associated with crack damage, projected to the loading direction.

The ultimate strength of a plate with cracking damage can then be calculated as follows:

$$\sigma_{xu} = \frac{A_{xo} - A_{xc}}{A_{xo}} \sigma_{xuo} \tag{9.40a}$$

$$\sigma_{yu} = \frac{A_{yo} - A_{yc}}{A_{yo}} \sigma_{yuo} \tag{9.40b}$$

where $A_{xo} = bt$, $A_{xc} = tc\sin\theta$, $A_{yo} = at$, $A_{yc} = tc\cos\theta$, and t is the plate thickness.

For edge shear loading, the ultimate shear strength of a cracked plate may be given by (Paik et al. 2003)

$$\tau_u = \frac{1}{2}\left(\frac{A_{xo} - A_{xc}}{A_{xo}} + \frac{A_{yo} - A_{yc}}{A_{yo}}\right)\tau_{uo} \tag{9.40c}$$

where τ_u and τ_{uo} are the ultimate shear strengths of a cracked or uncracked plate.

The crack damage model of Equation (9.39) can be applied to the ultimate strength calculations of a stiffened panel with cracking damage in longitudinal axial tension, as shown in Figure 9.21b, as follows:

$$\sigma_u = \frac{\left(b - c_p\right)t\sigma_{Yp} + \left(h_w - c_s\right)t_w\sigma_{Ys}}{bt + h_w t_w} \tag{9.41}$$

where c_p is the projected crack length for the plating, c_s is the projected crack length for the stiffener, $\sigma_{Yp} = \sigma_Y$ is the yield strength of the plating, and σ_{Ys} is the yield strength of the stiffener.

9.7.2 A Cracked Plate in Axial Tension

The applicability of Equation (9.39) may be verified by comparison to experiments on plates with an existing crack in axial tensile loads (Paik et al. 2005). In the structural testing related to residual strength, a small hole is first created mechanically at either the center or the edge of the plate, and axial fatigue loading is then applied in the plane of the plate until a crack of a desired size is achieved. The aim of this process is to embody the fatigue crack-like damage in the plate. Finally, controlled displacements that correspond to different levels of monotonic uniaxial tensile loads are applied and are progressively increased in a quasi-static condition until the cracked plate is split into two pieces, as shown in Figure 9.23. Because of the thin plates and the quasi-static loading rates involved, the overall test plate behavior in these room temperature tests carried out under controlled displacement conditions essentially remains ductile.

Figure 9.24 shows the variation of the ultimate tensile strength of plates as a function of the crack length as obtained by the experiments and by the simplified model noted earlier, that is, reducing the cross-sectional area to account for the crack damage.

(a)   (b)

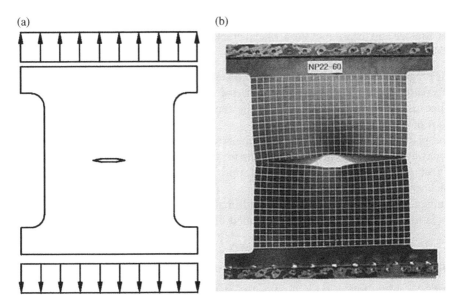

Figure 9.23 A typical pattern of the crack extension immediately before the plate is split into two pieces in axial tension.

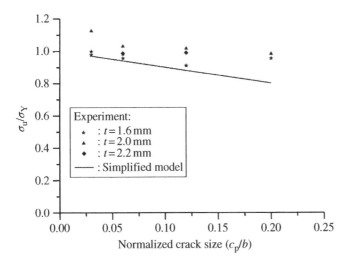

Figure 9.24 Variation of the ultimate tensile strength as a function of the crack length, as obtained by experiment and by the simplified model, with σ_Y being the measured yield stress.

It is evident that the simplified model provides adequate results at some pessimistic side. However, there is an apparent tendency for the experimental data points to lie above the simplified model prediction, as shown in Figure 9.23, and the degree of deviation seems to increase with the plate thickness. This perhaps is a manifestation of tearing and/or strain hardening.

In some of the aforementioned tests, carried out under monotonically increasing tensile loads, the size and location of the artificial cracks and the plate thickness were varied. Some additional insights so obtained are as follows:

1) As the crack length increases, the ultimate tensile strength and the elongation to failure decrease significantly because the cross-sectional area is reduced by the crack and also because the initial crack grows as the tensile load increases, that is, a certain amount of tearing occurs.
2) As the plate thickness increases, the elongation slightly increases. Although more data may be necessary to generalize results for thin plates to relatively thicker plates, the effect of plate thickness on strength is approximately supposed to be proportional to thickness.
3) The effect of the crack location on the ultimate tensile strength of the overall test plates is not significant under the controlled displacement conditions used.

9.7.3   A Cracked Stiffened Panel in Axial Tension

To validate the applicability of the simplified model for a stiffened panel with an existing crack under axial tensile loads, the nonlinear finite element solutions are compared with the predictions from Equation (9.41). The panel dimensions, with the nomenclature shown in Figure 9.21b, are $a = 1600$ mm, $b = 400$ mm, $t = 15$ mm, $h_w = 150$ mm, and $t_w = 12$ mm, and the material properties are $\sigma_{Yp} = \sigma_{Ys} = 249.7$ MPa, $E = 202.2$ GPa, and $v = 0.3$. Two sets of crack lengths are considered, namely, $2c_p = 50$ mm or 150 mm at the plating (i.e., $c_p = 25$ mm or 75 mm on each side of the stiffener) and $c_s = 25$ mm or 75 mm at the stiffener.

In the nonlinear finite element method analysis under displacement control, two types of panel material models are used: (i) the stress–strain relationship of the panel material obtained by the tensile coupon test is used, meaning that the effects of strain hardening, necking, and ductile fracture are considered to the extent possible, and (ii) the elastic–perfectly plastic material model is used by neglecting the effect of strain hardening but taking into account ductile fracture. For all analyses, the effects of crack extension by tearing are taken into consideration as the external tensile loads monotonically increase.

Figure 9.25 shows a typically deformed shape immediately before the panel is entirely fractured. Figure 9.26 shows the nonlinear finite element method solutions for the average stress–strain relation for the stiffened panel under monotonically applied axial tensile loads until the panel is broken into two pieces.

Applying the simplified model indicated in Equation (9.41), the ultimate tensile strength of the stiffened panel can be predicted by $P_u = (b - 2c_p)t\sigma_{Yp} + (h_w - c_s)t_w\sigma_{Ys} = 3183.7$ kN or $\sigma_u = 230.7$ MPa for $c_p = c_s = 25$ mm, and $P_u = 2659.3$ kN or $\sigma_u = 192.7$ MPa for $c_p = c_s = 75$ mm. Table 9.7 compares the solutions by the simplified model with the nonlinear finite element method solutions. Although the ultimate strengths predicted by the simplified model are between 74 and 78% of the nonlinear finite element method solutions for material model (i) noted earlier, they are between 91 and 94% of the nonlinear finite element method solutions for material model (ii). The pessimism of the simplified model of Equation (9.41) is partly because the effect of strain hardening is neglected and perhaps partly also due to stable crack extension. It is evident, at least in these illustrative cases, that the simplified crack model can be useful to predict the ultimate tensile strength of stiffened panels with existing crack damage, but on the pessimistic side.

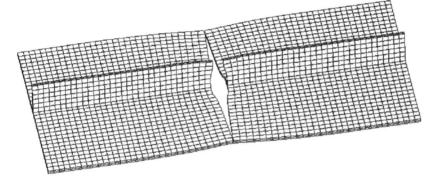

Figure 9.25 Deformed shape immediately before the entire fracture of the stiffened panel under monotonic tensile loads, as obtained by the nonlinear finite element method.

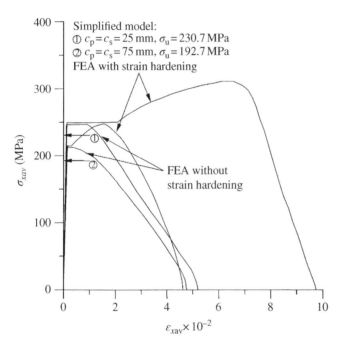

Figure 9.26 The elastic–plastic behavior of a stiffened panel with premised cracks and under monotonic tensile loads.

Table 9.7 Comparison of the simplified model solutions with FEA.

		σ_u (MPa)			
c_p (mm)	c_s (mm)	(A)	(B)	$\dfrac{(A)}{(B)}$	Effect of strain hardening
25	25	230.7	311.4	0.741	Included
			246.8	0.935	Not included
75	75	192.7	247.8	0.778	Included
			212.9	0.905	Not included

Note: (A), simplified model; (B), FEM. The simplified model does not account for the effect of strain hardening.

9.7.4 A Cracked Plate in Axial Compression

The applicability of Equation (9.39) for a cracked plate under axial compressive loads is validated by an experiment performed by Paik et al. (2005). The tests were undertaken on box column models with premised cracks, as shown in Figure 9.27. The test structure is fabricated by tack welding with four sheets of similar steel plates as an assembly. A set of the same through-thickness cracking is artificially made in each of the four plates of the test structure.

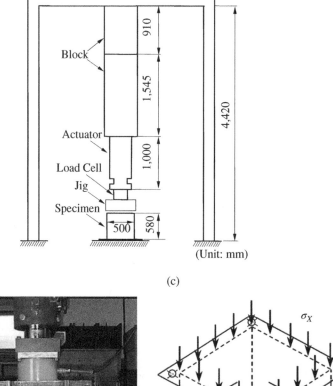

Figure 9.27 The ultimate strength test setup of the box type plated structure with cracking damage and under axial compressive loads: (a) a schematic view of the test setup; (b) a photo of the test setup; (c) a schematic of the test structure.

Figure 9.28 shows a schematic of the test structure with cracking damage and under axial compressive loads, which includes the structural dimensions. Three types of crack locations are considered, namely, VC-Center, VC-Edge(1), and VC-Edge(2). The VC-Center represents that the crack is located at the plate center. VC-Edge(1) and VC-Edge(2) have the edge cracking where VC-Edge(1) has a crack on one side of the plate and VC-Edge(2) has a crack on each side of the plate. It is considered that the ultimate strength behavior of a plate with cracking damage may also be affected by the size of gap between crack faces, denoted by G. Therefore, two types of crack gap sizes, namely, 0.3 and 3.0 mm, are considered in the test. The crack with the smaller gap (i.e., 0.3 mm) is made by wire cutting method, while it is made by plasma cutting for crack of 3.0 mm gap. The material of the test structure is mild steel. Figure 9.29 shows the engineering stress

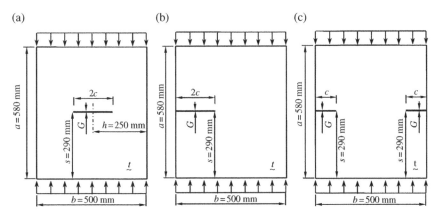

Figure 9.28 Various crack locations in the test structure: (a) vertical crack-center (VC-Center); (b) vertical crack-edge(1) (VC-Edge(1)); (c) vertical crack-edge(2) (VC-Edge(2)).

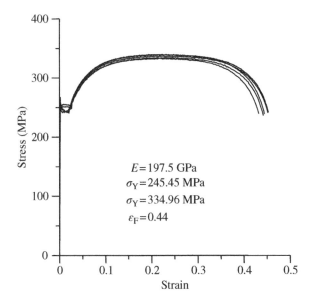

$$E = 197.5 \text{ GPa}$$
$$\sigma_Y = 245.45 \text{ MPa}$$
$$\sigma_Y = 334.96 \text{ MPa}$$
$$\varepsilon_F = 0.44$$

Figure 9.29 Engineering stress-engineering strain curves of the material as obtained by the tensile coupon test (ε_F, fracture strain; σ_T, ultimate tensile stress; σ_Y, yield stress; E, elastic modulus).

versus engineering strain curves of the material as obtained by tensile coupon test. Table 9.8 indicates the maximum initial deflections measured for each of the four plates in the test structure and Table 9.9 identifies the test models.

Figure 9.30 shows the average axial compressive stress–strain curves of the test structures with varying the crack sizes or locations obtained by the experiment. An average value of the ultimate compressive strength of the four individual plates making up the

Table 9.8 Maximum initial deflection measured for each of the four plates in the test structure.

Crack type	A1 (mm)	A2 (mm)	A3 (mm)	A4 (mm)	Average
Intact-1	1.15	1.88	2.81	1.26	1.78
Intact-2	1.10	0.80	0.96	1.80	1.17
VC-Center-0.3-15-1	1.71	2.73	0.68	1.93	1.76
VC-Center-0.3-15-2	2.10	1.50	1.50	1.52	1.66
VC-Center-0.3-30-1	0.50	1.15	0.95	0.79	0.85
VC-Center-0.3-30-2	1.41	1.05	0.90	0.65	1.00
VC-Center-3.0-50	1.95	0.34	0.90	0.53	0.93
VC-Edge(1)-3.0-15	1.06	2.03	1.20	2.21	1.63
VC-Edge(1)-3.0-30	0.58	1.59	1.60	3.38	1.79
VC-Edge(1)-3.0-50	2.40	1.70	0.58	1.76	1.61
VC-Edge(2)-0.3-30	2.1	1.75	3.10	2.28	2.31
VC-Edge(2)-3.0-30	0.83	0.35	1.38	1.1	0.92
VC-Edge(2)-3.0-50	0.80	1.68	0.70	0.80	1.00

Note: A1, maximum initial deflection of plate 1; A2, maximum initial deflection of plate 2; A3, maximum initial deflection of plate 3; A4, maximum initial deflection of plate 4.

Table 9.9 Identification number of the test structures with cracking damage.

| Location of crack | Gap of crack, G (mm) | Size of crack | | |
		0.15b	0.3b	0.5b
Center	0.3	VC-Center-0.3-15	VC-Center-0.3-30	–
	3.0	–	–	VC-Center-3.0-50
Edge (one side)	0.3	–	–	–
	3.0	VC-Edge(1)-3.0-15	VC-Edge(1)-3.0-30	VC-Edge(1)-3.0-50
Edge (both side)	0.3	–	VC-Edge(2)-0.3-30	–
	3.0	–	VC-Edge(2)-3.0-30	VC-Edge-3.0-50

(a)

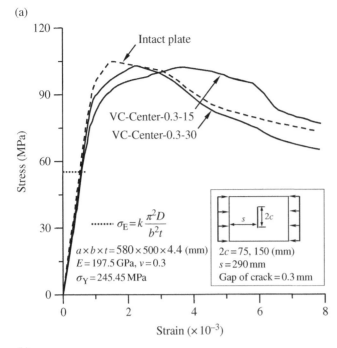

(b)

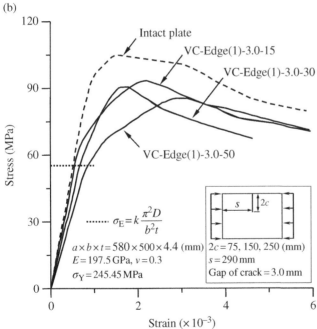

Figure 9.30 Average axial compressive stress–strain curves of the test structures: (a) center crack of varying crack sizes (gap size = 0.3 mm); (b) one side edge crack of varying crack sizes (gap size = 3.0 mm); (c) both side edge crack of varying crack sizes (gap size = 3.0 mm); (d) both side edge crack of varying gap sizes (2c = 150 mm); (e) edge crack on one or both sides (gap size = 3.0 mm); (f) varying locations of edge crack (gap size = 3.0 mm).

(c)

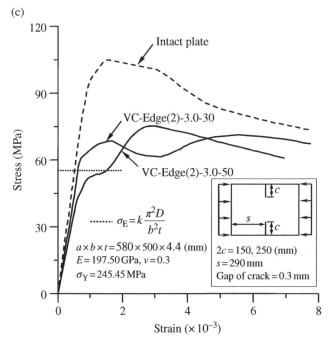

(d)

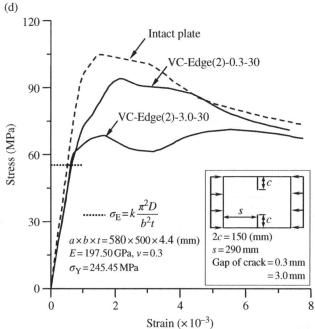

Figure 9.30 (Continued)

(e)

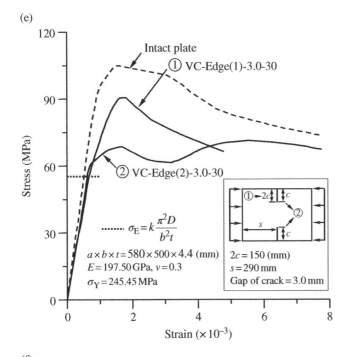

(f)

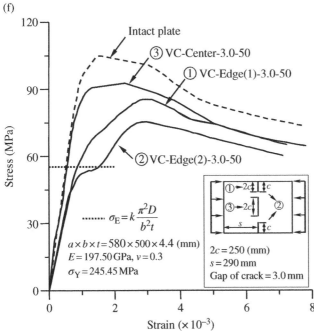

Figure 9.30 (Continued)

Table 9.10 Ultimate strength of cracked plates obtained by the experiment.

Crack type	σ_{xu} (MPa)	σ_{xu}/σ_Y	σ_{xu}/σ_{xuo}
Intact	105.3	0.429	1.0
VC-Center-0.3-15	102.27	0.417	0.971
VC-Center-0.3-30	102.89	0.419	0.977
VC-Center-3.0-50	92.65	0.377	0.880
VC-Edge(1)-3.0-15	93.38	0.380	0.887
VC-Edge(1)-3.0-30	90.55	0.369	0.860
VC-Edge(1)-3.0-50	84.40	0.344	0.802
VC-Edge(2)-0.3-30	94.11	0.383	0.894
VC-Edge(2)-3.0-30	68.60	0.279	0.651
VC-Edge(2)-3.0-50	53.64	0.219	0.509

Note: σ_{xuo} and σ_{xu} are the ultimate compressive strength of the intact or cracked structures.

test structure is then calculated by the applied loads divided by the total cross-sectional area of the structure. Table 9.10 summarizes the ultimate strength of cracked plates with varying crack size and location, obtained from the experiment.

The ultimate strength predictions for the test structures are made using Equation (9.39). Figure 9.31a shows the ultimate compressive strength reduction characteristics of plates due to varying crack size, obtained from the experiment. In Figure 9.31b, a comparison of the ultimate compressive strength of cracked plates is made between the experiment and the simplified formula of Equation (9.39). Figure 9.32 shows a sample finite element model for a plate with edge crack at one side and under axial compressive loads. Figure 9.33 compares the predictions of Equation (9.39) with the experimental results and nonlinear finite element method solutions for a plate with a single crack or multiple cracks. It is apparent that Equation (9.39) reasonably well predicts the ultimate compressive strength of cracked plates at somewhat pessimistic side.

Interested readers may also refer to Paik (2008, 2009), Wang et al. (2009), Shi and Wang (2012), Rahbar-Ranji and Zarookian (2014), Underwood et al. (2015), Wang et al. (2015), Cui et al. (2016, 2017), and Shi et al. (2017), among others.

9.7.5 A Cracked Plate in Edge Shear

To validate the applicability of the simplified model, nonlinear finite element method solutions are compared with the predictions from Equation (9.40c). Figure 9.34 shows the variation of the ultimate shear strength of a cracked plate as a function of the crack length as obtained by the nonlinear finite element method solutions and by the simplified model of Equation (9.40c), that is, reducing the cross-sectional area to account for the crack damage. It is evident from Figure 9.34 that the simplified model provides adequate results at some pessimistic side although there is an apparent

(a)

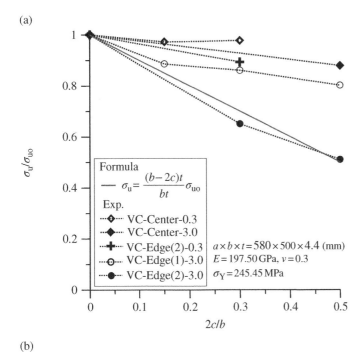

(b)

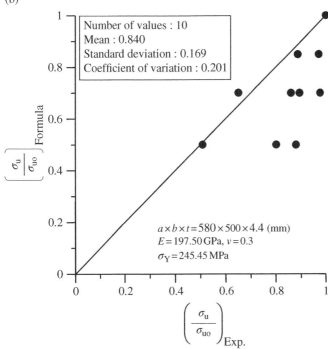

Figure 9.31 (a) The ultimate compressive strength reduction characteristics of steel plates as a function of the crack size; (b) a comparison between the experiment and Equation (9.39).

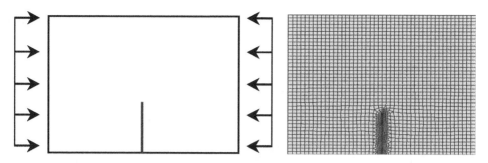

Figure 9.32 A sample finite element mesh in a plate with one edge crack and under axial compression.

(a)

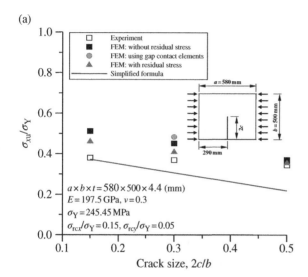

(b)

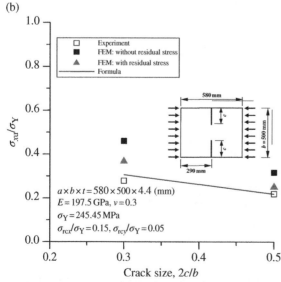

Figure 9.33 Variation of the ultimate compressive strength for a cracked plate, plotted versus the crack size (a) with a single edge crack and (b) with multiple cracks.

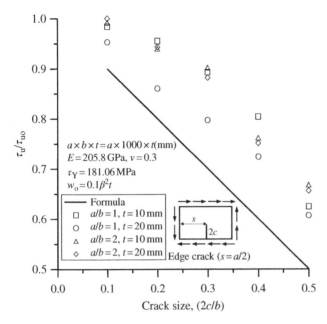

Figure 9.34 Variation of the ultimate shear strength of a cracked plate, plotted versus the crack size with varying the plate thickness and aspect ratio (solid line: Equation (9.40c), symbols: nonlinear finite element method solutions, $\beta = (b/t)\sqrt{\sigma_Y/E}$, the total length of crack is denoted by 2c).

tendency for the nonlinear finite element method solutions to lie above the simplified model prediction.

For the ultimate shear strength of a stiffened panel with multiple cracks, interested readers may refer to Wang et al. (2015), among others.

References

Anderson, T.L. (1995). *Fracture mechanics: fundamentals and applications*. Second Edition, CRC Press, London.

Barenblatt, G.I. (1962). The mathematical theory of equilibrium cracks in brittle fracture. *Advances in Applied Mechanics*, **7**: 55–129.

Barrette, M., Berry, A. & Beslin, O. (2000). Vibration of stiffened plates using hierarchical trigonometric functions. *Journal of Sound and Vibration*, **235**(8): 727–747.

Barsom, J.M. (ed.) (1987). *Fracture mechanics retrospective: early classic papers (1913–1965)*, The American Society of Testing and Materials (RPS), **1**. ASTM, Philadelphia, PA.

Begley, J.A. & Landes, J.D. (1972). *The J-integral as a fracture criterion*, ASTM STP, **514**. The American Society for Testing and Materials, Philadelphia, PA, 1–20.

Beslin, O. & Nicolas, J. (1997). A hierarchical functions set for predicting very high order plate bending modes with any boundary conditions. *Journal of Sound and Vibration*, **202**(5): 633–655.

Broek, D. (1986). *Elementary engineering fracture mechanics.* Martinus Nijhoff, Dordrecht/ Boston/Lancaster.

BS (British Standards Institution) (1980). *Guidance on some methods for the derivation of acceptance levels for defects in fusion welded joints*, PD, **6493**. British Standards Institution, London.

Burdekin, F.M. & Dawes, M.G. (1971). Practical use of linear elastic and yielding fracture mechanics with particular reference to pressure vessels. *Proceedings of the Institute of Mechanical Engineers Conference*, London, May: 28–37.

Cui, C., Yang, P., Li, C. & Xia, T. (2017). Ultimate strength characteristics of cracked stiffened plates subjected to uniaxial compression. *Thin-Walled Structures*, **113**: 27–39.

Cui, C., Yang, P., Xia, T. & Du, J. (2016). Assessment of residual ultimate strength of cracked steel plates under longitudinal compression. *Ocean Engineering*, **121**: 174–183.

Dawes, M.G. (1974). Fracture control in high yield strength weldments. *Welding Journal*, **53**: 369–380.

Donahue, R.J., Clark, H.M., Atanmo, P., Kumble, R. & McEvily, A.J. (1972). Crack opening displacement and the rate of fatigue crack growth. *International Journal of Fracture Mechanics*, **8**: 209–219.

Dugdale, D.S. (1960). Yielding of steel sheets containing slits. *Journal of the Mechanics and Physics of Solids*, **8**: 100–108.

Erdogan, F. (1963). Stress intensity factors. *Journal of Applied Mechanics*, **50**: 992–1002.

Forman, R.G., Kearney, V.E. & Engle, R.M. (1967). Numerical analysis of crack propagation in cyclic-loaded structures. *Journal of Basic Engineering*, **89**: 459–464.

Griffith, A.A. (1920). The phenomena of rupture and flow in solids. *Philosophical Transactions, Series A*, **221**: 163–198.

Hutchinson, J.W. (1968). Singular behavior at the end of a tensile crack in a hardening material. *Journal of the Mechanics and Physics of Solids*, **16**:13–31.

Hutchinson, J.W. & Paris, P.C. (1979). *Stability analysis of J-controlled crack growth*, ASTM STP, **668**. The American Society for Testing and Materials, Philadelphia, PA, 37–64.

Irwin, G.R. (1948). *Fracture dynamics: fracturing of metals.* The American Society for Metals, Cleveland, OH, 147–166.

Irwin, G.R. (1956). Onset of fast crack propagation in high strength steel and aluminum alloys. *Singapore Research Conference Proceedings*, **2**: 289–305.

Irwin, G.R. (1957). Analysis of stresses and strains near the end of a crack traversing a plate. *Journal of Applied Mechanics*, **24**: 361–364.

Kanninen, M.F. & Popelar, C.H. (1985). *Advanced fracture mechanics.* Oxford University Press, New York.

Lotsberg, I. (2016). *Fatigue design of marine structures.* Cambridge University Press, Cambridge.

Lui, F.L. (2001). Differential quadrature element method for buckling analysis of rectangular Mindlin plates having discontinuities. *International Journal of Solids and Structures*, **38**: 2305–2321.

Machida, S. (1984). Ductile Fracture Mechanics. Nikkan Kogyo Shimbunsha (Daily Engineering Newspaper Company), Tokyo (in Japanese).

McEvily, A.J. & Groeger, J. (1977). On the threshold for fatigue-crack growth. The Fourth International Conference on Fracture, University of Waterloo Press, Waterloo, Ontario, **2**: 1293–1298.

Mott, N.F. (1948). Fracture of metals: theoretical considerations. *Journal of Engineering*, **165**: 16–18.

Murakami, Y. (ed.) (1987). *Stress intensity factors handbook*. Pergamon Press, New York.

Orowan, E. (1948). Fracture and strength of solids. *Reports on Progress in Physics*, **XII**: 185–232.

Paik, J.K. (2008). Residual ultimate strength of steel plates with longitudinal cracks under axial compression: experiments. *Ocean Engineering*, **35**(17–18): 1775–1783.

Paik, J.K. (2009). Residual ultimate strength of steel plates with longitudinal cracks under axial compression: nonlinear finite element method investigations. *Ocean Engineering*, **36**(3–4): 266–276.

Paik, J.K. & Satish Kumar, Y.V. (2006). Ultimate strength of stiffened panels with cracking damage under axial compression or tension. *Journal of Ship Research*, **50**(3): 231–238.

Paik, J.K., Satish Kumar, Y.V. & Lee, J.M. (2005). Ultimate strength of cracked plate elements under axial compression or tension. *Thin-Walled Structures*, **43**: 237–272.

Paik, J.K., Wang, G., Thayamballi, A.K., Lee, J.M. & Park, Y.I. (2003). Time-dependent risk assessment of aging ships accounting for general/pit corrosion, fatigue cracking and local denting damage. *Transactions of the Society of Naval Architects and Marine Engineers*, **111**: 159–197.

Paris, P.C. & Erdogan, F. (1960). A critical analysis of crack propagation law. *Journal of Basic Engineering*, **85**(4): 528–534.

Paris, P.C., Gomez, M.P. & Anderson, W.P. (1961). A rational analytical theory of fatigue. *The Trend in Engineering*, **13**: 9–14.

Paris, P.C., Tada, H., Zahoor, A. & Ernst, H.A. (1979). *Instability of the tearing mode of elastic–plastic crack growth*, ASTM STP, **668**. The American Society for Testing and Materials, Philadelphia, PA, 5–36 and 251–265.

Priddle, E.K. (1976). High cycle fatigue crack propagation under random and constant amplitude loadings. *International Journal of Pressure Vessels and Piping*, **4**: 89–117.

Rahbar-Ranji, A. & Zarookian, A. (2014). Ultimate strength of stiffened plates with a transverse crack under uniaxial compression. *Ships and Offshore Structures*, **10**(4): 416–425.

Rice, J.R. (1968). A path independent integral and the approximate analysis of strain concentrations by notches and cracks. *Journal of Applied Mechanics*, **35**: 379–386.

Rice, J.R. & Rosengren, G.F. (1968). Plane strain deformation near a crack tip in a power-law hardening material. *Journal of the Mechanics and Physics of Solids*, **16**: 1–12.

Riks, E., Rankin, C.C. & Bargon, F.A. (1992). Buckling behavior of a central crack in a plate under tension. *Engineering Fracture Mechanics*, **43**: 529–548.

Rooke, D.R. & Cartwright, D.J. (1976). *Compendium of stress intensity factors*. Hillington, Uxbridge.

Roy, Y.A., Shastry, B. P. & Rao, G.V. (1990). Stability of square plates with through transverse cracks. *Computers & Structures*, **36**: 387–388.

Satish Kumar, Y.V. & Mukhopadhyay, M. (2000). A new triangular stiffened plate element for laminate analysis. *Composites Science and Technology*, **60**(6): 935–943.

Satish Kumar, Y.V. & Paik, J.K. (2004). Buckling analysis of cracked plates using hierarchical trigonometric functions. *Thin-Walled Structures*, **42**: 687–700.

Schijve, J. (1979). Four lectures on fatigue crack growth. *Engineering Fracture Mechanics*, **11**: 167–221.

Shaw, D. & Huang, Y.H. (1990). Buckling behavior of a central cracked thin plate under tension. *Engineering Fracture Mechanics*, **35**: 1019–1027.

Shi, G.J. & Wang, D.Y. (2012). Residual ultimate strength of open box girders with cracked damage. *Ocean Engineering*, **43**: 90–101.

Shi, X.H., Zhang, J. & Guedes Soares, C. (2017). Experimental study on collapse of cracked stiffened plate with initial imperfections under compression. *Thin-Walled Structures*, **114**: 39–51.

Shih, C.F. (1981). Relationship between the J-integral and the crack opening displacement for stationary and extending cracks. *Journal of the Mechanics and Physics of Solids*, **29**: 305–326.

Shih, C.F., deLorenzi, H.G. & Andrews, W.R. (1977). Studies on crack initiation and stable crack growth. *Proceedings of the International Symposium on Elastic-plastic Fracture, The American Society for Testing and Materials*, Atlanta, GA, November: 64–120.

Shih, C.F. & Hutchinson, J.W. (1976). Fully plastic solutions and large scale yielding estimates for plane stress crack problems. *Journal of Engineering Materials and Technology*, **98**: 289–295.

Sih, G.C. (1973). *Handbook of stress intensity factors*. Lehigh University, Bethlehem, PA.

Stahl, B. & Keer, M. (1972). Vibration and stability of cracked rectangular plates. *Internal Journal of Solids and Structures*, **8**: 69–92.

Tada, H., Paris, P.C. & Irwin, G.R. (1973). *Stress analysis of cracks handbook*. Del Research Corporation, Hellertown, PA.

Underwood, J.M., Sobey, A.J., Blake, J.I.R. & Shenoi, R.A. (2015). Ultimate collapse strength assessment of damaged steel plated grillages. *Engineering Structures*, **99**: 517–535.

Walker, K. (1970). *The effect of stress ratio during crack propagation and fatigue for 2024-T3 and 7075-T6 aluminum*, ASTM STP, **462**. The American Society for Testing and Materials, Philadelphia, PA, 1–14.

Wang, F., Cui, W.C. & Paik, J.K. (2009). Residual ultimate strength of structural members with multiple crack damage. *Thin-Walled Structures*, **47**: 1439–1446.

Wang, F., Paik, J.K., Kim, B.J., Cui, W.C., Hayat, T. & Ahmad, B. (2015). Ultimate shear strength of intact and cracked stiffened panels. *Thin-Walled Structures*, **88**: 48–57.

Wells, A.A. (1961). Unstable crack propagation in metals: cleavage and fast fracture. *Proceedings of the Crack Propagation Symposium*, Cranfield, UK, Paper No. 84: 210–230.

Wells, A.A. (1963). Application of fracture mechanics at and beyond general yield. British Welding Research Association, Report No. M13/63.

Westergaard, H.M. (1939). Bearing pressures and cracks. *Journal of Applied Mechanics*, **6**: 49–53.

Williams, M.L. (1957). On the stress distribution at the base of a stationary crack. *Journal of Applied Mechanics*, **24**: 109–114.

10

Structural Impact Mechanics

10.1 Fundamentals of Structural Impact Mechanics

Structures can be subjected to dynamic or impact loads during service. Any loading that gives rise to a time-dependent structural response is in this chapter termed dynamic or impact loading. Three types of dynamic or impact loadings are normally considered: impact, dynamic pressure, and impulsive loading (Jones 2012).

In some cases, objects may be dropped from a crane, for instance, onto the deck plating of an offshore platform. The striking mass dropped from a height, h, is often large and travels at a relatively low velocity, $V_0 = \sqrt{(2gh)}$, which may be around 10–15 m/s. In other cases, a mass, W, may be struck at much higher velocities as a result of a gas explosion on a land-based structure or an offshore platform. This class of loading is typically called impact, and the total impact energy of a striking mass with an initial velocity, V_0, is given by the sum of the initial kinetic energy ($WV_0^2/2$) plus the additional potential energy of the striking mass traveling through the permanent displacement of the struck structure. Water impacts and explosive events (blast) that give rise to a pressure–time history acting on the exposed area of a structure are termed "dynamic pressure." When the magnitude of the dynamic pressure is great and the duration is very short, the dynamic pressure can be idealized as impulsive loading.

The mechanical properties of steels or aluminum alloys are significantly affected by the loading speed or strain rate, $\dot{\varepsilon}$, which is defined as a relevant ratio of the loading speed to the structural displacement measured between two reference points, that is, $\dot{\varepsilon} = d\varepsilon/dt$, where ε is the strain and t is time. Table 10.1 indicates a classification of dynamic or impact loading modes as a function of the strain rate.

Three major differences between static/quasistatic and dynamic/impact loading cases are usually evident. The first difference regards the stress field. In an impact loading situation, tensile stresses can occur even under compressive far-field loading, and stress concentrations can occur even without notches. The second difference is that the dynamic/impact structural behavior can vary as a function of the strain rate. The first and second aspects always interact. The third difference regards the failure pattern. Under impact loading, brittle fracture is a greater possibility for steels or aluminum alloys that are predominantly ductile under static/quasistatic loading because the energy absorption capacity by ductile yielding decreases at high strain rates after an increase in the yield strength but a decrease in the fracture strain.

Ultimate Limit State Analysis and Design of Plated Structures, Second Edition. Jeom Kee Paik.
© 2018 John Wiley & Sons Ltd. Published 2018 by John Wiley & Sons Ltd.

Table 10.1 Dynamic modes of loading versus the strain rate (Hayashi & Tanaka 1988).

Strain rate $\dot{\varepsilon}$ (1/s)	$<10^{-5}$	10^{-5} to 10^{-1}	10^{-1} to $10^{1.5}$	$10^{1.5}$ to 10^4	$>10^4$
Dynamic loading mode	Creep	Static or quasistatic	Dynamic	Impact	Hypervelocity impact
Examples	Constant loading machine	Dead or live loading on a ship hull girder	Impulse pressure effects on high-speed craft, wave breaking loads	Explosion, ship collision	Bombing

When the externally applied dynamic (e.g., kinetic) energy is great, a dynamic plastic structural problem cannot be adequately solved using quasistatic analysis methods based on "equivalent" static loading and considering dynamic magnification factors. The general procedure for analysis of the dynamic plastic behavior of structures in such cases is similar to that used for the static behavior, except that a kinematically admissible velocity field that represents the structure's motion should be considered in the dynamic problems instead of a statically admissible deformation field. The dynamic governing differential equations, which involve an inertial term in addition to the static equilibrium-related terms, will then be solved to satisfy the initial and boundary conditions. The characteristics of the static collapse mode are often helpful for establishment of a kinematically admissible velocity field. Solutions are then found that satisfy the yield condition associated with the presumed velocity field.

It can be said with reasonable certainty that the elastic effects may be disregarded when the external dynamic energy (e.g., kinetic energy) is significantly greater than the maximum amount of strain energy that can be absorbed in a wholly elastic manner, provided that the duration of an impact loading pulse is sufficiently short compared with the structure's natural elastic period. This implies that an approximation using rigid-perfectly plastic material (instead of elastic–perfectly plastic material or elastic–plastic material including strain hardening and necking) may be used to examine the dynamic plastic response (Johnson 1972, Jones 1997, 2012).

The limit state design of plated structures under impact loading can also be undertaken following Equation (1.17). The design procedure in this case is, however, better formulated in terms of energy-related parameters, unlike the design of structures for static or quasistatic loading, which is based on the load effects and the corresponding load-carrying capacities. For instance, the demand in Equation (1.17) may be defined as the design loss of kinetic energy, whereas the capacity may be the strain energy absorption capability in the limit state design of plated structures under impact loads. This chapter describes the fundamentals of structural impact mechanics. It is noted that the theories and methodologies described in this chapter can be commonly applied to both steel- and aluminum-plated structures.

10.2   Load Effects Due to Impact

The load effects (i.e., stresses) of structures under static or quasistatic loading are normally calculated via the classical theory of structural mechanics or linear elastic finite element analysis. Under impact loading, however, the methods for the analysis of load effects are entirely different from those for static loading.

   This section describes some related considerations. Essentially, under dynamic loading, a structure not only deforms but can also vibrate, implying that the load effects are time dependent and that dynamic amplification of load effects is possible. Because in many cases the response is transient, many structures can tolerate dynamic loads that far exceed the same magnitude of static loads. Both simplified and more sophisticated methods exist for calculation of the structural response under dynamic loading. Finite element methods are useful for calculation of the dynamic structural response, taking account of material and geometric nonlinearities, including plasticity and buckling. When using simplified methods, the local strength aspects themselves are often characterized using methods based on plastic theory.

   Now consider the case of a rigid body with a mass, W, that falls freely as shown in Figure 10.1, resulting in axial impact loading to an elastic rod. The kinetic energy loss, E_k, for the falling body is given by

$$E_k = Wg(h + \delta) \tag{10.1a}$$

where g is the acceleration of gravity, h is the initial height of the falling object, and δ is the displacement.

   The strain energy, E_s, absorbed by the elastic rod is calculated by

$$E_s = \frac{EA}{2}\left(\frac{\delta}{L}\right)^2 \tag{10.1b}$$

where A is the cross-sectional area, L is the initial length of the elastic rod, and E is Young's modulus.

Figure 10.1 Impact by freefall.

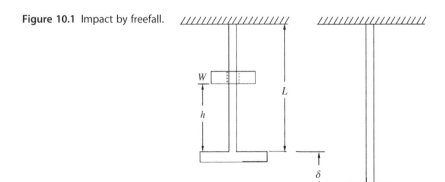

By the principle of energy conservation, $E_k = E_s$, and thus the displacement, δ, is determined. Therefore, the impact stress, σ, can be obtained as follows:

$$\sigma = \frac{E\delta}{L} = \frac{Wg}{A}\left(1 + \sqrt{1 + \frac{2EAh}{WgL}}\right) \tag{10.2}$$

The impact stress is time variant in nature. During contact of the falling object with the bottom plate, the falling object will be subjected to reaction forces from the rod and gravitational forces. The governing differential equation for the motion of the falling object is given by neglecting the effect of the rod mass as follows:

$$W\frac{d^2x}{dt^2} = Wg - A\sigma = Wg - \frac{EAx}{L} \tag{10.3}$$

where x represents the displacement of the rod. If the origin of the time, t, is taken as the time at which the falling object first contacts the bottom plate, the initial conditions for the impact are given by

$$x = 0, \quad \frac{dx}{dt} = V_0 \text{ at } t = 0 \tag{10.4}$$

where $V_0 = \sqrt{(2gh)}$ is the velocity of freefall.

The solution of Equation (10.3) is then obtained by considering the initial conditions, Equation (10.4), as follows:

$$x = \frac{WgL}{EA}\left(1 - \cos\phi + \sqrt{\frac{2EAh}{WgL}}\sin\phi\right) \tag{10.5}$$

where $\phi = [EA/(WL)]^{1/2}t$.

Therefore, the time-variant impact stress, σ, is given using Equation (10.5) as follows:

$$\sigma = \frac{Ex}{L} = \frac{Wg}{A}\left\{1 + \sqrt{1 + \frac{2EAh}{WgL}}\sin\left[\sqrt{\frac{EA}{WL}}(t - T)\right]\right\} \tag{10.6}$$

where

$$T = \sqrt{\frac{WL}{EA}}\arctan\sqrt{\frac{WgL}{2EAh}}$$

It is realized from Equation (10.6) that the impact stress develops in a sinusoidal form with regard to time and takes the maximum value equal to Equation (10.2) at the time, t_0, which is defined by

$$t_0 = \frac{\pi}{2}\sqrt{\frac{WL}{EA}} + T \tag{10.7}$$

It has been assumed here that impact loads may be transferred to the fixed end of the rod immediately after the falling object strikes the bottom plate, and the rod would then deform in a uniform manner along its length. Also, the effect of the stress wave propagation is neglected. For higher-speed impact loading, however, the latter assumption may not be appropriate because the propagation, reflection, refraction, and interference of the stress wave may play important roles.

Related to the effects of the stress wave propagation, a simple example is now considered as studied by Hayashi and Tanaka (1988). When a rigid body with a velocity of V_0 strikes a column in the length direction, as shown in Figure 10.2, the resulting time-variant impact stress is estimated with consideration of the effects of stress wave propagation.

Denote the propagation velocity of the stress wave and the velocity of a particle within the impacted body by V_p and V_m, respectively. The stress wave propagates from point B to point C, whereas point B deforms to point B′ at time $t = t^*$. In this case, the strain, ε, of the column at $t = t^*$ can be given, with the compressive strain taken as positive as follows:

$$\varepsilon = \frac{BB'}{BC} = \frac{V_m}{V_p} \tag{10.8}$$

During the strike, the momentum imparted to the struck body is given by $\rho_0 A V_p t^* V_m$ because the part B′C of the column with cross-sectional area A moves with velocity V_m for the period of t^*, and the pulse that acts on the column during the same period is calculated by $A\sigma t^*$. Because of the equilibrium between the two quantities, that is, $\rho_0 A V_p t^* V_m = A\sigma t^*$, the impact stress, σ, is obtained by

$$\sigma = \rho_0 V_m V_p \tag{10.9a}$$

where ρ_0 is the density of the column before it is struck.

Using the strain given by Equation (10.8), the impact stress is also given by

$$\sigma = E\varepsilon = E\frac{V_m}{V_p} \tag{10.9b}$$

Figure 10.2 An example to illustrate the effects of impact stress wave propagation.

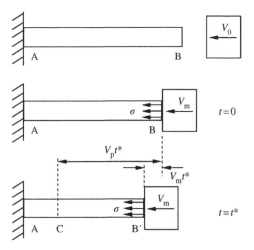

The propagation velocity of the stress wave can then be obtained from Equations (10.9a) and (10.9b) as follows:

$$V_{\mathrm{p}} = \left(\frac{E}{\rho_0}\right)^{1/2} \tag{10.10}$$

Substitution of Equation (10.10) into Equation (10.9b) yields

$$\sigma = (E\rho_0)^{1/2} V_{\mathrm{m}} \tag{10.11}$$

In the aforementioned, it is noted that the particle velocity, V_{m}, does not necessarily equal the initial velocity, V_0, of the striking body. It is apparent that Equation (10.11) takes an entirely different expression from that of Equation (10.6) because Equation (10.11) represents the impact stress propagation immediately after striking, whereas Equation (10.6) indicates the stress after the stress wave ceases.

For a more elaborate treatment of the stress wave effects in structural members under impact loading, interested readers may refer to Karagiozova and Jones (1998), Kolsky (1963), or Hayashi and Tanaka (1988), among others.

It is not always straightforward to calculate the load effects of structures under impact loading because propagation of the stress wave can play an important role. In the context of Equation (1.17), therefore, an easier alternative is to use the loss of kinetic energy instead of the load effects as the measure of "demand," whereas the energy absorption capability is used as the measure of "capacity."

10.3   Material Constitutive Equation of Structural Materials Under Impact Loading

The stresses in structural materials under a given magnitude of impact loading in the elastic–plastic regime increase as the strain rate increases. The mechanism of plastic deformation under impact loading in such a case may be classified into four regions, as shown in Figure 10.3 (Perzyna 1974, Rosenfield & Hahn 1974).

In region I, the yield (or flow) stress is not affected by the strain rate or temperature. In region II, the yield (or flow) stress increases as the strain rate increases. In region III, that is, at low temperatures, the effect of the strain rate on the yield (or flow) stress becomes more moderate. Region III is distinguished from region II by a boundary known as the twinning mode of the plastic deformation mechanism due to the low temperature. In region IV, the yield (or flow) stress is extremely sensitive to the strain rate.

Figure 10.4 shows a sample relationship between the dynamic shear yield stress and the shear strain rate for mild steels, varying the strain rate and temperature (Campbell & Ferguson 1970).

10.3.1   The Malvern Constitutive Equation

The stress–strain relationship for structural steels under impact loading in a one-dimensional form, which is typically applied in region II, may be given by (Malvern 1969)

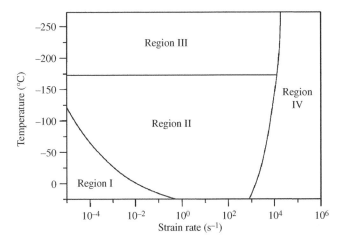

Figure 10.3 Classification of plastic deformation mechanisms as a function of strain rate and temperature.

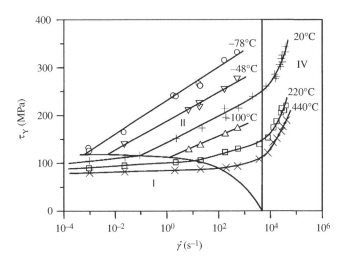

Figure 10.4 Variation of the dynamic shear yield stress of mild steel as a function of shear strain rate and temperature (symbols denote experimental results after Campbell & Ferguson 1970).

$$\sigma = f(\varepsilon) + c_1 \ln\left(1 + c_2 \dot{\varepsilon}_p\right) \qquad (10.12)$$

where $\sigma = f(\varepsilon)$ is the stress–strain relationship of material under static loading, $\dot{\varepsilon}_p$ is the plastic strain rate, and c_1 and c_2 are material constants.

The plastic strain rate can then be obtained from Equation (10.12) as follows:

$$\dot{\varepsilon}_p = \frac{1}{c_2} \left\{ \exp\left[\frac{\sigma - f(\varepsilon)}{c_1}\right] - 1 \right\} \qquad (10.13)$$

It is realized from Equation (10.13) that the plastic strain rate represents the effect of the excess stress between the dynamic stress, σ, and the static stress, $f(\varepsilon)$. As the simplest expression, the excess stress may sometimes be expressed with a linear function as follows:

$$E\dot{\varepsilon}_p = c_3[\sigma - f(\varepsilon)] \tag{10.14}$$

where E is the elastic modulus and c_3 is the material constant.

Because the elastic component, $\dot{\varepsilon}_e$, of the strain rate has a linear relationship with the corresponding stress $\dot{\sigma}$, that is, $E\dot{\varepsilon}_p = \dot{\sigma}$, and the total strain is a sum of the elastic and plastic components, the following equation is obtained:

$$E\dot{\varepsilon} = E(\dot{\varepsilon}_e + \dot{\varepsilon}_p) = \dot{\sigma} + c_3[\sigma - f(\varepsilon)] \tag{10.15}$$

Equation (10.15) is sometimes called the Malvern constitutive equation and is widely used for the response analysis of structures under impact loading, primarily representing region II in Figure 10.3. The dynamic yield strength and fracture strain are thus the two primary influential parameters in analysis of the crashworthiness of structures under impact loading. The dynamic effect on the material properties is associated with the strain rate, as previously noted.

10.3.2 Dynamic Yield Strength: The Cowper–Symonds Equation

The dynamic yield strength of the material may be expressed as follows (Karagiozova & Jones 1997):

$$\frac{\sigma_{Yd}}{\sigma_Y} = f(\dot{\varepsilon})g(\varepsilon) \tag{10.16}$$

where σ_Y and σ_{Yd} are the static or dynamic yield stresses, respectively, $f(\dot{\varepsilon})$ is a function of the strain rate sensitivity effect, $g(\varepsilon)$ is a material strain-hardening function, and $\dot{\varepsilon}$ is the strain rate.

If the effect of strain hardening is neglected, $g(\varepsilon) = 1$ can be assumed. The strain rate sensitivity parameter, $f(\dot{\varepsilon})$, is often given using the Cowper–Symonds equation (Cowper & Symonds 1957) as follows:

$$\frac{\sigma_{Yd}}{\sigma_Y} = 1.0 + \left(\frac{\dot{\varepsilon}}{C}\right)^{1/q} \tag{10.17}$$

where C and q are coefficients to be determined based on test data, as indicated in Table 10.2.

Figure 10.5 plots the Cowper–Symonds equation together with the relevant coefficients for mild or high-tensile steels when $g(\varepsilon) = 1$. Figure 10.6 shows the effects of both the strain rate and the temperature on the yield strength of a higher-tensile steel. It is evident from this figure that the material yield strength increases as the strain rate increases and the temperature decreases. Also, for higher-tensile steel, the percentage increase of σ_{Yd}/σ_Y is less than that for mild steel. Figure 10.7 shows the effects of the strain rate and low temperature on the dynamic yield stress for mild steel, high-tensile steel, aluminum alloy, and stainless steel (Paik et al. 2017); the references shown are within the figure after Paik et al. (2017).

Table 10.2 Sample coefficients for the Cowper–Symonds constitutive equation associated with the dynamic yield stress.

Material	C (1/s)	q	Reference
Mild steel	40.4	5	Cowper and Symonds (1957)
High-tensile steel	3200	5	Paik and Chung (1999)
Aluminum alloy	6500	4	Bodner and Symonds (1962)
	10.39×10^{10}	10.55	Hsu and Jones (2004)
	610.4	3.6	Paik et al. (2017)
α-Titanium (Ti 50A)	120	9	Symonds and Chon (1974)
Stainless steel	100	10	Forrestal and Sagartz (1978)
	3000	2.8	Paik et al. (2017) for room temperature
	35.9	1.5	Paik et al. (2017) for low temperature

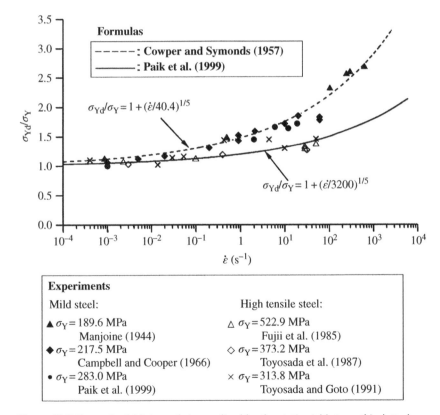

Figure 10.5 Dynamic yield strength (normalized by the static yield strength) plotted versus strain rate for mild and high-tensile steels (references shown within the figure after Paik et al. 1999).

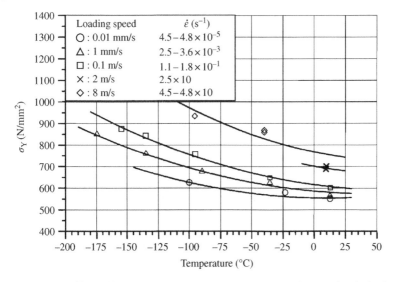

Figure 10.6 Effects of strain rate and temperature on the yield strength of a higher-tensile steel (Toyosada et al. 1987).

10.3.3 Dynamic Fracture Strain

Both crushing effects and the yield strength increase as the loading speed increases, whereas any fracture or tearing of steel (and the welded regions) in a structure tends to occur earlier. The following approximate formula, which is the inverse of the Cowper–Symonds equation for the dynamic yield stress, is then useful for estimation of the dynamic fracture strain as a function of the strain rate (Jones 1989), namely,

$$\frac{\varepsilon_{Fd}}{\varepsilon_F} = \xi \left[1.0 + \left(\frac{\dot{\varepsilon}}{C} \right)^{1/q} \right]^{-1} \tag{10.18}$$

where ε_F and ε_{Fd} are the static and dynamic fracture strains, respectively, and ξ is the ratio of the total energies to rupture for dynamic and static uniaxial loadings.

If the energy to failure is assumed to be invariant, that is, independent of $\dot{\varepsilon}$, then it may be taken that $\xi = 1$. Figure 10.8 plots Equation (10.18) with three sets of the coefficients together with experimental results for mild steels when $\xi = 1$. The expression in Equation (10.18) represents the decrease of the dynamic fracture strain with the increase in the strain rate, but the coefficients for the dynamic fracture strain differ from those for the dynamic yield strength. It is again evident that the strain rate is a primary parameter that affects the impact mechanics.

Figure 10.9 shows the effects of the strain rate and low temperature on the dynamic fracture strain for mild steel, high-tensile steel, aluminum alloy, and stainless steel (Paik et al. 2017); the references shown are within the figure after Paik et al. (2017). Figures 10.8 and 10.9 also show the best coefficients of the Cowper–Symonds equation for different materials associated with the dynamic fracture strain. It is evident that the coefficients for dynamic fracture strain differ from those for dynamic yield stress.

(a)

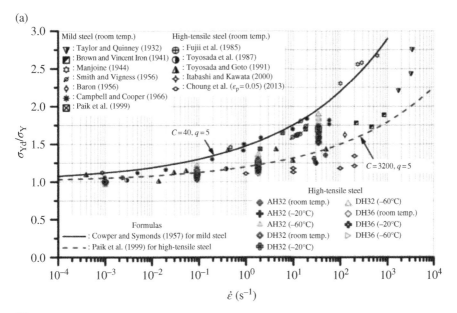

(b)

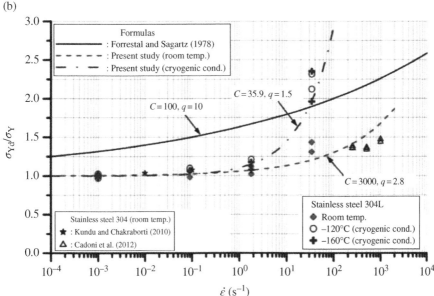

Figure 10.7 Effects of strain rates and room or low temperature on dynamic yield stress of (a) mild and high-tensile steel and (b) stainless steel 304L (Paik et al. 2017).

10.3.4 Strain-Hardening Effects

Two mathematical descriptions are relevant for the strain-hardening phenomenon, that is, the Hollomon equation (1945) and the Ludwik equation (1909). The Hollomon equation is a power law relationship between the true stress and the true plastic strain given by

$$\sigma = K\varepsilon_{\mathrm{p}}^{n} \tag{10.19}$$

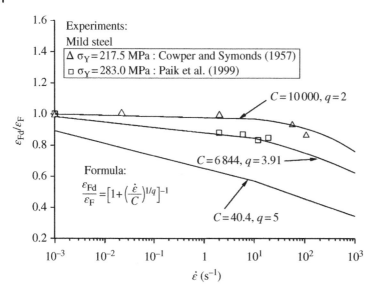

Figure 10.8 Dynamic fracture strain (normalized by the static fracture strain) versus the strain rate for mild steels.

where σ is the true stress, ε_p is the true plastic strain, K is the material coefficient, and n is the strain-hardening exponent.

The Ludwik equation is similar to the Hollomon equation, but it includes the yield strength as follows:

$$\sigma = \sigma_Y + K\varepsilon_p^n \tag{10.20}$$

where σ_Y is the yield strength of material.

While the strain-hardening exponent n usually lies in the range of 0.2–0.5, it can be expressed as follows:

$$n = \frac{d\log(\sigma)}{d\log(\varepsilon)} = \frac{\varepsilon \, d\sigma}{\sigma \, d\varepsilon} \tag{10.21}$$

Equation (10.21) can be evaluated from the slope of a $\log(\sigma) - \log(\varepsilon)$ plot, and it allows a determination of the rate of the strain hardening at a given (true) stress and (true) strain as follows:

$$\frac{d\sigma}{d\varepsilon} = n\frac{\sigma}{\varepsilon} \tag{10.22}$$

Figure 10.10 shows the test results associated with the material coefficient K and the strain-hardening exponent n for carbon steels (grade E) at low temperatures, where the thickness of tensile coupon specimens is 16 mm (Park 2015).

10.3.5 Inertial Effects

Inertial effects must sometimes be considered for the impact response of plated structures (Reid & Reddy 1983, Harrigan et al. 1999, Paik & Chung 1999, Karagiozova et al.

(a)

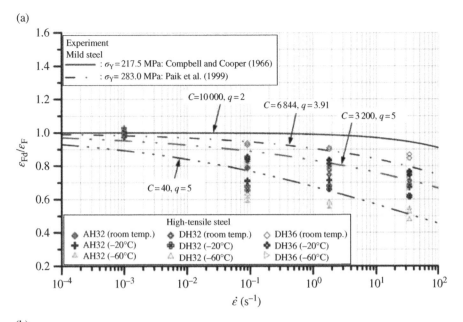

(b)

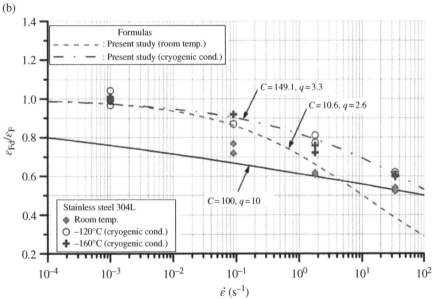

Figure 10.9 Effects of strain rates and room or low temperature on the dynamic fracture strain of (a) mild and high-tensile steel and (b) stainless steel 304L (Paik et al. 2017).

2000). Due to inertial effects and stress wave propagation phenomena within the structures during impact loading, the strain distribution (or deformation pattern) at any moment in time would be nonhomogeneous. It is typically considered that the inertial effects become more important when the strain rate is greater than about 0.1 s^{-1}.

To investigate the characteristics of inertial effects, Paik and Chung (1999) carried out a series of experiments on crushing square steel tubes and observed that as the maximum

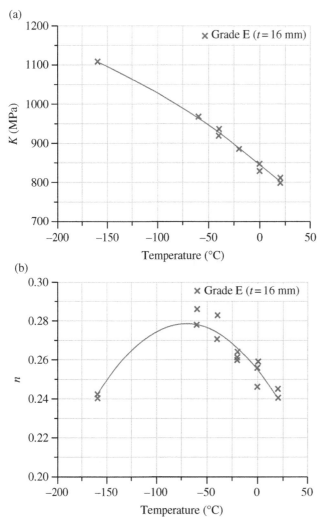

Figure 10.10 Variation of (a) the material coefficient K and (b) the strain-hardening exponent n, of the Hollomon equation for carbon steels (grade E) at low temperatures (symbol, test data; line, trend line) (Park 2015).

crushing indentation increased (indicating that the mass of the striking body and/or the impact speed increased, thus increasing the initial kinetic energy involved), the inertial effects also increased, but these effects could be ignored for strain rates of less than about $50 \, s^{-1}$.

10.3.6 Friction Effects

During an impact loading process, the influence of friction would normally be great when there is a relative velocity between the struck body and the indentor (or the striking body). This situation is often seen in ship grounding incidents when a ship with forward speed runs onto a rock pinnacle. In structural engineering problems where an object slides across a surface, the frictional force can be defined as product of the friction constant of the two materials and the normal force. The friction constant depends on various factors such as impact velocity, contact area, surface grain feature, moisture (wet or dry), and temperature.

10.4   Ultimate Strength of Beams Under Impact Lateral Loads

In the quasistatic case, it is considered that a beam under lateral load, q, collapses if a plastic hinge mechanism is formed at $q = q_c$ as described in Section 2.8. A beam made from a rigid-perfectly plastic material is assumed to remain rigid as long as the external lateral load is less than the static collapse lateral load, q_c. Hence static equilibrium cannot be achieved in the collapsed beam if the effects of strain-hardening or large deformations are neglected.

In contrast, when an external load larger than q_c is applied suddenly or impulsively, the beam deforms plastically and inertial forces are generated. If the external load pulse continues for a sufficient duration, the beam's lateral deflection will become excessive. However, if the external load pulse is removed soon enough or decays to a sufficiently small value after a certain time, part of the kinetic energy will be absorbed by the beam after permanent plastic deformations.

The governing differential equation of a plate-beam combination under dynamic lateral load, q, may be expressed by (Jones 2012)

$$EI_e \frac{\partial^4 w}{\partial x^4} = q - m \frac{\partial^2 w}{\partial t^2} \tag{10.23}$$

where $m = \rho A$ is the mass per unit length of beam, ρ is the density of the material, t is time, EI_e is the bending rigidity of the effective section beam, and A is the beam's cross-sectional area.

The elastic strain energy in a beam can be given from Equation (2.68) as follows:

$$U = \int_{Vol} \frac{\sigma_x^2}{2E} dVol \tag{10.24a}$$

The maximum possible amount of the elastic strain energy that a beam can absorb is obtained when σ_x reaches the equivalent yield stress, σ_{Yeq}, in the entire volume of the beam, namely,

$$U_{max} = \frac{\sigma_{Yeq}^2 LA}{2E} \tag{10.24b}$$

where σ_{Yeq} accounts for the effect of different yield stresses of the plating and stiffener, as indicated in Table 2.1, and L is the span of the beam.

Because local plastic deformations occur at smaller values of elastic strain energy, Equation (10.24b) represents the upper limit of the strain energy absorbed. The initial kinetic energy absorbed by the beam under a uniformly distributed impulsive velocity, V_0, may approximately be given by

$$E_k = \frac{1}{2} \rho LA V_0^2 \tag{10.25a}$$

For a beam struck by a total mass W with an initial velocity V_0, the initial kinetic energy is given by

$$E_k = \frac{1}{2}WV_0^2 \qquad (10.25b)$$

It appears that the rigid-perfectly plastic material approach may be relevant when the ratio of the initial kinetic energy to the maximum amount of the strain energy is greater than about 10, namely (Jones 1989),

$$\frac{E_k}{U_{max}} > 10 \qquad (10.26)$$

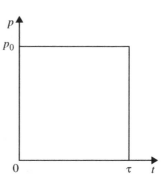

Figure 10.11 A schematic of a rectangular-shaped pressure pulse.

Figure 10.11 shows a schematic of a rectangular-type load pulse. The lateral pressure, p_0 ($p_0 = q_0/b$ for a plate–beam combination model), is dynamically applied at the beginning and is kept constant for a duration τ, after which the load is removed. The rectangular pressure pulse is often termed "impulsive loading" when $p_0/p_c \gg 1$ ($p_c = q_c/b$ for a plate–beam combination) (i.e., $\eta = p_0/p_c \to \infty$) and $\tau \neq 0$. In this situation, the following condition must be satisfied from the conservation of linear momentum at $t \neq 0$:

$$I = \int_0^\tau p(t)\, dt = p_0\tau = \mu V_0 \qquad (10.27)$$

where I is the pulse and μ is the mass per unit area.

It has been noted that the elastic effects are not important when the duration is sufficiently shorter than the corresponding elastic natural period, namely,

$$\frac{\tau}{T} \ll 1 \qquad (10.28a)$$

where T is the fundamental period of elastic vibration, which may be taken as follows:

$$T = \frac{2L^2}{\pi}\left(\frac{m}{EI_e}\right)^{1/2} \quad \text{for a beam simply supported at both ends} \qquad (10.28b)$$

$$T = \frac{2\pi L^2}{(4.73)^2}\left(\frac{m}{EI_e}\right)^{1/2} \quad \text{for a beam clamped at both ends} \qquad (10.28c)$$

The maximum permanent lateral deflections, w_p, of clamped beams under dynamic lateral loading with an initial impact velocity, V_0, may be calculated for rectangular cross sections with breadth B and thickness H, to account for the large-deflection effect, as follows (Jones 1997):

$$\frac{w_p}{H} = \frac{1}{2}\left[\left(1 + \frac{\lambda}{2\alpha}\right)^{1/2} - 1\right] \quad \text{for impact loading } (\alpha \ll 1) \qquad (10.29a)$$

$$\frac{w_p}{H} = \frac{1}{2}\left[\left(1 + \frac{3\lambda}{4}\right)^{1/2} - 1\right] \quad \text{for impulsive loading} \qquad (10.29b)$$

where $\lambda = \mu V_0^2 L^2 / (16 M_p H)$, $\alpha = \mu L / (2W)$, $M_p = \sigma_0 BH^2 / 4$, W is the impact mass, and σ_0 is the flow stress (refer to Section 10.8.1).

For plate–stiffener combination cross sections under dynamic loading, interested readers may refer to Schubak et al. (1989), Nurick et al. (1994), and Nurick and Jones (1995), among others.

10.5 Ultimate Strength of Columns Under Impact Axial Compressive Loads

The governing differential equation of the small-deformation theory for a column with initial deflection under impact axial compressive loading, P, may be expressed by (Jones 2012)

$$EI\frac{\partial^4 w}{\partial x^4} + P\frac{\partial^2 (w + w_0)}{\partial x^2} = -m\frac{\partial^2 w}{\partial t^2} \tag{10.30}$$

For a column simply supported at both ends, the lateral deflection functions may be assumed as follows:

$$w_0 = \delta_0 \sin\frac{\pi x}{L}, \quad w = \delta(t)\sin\frac{\pi x}{L} \tag{10.31}$$

where w_0 is the initial deflection, w is the added deflection due to applied loading, and δ_0 and δ are the initial and added deflection amplitudes, respectively.

Substitution of Equation (10.31) into Equation (10.30) results in

$$\frac{d^2\delta(t)}{dt^2} + C_1\delta(t) = C_2 \tag{10.32}$$

where

$$C_1 = \left(\frac{\pi}{L}\right)^2 \left(\frac{P_E}{m}\right)\left(1 - \frac{P}{P_E}\right), \quad C_2 = \left(\frac{\pi}{L}\right)^2 \frac{P}{m}\delta_0$$

$P_E = \pi^2 EI_e / L^2$ is the Euler buckling load.

It is obvious that the second-order differential equation, Equation (10.32), has a different form of solution depending on the sign of the coefficient C_1: if $P < P_E$ then $C_1 > 0$, and if $P > P_E$ then $C_1 < 0$.

10.5.1 Oscillatory Response

When $P < P_E$, the coefficient C_1 is positive, and thus Equation (10.32) has the solution as follows:

$$\delta(t) = \frac{C_2}{C_1}\left[1 - \cos\left(\sqrt{C_1}t\right)\right] \tag{10.33}$$

On substituting Equation (10.33) into Equation (10.31), the added deflection of the column is given, considering that the initial lateral velocity is zero (i.e., $d\delta/dt = 0$ at $t = 0$) and that the added deflection is zero (i.e., $\delta = 0$) at $t = 0$:

$$w = B_1(\tau)\delta_0 \sin\frac{\pi x}{L} \tag{10.34}$$

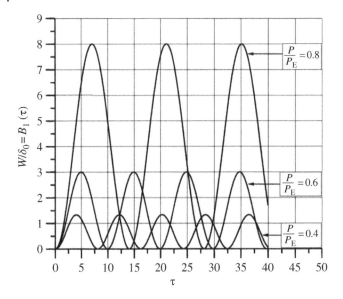

Figure 10.12 Oscillation of the added deflection of a column under dynamic axial compressive load, P, less than P_E versus time, as obtained from Equation (9.30) at $x = L/2$.

where

$$B_1(\tau) = \frac{P/P_E}{1 - P/P_E}\left\{1 - \cos\left[\left(1 - \frac{P}{P_E}\right)^{1/2}\tau\right]\right\}$$

$\tau = 2\pi t/T = [\pi^4 EI_e/(mL^4)]^{1/2}$ $t =$ non-dimensional time, where T is as defined in Equation (10.28).

Figure 10.12 plots Equation (10.34) at $x = L/2$, showing the variations of the maximum added deflections of the column versus non-dimensional time. It is seen from Figure 10.12 that the added deflection of a column under a dynamic axial compressive load, P, less than P_E varies cyclically with time, but the dynamic buckling phenomenon does not occur. The added deflection increases significantly as P approaches P_E and the period of oscillation becomes infinite at $P = P_E$.

10.5.2 Dynamic Buckling Response

If $P > P_E$, then $C_1 < 0$. In this case, the second-order differential equation, Equation (10.32), involves hyperbolic functions. For convenience, Equation (10.32) may be rewritten as follows:

$$\frac{d^2\delta(t)}{dt^2} - C_1^*\delta(t) = C_2 \tag{10.35}$$

where

$$C_1^* = \left(\frac{\pi}{L}\right)^2\left(\frac{P_E}{m}\right)\left(\frac{P}{P_E} - 1\right)$$

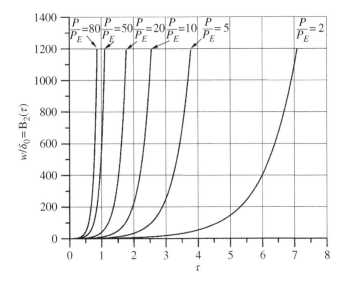

Figure 10.13 Dynamic buckling of a column under dynamic axial compressive load P larger than P_E, as obtained from Equation (10.37) at $x = L/2$.

The solution of the differential equation, Equation (10.31), is given by

$$\delta(t) = \frac{C_2}{C_1^*}\left[\cosh\left(\sqrt{C_1^*}t\right) - 1\right] \tag{10.36}$$

The added deflection of the column is then obtained by substituting Equation (10.36) into Equation (10.31) as follows:

$$w = B_2(\tau)\delta_0 \sin\frac{\pi x}{L} \tag{10.37}$$

where

$$B_2(\tau) = \frac{P/P_E}{P/P_E - 1}\left\{\cosh\left[\left(\frac{P}{P_E} - 1\right)^{1/2}\tau\right] - 1\right\}$$

Figure 10.13 plots Equation (10.37) at $x = L/2$. It is evident from Figure 10.13 that the column-added deflection significantly increases over time and becomes very large at a certain point in time. This phenomenon is sometimes called dynamic buckling.

10.6 Ultimate Strength of Plates Under Impact Lateral Pressure Loads

10.6.1 Analytical Formulations: Small-Deflection Theory

In a quasistatic loading condition, a plate under lateral pressure collapses if the applied pressure is greater than the collapse pressure load, p_c, as described in Chapter 4.

When the plate is subjected to impact pressure pulses applied for a short period, however, it may withstand the loads even if the initial peak pressure is greater than p_c.

The plate under such a pressure pulse initially deforms, but the mass of the plate has a certain amount of kinetic energy when the pressure is removed. This kinetic energy is then absorbed by the structure, which deforms accordingly. The motion of the plate ceases when all kinetic energies are dissipated as strain energy. In this process, the plate response is time dependent, and the inertial forces may play an important role in equilibrium equations.

The governing differential equations for the dynamic behavior of an element of a rectangular plate are given by accounting for the effects of inertial forces as follows (for the symbols used, Figure 10.14 may be referred to unless otherwise specified later):

$$\frac{\partial Q_x}{\partial x} + \frac{\partial Q_y}{\partial y} + p = \mu \frac{\partial^2 w}{\partial t^2} \tag{10.38a}$$

$$\frac{\partial M_{yx}}{\partial y} + \frac{\partial M_x}{\partial x} - Q_x = 0 \tag{10.38b}$$

$$\frac{\partial M_{xy}}{\partial x} - \frac{\partial M_y}{\partial y} + Q_y = 0 \tag{10.38c}$$

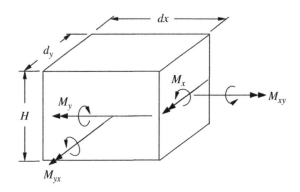

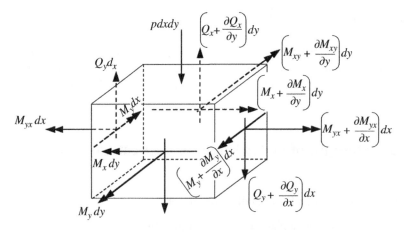

Figure 10.14 An infinitesimal element, *dx dy*, of a rectangular plate (*H* is the plate thickness).

where p is the applied pressure that is a function of t (time), Q_x and Q_y are the transverse shear forces, $\mu = \rho H$ is the mass of the plate per unit area, H is the plate thickness, and ρ is the material density. Also, bending moments per unit length, M_x and M_y, and the twisting moment per unit length, $M_{xy}(= -M_{yx})$, are given by

$$M_x = -D\left(\frac{\partial^2 w}{\partial x^2} + v\frac{\partial^2 w}{\partial y^2}\right) \tag{10.38d}$$

$$M_y = -D\left(\frac{\partial^2 w}{\partial y^2} + v\frac{\partial^2 w}{\partial x^2}\right) \tag{10.38e}$$

$$M_{xy} = -M_{yx} = D(1-v)\frac{\partial^2 w}{\partial x \partial y} \tag{10.38f}$$

where $D = EH^3/[12(1-v^2)]$ is the plate bending rigidity, E is Young's modulus, and v is Poisson's ratio.

Eliminating Q_x and Q_y from Equations (10.38a) to (10.38c) and considering Equation (10.38f), we have

$$\frac{\partial^2 M_x}{\partial x^2} - 2\frac{\partial^2 M_{xy}}{\partial x \partial y} + \frac{\partial^2 M_y}{\partial y^2} = -p + \mu\frac{\partial^2 w}{\partial t^2} \tag{10.38g}$$

Substitution of Equations (10.38d)–(10.38f) into Equation (10.38g) yields

$$\frac{\partial^4 w}{\partial x^4} + 2\frac{\partial^4 w}{\partial x^2 \partial y^2} + \frac{\partial^4 w}{\partial y^4} = \frac{1}{D}\left(p - \mu\frac{\partial^2 w}{\partial t^2}\right) \tag{10.38h}$$

Jones (2012) solved the governing differential equation, Equation (10.38g), for a simply supported square plate of $a/b = 1$ when a uniformly distributed pressure pulse with an initial value of p_0 was applied for duration τ, following the pulse profile as shown in Figure 10.11. When p_0 is less than twice the static collapse pressure, which is equal to $24\,M_p/b^2$, the response has two phases: the first phase coincides with the period of application of the pressure, and the second starts when the pressure is removed and ends when the kinetic energy of the plate becomes zero. The maximum permanent deflection, w_p, at the center of the plate is in this case given by

$$w_p = \frac{p_c \tau^2}{\mu}\eta(\eta-1) \text{ for } p_c \le p_0 \le 2p_c \tag{10.39}$$

where $\eta = p_0/p_c$, $p_c = 24\,M_p/b^2$, $M_p = \sigma_0 H^2/4$, and σ_0 is the flow stress.

As noted earlier, Equation (10.39) has been derived considering the dynamic pressure profile shown in Figure 10.11. It has been shown that the resulting cross-sectional forces (i.e., bending moments M_x, M_y) are "admissible," in the sense that the Tresca-type yield criterion, as indicated in Equation (1.31b), is not violated and also remain stationary. However, when the initial peak pressure, p_0, shown in Figure 10.11 is more than twice the static collapse pressure, the two-phase response is not statically admissible because the assumed velocity profile results in a violation of the assumed yield criterion. In this case, it is assumed that the response has three phases: the first coincides with the application of the pressure pulse, the second is characterized by moving hinge lines, and in the third phase, the plastic hinges become stationary. At the end of the third phase, the maximum lateral deflection of the square plate is given by

$$w_p = \frac{p_c \tau^2}{4\mu} \eta (3\eta - 2) \text{ for } p_0 > 2p_c \tag{10.40}$$

For both the two-phase response and the three-phase response noted earlier, the response time, T, is found to be equal to $\eta\tau$.

10.6.2 Analytical Formulations: Large-Deflection Theory

Equations (10.39) and (10.40) were derived under the assumption that the lateral deflections are not sufficiently large to cause any change to the plate's geometry, and the supports do not provide any resistance to the axial movement of the plate edges. However, when the deflections are relatively large and the supports resist the axial movement of the plate edges, membrane stresses develop, and the structure provides further resistance against the applied pressure. The resistance is experienced regardless of whether the pressure is applied statically or dynamically.

Jones (2012) proposes the rigid-plastic approach to compute the permanent deflection of beams and rectangular plates loaded by a pressure pulse, $p(t)$, taking into account the large-deflection effect. According to the Jones approach, if bending moments and membrane forces are developed within the plate, resulting from the axial restraint of the supports, the lateral deflection, w, obeys the following equation:

$$\int_A (p - \mu \ddot{w}) \dot{w} dA = \sum_{m=1}^{r} \int_{\ell_m} (M + Nw) \dot{\theta}_m d\ell_m \tag{10.41}$$

where μ is the mass of the plate per unit area, r is the number of hinge lines, ℓ_m is the length of the hinge line, θ_m is the relative angular rotation across a hinge line, and N and M are the membrane and bending forces, respectively, that act along the hinge lines. It has been assumed that the material is rigid perfectly plastic, and the loaded plate is divided into a number of rigid sections separated by straight-line hinges.

If it is further assumed that (i) the plate is clamped along its four edges, (ii) the material obeys the Tresca-type yield criterion, and (iii) shear forces do not affect yielding, then the maximum permanent deflection, w_p, of the plate is given by

$$w_p = H \frac{(3 - \xi_0) \left[\sqrt{1 + 2\eta(\eta - 1)(1 - \cos\gamma\tau)} - 1 \right]}{2[1 + (\xi_0 - 2)(\xi_0 - 1)]} \tag{10.42}$$

where

$$\xi_0 = \alpha \left(\sqrt{3 + \alpha^2} - \alpha \right), \quad \gamma^2 = \frac{96 M_p}{\mu b^2 H (3 - 2\xi_0)} \left(1 - \xi_0 + \frac{1}{2 - \xi_0} \right), \quad \alpha = \frac{b}{a}$$

a is the plate length, b is the plate breadth, $M_p = \sigma_0 H^2/4$, σ_0 is the flow stress, H is the plate thickness, and η is as defined in Equation (10.39).

If the load is impulsive, that is, $p \to \infty$ and $\tau \to 0$, Equation (10.42) can be rewritten as follows:

$$w_p = H \frac{(3 - \xi_0)\left\{ \sqrt{1 + (\lambda a^2/6)(3 - 2\xi_0)[1 - \xi_0 + 1/(2 - \xi_0)]} - 1 \right\}}{2[1 + (\xi_0 - 2)(\xi_0 - 1)]} \tag{10.43}$$

where $\lambda = \rho V_0^2 a^2/(\sigma_0 H^2)$. The velocity, V_0, is calculated from the relationship $p_0 \tau = \mu V_0 = I$, where I is the pulse, refer to Equation (10.27).

Using the same approach, similar equations can be derived for simply supported plates. In this case, the maximum permanent deflection is given by

$$w_p = H \frac{(3 - \xi_0)\left[\sqrt{1 + 2\eta(\eta - 1)(1 - \cos\gamma\tau)} - 1 \right]}{4[1 + (\xi_0 - 2)(\xi_0 - 1)]} \tag{10.44}$$

To account for the effect of the strain rate on yield stress, Symonds and Jones (1972) suggested a correction factor, f, that will be multiplied by the flow stress, σ_0, in either Equation (10.43) or (10.44), namely,

$$f = 1 + \left(\frac{H^2 \lambda^{3/2}}{6Ca^3} \sqrt{\frac{\sigma_0}{\mu}} \right)^{1/q} \tag{10.45}$$

where C and q are as defined in Equation (10.17) or Table 10.2.

Yu and Chen (1992) showed that in a case of intense impulse, it is essential to consider the effect of traveling hinges, that is, to update the mode of the assumed collapse mechanism during the response. A similar approach was developed by Shen (1997) for the dynamic response of a thin rectangular plate struck transversely (laterally) by a wedge. Chen (1993) applied the rigid-plastic theory to derive closed-form expressions to predict the permanent deflections of rectangular plates under impulsive loads, using two types of deflection patterns: roof-shaped and sinusoidal patterns. It was shown that the sinusoidal shape function gives larger permanent deflections than the roof-shaped deformation function.

The insights noted earlier are based on the assumption that the material is rigid perfectly plastic, that is, that neither elastic deformations nor strain-hardening effects occur. This may cause overestimation of permanent deflections, whereas the influence of strain rate sensitivity, which tends to reduce the permanent deflections, may be accounted for with the Cowper–Symonds formula, Equation (10.17).

10.6.3 Empirical Formulations

Based on existing experimental data for plates under impact lateral pressure loading, numerous empirical formulations were derived by curve fitting for prediction of the permanent lateral deflection of clamped rectangular plates as follows (Nurick & Martin 1989):

$$\frac{w_p}{H} = 0.471\phi_r + 0.001 \tag{10.46a}$$

(and Saitoh et al. (1995)):

$$\frac{w_p}{H} = 0.593\phi_r + 1.38 \tag{10.46b}$$

where $\phi_r = I/\left(2H^2\sqrt{ba\rho\sigma_0}\right)$, ρ is the density of the material, and I is the pulse.

10.7 Ultimate Strength of Stiffened Panels Under Impact Lateral Loads

The literature on the collapse behavior of stiffened panels subjected to impact lateral loads is very meager because of the complexity of the structural response phenomena involved. There are, however, a few closed-form formulas or analytical methods available that may help to predict stress levels or deflections of stiffened panels under impact lateral loads (Schubak et al. 1989, Nurick et al. 1994, 1995, Nurick & Jones 1995). The effects of impact loading on the elastic–plastic structural response of stiffened panels have mainly been investigated experimentally (Jones et al. 1991, Saitoh et al. 1995) or numerically (Smith 1989, Rudrapatna et al. 2000).

A simple formula to predict the damage to a stiffened panel under impact lateral loads was derived by Woisin (1979). He suggested that in the case of ship–ship collisions, the energy absorbed by the impacted and damaged plating in MJ equals $0.5 \sum_i h_i t_i^2$, where h_i (m) is the height of the broken or heavily deformed side shell plating or longitudinal bulkhead of constant thickness t_i (cm). The Woisin formula was derived on the basis of his own test results obtained using relatively large-scale collision test models. Woisin (1990) later modified his formula and suggested that the energy absorbed by a damaged plate under impact is better represented by $0.2 \sum_i h_i t_i d_i$, where d_i (m) is the distance between horizontal structural elements such as decks or stringers.

Jones et al. (1991) carried out a series of impact tests on steel and aluminum grillages (or cross-stiffened panels). The impactor had a mass of 3 kg and an impact speed between 3 and 7 m/s. They also predicted deformations using a quasistatic analysis, which showed fair agreement with the test results. The conclusion drawn from the study of Jones et al. (1991) strengthens the opinion that relatively low-velocity impact may be treated in a quasistatic manner as long as the strain-hardening effect is taken into account.

10.8 Crushing Strength of Plate Assemblies

10.8.1 Fundamentals of Crushing Behavior

Consider a plated structure under predominantly compressive loads as shown in Figure 10.15. Figure 10.16 represents a typical history of the resulting load versus displacement curve. As the compressive load increases, the structure eventually reaches the ultimate strength, which is the first peak in Figure 10.16. If the displacement continues to increase, the internal load will decrease rapidly. During the unloading process, some parts of the structure may bend or stretch. A lobe emerges, and the walls begin to fold. As the deformation continues, walls make contact with each other, which ends the first fold and initiates a new fold.

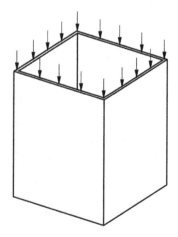

Figure 10.15 A plated structure under predominantly axial compressive loads.

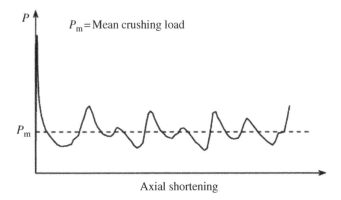

Figure 10.16 Crushing response of a thin-walled structure under predominantly compressive loads.

The internal load increases until the adjacent walls buckle. The structure begins to fold in a manner similar to the previous one. This process repeats itself until the entire structure is completely crushed. The completely folded structure then behaves as a rigid body until gross yielding occurs by compression. Each trough–peak pair in Figure 10.16 is associated with the formation of one structural fold.

Figure 10.17 shows a crushed tube after cutting off half of the structure. It is apparent that many folds form during the crushing process. The formation of these folds results in large axial compressive displacements. The folds usually develop sequentially from one end of the tube, so the phenomenon is known as progressive crushing. A ship's bow may possibly be designed to be crushable to achieve high energy absorption capacity in a collision accident.

In a usual loading condition in which the deformation is relatively small, the designer's primary concern is the structure's ultimate strength, or the initial peak load in Figure 10.16. In an accidental loading condition, however, the energy absorption

Figure 10.17 A plated structure crushed under axial compressive loads and cut at its midsection.

capability is the more likely concern. The peak load of the structure is not always of primary interest, and the analysis of the detailed crushing behavior is not an easy task.

As long as the energy absorption capacity is of primary concern, a convenient alternative is to predict the structure's "mean crushing load," which represents a mean value of fluctuating loads as shown in Figure 10.16. With this mean crushing load and the crushing displacement known, the absorbed energy can be calculated by multiplying these two values, which are approximately equal to the area below the corresponding load–displacement curve.

The insights gained from experiments on crushing thin-walled structures lead to an approximation widely applied in theoretical computations of the mean crushing strength: one structural fold forms at a time (Alexander 1960) so that one such fold can be analyzed independently. The influence of adjacent structures is therefore neglected. By means of the rigid-plastic theory, the mean crushing strength of structural elements can be estimated based on the kinematically admissible folding mechanisms. The response of complex plated structures is then calculated from the assemblage of such individual elements.

For the solution of crushing problems of thin-walled structures, in which a plane stress state can be assumed and the material is often modeled as a perfectly plastic material with a well-defined flow stress, σ_0, the classical theorem of upper bound plasticity is commonly used. Under this theorem, if the work rate of a structural system under the applied loads during any kinematically admissible collapse of the structure becomes equal to the corresponding internal energy dissipation rate, this system under the applied loads will be at the point of collapse.

The analytical approach to calculate the crushing strength is typically based on the introduction of rigid-plastic collapse mechanisms into the basic structural unit. Two lines of thought currently exist regarding the modeling technique for a structure (Paik & Wierzbicki 1997). One technique, the intersecting unit method (Amdahl 1983, Wierzbicki 1983, Pedersen et al. 1993), aims to model a structure as an assembly of typical intersecting units, such as L, T, Y, and X (or cruciform) sections. The other technique, the individual plate unit method (Murray 1983, Paik & Pedersen 1995), aims to model a structure as a collection of individual plate units. The intersecting unit method allows for several possible crushing mechanisms, and crushing occurs in the mode that gives the lowest crushing strength.

For a complex plated structure under accidental crushing loads, the reaction force versus crushing displacement relationship is obtained by computing the mean crushing loads and the corresponding crushed distance as the striking body crushes into the struck body. The mean crushing strength may be obtained as a sum of the mean crushing strengths for individual elements. The structure's energy absorption capability is then calculated by integrating the area below the reaction force versus crushing displacement curve. In the following two sections, we present some useful analytical expressions for the mean crushing strength characteristics of individual plate units and intersecting elements.

Although the derivation of all mean crushing strength formulations noted later is undertaken in a static loading condition, the effect of impact loading can be approximately accounted for with the use of the dynamic yield stress, which accommodates the effects of strain rate sensitivity using Equation (10.16) in place of the static yield stress. Strain-hardening effects may also be approximately included by using the so-called flow stress, σ_0, which is defined as the average of the yield stress, σ_Y, and the ultimate tensile stress, σ_T, given by $\sigma_0 = (\sigma_Y + \sigma_T)/2$.

10.8.2   A Plate

A plated structure may be regarded as an assembly of individual plate elements, as shown in Figure 10.18. When thin-walled structures made up of such elements are subjected to axial compressive loads in one of the plate directions, as shown in Figure 10.15, the other (unloaded) edges of the plate elements are usually connected with those of the surrounding structures. When at least two such plate edges meet, they can restrain each other and may even remain straight. Depending on the condition at the plate edges, the structure may have two different folding modes, as shown in Figures 10.19 and 10.20 (Murray 1983, Paik & Pedersen 1995): mode I, in which one unloaded edge remains straight and the other is free to deform, and

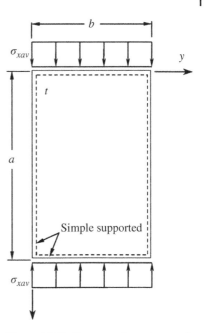

Figure 10.18 Geometry, boundary, and loading of a rectangular plate element.

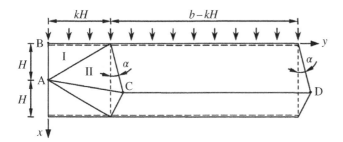

Figure 10.19 A schematic for the "one edge straight/one edge free" folding mechanism (mode I).

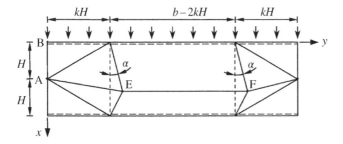

Figure 10.20 A schematic for the "both edges straight" folding mechanism (mode II).

mode II, in which both unloaded edges remain straight or the bending deformation of the edges is small.

For the folding mechanisms shown in Figures 10.19 and 10.20, the plastic energy is mainly dissipated by in-plane deformations in triangular regions and the horizontal/vertical hinge lines. The contributions from plastic bending in the plastic hinges and membrane stretching in these triangular regions may be summed and then divided by the effective crushing length to yield the mean crushing strength. Using such a procedure based on the rigid-plastic theory, Paik and Pedersen (1995) derived the following formulations for the mean crushing strength of unstiffened plate elements, namely,

$$\frac{\sigma_{xm}}{\sigma_0} = \frac{1}{\eta_x}\left(1.0046\sqrt{\frac{t}{b}} + 0.1332\frac{t}{b}\right) \quad \text{for mode I} \tag{10.47a}$$

$$\frac{\sigma_{xm}}{\sigma_0} = \frac{1}{\eta_x}\left(1.4206\sqrt{\frac{t}{b}} + 0.2665\frac{t}{b}\right) \quad \text{for mode II} \tag{10.47b}$$

where σ_{xm} is the mean crushing strength (stress) in the x direction, σ_0 is the flow stress ($\sigma_0 = \sigma_Y$ is taken when the strain-hardening effect is ignored), t is the plate thickness, b is the plate width along the loaded edge, and η_x is the normalized effective crushing length, taken as 0.728.

For a continuous plated structure, the use of the mean crushing strength formula for mode II, that is, Equation (10.47b), may be more relevant because the unloaded plate edges may in such cases remain straight.

10.8.3 A Stiffened Panel

Many engineering structures use stiffened panels. In terms of energy absorption capacity, a stiffened panel with stiffeners in the longitudinal direction parallel to the compressive loading direction can be approximately replaced by an unstiffened plate with equivalent wall thickness (Paik & Wierzbicki 1997). The equivalent wall thickness is an increased thickness that accounts for the cross-sectional area of both the plate and the stiffeners as follows:

$$t_{xeq} = t + \frac{A_{sx}}{b} \tag{10.48}$$

where t_{xeq} is the equivalent wall thickness of a stiffened panel in the x direction, A_{sx} is the cross-sectional area of the stiffeners in the x direction, and b is the spacing of the longitudinal stiffeners.

The contribution from transverse stiffeners in such cases can be neglected. The presence of longitudinal stiffeners in the direction of compressive loading usually has a significant effect (i.e., reduces) the effective crushing length because the stiffeners work against and thus disturb the folding process. Based on the crushing test data for thin-walled structures with stiffeners, Paik et al. (1996) derived an empirical formula by curve fitting to predict the effective crushing length as follows:

$$\eta_x = \begin{cases} 0.728 & \text{for } 0 < t_{xeq}/b \le 0.0336 \\ 704.49(t_{xeq}/b)^2 - 81.22 t_{xeq}/b + 2.66 & \text{for } 0.0336 < t_{xeq}/b < 0.055 \\ 0.324 & \text{for } 0.055 \le t_{xeq}/b \end{cases} \tag{10.49}$$

Because Equation (10.49) is based on the crushing test data for stiffened square tubes, it may be available for a stiffened panel with the unloaded edges simply supported. The longitudinal mean crushing strength, σ_{xm}, of a stiffened panel with longitudinal stiffeners is then predicted from Equations (10.47) but using the applicable effective crushing length and the equivalent wall thickness for the stiffened panel as follows:

$$\frac{\sigma_{xm}}{\sigma_0} = \frac{1}{\eta_x}\left(1.0046\sqrt{\frac{t_{xeq}}{b}} + 0.1332\frac{t_{xeq}}{b}\right) \text{ for mode I} \tag{10.50a}$$

$$\frac{\sigma_{xm}}{\sigma_0} = \frac{1}{\eta_x}\left(1.4206\sqrt{\frac{t_{xeq}}{b}} + 0.2665\frac{t_{xeq}}{b}\right) \text{ for mode II} \tag{10.50b}$$

A similar expression may be relevant for prediction of the transverse mean crushing strength, σ_{ym}, of a stiffened panel as follows:

$$\frac{\sigma_{ym}}{\sigma_0} = \frac{1}{\eta_y}\left(1.0046\sqrt{\frac{t_{yeq}}{a}} + 0.1332\frac{t_{yeq}}{a}\right) \text{ for mode I} \tag{10.51a}$$

$$\frac{\sigma_{ym}}{\sigma_0} = \frac{1}{\eta_y}\left(1.4206\sqrt{\frac{t_{yeq}}{a}} + 0.2665\frac{t_{yeq}}{a}\right) \text{ for mode II} \tag{10.51b}$$

where

$$\eta_y = \begin{cases} 0.728 & \text{for } 0 < t_{yeq}/a \leq 0.0336 \\ 704.49\left(t_{yeq}/a\right)^2 - 81.22 t_{yeq}/a + 2.66 & \text{for } 0.0336 < t_{yeq}/a < 0.055 \\ 0.324 & \text{for } 0.055 \leq t_{yeq}/a \end{cases}$$

$t_{yeq} = t + A_{sy}/a$ is the equivalent wall thickness of a stiffened panel in the y direction, A_{sy} is the cross-sectional area of the stiffeners in the y direction, and a is the spacing of the transverse stiffeners.

Under shearing force, the plate element can crush. In this case, it is assumed that the mean crushing strength, τ_m, of an unstiffened or stiffened panel equals the shear flow stress as follows:

$$\tau_m = \tau_0 = \frac{\sigma_0}{\sqrt{3}} \tag{10.52}$$

In combined loads, the following interactive relationship for crushing is suggested as a function of the average stresses acting on the panel element (Paik & Pedersen 1996):

$$f_c = \left(\frac{\sigma_{xav}}{\sigma_{xm}}\right)^2 + \left(\frac{\sigma_{yav}}{\sigma_{ym}}\right)^2 + \left(\frac{\tau_{av}}{\tau_m}\right)^2 - 1 = 0 \tag{10.53}$$

where f_c is the crushing function and σ_{xav}, σ_{yav}, and τ_{av} are average stress components.

10.8.4 An Inclined Plate

When the thin-walled structure with inclined sides is subjected to axial compressive loads as shown in Figure 10.21, the mean crushing strength may approximately be

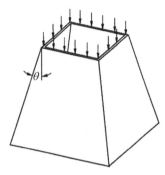

Figure 10.21 A plated structure under axial compressive loads in the vertical direction, with sides inclined at an angle.

calculated from Equations (10.47) or (10.50) but considering the effect of inclination as follows:

$$\sigma_{xm}^* = \frac{1}{\cos\theta}\sigma_{xm} \tag{10.54}$$

where σ_{xm}^* is the mean crushing strength for inclined loading and θ is the angle between the structure and the loading direction.

As previously noted, the dynamic loading effect may be approximately accounted for by replacing the flow or yield stress with the dynamic flow or yield stress, which is determined from Equation (10.16) or (10.17).

10.8.5 L-, T-, and X-Shaped Plate Assemblies

In contrast to the modeling presented in Section 10.8.2, a thin-walled structure may also be modeled as an assembly of intersecting elements such as L-, T-, or X-shaped elements (Paik & Wierzbicki 1997), as shown in Figure 10.22.

For L-shaped intersecting elements, two basic folding modes are relevant: the so-called quasi-inextensional and extensional modes, as shown in Figure 10.23 (Abramowicz & Wierzbicki 1989). The former mode consists of four trapezoidal elements that undergo rigid-body motions and are separated by plastic hinges. The horizontal plastic hinges are stationary. The vertical plastic hinges travel in the plate elements. These hinges are formed where the material is bent and re-bent.

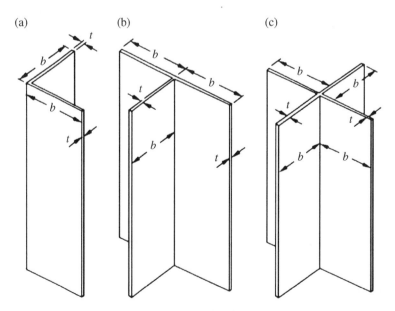

Figure 10.22 Intersecting plate element modeling of plated structures: (a) L-shape; (b) T-shape; (c) X-shape.

(a)   (b)

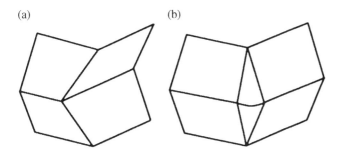

Figure 10.23 Schematic representation of the two basic folding modes for the L-shaped intersecting plate assembly: (a) quasi-inextensional mode; (b) extensional mode.

Although the structure may tend to follow an inextensional deformation mode, there may inevitably be extensions in some areas of the plate, as shown in Figure 10.23b. In this specific case, the extensional deformations in the circumferential directions are accommodated in a local zone of the vertical hinges. The extensional folding mode also consists of four trapezoidal elements that bend about the horizontal plastic hinges. As the vertical plastic hinges moves, the shape of the four elements distorts, and the material is stretched around the vertical connection.

Although an L-shaped structure under crushing loads involves either a quasi-inextensional or extensional mode, complex intersecting elements with T- or X-shapes exhibit many more different folding patterns. These complicated symmetric and asymmetric and mixed folding patterns may be the results of different combinations of the two basic folding modes earlier. Any one such particular combination is generally triggered by the geometry of the structure, the shape of the initial buckling mode, and the initial imperfections.

Ohtsubo and Suzuki (1994), among others, derived the mean crushing strength formulations for L-, T-, and X-shaped intersecting elements as follows:

$$\frac{\sigma_m}{\sigma_0} = \frac{1.5165}{\eta} \left(\frac{t}{b}\right)^{2/3} \quad \text{for L-shaped elements} \tag{10.55a}$$

$$\frac{\sigma_m}{\sigma_0} = \frac{1.1573}{\eta} \left(\frac{t}{b}\right)^{2/3} \quad \text{for T-shaped elements} \tag{10.55b}$$

$$\frac{\sigma_m}{\sigma_0} = \frac{1}{\eta} \left(1.2499\sqrt{\frac{t}{b}} + 0.2493\frac{t}{b}\right) \quad \text{for X-shaped elements} \tag{10.55c}$$

where σ_m is the mean crushing strength and η is the normalized effective crushing length, which may be taken as $\eta = 0.728$ for the unstiffened elements.

The effects of stiffeners, dynamic loading, or inclined loading may approximately be accounted for, as previously presented in Section 10.8.2.

10.9 Tearing Strength of Plates and Stiffened Panels

10.9.1 Fundamentals of Tearing Behavior

When a ship with a forward speed runs aground on a rock, its bottom may be torn from the initial impact. If the kinetic energy is not entirely spent during the initial impact, the ship will run atop the rock for a distance. As a result, the grounding damage on the bottom can become a long gash that extends tens or even hundreds of meters in length.

Similarly, when a ship collides with the side structure of another ship, the deck structure of the struck ship may be cut and penetrated by the striking bow. Steel is torn, separated, and bent in a manner similar to a ship bottom in a grounding accident. The plate tearing also plays an important role in absorbing impact energy in a collision accident.

For purposes of investigating structural strength under such conditions, many experiments were undertaken in the 1980s by dropping a heavy wedge into a vertical or near-vertical steel plate in a drop-hammer rig (Vaughan 1980, Woisin 1982, Jones & Jouri 1987). Quasistatic tests were also carried out in the 1990s when wedges were pushed very slowly into plates (Lu & Calladine 1990, Wierzbicki & Thomas 1993, Paik 1994). A quasistatic test has the advantage of continuous recording of various features.

Figure 10.24 shows a schematic of a cutting test setup in which a sharp wedge is pushed into a steel plate. The wedge in such a case is generally made to be rigid so that the impact energy is entirely absorbed by the steel plate. As the wedge is pushed into the plate, the plate buckles and bends out of plane. The load increases to a peak and then declines, but there is no separation of the material. Eventually, as the wedge pushes further, cutting commences, and loading picks up again. The plate is torn apart in front of the wedge tip in the transverse direction. The separated material then bends over, forming two curls or flaps. The wedge keeps pushing the curved plate flaps, which roll up in the wake of the wedge. Near the wedge tip, the plate also develops a global deformation pattern, where the plate deforms out of the plane and separates. Under some circumstances, the plate may bend in the opposite direction, and the curls reverse. Figure 10.25 shows plates cut by wedges, representing tearing and curling of the plate.

In such a process, there are several distinct mechanisms by which the energy is absorbed, namely, tearing, bending, and friction. In the vicinity of the wedge tip, the material is stretched transversely; the stress state there is mainly a result of membrane stretching. In the wake of the wedge, the plate bends out of plane; the stress state there is primarily due to plastic bending. As the wedge moves, its side makes contact with the plate, and friction builds up and consumes a portion of the energy.

The following sections present analytical and empirical formulations for the relationship between the tearing force, F, and penetration, ℓ. All these formulations are derived in a quasistatic loading condition, but the dynamic effect may approximately be accounted for by including the effects of strain rate sensitivity on the material yield stress.

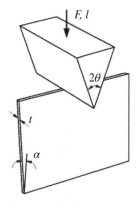

Figure 10.24 A schematic of a cutting test setup for a plate.

(a)

(b)

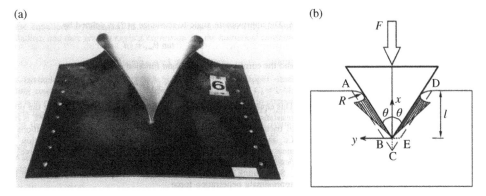

Figure 10.25 (a) A typical picture of tearing and curling of a plate upon being cut by a wedge (Thomas 1992); (b) a schematic of the analytical model on plate cutting by a rigid wedge (Zhang 2002).

10.9.2 Analytical Formulations

There are two distinct plastic deformation processes in play: near-tip tearing and global bending in the far field. Figure 10.26 shows a schematic of the plate damage, including tearing and curling considered by Wierzbicki and Thomas (1993).

The work involved in the bending of the two curls or flaps in the far field includes contributions from both a continuous velocity field and a discontinuous velocity field. The integration of the continuous deformation field is performed over the plastically deforming zone, whereas the contribution from a discontinuous field is summed over a finite number of straight-line segments. These discontinuous velocity fields are related to local plastic hinges.

The calculation of plastic membrane work near the crack tip can be based on either a traditional rigid-plastic approach or a fracture mechanics approach. In the rigid-plastic approach, such as that of Ohtsubo and Wang (1995), the membrane work to stretch the material is integrated over a continuous plastically deforming field. The extent of this membrane stretched plate may be determined using a criterion based on critical rupture strain (Zhang 2002). The fracture mechanics approach (e.g., Wierzbicki & Thomas 1993, Simonsen & Wierzbicki 1998) describes the local stress state using relevant parameters,

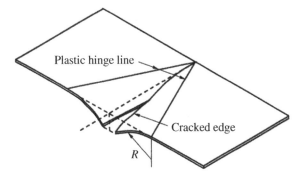

Figure 10.26 A schematic of the damage of a plate cut by a wedge.

such as the crack tip opening displacement (CTOD), and calculates the work required to propagate the crack as a function of the CTOD parameter (see Chapter 9).

Membrane stretching and curl bending are not independent; they are related through a single geometric parameter. This parameter is denoted by R, as depicted in Figure 10.26, which represents the instantaneous bending or rolling radius of the cylindrical flaps in the wake of the wedge. Specifically, the far-field bending work is inversely proportional to R, whereas the membrane work in the near-tip zone increases with the rolling radius.

The analytical expression for the plate tearing force versus the cutting length derived by Ohtsubo and Wang (1995), among others, using the rigid-plastic method is as follows:

$$F = 1.51\sigma_0 t^{1.5} l^{0.5} (\sin\theta)^{0.5} \left(1 + \frac{\mu}{\tan\theta}\right) \tag{10.56}$$

where σ_0 is the flow stress, t is the plate thickness, l is the tearing length, 2θ is the spreading angle of the wedge, and μ is the friction coefficient.

Using the fracture mechanics approach, Wierzbicki and Thomas (1993), among others, derived the following expression:

$$F = 1.67\sigma_0 (\delta_t)^{0.2} t^{1.6} l^{0.4} \frac{1}{(\cos\theta)^{0.8}} \left[(\tan\theta)^{0.4} + \frac{\mu}{(\tan\theta)^{0.6}} \right] \tag{10.57}$$

where δ_t is the CTOD parameter as described in Chapter 9.

Considering the critical rupture strain to membrane stretching, Zhang (2002) derived a semi-analytical expression as follows:

$$F = 1.942\sigma_0 t^{1.5} l^{0.5} \varepsilon_f^{0.25} (\tan\theta)^{0.5} \left(1 + \frac{\mu}{\tan\theta}\right) \tag{10.58}$$

where ε_f is the critical rupture strain.

Such analytical formulations noted earlier accommodate the coefficient of friction and the wedge's spreading angle as parameters of influence. In most laboratory experiments, there is no sign of a crack extending in front of the wedge tip, and local necking of the plate material is observed in that area before the material is separated. It is thus generally recognized that the material in front of the wedge may be torn apart rather than cut by the wedge, yet such a mechanism is very difficult to include precisely in an analytical model. The material separation in front of the wedge is due to ductile failure. The rigid-plastic analysis approach describes a local deformation zone near the wedge tip, whereas the fracture mechanics approach explains the mechanism that drives the crack. Both approaches calculate the internal energy absorbed in the local area in the vicinity of the wedge tip, and the two estimates can be similar. In fact, the tearing force depends very weakly on the value of δ_t, the CTOD parameter, because the exponent involved is 0.2, as per Equation (10.57).

Once the relationship between the tearing force, F, and the cutting length, ℓ, is known as the wedge is pushed into the plate, the strain energy, W, that is absorbed until $\ell = \ell_m$ is reached can be calculated by integrating the area below the F–ℓ curve as follows:

$$W = \int_0^{\ell_m} F \, d\ell \tag{10.59}$$

10.9.3 Empirical Formulations

Based on mechanical test results, empirical formulations for the tearing force may also be derived via dimensional analysis. The tearing force, F, depends on the wedge's geometric parameters, the thickness of the plate, and the tearing length. It also depends on the material yield stress, because plastic deformation is evident in the process of penetration. The parameter of Young's modulus is not included because deep penetration is concerned, whereas elastic deformation is limited to the stage up to the initial buckling.

If the dimensions of the problem are taken as the tearing load and tearing length, there are two different dimensionless parameters involved. If two other parameters, that is, the plate thickness and the yielding stress, are included, the problem involves four variables. The so-called Buckingham's Pi rule (Buckingham 1914, Jones 2012) tells us that a four-variable problem has two independent dimensionless groups. One such expression that satisfies the rule is

$$\frac{F}{\sigma_0 t^2} = C \left(\frac{\ell}{t}\right)^n \tag{10.60}$$

where F is the tearing force, σ_0 is the flow stress, t is the plate thickness, ℓ is the tearing length, and C and n are constants. The value of C depends on many factors, including the geometry of the wedge and the parameters for friction. The value of n reflects the interdependence of the major energy absorption mechanisms.

Based on curve fitting of their test results for unstiffened high tensile steel plates cut by a rigid wedge, Lu and Calladine (1990) found that the value of n is between 0.2 and 0.4. The thickness of their test plates was in the range of 0.7 and 2 mm, and $\alpha = 0°$ and $2\theta = 20$ and $40°$. They further simplified the expression by using a single value of $n = 0.3$ and worked out the corresponding best-fitting values for C. Their resulting empirical relation is as follows:

$$F = C\sigma_0 t^{1.7}\ell^{0.3} \quad \text{for} \quad 5 \le \frac{\ell}{t} \le 150 \tag{10.61}$$

where C is a constant that is dependent on the materials or test conditions.

Paik (1994) performed a series of cutting tests on high tensile steel panels with longitudinal stiffeners. Paik's test panels were 3.4–7.8 mm thick, and $\alpha = 0°$ and $2\theta = 15, 30, 45,$ and $60°$. This test series demonstrated a dependence of the value C in Equation (10.60) on the wedge's geometrical parameter in such a case. A least-squares best fit to the experimental data provides the following expression:

$$F = 1.5C\sigma_0 t_{eq}^{1.5}\ell^{0.5} \tag{10.62a}$$

where t_{eq} is the equivalent plate thickness, as defined in Equation (10.44), and C is a parameter that accounts for the influences of a wedge's geometry, which is a function of the spreading angle of the wedge:

$$C = 1.112 - 1.156\theta + 3.760\theta^2 \tag{10.62b}$$

where θ is as defined in Figure 10.24 (in rad).

It is noted that although Equation (10.61) is used for unstiffened plates, Equation (10.62) will be used for longitudinally stiffened panels. Although Equations (10.61) and (10.62) are based on the test results for relatively thin panels, they may be approximately applied to thicker plates.

Figure 10.27 compares between the plate tearing force expressions for a given case. In this comparison, a mild steel plate with $t = 8$ mm and $\sigma_0 = 270$ MPa is considered to be cut by a rigid wedge. The coefficient of friction is assumed to be 0.25, and the critical rupture strain is taken as $\varepsilon_f = 0.25$. Although the Paik formula is basically applicable to cutting of longitudinally stiffened panels, it is applied with $t_{eq} = t$ in this calculation.

It is seen from Figure 10.27 that good agreement between all methods is achieved when the wedge angle is relatively small, that is, $2\theta = 40°$, whereas some differences appear when the wedge angle is relatively large, that is, $2\theta = 90°$. For the case of a larger wedge

(a)

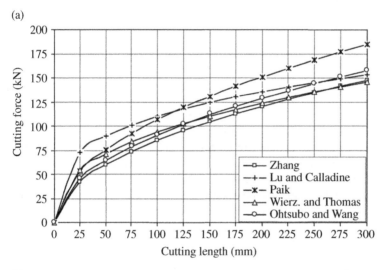

(b)

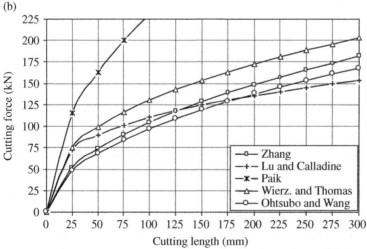

Figure 10.27 A comparison of cutting forces calculated by different methods for wedge angle of (a) $2\theta = 40°$ and (b) $2\theta = 90°$, undertaken by Zhang (2002).

angle that has a relatively large wedge width far behind the wedge tip in comparison with stiffener spacing, the Paik formula predicts larger tearing forces because it is mainly derived on the basis of longitudinally stiffened panels, whereas the rest of the formulas are only based on unstiffened plates.

10.9.4 Concertina Tearing

When a stiffened panel is cut by a wedge, the plate may crack in the vicinity of stiffening members. Two or more cracking lines advance with the wedge; there is no crack directly in front of the wedge tip, and the plate is folded like an accordion. This type of tearing process is sometimes termed "concertina tearing." Figure 10.28 shows a photo of concertina tearing on a stiffened panel cut by a wedge.

Concertina tearing is similar to the usual tearing process in the sense that the material is separated. It is also similar to the crushing of plating in the sense that the structure is folded. Applying the rigid-plastic theory, Wierzbicki (1995) derived an analytical formula of the mean concertina tearing load for a steel plate. In this approach, the length of the folds is treated as a parameter in the relationship between the tearing load and the indentation, thus linking membrane stretching and bending at the plastic hinges. The mean concertina tearing load formula is then derived by minimizing the load with regard to this parameter. The Wierzbicki formula of the mean concertina tearing load is given by

$$F_{\mathrm{m}} = \frac{1}{\lambda}\left(3.25\sigma_0 b^{0.33} t^{1.67} + 2Rt\right) \tag{10.63}$$

where F_{m} is the mean concertina tearing load, σ_0 is the flow stress, t is the plate thickness, b is the width of the folded plate, λ is the factor for effective crushing length, and R is the fracture parameter.

Figure 10.28 A concertina tearing.

10.10   Impact Perforation of Plates

The perforation of structural members by various objects has been of interest to the engineering community for over 300 years and to the military establishment for even longer (Johnson 2001, Blyth & Atkins 2002). This apparently simple topic is, in fact, a complex one because of the plethora of variables.

The striking mass (missile, striker or projectile) is characterized by many parameters including its shape, mass, impact velocity, structural strength, and hardness, though, for simplicity, the mass is often taken as rigid. The geometrical shapes range from military weapons having highly specified shapes to amorphous objects, which emanate from accidental explosions of pressure vessels or are dropped onto structures. The impact velocities can embrace several orders of magnitude and might elicit a range of markedly different responses. The target itself can have different geometries and be made from a wide range of materials, some ductile, others not, and having various strain-hardening and strain rate sensitive characteristics. The mass could strike at an oblique angle at any location on a plate, for example, near a hard point, impact velocities, etc. Thus, experimental data is available only for small subsets of the entire field that contains, as a consequence, many lacunae.

However, empirical equations have been developed largely on the basis of limited experimental data. These equations are strictly restricted to the range of applicability of the experimental test parameters, but they are often used outside these restrictions in design situations and for hazard assessments. It is clear that empirical equations are valuable for securing quick estimates for the perforation energies of plating. In the fullness of time, numerical schemes have been developed as design tools, but they do require considerable input from experimental tests on the strain rate, temperature, and failure characteristics of the target and missile materials (Borvik et al. 2009, Rusinek et al. 2009).

Impact velocities associated with missiles are higher than about 100 m/s. The perforation of plating in the lower velocity range is associated with dropped objects and large relatively slow moving fragments from explosive and other dynamic incidents. It is recognized that the impact perforation of thick and thin plates needs to be distinguished at high velocities (e.g., hypervelocity impact (Anderson 2001)), for which local effects, including possible temperature related phenomena, dominate global effects, which are often neglected entirely.

To discriminate more clearly between thin and thick plates, Backman and Goldsmith (1978) have examined the number of elastic wave transversals, n, through a target thickness, H, during the single traverse of an elastic wave along the length, L, of a missile as follows:

$$n = \frac{C_t\,L}{C_m\,H} \tag{10.64}$$

where C_t and C_m are the elastic wave speeds in the target and missile, respectively, L is the length of projectile, and H is the plate thickness.

If $n \gg 1$, then it is assumed that the elastic stress state in a target underneath the missile would be uniform across the target thickness before an elastic stress wave reaches the end of the missile. In this case, such a target could be considered as thin. An actual impact problem with inelastic behavior and perforation would be much more complex than this simplified analysis, but Backman and Goldsmith (1978) suggested that plate impact problems with $n > 5$, approximately, could be regarded as thin, those with $n < 1$ as thick, and $1 < n < 5$ for plates having an intermediate thickness.

Equation (10.64) is independent of the impact velocity, V_o, which is an important parameter in a dynamic problem. However, Johnson (1972) has introduced the damage number, Φ, as follows:

$$\Phi = \frac{\rho V_o^2}{\sigma} \tag{10.65}$$

where ρ is the density of plate material, σ is the mean dynamic flow stress, and V_o is the impact velocity.

By rearranging Equation (10.65) as $\rho V_o^2 = \sigma \Phi$, it is evident that Φ gives an estimate of the order of the strain in the region where severe plastic deformation occurs. Equation (10.65) might be used to suggest the phenomenon, which most likely governs the response of a particular impact problem. Another way to characterize a missile-plate response is to compare the time it takes for a plastic hinge to travel across a plate (t_h) with the time taken for a perforation failure to develop (t_p). A rough estimate $t_p = 2H/V_o$ is assumed for the perforation time of a projectile traveling with an impact velocity V_o, and an average velocity $V_o/2$, to travel one plate thickness H. A rigid-perfectly plastic analysis for a circular plate of radius R struck at the center by a mass with a cylindrical body and a flat face is developed in reference (Liu & Jones 1996), but a numerical analysis is required to obtain the hinge speed and, therefore, the time t_h.

This is also true for a similar analysis by Florence (1977) for the dynamic pressure load acting over a central circular area of a circular plate. Thus, an estimate is made for t_h by evaluating the time taken for a plastic hinge to travel across a simply supported circular plate, which is loaded impulsively over the entire plate. This expression when divided by t_p gives

$$\frac{t_h}{t_p} = \frac{\Phi}{6}\left(\frac{R}{H}\right)^2 \tag{10.66}$$

where R is the radius of a circular plate.

Equation (10.66) indicates that if $t_h \ll t_p$, then the entire plate would participate in the perforation process, that is, the global deformations of a plate would be important throughout the response. On the other hand, if $t_h \gg t_p$, then the perforation process would be highly localized with insufficient time for global effects to develop.

For an example of a steel plate, $t_h/t_p = 0.033$ gives $\Phi = 2 \times 10^{-3}$ when $R/H = 10$ with $\rho = 7850 \text{ kg/m}^3$, $\sigma = 392.5 \text{ MPa}$, and $V_o = 10 \text{ m/s}$. This suggests that global effects (e.g., membrane forces) would be an important aspect of the response because they would have developed well before perforation had occurred. If $V_o = 100 \text{ m/s}$ for the same plate, then $t_h/t_p = 3.33$, so that Equation (10.66) suggests that both local and global effects would be important during the perforation process. At $V_o = 1000 \text{ m/s}$, $t_h/t_p = 333$, which implies that the perforation for this plate is highly localized, because there is insufficient time for any global deformations to develop, and for the disturbance to be transmitted away from the vicinity of the impact site through the action of the plastic hinge movement.

For another example of an aluminum alloy plate, $t_h/t_p = 0.016$ gives $\Phi = 10^{-3}$ when $R/H = 10$ with $\rho = 2720 \text{ kg/m}^3$, $\sigma = 272 \text{ MPa}$, and $V_o = 10 \text{ m/s}$. If $V_o = 100 \text{ m/s}$ for the same plate, then $t_h/t_p = 1.6$, so that Equation (10.66) suggests that both local and global effects would be important during the perforation process. At $V_o = 1000$ m/s, $t_h/t_p = 166$, which also implies that the perforation for this plate is highly localized.

The perforation velocity is taken as the average between the maximum velocity that does not cause a projectile to pass through a plate and the minimum recorded velocity, which causes perforation of a plate. The threshold value of the perforation velocity is the limiting case with no residual, or exit, velocity of a projectile after perforation. However, in some experimental studies, perforation is taken to occur when the distal side of a plate is cracked.

The low-velocity behavior is related to impact velocities causing perforation up to about 20 m/s, which can be regarded as a quasistatic response for which global effects are significant. This topic is of interest for drop weight loading incidents and for the effect of large fragments emanating from explosive failures of pipelines and pressure vessels. For high-impact velocities, say, above about 300 m/s, then local effects generally dominate the response with less important global effects. In moderate-impact velocities in the approximate intermediate velocity range 20–300 m/s, both global and local effects control the response. The influence of global effects decreases as the impact velocity increases within this range, as indicated by the predictions of Equation (10.66) for t_h/t_p.

For more elaborate descriptions, interested readers may refer to Backman and Goldsmith (1978), Goldsmith (1999), Corbett et al. (1996), and Jones and Paik (2012, 2013), among others. Several existing and recently developed empirical equations are compared with some recent experimental results reported for mild steel and stainless steel plates in Jones and Paik (2012) and for aluminum plates in Jones and Paik (2013).

10.11 Impact Fracture of Plates and Stiffened Panels at Cold Temperature

Plated structures are likely to be subjected to impact loads at cold temperature. Paik et al. (2011) investigated the effects of low temperature (−40°C and −60°C) on the crushing response of steel-plated structures. Dipaolo and Tom (2009) examined the same topic at −45°C. McGregor et al. (1993) studied the crushing characteristics of aluminum-plated structures and found that the average crushing force of hexagonal aluminum box sections increased as the temperature decreased (from room temperature to −40°C). With respect to impact loads at low temperatures, Min et al. (2012) conducted an experiment associated with the plastic deformation of steel-plated structures subjected to impact loads (~5–5.5 m/s) and performed comparative studies through numerical analysis, where the experiment was conducted at −30°C and −50°C. Manjunath and Surendran (2013) studied dynamic fracture toughness of aluminum 6063 with multilayer composite patching at lower temperatures. Kim et al. (2016) performed an experimental and numerical study to examine the nonlinear impact response of steel-plated structures in an Arctic environment (at −60°C), involving buckling, yielding, crushing, and brittle fracture.

Figure 10.29 shows selected pictures representing the brittle fracture response of a cross-stiffened steel panel (with two flat-bar stiffeners) under centrally concentrated impact loading at −60°C, obtained by a dropped-object test (Kim et al. 2016).

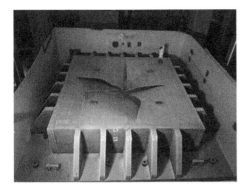
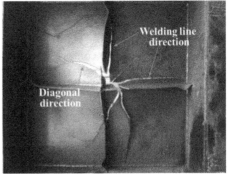

Figure 10.29 Impact fracture response of a steel-stiffened panel at cold temperature (Kim et al. 2016).

It is apparent from this figure that a metal structure exposed to cold temperature can suffer brittle fracture failure under impact loading in contrast to a structure at room temperature that may show ductile fracture or rupture. It is also observed that the aspects of crack propagation are quite complex in association with the welding line or in an arbitrary direction. Figure 10.29 also shows the local buckling (flexural-torsional buckling) failure that has occurred in the stiffeners.

10.12   Ultimate Strength of Plates Under Impact Axial Compressive Loads

Paik and Thayamballi (2003) provided an experimental data on the ultimate strength of steel plates under impact axial compressive loads. A series of impact collapse tests were carried out on a steel square plate ($a \times b \times t$ = 500 mm × 500 mm × 1.6 mm with a yield stress (σ_Y) of 251.8 MPa and Young's modulus (E) of 198.5 GPa) under axial compressive loads, varying the loading speed V_o in the range of 0.05–400 mm/s, which corresponds to strain rates in the range of 10^{-4} to 0.8 s^{-1}. where the strain rate $\dot{\varepsilon}$ is approximately determined as $\dot{\varepsilon} = V_o/a$. Based on the test data, the effect of loading speed (or strain rate) on the plate ultimate strength was investigated. Relevant useful formulations to predict the impact ultimate compressive strength of plates were also derived.

Figure 10.30 shows the ultimate strength behavior of the plate under impact axial compressive load, varying the loading speed. Figure 10.31 shows the impact ultimate compressive strength of the plate as a function of the strain rate. It is interesting to note that the impact ultimate compressive strength of a plate may be calculated using the dynamic yield strength of material that can be defined from the Cowper–Symonds equation described in Section 10.3.2. For a direct calculation, Paik and Thayamballi (2003)

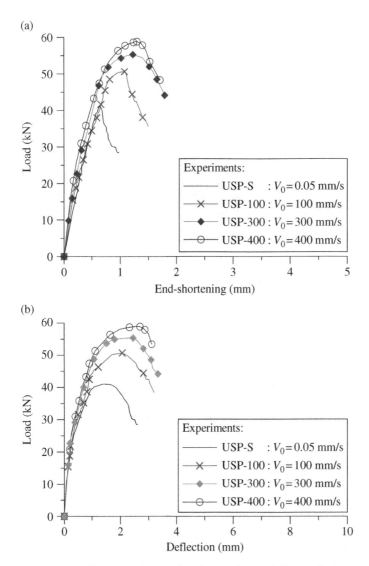

Figure 10.30 The ultimate strength behavior of a steel plate under impact compressive load, plotted versus (a) the end shortening and (b) the lateral deflection, as obtained by the experiment (Paik & Thayamballi 2003).

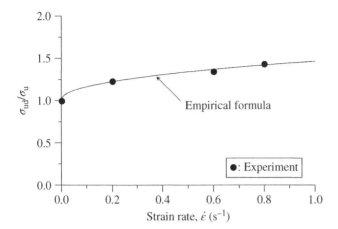

Figure 10.31 The impact ultimate compressive strength of a steel plate, plotted versus the strain rate.

derived an empirical formulation that is similar to the Cowper–Symonds equation in expression, Equation (10.17), as follows:

$$\frac{\sigma_{ud}}{\sigma_u} = 1.0 + \left(\frac{\dot{\varepsilon}}{D}\right)^{1/q} = 1.0 + \left(\frac{V_o}{aD}\right)^{1/q} \tag{10.67}$$

where σ_{ud} is the dynamic ultimate compressive strength of a plate, σ_u is the quasistatic ultimate compressive strength of a plate, $\dot{\varepsilon} = V_o/a$, V_o is the loading speed, a is the plate length, and D and q are the coefficients that are taken as $D = 5.41 s^{-1}$ and $q = 2.21$ for a steel plate. Figure 10.31 confirms the validity of Equation (10.67).

10.13 Ultimate Strength of Dented Plates

Plated structures may suffer mechanical damage in many ways. In offshore platforms, local denting damage occurs in deck plate panels subjected to impacts due to objects dropped from a crane. Inner bottom plate panels of cargo holds of bulk carriers often suffer mechanical damage by mishandled loading or unloading of cargoes; iron ore cargo strikes the plate panels during loading, and excavator hits the inner bottom plate panels in unloading of bulk cargoes such as coal or iron ore. While such mechanical damage may involve various features such as denting, cracking, residual stresses or strains due to plastic deformation, and coating damage, the load-carrying capacity of dented plate panels can be reduced.

As shown in Figure 10.32, local denting damage of plate panels on striking or being struck by an obstacle may depend on the shape and sharpness of the obstruction, among other factors such as mass and impact velocity. The formation of impact damage can be not only localized dent together with global deformation but also perforation or tearing, the latter being described in Section 10.10.

Figure 10.33 defines the geometric parameters for two types of typical dent shapes, namely, spherical shape and conical shape. The center of the localized dent is located

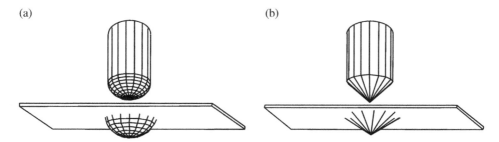

Figure 10.32 A schematic of local denting damage for a plate struck by (a) a spherical obstacle and (b) a conical obstacle.

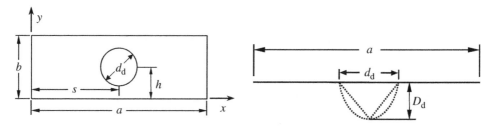

Figure 10.33 Geometric parameters of local denting damage.

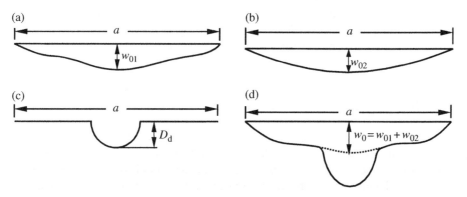

Figure 10.34 A schematic of spherical shape of denting damage: (a) the welding induced initial deflection; (b) the global deformation by denting; (c) the local spherical dent; (d) the combined initial deflection and denting damage.

in s in the x coordinate and h in the y coordinate. In general, denting causes both local dent and global deflection as shown in Figures 10.34 and 10.35. It is considered that the latter type of damage (global deflection) may affect the plate collapse behavior as if post-weld initial deflection does. In fact, global deflection due to denting will be inherently superposed by welding induced initial deflections. In this regard, global deflection by denting may be treated as a type of welding induced initial deflection in the plate strength calculations.

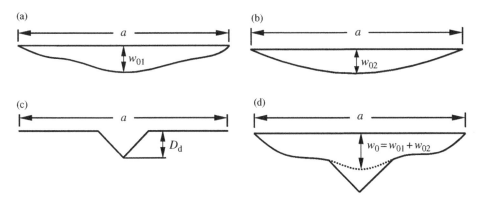

Figure 10.35 A schematic of conical shape of denting damage: (a) the welding induced initial deflection; (b) the global deformation by denting; (c) the local spherical dent; (d) the combined initial deflection and denting damage.

10.13.1 A Dented Plate in Axial Compression

Paik et al. (2003) investigated the ultimate strength behavior of plates under axial compressive loads using the nonlinear finite element method, where the plate is 2400 mm long and 800 mm wide, while the plate thickness t is varied. The material of the plate is high tensile steel with a yield strength of $\sigma_Y = 352.8$ MPa, an elastic modulus of $E = 205.8$ GPa, and Poisson ratio's of $\nu = 0.3$. The plate is considered to have the welding induced initial deflection (buckling mode component) of $w_{0pl} = 0.1\beta^2 t$ where $\beta = (b/t)\sqrt{\sigma_Y/E}$ is the plate slenderness ratio. The plate is assumed to be simply supported at all four edges.

Figure 10.36 shows selected illustrations of a plate with spherical shape or conical shape of local denting damage. Figure 10.37 shows the deformed shape of the plate with or without denting immediately after the ultimate strength is reached in axial compression. Figure 10.38 shows membrane stress distribution inside the plate at the ultimate limit state. It is observed from these figures that the deflected patterns and membrane stress distributions of dented plates are similar to those of un-dented (intact) plates as long as dent damage is limited to some local region.

Figures 10.39 and 10.40 show the nonlinear finite element method solutions for the effects of dent parameters (dent depth, diameter, and location) and plate aspects (thickness and aspect ratio) on the ultimate compressive strength behavior of dented plates. It is found that the ultimate compressive strength decreases significantly as the depth and/ or diameter of local denting increases. Also, it is apparent that the collapse behavior for spherical dent is similar to that for conical dent, while the former case is more likely to reduce the load-carrying capacity than the latter as long as the depth and diameter of denting are the same. It is also observed that the plate falls in the worst situation in terms of the load-carrying capacity when the local denting is located at the plate center rather than other places.

It is evident that the size (depth, diameter) and location of local denting are generally quite sensitive to the normalized ultimate compressive strength, while the influence of the dent depth on the plate ultimate compressive strength is not significant as long as the

(a)

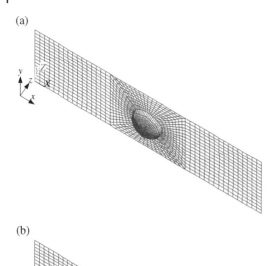

(b)

Figure 10.36 Selected illustrations of a plate with (a) the spherical shape of local denting and (b) the conical shape of local denting.

dent diameter is small. Also, since the plate collapse behavior for spherical dent is similar to that for conical dent and the former is slightly worse than the latter, the spherical dent may be taken as a representative of local dent shape for the purpose of the plate ultimate strength prediction, regardless of the actual shape of denting. It is seen from Figure 10.40f that as the dent location becomes closer to the unloaded plate edges the ultimate strength is decreased by 20%, compared with that of the dent located at the plate center.

As described in Section 4.12, the ultimate compressive strength of a dented plate may be predicted by an empirical formulation based on the strength reduction factor approach, that is, $R_{xu} = \sigma_{xu}/\sigma_{xuo}$, where R_{xu} is the strength reduction factor. In this case, the plate thickness and the aspect ratio are not influential parameters to the strength reduction factor because they are already implemented in the determination of σ_{xuo}:

$$\frac{\sigma_{xu}}{\sigma_{xuo}} = C_3 \left[C_1 \ln\left(\frac{D_d}{t}\right) + C_2 \right] \tag{10.68a}$$

Figure 10.37 Deflected shape of a plate with or without denting at the ultimate limit state under axial compressive loads: (a) an un-dented plate; (b) a dented plate with spherical type.

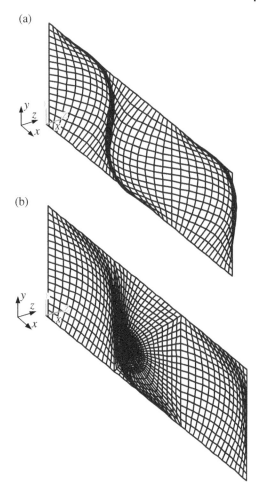

(a)

(b)

where σ_{xu} and σ_{xuo} are the ultimate compressive strengths of dented or un-dented (intact) plates, respectively, and C_1, C_2, and C_3 are coefficients that are empirically determined by curve fitting of the computed results, when $d_d/b < 1$, as follows:

$$C_1 = -0.042\left(\frac{d_d}{b}\right)^2 - 0.105\left(\frac{d_d}{b}\right) + 0.015 \qquad (10.68b)$$

$$C_2 = -0.138\left(\frac{d_d}{b}\right)^2 - 0.302\left(\frac{d_d}{b}\right) + 1.042 \qquad (10.68c)$$

(a)

(b)

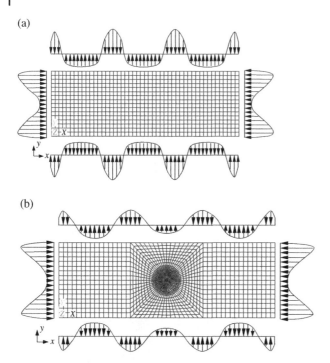

Figure 10.38 Membrane stress distribution with or without denting at the ultimate limit state under axial compressive loads for (a) an un-dented plate and (b) a dented plate with spherical type.

$$C_3 = -1.44 \left(\frac{H}{b}\right)^2 + 1.74 \left(\frac{H}{b}\right) + 0.49 \qquad (10.68\text{d})$$

with $H = h$ for $h \le \dfrac{b}{2}$ and $H = b - h$ for $h > \dfrac{b}{2}$.

Figure 10.41 confirms that Equation (10.68) gives reasonably accurate results for the plate ultimate strength reduction factor compared with the nonlinear finite element method solutions. It is interesting to note that the ultimate compressive strength of a dented plate with large dent depth is close to that of a perforated plate described in Section 4.10, as shown in Figure 10.41c.

Interested readers may also refer to Saad-Eldeen et al. (2015, 2016), Raviprakash et al. (2012), Xu and Guedes Soares (2013, 2015), and Li et al. (2014), among others. For a dented plate in axial tension, Li et al. (2015) may be referred to.

10.13.2 A Dented Plate in Edge Shear

Paik (2005) investigated the ultimate strength behavior of a dented plate in edge shear using the nonlinear finite element method. The same plate presented in Section 10.13.1 is used. Figure 10.42 shows the deformed shape of a plate with or without local dent immediately after the ultimate strength is reached in edge shear.

(a)

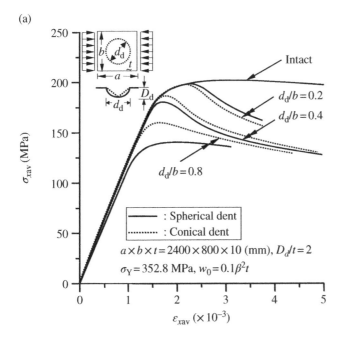

(b)

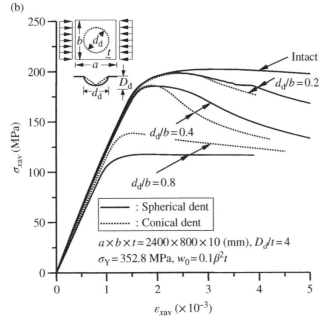

Figure 10.39 The ultimate compressive strength behavior of a dented plate with $t = 10$ mm, plotted versus (a) $D_d/t = 2$; (b) $D_d/t = 4$; (c) $D_d/t = 10$ and (d) $h/b = 0.25$–0.5.

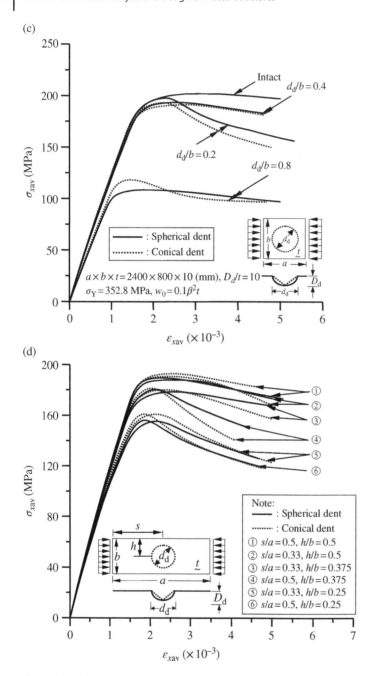

Figure 10.39 (Continued)

(a)

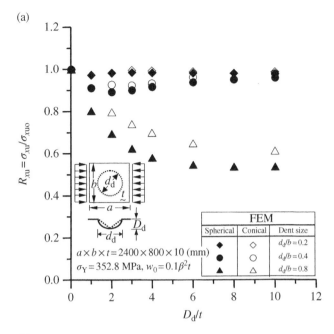

(b)

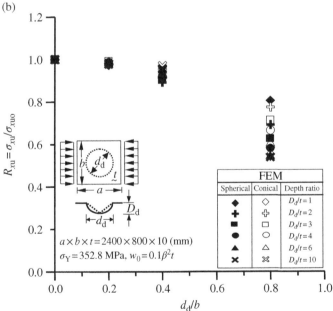

Figure 10.40 The effect of various dent parameters on the ultimate compressive strength of a dented plate, plotted versus (a) the dent depth; (b) the dent diameter; (c) the plate thickness with dent diameter; (d) the plate aspect ratio with dent diameter for a thin plate; (e) the plate aspect ratio with dent diameter for a thick plate and (f) the dent location.

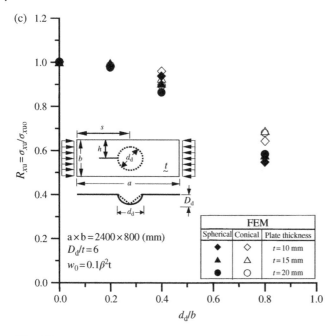

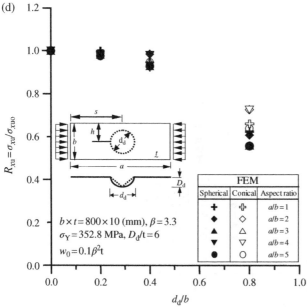

Figure 10.40 (Continued)

(e)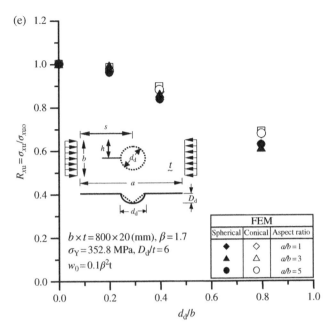

(f)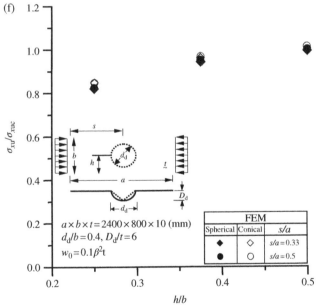

Figure 10.40 (Continued)

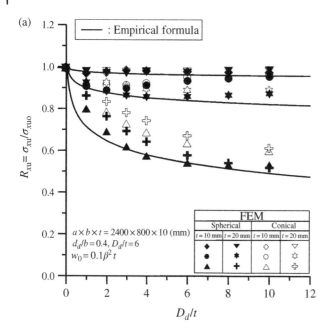

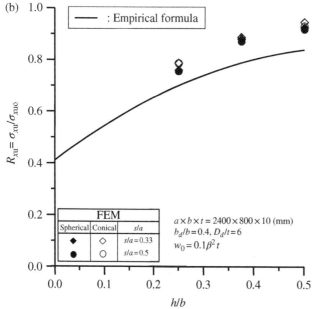

Figure 10.41 The validity of the empirical formulation for the ultimate compressive strength prediction of a dented plate, plotted versus (a) the dent depth; (b) the dent location and (c) the dent diameter.

(c)

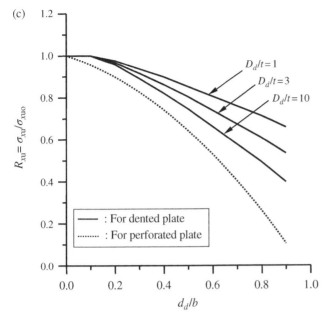

Figure 10.41 (Continued)

Figure 10.43 shows the ultimate shear strength behavior of a plate with spherical or conical shape of local denting, varying the dent size (i.e., depth and diameter) and location. It is found that the ultimate shear strength decreases significantly as the depth and/ or diameter of local denting increases. Also, it is apparent that the collapse behavior for spherical dent is similar to that for conical dent, while the former case is more likely to reduce the load-carrying capacity than the latter as long as the depth and diameter of denting are the same. Similar to the plate in axial compression, the plate falls in the worst situation in terms of the load-carrying capacity when the local denting is located at the plate center rather than other places.

Figure 10.44 shows the effect of dent parameters (dent depth, diameter, and location) and plate aspects (thickness and aspect ratio) on the ultimate shear strength of a dented plate. Similar to the plate in axial compression, the size (depth, diameter) and location of local denting are generally quite sensitive to the normalized ultimate shear strength, while the influence of the dent depth on the plate ultimate shear strength is not significant as long as the dent diameter is small. Also, as the plate collapse behavior for spherical dent is similar to that for conical dent and the former is slightly worse than the latter, the spherical dent may be taken as a representative of local dent shape for the purpose of the plate ultimate shear strength prediction, regardless of the actual shape of denting.

As described in Section 4.12, the ultimate shear strength of a dented plate may be predicted by an empirical formulation based on the strength reduction factor approach, that is, $R_\tau = \tau_u / \tau_{uo}$, where R_τ is the strength reduction factor. In this case, the plate thickness and the aspect ratio are not influential parameters to the

(a)

(b)

(c)

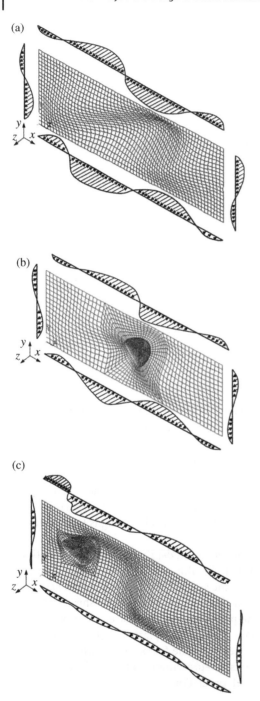

Figure 10.42 Deformed shape and membrane stress distribution of a plate immediately after the ultimate limit state is reached in edge shear: (a) without dent; (b) with the central dent; (c) with the side dent.

(a)

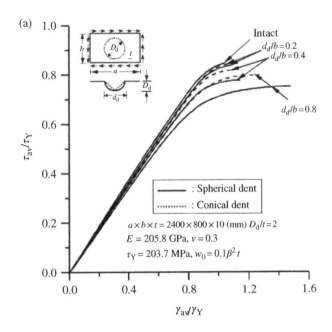

(b)

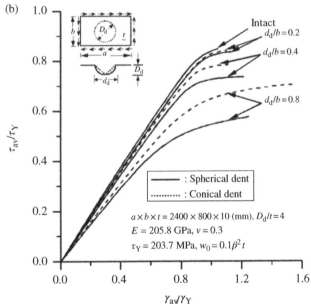

Figure 10.43 The ultimate shear strength behavior of a dented plate, plotted versus (a) $D_d/t = 2$; (b) $D_d/t = 4$; (c) $D_d/t = 10$ and (d) the dent location.

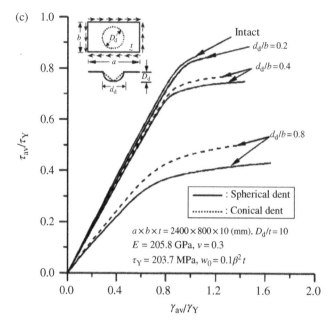

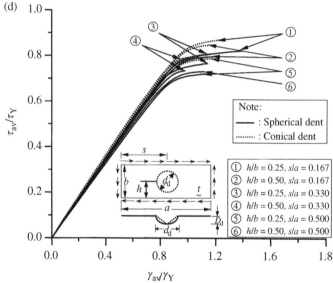

Figure 10.43 (Continued)

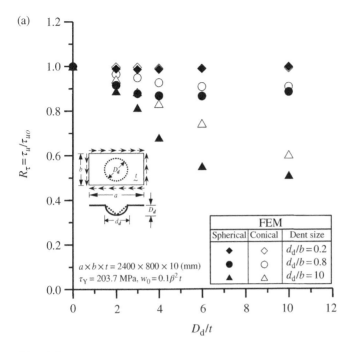

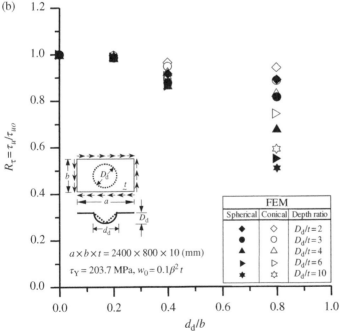

Figure 10.44 The effect of various dent parameters on the ultimate shear strength of a dented plate, plotted versus (a) the dent depth; (b) the dent diameter; (c) the plate thickness with dent diameter; (d) the plate aspect ratio with dent diameter for a thin plate and (e) the plate aspect ratio with dent diameter for a thick plate; (f) the dent location.

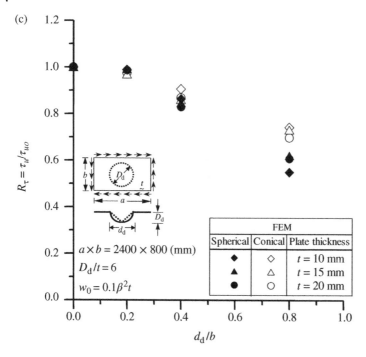

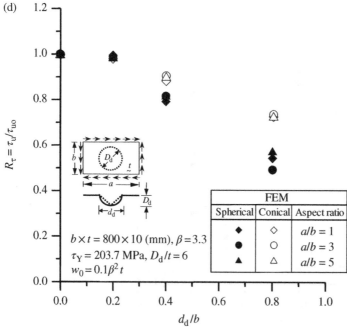

Figure 10.44 (Continued)

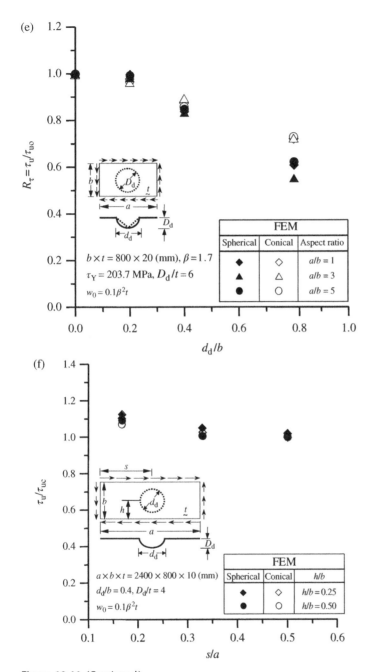

(e)

$b \times t = 800 \times 20$ (mm), $\beta = 1.7$
$\tau_Y = 203.7$ MPa, $D_d/t = 6$
$w_0 = 0.1\beta^2 t$

	FEM		
	Spherical	Conical	Aspect ratio
	◆	◇	$a/b = 1$
	▲	△	$a/b = 3$
	●	○	$a/b = 5$

$R_\tau = \tau_u/\tau_{uo}$ vs d_d/b

(f)

$a \times b \times t = 2400 \times 800 \times 10$ (mm)
$d_d/b = 0.4$, $D_d/t = 4$
$w_0 = 0.1\beta^2 t$

	FEM		
	Spherical	Conical	h/b
	◆	◇	$h/b = 0.25$
	●	○	$h/b = 0.50$

τ_u/τ_{uc} vs s/a

Figure 10.44 (Continued)

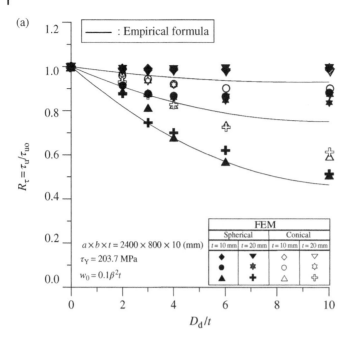

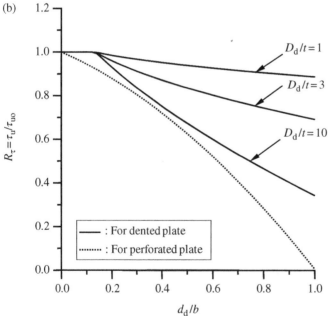

Figure 10.45 The validity of the empirical formulation for the ultimate shear strength prediction of a dented plate, plotted versus (a) the dent depth and (b) the dent diameter.

strength reduction factor because they are already implemented in the determination of τ_{uo}:

$$\frac{\tau_u}{\tau_{uo}} = \begin{cases} C_1\left(\dfrac{D_d}{t}\right)^2 - C_2\left(\dfrac{D_d}{t}\right) + 1 & \text{for } 1 < \dfrac{D_d}{t} \leq 10 \\ 100C_1 - 10C_2 + 1 & \text{for } 10 > \dfrac{D_d}{t} \end{cases} \tag{10.69a}$$

where τ_u and τ_{uo} are the ultimate shear strengths of dented or un-dented (intact) plates, respectively, and C_1 and C_2 are the coefficients that may be empirically determined by regression analysis of the computed results as follows:

$$C_1 = 0.0129\left(\frac{d_d}{b}\right)^{0.26} - 0.0076 \tag{10.69b}$$

$$C_2 = 0.1888\left(\frac{d_d}{b}\right)^{0.49} - 0.07 \tag{10.69c}$$

Figure 10.45 confirms that Equation (10.69) gives reasonably accurate results of the plate ultimate shear strength at somewhat pessimistic side, compared with the nonlinear finite element solutions. It is also interesting to note that the ultimate shear strength of a dented plate with large dent depth is close to that of a perforated plate described in Section 4.10.

References

Abramowicz, W. & Wierzbicki, T. (1989). Axial crushing of multi-corner sheet metal columns. *Journal of Applied Mechanics*, **156**: 113–119.

Alexander, J.M. (1960). An approximate analysis of the collapse of thin cylindrical shells under axial loading. *Quarterly Journal of Mechanics and Applied Mathematics*, **13**(1): 10–15.

Amdahl, J. (1983). Energy absorption in ship-platform impacts. Division of Marine Structures, University of Trondheim, Report No. UR-83-34, September.

Anderson, C.E. (2001). Proceedings of hypervelocity impact symposium. *International Journal of Impact Engineering*, **26**: 1–890.

Backman, M.E. & Goldsmith, W. (1978). The mechanics of penetration of projectiles into targets. *International Journal of Engineering Science*, **16**(1): 1–99.

Blyth, P.H. & Atkins, A.G. (2002). Stabbing of metal sheets by a triangular knife (an archaeological investigation). *International Journal of Impact Engineering*, **27**(4): 459–473.

Bodner, S.R. & Symonds, P.S. (1962). Experimental and theoretical investigation of the plastic deformation of cantilever beams subjected to impulsive loading. *Journal of Applied Mechanics*, **29**: 719–728.

Borvik, T., Forrestal, M.J., Hopperstad, O.S., Warren, T.L. & Langseth, M. (2009). Perforation of AA 5083-H116 aluminium plates with conical-nose steel projectiles—calculations. *International Journal of Impact Engineering*, **36**(3): 426–437.

Buckingham, E. (1914). On physically similar systems: illustrations of the use of dimensional equations. *Physics Review*, **4**: 347–350.

Campbell, J.D. & Ferguson, W.G. (1970). The temperature and strain-rate dependence of the shear strength of mild steel. *Philosophical Magazine*, **21**: 63.

Chen, W. (1993). A new bound solution for quadrangular plates subjected to impulsive loads. *Proceedings of the 3rd International Offshore and Polar Engineering Conference*, Singapore, June 6–11, IV: 702–708.

Corbett, G.G., Reid, S.R. & Johnson, W. (1996). Impact loading of plates and shells by free-flying projectiles: a review. *International Journal of Impact Engineering*, **18**(2): 141–230.

Cowper, G.R. & Symonds, P.S. (1957). Strain-hardening and strain-rate effects in the impact loading of cantilever beams. Technical Report No. 28, Division of Applied Mathematics, Brown University, September.

Dipaolo, B.P. & Tom, J.G. (2009). Effects of ambient temperature on a quasi-static axial-crush configuration response of thin-wall, steel box components. *Thin-Walled Structures*, **47**: 984–997.

Florence, A.L. (1977). Response of circular plates to central pulse loading. *International Journal of Solids and Structures*, **13**(11): 1091–1102.

Forrestal, M.J. & Sagartz, M.J. (1978). Elastic-plastic response of 304 stainless steel beams to impulse loads. *Journal of Applied Mechanics*, **45**: 685–687.

Goldsmith, W. (1999). Non-ideal projectile impact on targets. *International Journal of Impact Engineering*, **22**(2–3): 95–395.

Harrigan, J.J., Reid, S.R. & Peng, C. (1999). Inertia effects in impact energy absorbing materials and structures. *International Journal of Impact Engineering*, **22**(9): 955–979.

Hayashi, T. & Tanaka, Y. (1988). Impact engineering. Nikkan Kogyo Simbunsha (Daily Engineering Newspaper Company), Tokyo (in Japanese).

Hollomon, J.H. (1945). Tensile deformation. *Transactions of the Metallurgical Society of American Institute of Mining, Metallurgical, and Petroleum Engineers*, **162**: 268–290.

Hsu, S.S. & Jones, N. (2004). Dynamic axial crushing of aluminum alloy 6063-T6 circular tubes. *Latin American Journal of Solids and Structures*, **1**(3): 277–296.

Johnson, W. (1972). *Impact strength of materials*. Edward Arnold, London.

Johnson, W. (2001). *Collected works on Benjamin Robins and Charles Hutton*. Phoenix Publishing House Pvt Ltd, New Delhi.

Jones, N. (1989). On the dynamic inelastic failure of beams. In *Structural failure*, Edited by Wierzbicki, T. & Jones, N., John Wiley & Sons, Inc., New York, 133–159.

Jones, N. (1997). Dynamic plastic behaviour of ship and ocean structures. *Transactions of the Royal Institution of Naval Architects*, **139**: 65–97.

Jones, N. (2012). *Structural impact*. Second Edition, Cambridge University Press, Cambridge.

Jones, N. & Jouri, W.S. (1987). A study of plate tearing for ship collision and grounding damage. *Journal of Ship Research*, **31**: 253–268.

Jones, N., Liu, T., Zheng, J.J. & Shen, W.Q. (1991). Clamped beam grillages struck transversely by a mass at the center. *International Journal of Impact Engineering*, **11**: 379–399.

Jones, N. & Paik, J.K. (2012). Impact perforation of aluminum alloy plates. *International Journal of Impact Engineering*, **48**: 46–53.

Jones, N. & Paik, J.K. (2013). Impact perforation of steel plates. *Ships and Offshore Structures*, **8**(5): 579–596.

Karagiozova, D., Alves, M. & Jones, N. (2000). Inertia effects in axisymmetrically deformed cylindrical shells under axial impact. *International Journal of Impact Engineering*, **24**: 1083–1115.

Karagiozova, D. & Jones, N. (1997). Strain-rate effects in the dynamic buckling of a simple elastic-plastic model. *Journal of Applied Mechanics*, **64**: 193–200.

Karagiozova, D. & Jones, N. (1998). Stress wave effects on the dynamic axial buckling of cylindrical shells under impact. In *Structures under shock and impact V*, Edited by Jones, N., Computational Mechanics Publications, Southampton, 201–210.

Kim, K.J., Lee, J.H., Park, D.K., Jung, B.G., Han, X. & Paik, J.K. (2016). An experimental and numerical study on nonlinear impact response of steel-plated structures in an Arctic environment. *International Journal of Impact Engineering*, **93**: 99–115.

Kolsky, H. (1963). *Stress waves in solids*. Dover, New York.

Li, Z.G., Zhang, M.Y., Liu, F., Ma, C.S., Zhang, J.H., Hu, Z.M., Zhang, J.Z. & Zhao, Y.N. (2014). Influence of dent on residual ultimate strength of 2024-T3 aluminum alloy plate under axial compression. *Transactions of Nonferrous Metals Society of China*, **24**(10): 3084–3094.

Li, Z.G., Zhang, D.N., Peng, C.L., Ma, C.S., Zhang, J.H., Hu, Z.M., Zhang, J.Z. & Zhao, Y.N. (2015). The effect of local dents on the residual ultimate strength of 2024-T3 aluminum alloy plate used in aircraft under axial tension tests. *Engineering Failure Analysis*, **48**: 21–29.

Liu, J.H. & Jones, N. (1996). Shear and bending response of a rigid-perfectly plastic circular plate struck transversely by a mass. *Mechanics Based Design of Structures and Machines*, **24**(3): 361–388.

Lu, G. & Calladine, C.R. (1990). On the cutting of a plate by a wedge. *International Journal of Mechanical Science*, **32**: 293–313.

Ludwick, P. (1909). *Elemente der technologischen mechanic*. Springer Verlag, Berlin.

Malvern, L.E. (1969). *Introduction to the mechanics of continuous media*. Prentice Hall, Englewood Cliffs, NJ.

Manjunath, G.L. & Surendran, S. (2013). Dynamic fracture toughness of aluminium 6063 with multilayer composite patching at lower temperatures. *Ships and Offshore Structures*, **8**(2): 163–175.

McGregor, I.J., Meadows, D.J., Scott, C.E. & Seeds, A.D. (1993). Impact performance of aluminium structures. In *Structural crashworthiness and failure*, Edited by Jones, N. & Wierzbicki, T., Alcan International Limited, Banbury Laboratory, Oxfordshire, 385–421.

Min, D.K., Shin, D.W., Kim, S.H., Heo, Y.M. & Cho, S.R. (2012). On the plastic deformation of polar-class ship's single frame structures subjected to collision loadings. *Journal of the Society of Naval Architects of Korea*, **49**: 232–238.

Murray, N.W. (1983). The static approach to plastic collapse and energy dissipation in some thin-walled steel structures. In *Structural crashworthiness*, Edited by Jones, N. & Wierzbicki, T., Butterworths, London, 44–65.

Nurick, G.N. & Jones, N. (1995). Prediction of large inelastic deformations of T-beams subjected to uniform impulsive loads. In *High strain rate effects on polymer metal and ceramic matrix composites and other advanced materials*, vol. **48**, Edited by Rajapakse, Y., The American Society of Mechanical Engineers, New York, 127–153.

Nurick, G.N., Jones, N. & von Alten-Reuss, G.V. (1994). Large inelastic deformations of T-beams subjected to impulsive loads. *Proceedings of the 3rd International Conference on Structures Under Shock and Impact, Computational Mechanics Publications*, Southampton and Boston, MA: 191–206.

Nurick, G.N. & Martin, J.B. (1989). Deformation of thin plates subjected to impulsive loading—a review. Part II experimental studies. *International Journal of Impact Engineering*, **8**: 171–186.

Nurick, G.N., Olson, M.D., Fagnan, J.R. & Levin, A. (1995). Deformation and tearing of blast loaded stiffened square plates. *International Journal of Impact Engineering*, **16**: 273–291.

Ohtsubo, H. & Suzuki, K. (1994). The crushing mechanics of bow structures in head-on Collision (1st report). *Journal of the Society of Naval Architects of Japan*, **176**: 301–308 (in Japanese).

Ohtsubo, H. & Wang, G. (1995). An upper-bound solution to the problem of plate tearing. *Journal of Marine Science and Technology*, **1**: 46–51.

Paik, J.K. (1994). Cutting of a longitudinally stiffened plate by a wedge. *Journal of Ship Research*, **38**(4): 340–348.

Paik, J.K. (2005). Ultimate strength of dented steel plates under edge shear loads. *Thin-Walled Structures*, **43**: 1475–1492.

Paik, J.K. & Chung, J.Y. (1999). A basic study on static and dynamic crushing behavior of a stiffened tube. *Transactions of the Korean Society of Automotive Engineers*, **7**(1): 219–238 (in Korean).

Paik, J.K., Chung, J.Y., Choe, I.H., Thayamballi, A.K., Pedersen, P.T. & Wang, G. (1999). On the rational design of double hull tanker structures against collision. *Transactions of the Society of Naval Architects and Marine Engineers*, **107**: 323–363.

Paik, J.K., Chung, J.Y. & Chun, M.S. (1996). On quasi-static crushing of a stiffened square tube. *Journal of Ship Research*, **40**(3): 258–267.

Paik, J.K., Kim, K.J., Lee, J.H., Jung, B.G. & Kim, S.J. (2017). Test database of the mechanical properties of mild, high-tensile and stainless steel and aluminum alloy associated with cold temperatures and strain rates. *Ships and Offshore Structures*, **12**(S1): S230–S256.

Paik, J.K., Kim, B.J., Park, D.K. & Jang, B.S. (2011). On quasi-static crushing of thin-walled steel structures in cold temperature: experimental and numerical studies. *International Journal of Impact Engineering*, **38**: 13–28.

Paik, J.K., Lee, J.M. & Kee, D.H. (2003). Ultimate strength of dented steel plates under axial compressive loads. *International Journal of Mechanical Sciences*, **45**: 433–448.

Paik, J.K. & Pedersen, P.T. (1995). Ultimate and crushing strength of plated structures. *Journal of Ship Research*, **39**(3): 259–261.

Paik, J.K. & Pedersen, P.T. (1996). Modeling of the internal mechanics in ship collisions. *Ocean Engineering*, **23**(2): 107–142.

Paik, J.K. & Thayamballi, A.K. (2003). An experimental investigation on the dynamic ultimate compressive strength of ship plating. *International Journal of Impact Engineering*, **28**: 803–811.

Paik, J.K. & Wierzbicki, T. (1997). A benchmark study on crushing and cutting of plated structures. *Journal of Ship Research*, **41**(2): 147–160.

Park, D.K. (2015). Nonlinear structural response analysis of ship and offshore structures in low temperature. Ph.D. Thesis, Department of Naval Architecture and Ocean Engineering, Pusan National University, Busan.

Pedersen, P.T., Valsgaard, S., Olsen, D. & Spangenberg, S. (1993). Ship impacts—bow collisions. *International Journal of Impact Engineering*, **13**: 163–187.

Perzyna, P. (1974). The constitutive equations describing thermo-mechanical behavior of materials at high rates of strain. In *Mechanical properties at high rates of strain*, Conference Series, vol. **21**, Edited by Harding, J.E., American Institute of Physics, New York, 138–153.

Raviprakash, A.V., Prabu, B. & Alagumurthi, N. (2012). Residual ultimate compressive strength of dented square plates. *Thin-Walled Structures*, **58**: 32–39.

Reid, S.R. & Reddy, T.Y. (1983). Experimental investigation of inertia effects in one-dimensional metal ring systems subjected to end impact—part I fixed-ended systems. *International Journal of Impact Engineering*, **1**(1): 85–106.

Rosenfield, A.R. & Hahn, G.T. (1974). Numerical descriptions of the ambient low-temperature and high strain rate flow and fracture behavior of plain carbon steel. *Transactions of the American Society of Mechanical Engineers*, **59**: 138.

Rudrapatna, N.S., Vaziri, R. & Olson, M.D. (2000). Deformation and failure of blast loaded stiffened plates. *International Journal of Impact Engineering*, **24**: 457–474.

Rusinek, A., Rodriguez-Martinez, J.A., Zaera, R., Klepaczko, J.R., Arias, A. & Sauvelet, C. (2009). Experimental and numerical study on the perforation process of mild steel sheets subjected to perpendicular impact by hemispherical projectiles. *International Journal of Impact Engineering*, **36**(4): 565–587.

Saad-Eldeen, S., Garbatov, Y. & Guedes Soares, C. (2015). Stress-strain analysis of dented rectangular plates subjected to uni-axial compressive loading. *Engineering Structures*, **99**: 78–91.

Saad-Eldeen, S., Garbatov, Y. & Guedes Soares, C. (2016). Ultimate strength analysis of highly damaged plates. *Marine Structures*, **45**(1): 63–85.

Saitoh, T., Yosikawa, T. & Yao, H. (1995). Estimation of deflection of steel panel under impulsive loading. *Transactions of the Society of Mechanical Engineers of Japan*, **61**(590): 2241–2246 (in Japanese).

Schubak, R.B., Anderson, D.L. & Olson, M.D. (1989). Simplified dynamic analysis of rigid-plastic beams. *International Journal of Impact Engineering*, **8**(1): 27–42.

Shen, W.Q. (1997). Dynamic response of rectangular plates under drop mass impact. *International Journal of Impact Engineering*, **19**(3): 207–229.

Simonsen, B.C. & Wierzbicki, T. (1998). Plasticity, fracture, friction in steady-state plate cutting. *International Journal of Impact Engineering*, **21**(5): 387–411.

Smith, C.S. (1989). Behavior of composite and metallic superstructures under blast loading. In *Structural failure*. John Wiley & Sons, Inc., New York, 435–462.

Symonds, P.S. & Chon, C.T. (1974). Approximation techniques for impulsive loading of structures of time-dependent plastic behaviour with finite-deflections. In *Mechanical properties of materials at high rates of strain*, Conference Series, vol. **21**, Edited by Harding, J.E., American Institute of Physics, New York, 299–316.

Symonds, P.S. & Jones, N. (1972). Impulsive loading of fully clamped beams with finite plastic deflections. *International Journal of Mechanical Sciences*, **14**: 49–69.

Thomas, P.F. (1992). Application of plate cutting mechanics to damage prediction in ship grounding. MIT–Industry Joint Program on Tanker Safety, Report No. 8, Department of Ocean Engineering, Massachusetts Institute of Technology, Cambridge, MA.

Toyosada, M., Fujii, E., Nohara, K., Kawaguchi, Y., Arimochi, K. & Isaka, K. (1987). The effect of strain rate on critical CTOD and J-integral. *Journal of the Society of Naval Architects of Japan*, **161**: 343–356 (in Japanese).

Vaughan, H. (1980). The tearing of mild steel plate. *Journal of Ship Research*, **24**(2): 96–100.

Wierzbicki, T. (1983). Crushing behavior of plate intersections. *Proceedings of the 1st International Symposium on Structural Crashworthiness*, University of Liverpool, September: 66–95.

Wierzbicki, T. (1995). Concertina tearing of metal plates. *International Journal of Solid Structures*, **19**: 2923–2943.

Wierzbicki, T. & Thomas, P. (1993). Closed-form solution for wedge cutting force through thin metal sheets. *International Journal of Mechanical Science*, **35**: 209–229.

Woisin, G. (1979). Design against collision. *Schiff & Hafen*, **31**: 1059–1069.

Woisin, G. (1982). Comments on Vaughan: the tearing strength of mild steel plate. *Journal of Ship Research*, **26**: 50–52.

Woisin, G. (1990). Analysis of the collisions between rigid bulb and side shell panel. *Proceedings of the 7th International Symposium on Practical Design of Ships and Mobile Units (PRADS'90)*, The Hague: 165–172.

Xu, M.C. & Guedes Soares, C. (2013). Assessment of residual ultimate strength for side dented stiffened panels subjected to compressive loads. *Engineering Structures*, **49**: 316–328.

Xu, M.C. & Guedes Soares, C. (2015). Effect of a central dent on the ultimate strength of narrow stiffened panels under axial compression. *International Journal of Mechanical Sciences*, **100**: 68–79.

Yu, T.X. & Chen, F.L. (1992). The large deflection dynamic plastic response of rectangular plates. *International Journal of Impact Engineering*, **12**(4): 603–616.

Zhang, S. (2002). Plate tearing and bottom damage in ship grounding. *Marine Structures*, **15**: 101–117.

11

The Incremental Galerkin Method

11.1 Features of the Incremental Galerkin Method

This chapter describes the incremental Galerkin method, which is a semi-analytical method for analysis of the elastic–plastic large-deflection behavior of steel or aluminum plates and stiffened panels up to their ultimate limit state (ULS). This method is designed to accommodate the geometric nonlinearity associated with buckling via an analytical procedure, whereas a numerical procedure accounts for the material nonlinearity associated with plasticity (Paik et al. 2001, Paik & Kang 2005).

The method is unique in its use to analytically formulate the incremental forms of nonlinear governing differential equations for elastic large-deflection plate theory. After solving these incremental governing differential equations using the Galerkin approach (Fletcher 1984), a set of easily solved linear first-order simultaneous equations for the unknowns is obtained, which facilitates a reduction in the computational effort.

It is normally difficult, but not impossible, to formulate the nonlinear governing differential equations to represent both geometric and material nonlinearities for plates and stiffened panels. A major source of difficulty is that an analytical treatment of plasticity with increases in the applied loads is quite cumbersome. An easier alternative is to deal with the progress of the plasticity numerically.

The benefits of this method are to provide excellent solution accuracy with great savings in computational effort and to handle in the analysis the combined loading for all potential load components, including biaxial compression or tension, biaxial in-plane bending, edge shear, and lateral pressure loads. The effects of initial imperfections in the form of initial deflection and welding induced residual stresses are also considered. The present theory can be applied to both steel and aluminum plate panels.

11.2 Structural Idealizations of Plates and Stiffened Panels

In this section, some of the more important basic hypotheses used to formulate the incremental Galerkin method for computation of the elastic–plastic large-deflection behavior of plate and stiffened panels are described:

1) The plate panel is made of isotropic homogeneous steel or aluminum alloys with Young's modulus of E and Poisson's ratio of ν. For a stiffened panel, Young's

Ultimate Limit State Analysis and Design of Plated Structures, Second Edition. Jeom Kee Paik.
© 2018 John Wiley & Sons Ltd. Published 2018 by John Wiley & Sons Ltd.

modulus of the plate part between stiffeners is the same as that of the stiffeners, but the yield stress of the plate part can differ from that of the stiffeners.

2) The length and breadth of the plate are a and b, respectively, as shown in Figure 11.1a. The plate thickness is t.

3) The spacing of the stiffeners or the breadth of the plating between stiffeners can differ as shown in Figure 11.1b.

4) The edge of the panel can be simply supported, clamped, or some combination of the two.

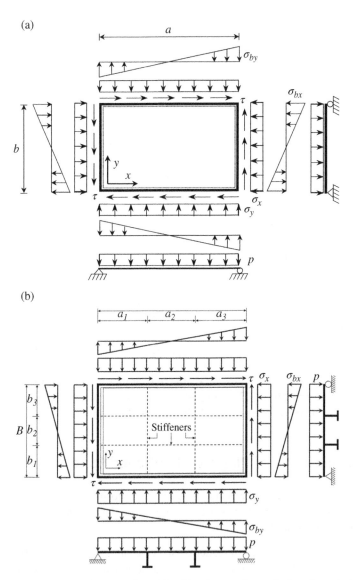

Figure 11.1 Application of combined in-plane and out-of-plane loads in (a) a plate and (b) a stiffened panel.

5) The panel is normally subjected to combined loads. Several potential load components act on the panel: biaxial compression or tension, edge shear, biaxial in-plane bending moment, and lateral pressure loads, as shown in Figure 11.1.

6) The applied loads are increased incrementally.

7) The shape of the initial deflection in the plate panel is normally complex, but it can be expressed with a Fourier series function. For a stiffened panel, the plate part between stiffeners may have the same set of local plate initial deflections, whereas the stiffeners may have a different set of global column-type initial deflections.

8) Due to the welding along the panel edges and at the intersections between the lower part of the stiffener web and parent plate, the panel has welding induced residual stresses. These can develop in the plate part in both *x* and *y* directions, as welding is normally carried out in these two directions. As shown in Figure 11.2, the distribution of welding induced residual stresses for the plate part between stiffeners is idealized to be composed of two stress blocks, that is, compressive and tensile

Figure 11.2 Idealized welding induced residual stress distribution inside (a) the plating and (b) the stiffeners.

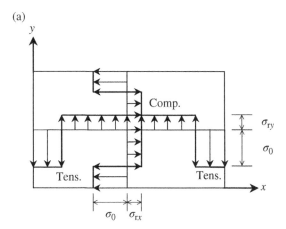

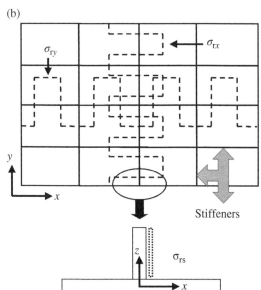

(a)

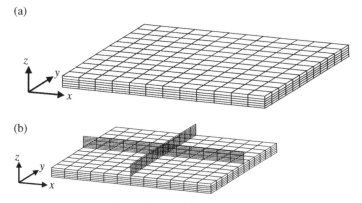

(b)

Figure 11.3 Example subdivision of mesh regions used for the treatment of plasticity for (a) a plate and (b) a stiffened panel (note that geometric nonlinearity is handled analytically).

residual stress blocks, as described in Section 1.7. It is assumed that the stiffener webs have uniform compressive residual stresses "equivalent" to that shown in Figure 11.2.

9) For evaluation of the plasticity, it is assumed that the panel is composed of a number of membrane fibers in the x and y directions. Each membrane fiber is considered to have a number of layers in the z direction, as shown in Figure 11.3.

10) It is recognized that the strength of welded aluminum alloys in the softened zone may be recovered by natural aging over a period of time (Lancaster 2003), but the ultimate strength of welded aluminum alloy panels may be reduced by softening phenomenon in the heat-affected zone as far as the material strength is not recovered. The effect of softening is accounted for using the technique noted in item 9.

11.3 Analysis of the Elastic–Plastic Large-Deflection Behavior of Plates

11.3.1 The Traditional Approach

As described in Chapter 4, the elastic large-deflection behavior of plates with initial deflections is governed by two differential equations: one that represents the equilibrium condition and one that represents the compatibility condition (Marguerre 1938). These equations are as follows:

$$\Phi = D\left(\frac{\partial^4 w}{\partial x^4} + 2\frac{\partial^4 w}{\partial x^2 \partial y^2} + \frac{\partial^4 w}{\partial y^4}\right)$$

$$-t\left[\frac{\partial^2 F}{\partial y^2}\frac{\partial^2 (w+w_0)}{\partial x^2} + \frac{\partial^2 F}{\partial x^2}\frac{\partial^2 (w+w_0)}{\partial y^2} - 2\frac{\partial^2 F}{\partial x \partial y}\frac{\partial^2 (w+w_0)}{\partial x \partial y} + \frac{p}{t}\right] = 0$$

(11.1a)

$$\frac{\partial^4 F}{\partial x^4} + 2\frac{\partial^4 F}{\partial x^2 \partial y^2} + \frac{\partial^4 F}{\partial y^4} - E\left[\left(\frac{\partial^2 w}{\partial y \partial x}\right)^2 - \frac{\partial^2 w}{\partial x^2}\frac{\partial^2 w}{\partial y^2}\right.$$

$$\left. + 2\frac{\partial^2 w_0}{\partial x \partial y}\frac{\partial^2 w}{\partial x \partial y} - \frac{\partial^2 w_0}{\partial x^2}\frac{\partial^2 w}{\partial y^2} - \frac{\partial^2 w}{\partial x^2}\frac{\partial^2 w_0}{\partial y^2}\right] = 0 \tag{11.1b}$$

where $D = Et^3/[12(1 - v^2)]$ and the lateral pressure p can be varying over the plate as $p = p(x,y)$.

By using Airy's stress function, F, the stress components at a certain location inside the plate may be calculated as follows:

$$\sigma_x = \frac{\partial^2 F}{\partial y^2} - \frac{Ez}{1-v^2}\left(\frac{\partial^2 w}{\partial x^2} + v\frac{\partial^2 w}{\partial y^2}\right) \tag{11.2a}$$

$$\sigma_y = \frac{\partial^2 F}{\partial x^2} - \frac{Ez}{1-v^2}\left(\frac{\partial^2 w}{\partial y^2} + v\frac{\partial^2 w}{\partial x^2}\right) \tag{11.2b}$$

$$\tau = \tau_{xy} = -\frac{\partial^2 F}{\partial x \partial y} - \frac{Ez}{2(1+v)}\frac{\partial^2 w}{\partial x \partial y} \tag{11.2c}$$

Also, the corresponding strain components at a certain location inside the plate are given by

$$\varepsilon_x = \frac{\partial u}{\partial x} + \frac{1}{2}\left(\frac{\partial w}{\partial x}\right)^2 + \frac{\partial w \partial w_0}{\partial x \, \partial x} - z\frac{\partial^2 w}{\partial x^2} \tag{11.3a}$$

$$\varepsilon_y = \frac{\partial v}{\partial y} + \frac{1}{2}\left(\frac{\partial w}{\partial y}\right)^2 + \frac{\partial w \partial w_0}{\partial y \, \partial y} - z\frac{\partial^2 w}{\partial y^2} \tag{11.3b}$$

$$\gamma_{xy} = \frac{\partial u}{\partial y} + \frac{\partial v}{\partial x} + \frac{\partial w \partial w}{\partial x \, \partial y} + \frac{\partial w_0 \partial w}{\partial x \, \partial y} + \frac{\partial w \partial w_0}{\partial x \, \partial y} - 2z\frac{\partial^2 w}{\partial x \partial y} \tag{11.3c}$$

where u and v are the axial displacements in the x and y directions, respectively.

Each strain component noted previously is expressed as a function of stress components as follows:

$$\varepsilon_x = \frac{1}{E}\left(\sigma_x - v\sigma_y\right) \tag{11.4a}$$

$$\varepsilon_y = \frac{1}{E}\left(\sigma_y - v\sigma_x\right) \tag{11.4b}$$

$$\gamma_{xy} = \frac{2(1+v)}{E}\tau_{xy} \tag{11.4c}$$

In solving the nonlinear governing differential equations, Equations (11.1a) and (11.1b), by the Galerkin method, the added deflection, w, and initial deflection, w_0, can be assumed as follows:

$$w = \sum_{m=1}^{\infty}\sum_{n=1}^{\infty} A_{mn} f_m(x) g_n(y) \tag{11.5a}$$

$$w_0 = \sum_{m=1}^{\infty}\sum_{n=1}^{\infty} A_{0mn} f_m(x) g_n(y) \tag{11.5b}$$

where $f_m(x)$ and $g_n(y)$ are functions that satisfy the boundary conditions for the plate and A_{mn} and A_{0mn} are the unknown and known deflection coefficients, respectively.

Upon substituting Equations (11.5a) and (11.5b) into Equation (11.1b) and solving for the stress function, F, the particular solution, F_P, may be expressed as follows:

$$F_P = \sum_{r=1}\sum_{s=1} K_{rs} p_r(x) q_s(y) \tag{11.6}$$

where K_{rs} are the coefficients of second-order functions with regard to the unknown deflection coefficients A_{mn}.

Including the applied loading, the complete stress function, F, may be given by

$$F = F_H + \sum_{r=1}\sum_{s=1} K_{rs} p_r(x) q_s(y) \tag{11.7}$$

where F_H is the homogeneous solution of the stress function that satisfies the applied loading condition.

To compute the unknown coefficients A_{mn}, one may use the Galerkin method for the equilibrium equation, Equation (11.1a), resulting in the following equation:

$$\iiint \Phi f_r(x) g_s(y) d\mathrm{Vol} = 0, \quad r = 1,2,3,\ldots, \quad s = 1,2,3,\ldots \tag{11.8}$$

Substituting Equations (11.5a), (11.5b), and (11.7) into Equation (11.8) and performing the integration over the whole volume of the plate, a set of third-order simultaneous equations with regard to the unknown coefficients, A_{mn}, is obtained.

Solving the simultaneous equations to obtain the coefficients A_{mn} normally requires an iteration process. As the solution of each coefficient should be unique, one must correctly select one of the three solutions obtained for each coefficient. Unfortunately, it is not always easy to solve a set of such third-order simultaneous equations, especially when the number of unknown coefficients, A_{mn}, becomes large.

11.3.2 The Incremental Approach

An incremental approach is possible to more efficiently solve the nonlinear governing differential equations for plates subjected to combined loads (Ueda et al. 1987). First, the incremental forms of the governing differential equations for plates must be formulated. After analytically solving these incremental governing differential equations using the Galerkin method, a set of linear (i.e., first-order) simultaneous equations for the unknowns (which can be easily solved) is obtained. Such a method normally drastically reduces the computational effort. Another benefit is that the solution is uniquely determined, unlike the traditional approach described in Section 11.3.1.

In the following section, the incremental forms of governing differential equations for plates are derived. First, it is assumed that the load is applied incrementally. At the end of the $(i-1)$th load increment step, the deflection and stress functions can be denoted by w_{i-1} and F_{i-1}, respectively. In the same manner, the deflection and stress functions at the end of the ith load increment step are denoted by w_i and F_i, respectively.

Therefore, the equilibrium equation, Equation (11.1a), and the compatibility equation, Equation (11.1b), at the end of the $(i-1)$th load increment step are written as follows:

$$\Phi_{i-1} = D\left(\frac{\partial^4 w_{i-1}}{\partial x^4} + 2\frac{\partial^4 w_{i-1}}{\partial x^2 \partial y^2} + \frac{\partial^4 w_{i-1}}{\partial y^4}\right) - t\left[\frac{\partial^2 F_{i-1}}{\partial y^2}\frac{\partial^2(w_{i-1}+w_0)}{\partial x^2}\right.$$

$$+\frac{\partial^2 F_{i-1}}{\partial x^2}\frac{\partial^2(w_{i-1}+w_0)}{\partial y^2} - 2\frac{\partial^2 F_{i-1}}{\partial x \partial y}\frac{\partial^2(w_{i-1}+w_0)}{\partial x \partial y} + \left.\frac{p_{i-1}}{t}\right] = 0 \tag{11.9a}$$

$$\frac{\partial^4 F_{i-1}}{\partial x^4} + 2\frac{\partial^4 F_{i-1}}{\partial x^2 \partial y^2} + \frac{\partial^4 F_{i-1}}{\partial y^4} - E\left[\left(\frac{\partial^2 w_{i-1}}{\partial y \partial x}\right)^2 - \frac{\partial^2 w_{i-1}}{\partial x^2}\frac{\partial^2 w_{i-1}}{\partial y^2}\right.$$

$$+ 2\frac{\partial^2 w_0}{\partial x \partial y}\frac{\partial^2 w_{i-1}}{\partial x \partial y} - \frac{\partial^2 w_0}{\partial x^2}\frac{\partial^2 w_{i-1}}{\partial y^2} - \left.\frac{\partial^2 w_{i-1}}{\partial x^2}\frac{\partial^2 w_0}{\partial y^2}\right] = 0 \tag{11.9b}$$

In the same manner, Equations (11.1a) and (11.1b) at the *i*th load increment step are given by

$$\Phi_i = D\left(\frac{\partial^4 w_i}{\partial x^4} + 2\frac{\partial^4 w_i}{\partial x^2 \partial y^2} + \frac{\partial^4 w_i}{\partial y^4}\right) - t\left[\frac{\partial^2 F_i}{\partial y^2}\frac{\partial^2(w_i+w_0)}{\partial x^2}\right.$$

$$+\frac{\partial^2 F_i}{\partial x^2}\frac{\partial^2(w_i+w_0)}{\partial y^2} - 2\frac{\partial^2 F_i}{\partial x \partial y}\frac{\partial^2(w_i+w_0)}{\partial x \partial y} + \left.\frac{p_i}{t}\right] = 0 \tag{11.10a}$$

$$\frac{\partial^4 F_i}{\partial x^4} + 2\frac{\partial^4 F_i}{\partial x^2 \partial y^2} + \frac{\partial^4 F_i}{\partial y^4} - E\left[\left(\frac{\partial^2 w_i}{\partial y \partial x}\right)^2 - \frac{\partial^2 w_i}{\partial x^2}\frac{\partial^2 w_i}{\partial y^2}\right.$$

$$+ 2\frac{\partial^2 w_0}{\partial x \partial y}\frac{\partial^2 w_i}{\partial x \partial y} - \frac{\partial^2 w_0}{\partial x^2}\frac{\partial^2 w_i}{\partial y^2} - \left.\frac{\partial^2 w_i}{\partial x^2}\frac{\partial^2 w_0}{\partial y^2}\right] = 0 \tag{11.10b}$$

It is assumed that the added deflection, w_i, and stress function, F_i, at the end of the *i*th load increment step are calculated by

$$w_i = w_{i-1} + \Delta w \tag{11.11a}$$

$$F_i = F_{i-1} + \Delta F \tag{11.11b}$$

where Δw and ΔF are the increments of the deflection and stress functions, respectively, and the prefix Δ indicates the increment for the variable.

Substituting Equations (11.11a) and (11.11b) into Equations (11.10a) and (11.10b) and subtracting Equation (11.9a) from Equation (11.10a) or Equation (11.9b) from Equation (11.10b), respectively, the necessary incremental forms of governing differential equations emerge as follows:

$$\Delta\Phi = D\left(\frac{\partial^4 \Delta w}{\partial x^4} + 2\frac{\partial^4 \Delta w}{\partial x^2 \partial y^2} + \frac{\partial^4 \Delta w}{\partial y^4}\right) - t\left[\frac{\partial^2 F_{i-1}}{\partial y^2}\frac{\partial^2 \Delta w}{\partial x^2} + \frac{\partial^2 \Delta F}{\partial y^2}\frac{\partial^2(w_{i-1}+w_0)}{\partial x^2}\right.$$

$$+\frac{\partial^2 F_{i-1}}{\partial x^2}\frac{\partial^2 \Delta w}{\partial y^2} + \frac{\partial^2 \Delta F}{\partial x^2}\frac{\partial^2(w_{i-1}+w_0)}{\partial y^2} - 2\frac{\partial^2 F_{i-1}}{\partial x \partial y}\frac{\partial^2 \Delta w}{\partial x \partial y}$$

$$- 2\frac{\partial^2 \Delta F}{\partial x \partial y}\frac{\partial^2(w_{i-1}+w_0)}{\partial x \partial y} + \left.\frac{\Delta p}{t}\right] = 0$$

$$\tag{11.12a}$$

$$\frac{\partial^4 \Delta F}{\partial x^4} + 2\frac{\partial^4 \Delta F}{\partial x^2 \partial y^2} + \frac{\partial^4 \Delta F}{\partial y^4} - E\left[2\frac{\partial^2 (w_{i-1} + w_0)}{\partial x \partial y}\frac{\partial^2 \Delta w}{\partial x \partial y}\right.$$

$$\left. - \frac{\partial^2 (w_{i-1} + w_0)}{\partial x^2}\frac{\partial^2 \Delta w}{\partial y^2} - \frac{\partial^2 \Delta w}{\partial x^2}\frac{\partial^2 (w_{i-1} + w_0)}{\partial y^2}\right] = 0 \tag{11.12b}$$

where the terms of very small quantities with an order higher than second order of the increments Δw and ΔF have been neglected.

At the end of the $(i-1)$th load increment step, the deflection, w_{i-1}, and the stress function, F_{i-1}, are obtained as follows:

$$w_{i-1} = \sum_{m=1}\sum_{n=1} A_{mn}^{i-1} f_m(x)g_n(y) \tag{11.13a}$$

$$F_{i-1} = F_H^{i-1} + \sum_{i=1}\sum_{j=1} K_{ij}^{i-1} p_i(x)q_j(y) \tag{11.13b}$$

where A_{mn}^{i-1} and K_{ij}^{i-1} are the known coefficients and F_H^{i-1} is a homogeneous solution for the stress function that satisfies the applied loading condition. The welding induced residual stresses can be included in the stress function, F_H^{i-1}, as initial stress terms.

The deflection increment, Δw, associated with the load increment at the ith step can be assumed as follows:

$$\Delta w = \sum_{k=1}\sum_{l=1} \Delta A_{kl} f_k(x)g_l(y) \tag{11.14}$$

where ΔA_{kl} is the unknown added deflection increment.

Substituting Equations (11.5b), (11.11a), and (11.14) into Equation (11.12b), the stress function increment, ΔF, can be obtained by

$$\Delta F = \Delta F_H + \sum_{i=1}\sum_{j=1} \Delta K_{ij} p_i(x)q_j(y) \tag{11.15}$$

where ΔK_{ij} are linear (i.e., first-order) functions in the unknown coefficients and ΔA_{kl}. ΔF_H is a homogeneous solution for the stress function increment that satisfies the applied loading condition.

To compute the unknown coefficients, ΔA_{kl}, the Galerkin method can then be applied to Equation (11.12a):

$$\iiint \Delta \Phi f_r(x)g_s(y)d\text{Vol} = 0, \quad r = 1,2,3,\dots, \quad s = 1,2,3,\dots \tag{11.16}$$

Substituting Equation (11.5b) and Equations (11.13)–(11.15) into Equation (11.16) and performing the integration over the entire volume of the plate, a set of linear simultaneous equations for the unknown coefficients, ΔA_{kl}, is obtained. Solving these linear simultaneous equations is normally easy. Having obtained ΔA_{kl}, one can then calculate Δw, that is, from Equation (11.14); ΔF, that is, from Equation (11.15); $w_i(= w_{i-1} + \Delta w)$, that is, from Equation (11.11a); and $F_i(= F_{i-1} + \Delta F)$, that is, from Equation (11.11b), at the end of the ith load increment step.

The elastic large-deflection behavior of the plate can be obtained by repeating the preceding procedure with increases in the applied loads. In this process, it is apparent that the load increments must be small to obtain more accurate solutions by avoiding non-equilibrating forces. As the computational effort required for this procedure is normally small, using smaller load increments does not usually lead to any severe penalties, unlike the usual nonlinear numerical methods.

11.3.3 Application to the Plates Simply Supported at Four Edges

The incremental Galerkin method can be applied to the nonlinear analysis of plates subject to various edge conditions: clamped, simply supported, or their combination. In the following, the formulations of the incremental Galerkin method for plates simply supported at all (four) edges are described in detail.

The simply supported edge conditions for the plate should satisfy

$$w = 0, \quad \frac{\partial^2 w}{\partial y^2} + v\frac{\partial^2 w}{\partial x^2} = 0 \quad \text{at} \quad y = 0,b \tag{11.17a}$$

$$\frac{\partial^2 w}{\partial y^2} = 0 \quad \text{at} \quad y = 0,b \tag{11.17b}$$

$$w = 0, \quad \frac{\partial^2 w}{\partial x^2} + v\frac{\partial^2 w}{\partial y^2} = 0 \quad \text{at} \quad x = 0,a \tag{11.17c}$$

$$\frac{\partial^2 w}{\partial y^2} = 0 \quad \text{at} \quad x = 0,a \tag{11.17d}$$

The Fourier series deflection functions that must satisfy the boundary conditions can be assumed as follows:

$$w_0 = \sum_{m=1}^{\infty}\sum_{n=1}^{\infty} A_{0mn}\sin\frac{m\pi x}{a}\sin\frac{n\pi y}{b} \tag{11.18a}$$

$$w_{i-1} = \sum_{m=1}^{\infty}\sum_{n=1}^{\infty} A_{mn}^{i-1}\sin\frac{m\pi x}{a}\sin\frac{n\pi y}{b} \tag{11.18b}$$

$$\Delta w = \sum_{k=1}^{\infty}\sum_{l=1}^{\infty} \Delta A_{kl}\sin\frac{k\pi x}{a}\sin\frac{l\pi y}{b} \tag{11.18c}$$

where $A_{0mn}\left(= A_{mn}^0\right)$ and A_{mn}^{i-1} are the known coefficients and ΔA_{kl} are the unknown coefficients to be calculated for the external load increments.

The condition of combined load application—biaxial loads, biaxial in-plane bending, edge shear, and lateral pressure loads—gives

$$\int_0^b \frac{\partial^2 F}{\partial y^2}t\,dy = P_x \quad \text{at} \quad x = 0,a \tag{11.19a}$$

$$\int_0^b \frac{\partial^2 F}{\partial y^2}t\left(y-\frac{b}{2}\right)dy = M_x \quad \text{at} \quad x = 0,a \tag{11.19b}$$

$$\int_0^a \frac{\partial^2 F}{\partial x^2}t\,dx = P_y \quad \text{at} \quad x = 0,b \tag{11.19c}$$

$$\int_0^a \frac{\partial^2 F}{\partial x^2} t \left(x - \frac{a}{2}\right) dx = M_y \quad \text{at} \quad x = 0, b \tag{11.19d}$$

$$\frac{\partial^2 F}{\partial x \partial y} = -\tau \quad \text{at all boundaries} \tag{11.19e}$$

where P_x and P_y are the axial loads in the x and y directions, respectively, and M_x and M_y are the in-plane bending moments in the x and y directions, respectively.

For the sake of simplicity in expressing the various functions, the following abbreviations are used from this point:

$$sx(m) = \sin\frac{m\pi x}{a}, \quad sy(n) = \sin\frac{n\pi y}{b}, \quad cx(m) = \cos\frac{m\pi x}{a}, \quad cy(n) = \cos\frac{n\pi y}{b}$$

To get the stress function increment, ΔF, substituting Equations (11.18a)–(11.18c) into Equation (11.12b) yields

$$\frac{\partial^4 \Delta F}{\partial x^4} + 2\frac{\partial^4 \Delta F}{\partial x^2 \partial y^2} + \frac{\partial^4 \Delta F}{\partial y^4} = \frac{E\pi^4}{4a^2 b^2} \sum_m \sum_n \sum_k \sum_l \Delta A_{kl} A_{mn}^{i-1}$$

$$\times \left[-(kn - ml)^2 cx(m-k)cy(n-l) + (kn + ml)^2 cx(m-k)cy(n+l) \right.$$
$$\left. + (kn + ml)^2 cx(m+k)cy(n-l) - (kn - ml)^2 cx(m+k)cy(n+l) \right] \tag{11.20}$$

A particular solution, ΔF_P, for the stress function increment can then be obtained as follows:

$$\Delta F_P = \sum_m \sum_n \sum_k \sum_l [B_1(m, n, k, l)cx(m-k)cy(n-l)$$

$$+ B_2(m, n, k, l)cx(m-k)cy(n+l) + B_3(m, n, k, l)cx(m+k)cy(n-l)$$

$$+ B_4(m, n, k, l)cx(m+k)cy(n+l)] \tag{11.21}$$

Substituting Equation (11.21) into Equation (11.20) by using $\Delta F = \Delta F_P$ yields the coefficients B_1–B_4 as follows:

$$B_1(m, n, k, l) = \frac{E\alpha^2 \pi^4}{4} \Delta A_{kl} A_{mn}^{i-1} \frac{-(kn - ml)^2}{\left[(m-k)^2 + \alpha^2(n-l)^2\right]^2} \tag{11.22a}$$

$$B_2(m, n, k, l) = \frac{E\alpha^2 \pi^4}{4} \Delta A_{kl} A_{mn}^{i-1} \frac{(kn + ml)^2}{\left[(m-k)^2 + \alpha^2(n+l)^2\right]^2} \tag{11.22b}$$

$$B_3(m, n, k, l) = \frac{E\alpha^2}{4} \Delta A_{kl} A_{mn}^{i-1} \frac{(kn - ml)^2}{(m+k)^2 + \alpha^2(n-l)^2} \tag{11.22c}$$

$$B_4(m, n, k, l) = \frac{E\alpha^2}{4} \Delta A_{kl} A_{mn}^{i-1} \frac{-(kn - ml)^2}{(m+k)^2 + \alpha^2(n+l)^2} \tag{11.22d}$$

where $\alpha = a/b$.

Substituting Equations (11.22a)–(11.22d) into Equation (11.21), ΔF_P can be written in a more simplified form as follows:

$$\Delta F_P = \frac{E\alpha^2\pi^4}{4}\sum_m\sum_n\sum_k\sum_l \Delta A_{kl}A_{mn}^{i-1}$$

$$\times \sum_{r=1}^{2}\sum_{s=1}^{2}(-1)^{r+s+1}h_1[(-1)^r k,(-1)^r l]cx[m+(-1)^r k][n+(-1)^s l]$$

(11.23)

where

$$h_1[\omega_1,\omega_2] = \frac{(-n\omega_1 + m\omega_2)^2}{\left[(m+\omega_1)^2 + \alpha^2(n+\omega_2)^2\right]^2}$$

$h_1 = 0$ if $m + \omega_1 = 0$ and $n + \omega_2 = 0$.

By considering the condition of load application, the homogeneous solution, ΔF_H, for the stress function increment is given by

$$\Delta F_H = \Delta P_x\frac{y^2}{2bt} + \Delta P_y\frac{x^2}{2at} - \Delta M_x\frac{y^2(2y-3b)}{b^3 t} - \Delta M_y\frac{x^2(2x-3a)}{a^3 t} - \Delta\tau_{xy}xy \quad (11.24)$$

The overall stress function increment may then be expressed by the sum of the particular and homogeneous solutions as follows:

$$\Delta F = \Delta F_P + \Delta F_H \quad (11.25)$$

In the same manner, to obtain the stress function, F_P^{i-1}, at the end of the $(i-1)$th load increment step, Equations (11.18a)–(11.18c) can be substituted into Equation (11.18b), resulting in the following equation:

$$\frac{\partial^4 F_{i-1}}{\partial x^4} + 2\frac{\partial^4 F_{i-1}}{\partial x^2\partial y^2} + \frac{\partial^4 F_{i-1}}{\partial y^4} = \frac{E\pi^4}{4a^2 b^2}\sum_m\sum_n\sum_k\sum_l \left(A_{mn}^{i-1}A_{kl}^{i-1} - A_{mn}^0 A_{kl}^0\right)$$

$$\times [ml(kn-ml)]cx(m-k)cy(n-l) + ml(kn+ml)cx(m-k)cy(n+l).$$
$$+ ml(kn+ml)cx(m+k)cy(n-l) + ml(kn-ml)cx(m+k)cy(n+l)]$$

(11.26)

A particular solution, F_P^{i-1}, for the stress function, $F_{i-1}(\equiv F^{i-1})$, can then be given by

$$F_P^{i-1} = \sum_m\sum_n\sum_k\sum_l [C_1(m,n,k,l)cx(m-k)cy(n-l)$$

$$+ C_2(m,n,k,l)cx(m-k)cy(n+l) + C_3(m,n,k,l)cx(m+k)cy(n-l)$$
$$+ C_4(m,n,k,l)cx(m+k)cy(n+l)]$$

(11.27)

The coefficients C_1–C_4 of Equation (11.27) can be determined by substituting Equation (11.27) into Equation (11.26) as $F_{i-1} = F_P^{i-1}$:

$$C_1(m,n,k,l) = \frac{E\pi^4}{4\alpha^2}\left(A_{mn}^{i-1}A_{kl}^{i-1} - A_{mn}^0 A_{kl}^0\right)\frac{ml(kn-ml)}{\left[(m-k)^2 + (n-l)^2/\alpha^2\right]^2} \quad (11.28a)$$

$$C_2(m,n,k,l) = \frac{E\pi^4}{4\alpha^2}\left(A_{mn}^{i-1}A_{kl}^{i-1} - A_{mn}^0 A_{kl}^0\right)\frac{ml(kn+ml)}{\left[(m-k)^2 + (n+l)^2/\alpha^2\right]^2} \tag{11.28b}$$

$$C_3(m,n,k,l) = \frac{E\pi^4}{4\alpha^2}\left(A_{mn}^{i-1}A_{kl}^{i-1} - A_{mn}^0 A_{kl}^0\right)\frac{ml(kn+ml)}{\left[(m+k)^2 + (n-l)^2/\alpha^2\right]^2} \tag{11.28c}$$

$$C_4(m,n,k,l) = \frac{E\pi^4}{4\alpha^2}\left(A_{mn}^{i-1}A_{kl}^{i-1} - A_{mn}^0 A_{kl}^0\right)\frac{ml(kn-ml)}{\left[(m+k)^2 + (n+l)^2/\alpha^2\right]^2} \tag{11.28d}$$

By substituting Equations (11.28a)–(11.28d) into Equation (11.27), F_P^{i-1} can be rewritten as follows:

$$F_P^{i-1} = \frac{E\pi^4}{4\alpha^2}\sum_m\sum_n\sum_k\sum_l\left(A_{mn}^{i-1}A_{kl}^{i-1} - A_{mn}^0 A_{kl}^0\right)$$

$$\times \sum_{r=1}^2\sum_{s=1}^2(-1)^{r+s}h_2[(-1)^r k,(-1)^r l]cx[m+(-1)^r k]cy[n+(-1)^s l] \tag{11.29}$$

where

$$h_2[\omega_1,\omega_2] = \frac{m\omega_2(n\omega_1 - m\omega_2)}{\left[(m+\omega_1)^2 + (n+\omega_2)^2/\alpha^2\right]^2}$$

$h_2 = 0$ if $m + \omega_1 = 0$ and $n + \omega_2 = 0$.

The homogeneous solution, F_H^{i-1}, can be expressed by considering the condition of load application as follows:

$$F_H^{i-1} = P_x^{i-1}\frac{y^2}{2bt} + \sigma_{rx}\frac{y^2}{2} + P_y^{i-1}\frac{x^2}{2at} + \sigma_{ry}\frac{x^2}{2}$$

$$- M_x^{i-1}\frac{y^2(2y-3b)}{b^3t} - M_y^{i-1}\frac{x^2(2x-3a)}{a^3t} - \tau_{xy}^{i-1}xy \tag{11.30}$$

where σ_{rx} and σ_{ry} are the welding induced residual stresses, as described in Section 1.7, that are included as initial stress terms.

Therefore, the stress function, F^{i-1}, at the end of the $(i-1)$th load increment step can be obtained by the sum of Equations (11.27) and (11.28) as follows:

$$F^{i-1} = F_P^{i-1} + F_H^{i-1} \tag{11.31}$$

A numerical technique can be used to efficiently calculate Equation (11.16). With the panel subdivided (meshed) into a number of regions in the x, y, and z directions, Equation (11.16) may then be expressed by

$$\sum_u\sum_v\sum_w\int_{a_u}^{a_{u+1}}\int_{b_v}^{b_{v+1}}\int_{t_w}^{t_{w+1}}\Delta\Phi(x,y,z)sx(r)sy(s)dxdydz = 0, \quad r=1,2,3,\dots, \quad s=1,2,3,\dots \tag{11.32}$$

where $\sum_u$, $\sum_v$, and $\sum_w$ indicate the summation for mesh regions in the x, y, and z directions, respectively.

Substitution of Equation (11.12a) into Equation (11.32) yields the following expression, given in detail:

$$
\sum_{u}\sum_{v}\sum_{w}\int_{a_u}^{a_{u+1}}\int_{b_v}^{b_{v+1}}\int_{t_w}^{t_{w+1}}\left[\frac{E}{1-v^2}\left(\frac{\partial^4\Delta w}{\partial x^4}+2\frac{\partial^4\Delta w}{\partial x^2\partial y^2}+\frac{\partial^4\Delta w}{\partial y^4}\right)z^2\right.
$$

$$
-\left(\frac{\partial^2 F_{i-1}}{\partial y^2}\frac{\partial^2\Delta w}{\partial x^2}+\frac{\partial^2\Delta F}{\partial y^2}\frac{\partial^2 w_{i-1}}{\partial x^2}-2\frac{\partial^2 F_{i-1}}{\partial x\partial y}\frac{\partial^2\Delta w}{\partial x\partial y}\right.
$$

$$
\left.\left.-2\frac{\partial^2\Delta F}{\partial x\partial y}\frac{\partial^2 w_{i-1}}{\partial x^2}+\frac{\partial^2 F_{i-1}}{\partial x^2}\frac{\partial^2\Delta w}{\partial y^2}+\frac{\partial^2\Delta F}{\partial x^2}\frac{\partial^2 w_{i-1}}{\partial y^2}\right)\right]sx(r)sy(s)dxdydz
$$

$$
-\sum_{u}\sum_{v}\int_{a_u}^{a_{u+1}}\int_{b_v}^{b_{v+1}}\Delta p sx(r)sy(s)dxdy=0
$$

(11.33)

where $D=\sum_u\int_{-t/2}^{t/2}Ezdz/\left(1-v^2\right)$ and $t=\sum_w\int_{-t/2}^{t/2}dz$ have been used. The lateral pressure loads are also considered as distributed on the surface of the plate, and the related integration is performed in the two directions in the xy plane; that is, no integration associated with the lateral pressure loads is undertaken for the z direction.

The integration of Equation (11.33) eventually results in a set of linear (i.e., first-order) simultaneous equations for the unknown coefficients, ΔA_{kl}. The equations can be written in matrix form as follows:

$$
\{\Delta P\}=\left(\left[P_O\right]+\left[K_B\right]+\left[K_M\right]\right)\{\Delta A\}
$$

(11.34)

where $\{\Delta P\}$ are the external load increments, $[P_O]$ is the stiffness matrix associated with initial stress (including welding induced residual stresses), $[K_B]$ is the bending stiffness matrix, $[K_M]$ is the stiffness matrix due to membrane action, and $\{\Delta A\}$ are unknown coefficients of deflection amplitudes.

11.3.4 Treatment of Plasticity

Thus far, the differential equations that govern the elastic large-deflection behavior of plates have been formulated and solved analytically, but the effects of plasticity have not been included. It is normally not straightforward to formulate governing differential equations to simultaneously represent both geometric and material nonlinearities, although it is not impossible for plates. As previously noted, a major source of difficulty is that an analytical treatment of plasticity with increases in the applied loads is very difficult. An easier alternative is to deal with the progress of the plasticity numerically.

In the incremental Galerkin method, therefore, the progress of plasticity with increases in the applied loads is treated with a numerical approach. For this purpose, the plate is subdivided into a number of mesh regions in the three directions, similar to the conventional finite element method, as shown in Figure 11.3a. The average membrane stress components for each mesh region can be calculated at every load increment step. Yielding for each mesh region is checked with the relevant yield criteria, such as the von Mises yield condition, as defined in Equation (1.31c), neglecting the strain-hardening effect:

$$\sigma_x^2 - \sigma_x\sigma_y + \sigma_y^2 + 3\tau^2 \geq \sigma_Y^2 \tag{11.35}$$

It is assumed that a plate element bounded by support members along its four edges is composed of a number of membrane strings (or fibers) in the plate length and breadth directions. Each fiber has a number of layers in the thickness direction. The membrane stresses for each region can be analytically calculated at every step of the load increments, and yielding at local regions of the plating can also be checked analytically. If any local region in the fiber yields, the fiber (i.e., string) will be cut so that the related membrane action is no longer available, resulting in a larger deflection. For the approximate evaluation of plasticity, it is assumed that the plate is composed of a number of membrane fibers in the x and y directions. Each membrane fiber is considered to have a number of layers in the z direction, as shown in Figure 11.3a.

As the applied loads increase, the stiffness matrices for the plate are redefined by considering the progress of plasticity. In Equation (11.34), the stiffness matrix associated with external loads should be calculated for the plate's whole volume regardless of the plasticity. However, the bending stiffness will be reduced by the plasticity if any mesh region yields. In the calculation (i.e., integration) of the bending stiffness matrix, therefore, the contribution to the yielded regions is removed.

As the plate is composed of a number of membrane strings (or fibers) in the two (i.e., x, y) directions, where each fiber has a number of layers in the z direction, the end condition for each fiber also satisfies the plate edge condition. In fact, due to the membrane action of the fibers, the occurrence of additional plate deflection may to some extent be disturbed, with further increases in the applied loads. However, if any local region in the fiber yields, the fiber (i.e., string) will be cut such that the membrane action is no longer available.

In calculating (and integrating) the stiffness matrix due to membrane action, therefore, the entire fiber associated with yielded regions is not included. It should be noted that a mesh region inside the plate may be common to two fibers, that is, in the x (i.e., length) and y (i.e., breadth) directions. In this case, the contribution from the two fibers (i.e., strings) should be removed in the calculation of the stiffness matrix associated with the membrane effects. The plate's stiffness will be progressively reduced by the large deflection and local yielding. The plate can be considered to have reached the ultimate strength when the plate stiffness eventually becomes zero (or negative).

11.4 Analysis of the Elastic–Plastic Large-Deflection Behavior of Stiffened Panels

11.4.1 The Traditional Approach

The elastic large-deflection response of stiffened panels with initial deflections is governed by two differential equations: one that represents the equilibrium condition and one that represents the compatibility condition. These equations are as follows:

$$\Phi = D\left(\frac{\partial^4 w}{\partial x^4} + 2\frac{\partial^4 w}{\partial x^2 \partial y^2} + \frac{\partial^4 w}{\partial y^4}\right)$$

$$-t\left[\frac{\partial^2 F}{\partial y^2}\frac{\partial^2(w+w_0)}{\partial x^2} + \frac{\partial^2 F}{\partial x^2}\frac{\partial^2(w+w_0)}{\partial y^2} - 2\frac{\partial^2 F}{\partial x \partial y}\frac{\partial^2(w+w_0)}{\partial x \partial y}\right]$$

$$+\sum_{ii=1}^{n_{sx}}\left[EI_{ii}\frac{\partial^4 w}{\partial x^4} - A_{ii}\left(\frac{\partial^2 F}{\partial y^2} - \nu\frac{\partial^2 F}{\partial x^2}\right)\frac{\partial^2(w+w_0)}{\partial x^2}\right]_{y=y_{ii}}$$

$$+\sum_{jj=1}^{n_{sy}}\left[EI_{jj}\frac{\partial^4 w}{\partial y^4} - A_{jj}\left(\frac{\partial^2 F}{\partial x^2} - \nu\frac{\partial^2 F}{\partial y^2}\right)\frac{\partial^2(w+w_0)}{\partial y^2}\right]_{x=x_{jj}} - p = 0 \tag{11.36a}$$

$$\frac{\partial^4 F}{\partial x^4} + 2\frac{\partial^4 F}{\partial x^2 \partial y^2} + \frac{\partial^4 F}{\partial y^4} - E\left[\left(\frac{\partial^2 w}{\partial y \partial x}\right)^2 - \frac{\partial^2 w}{\partial x^2}\frac{\partial^2 w}{\partial y^2} + 2\frac{\partial^2 w_0}{\partial x \partial y}\frac{\partial^2 w}{\partial x \partial y}\right.$$

$$\left.- \frac{\partial^2 w_0}{\partial x^2}\frac{\partial^2 w}{\partial y^2} - \frac{\partial^2 w}{\partial x^2}\frac{\partial^2 w_0}{\partial y^2}\right] = 0 \tag{11.36b}$$

where D and p are defined in Equation (11.1). I_{ii} is moment of inertia for the iith stiffener in the x direction and I_{jj} is moment of inertia for the jjth stiffener in the y direction.

By using Airy's stress function, the stress components at a certain position inside the panel may be expressed as follows:

$$\sigma_x = \frac{\partial^2 F}{\partial y^2} - \frac{Ez}{1-\nu^2}\left[\frac{\partial^2 w}{\partial x^2} + \nu\frac{\partial^2 w}{\partial y^2}\right] \tag{11.37a}$$

$$\sigma_y = \frac{\partial^2 F}{\partial x^2} - \frac{Ez}{1-\nu^2}\left(\frac{\partial^2 w}{\partial y^2} + \nu\frac{\partial^2 w}{\partial x^2}\right) \tag{11.37b}$$

$$\tau = \tau_{xy} = -\frac{\partial^2 F}{\partial x \partial y} - \frac{Ez}{2(1+\nu)}\frac{\partial^2 w}{\partial x \partial y} \tag{11.37c}$$

In solving Equations (11.36a) and (11.36b), by the Galerkin method, the added deflection w and initial deflection w_0 can be assumed as follows:

$$w = \sum_{m=1}^{\infty}\sum_{n=1}^{\infty} A_{mn}f_m(x)g_n(y) \tag{11.38a}$$

$$w_0 = \sum_{m=1}^{\infty}\sum_{n=1}^{\infty} A_{0mn}f_m(x)g_n(y) \tag{11.38b}$$

where $f_m(x)$ and $g_n(y)$ are functions that satisfy the boundary conditions for the panel and A_{mn} and A_{0mn} are unknown and known deflection coefficients, respectively.

Upon substituting Equations (11.38a) and (11.38b) into Equation (11.36b) and solving for the stress function F, the particular solution of F denoted by F_P may be expressed as follows:

$$F_P = \sum_{r=1}^{\infty}\sum_{s=1}^{\infty} K_{rs}p_r(x)q_s(y) \tag{11.39}$$

where K_{rs} are the coefficient of second-order functions with regard to the unknown deflection coefficients A_{mn}.

Including the condition of load application, the complete stress function F may be given by

$$F = F_H + \sum_{r=1}\sum_{s=1} K_{rs} p_r(x) q_s(y) \tag{11.40}$$

where F_H is a homogeneous solution of the stress function that satisfies the applied loading condition.

To compute the unknown coefficients A_{mn}, one may use the Galerkin method for the equilibrium Equation (11.36a), resulting in the following equation:

$$\iiint \Phi f_r(x) g_s(y) d\text{Vol} = 0, \quad r = 1,2,3,\ldots, \quad s = 1,2,3,\ldots \tag{11.41}$$

Substituting Equations (11.38a), (11.38b), and (11.40) into Equation (11.41) and performing the integration over the whole volume of the panel, a set of third-order simultaneous equations with regard to the unknown coefficients A_{mn} is obtained.

Solving the simultaneous equations to obtain the coefficients A_{mn} normally requires an iteration process. As the solution of each coefficient should be unique, one must correctly select one of the three solutions obtained for each coefficient. Unfortunately, it is not always easy to solve a set of such third-order simultaneous equations, especially when the number of unknown coefficients A_{mn} becomes large.

11.4.2 The Incremental Approach

The incremental forms of the governing differential equations for stiffened panels are derived. As the equilibrium equation for unstiffened plates readily results from that for stiffened panels by setting the properties of the stiffeners to zero, the derivation begins with the traditional governing differential equations for stiffened steel panels.

First, it is assumed that the load is applied incrementally. At the end of the $(i-1)$th load increment step, the deflection and stress function can be denoted by w_{i-1} and F_{i-1}, respectively. In the same manner, the deflection and stress function at the end of the ith load increment step are denoted by w_i and F_i, respectively.

The equilibrium equation, Equation (11.36a), and the compatibility equation, Equation (11.36b), are written at the end of the $(i-1)$th load increment step as follows:

$$\Phi_{i-1} = D\left(\frac{\partial^4 w_{i-1}}{\partial x^4} + 2\frac{\partial^4 w_{i-1}}{\partial x^2 \partial y^2} + \frac{\partial^4 w_{i-1}}{\partial y^4}\right)$$

$$-t\left[\frac{\partial^2 F_{i-1}}{\partial y^2}\frac{\partial^2 (w_{i-1}+w_0)}{\partial x^2} + \frac{\partial^2 F_{i-1}}{\partial x^2}\frac{\partial^2 (w_{i-1}+w_0)}{\partial y^2} - 2\frac{\partial^2 F_{i-1}}{\partial x\partial y}\frac{\partial^2 (w_{i-1}+w_0)}{\partial x\partial y}\right]$$

$$+\sum_{ii=1}^{n_{sx}}\left[EI_{ii}\frac{\partial^4 w_{i-1}}{\partial x^4} - A_{ii}\left(\frac{\partial^2 F_{i-1}}{\partial y^2} - \nu\frac{\partial^2 F_{i-1}}{\partial x^2}\right)\frac{\partial^2 (w_{i-1}+w_0)}{\partial x^2}\right]_{y=y_{ii}}$$

$$+\sum_{jj=1}^{n_{sy}}\left[EI_{jj}\frac{\partial^4 w_{i-1}}{\partial y^4} - A_{jj}\left(\frac{\partial^2 F_{i-1}}{\partial x^2} - \nu\frac{\partial^2 F_{i-1}}{\partial y^2}\right)\frac{\partial^2 (w_{i-1}+w_0)}{\partial y^2}\right]_{x=x_{jj}} - p_{i-1} = 0$$

$$\tag{11.42a}$$

$$\frac{\partial^4 F_{i-1}}{\partial x^4} + 2\frac{\partial^4 F_{i-1}}{\partial x^2 \partial y^2} + \frac{\partial^4 F_{i-1}}{\partial y^4} - E\left[\left(\frac{\partial^2 w_{i-1}}{\partial y \partial x}\right)^2 - \frac{\partial^2 w_{i-1}}{\partial x^2}\frac{\partial^2 w_{i-1}}{\partial y^2} + 2\frac{\partial^2 w_0}{\partial x \partial y}\frac{\partial^2 w_{i-1}}{\partial x \partial y}\right.$$

$$\left. -\frac{\partial^2 w_0}{\partial x^2}\frac{\partial^2 w_{i-1}}{\partial y^2} - \frac{\partial^2 w_{i-1}}{\partial x^2}\frac{\partial^2 w_0}{\partial y^2}\right] = 0 \tag{11.42b}$$

In the same manner, the equilibrium equation and the compatibility equation are written at the ith load increment step as follows:

$$\Phi_i = D\left(\frac{\partial^4 w_i}{\partial x^4} + 2\frac{\partial^4 w_i}{\partial x^2 \partial y^2} + \frac{\partial^4 w_i}{\partial y^4}\right)$$

$$-t\left[\frac{\partial^2 F_i}{\partial y^2}\frac{\partial^2 (w_i + w_0)}{\partial x^2} + \frac{\partial^2 F_i}{\partial x^2}\frac{\partial^2 (w_i + w_0)}{\partial y^2} - 2\frac{\partial^2 F_i}{\partial x \partial y}\frac{\partial^2 (w_i + w_0)}{\partial x \partial y}\right]$$

$$+\sum_{ii=1}^{n_{sx}}\left[EI_{ii}\frac{\partial^4 w_i}{\partial x^4} - A_{ii}\left(\frac{\partial^2 F_i}{\partial y^2} - \nu\frac{\partial^2 F_i}{\partial x^2}\right)\frac{\partial^2 (w_i + w_0)}{\partial x^2}\right]_{y=y_{ii}}$$

$$+\sum_{jj=1}^{n_{sy}}\left[EI_{jj}\frac{\partial^4 w_i}{\partial y^4} - A_{jj}\left(\frac{\partial^2 F_i}{\partial x^2} - \nu\frac{\partial^2 F_i}{\partial y^2}\right)\frac{\partial^2 (w_i + w_0)}{\partial y^2}\right]_{x=x_{jj}} - p_i = 0 \tag{11.43a}$$

$$\frac{\partial^4 F_i}{\partial x^4} + 2\frac{\partial^4 F_i}{\partial x^2 \partial y^2} + \frac{\partial^4 F_i}{\partial y^4} - E\left[\left(\frac{\partial^2 w_i}{\partial y \partial x}\right)^2 - \frac{\partial^2 w_i}{\partial x^2}\frac{\partial^2 w_i}{\partial y^2} + 2\frac{\partial^2 w_0}{\partial x \partial y}\frac{\partial^2 w_i}{\partial x \partial y}\right.$$

$$\left. -\frac{\partial^2 w_0}{\partial x^2}\frac{\partial^2 w_i}{\partial y^2} - \frac{\partial^2 w_i}{\partial x^2}\frac{\partial^2 w_0}{\partial y^2}\right] = 0 \tag{11.43b}$$

It is assumed that the accumulated (total) deflection w_i and stress function F_i at the end of the ith load increment step are calculated by

$$w_i = w_{i-1} + \Delta w \tag{11.44a}$$

$$F_i = F_{i-1} + \Delta F \tag{11.44b}$$

where Δw and ΔF are the increments of deflection or stress function, respectively, where the prefix Δ indicates the increment for the variable.

Substituting Equations (11.44a) and (11.44b) into Equations (11.43a) and (11.43b) and subtracting Equation (11.42a) from Equation (11.43a) or Equation (11.42b) from Equation (11.43b), respectively, the necessary incremental forms of the governing differential equations emerge as follows:

$$\Delta\Phi = D\left(\frac{\partial^4 \Delta w}{\partial x^4} + 2\frac{\partial^4 \Delta w}{\partial x^2 \partial y^2} + \frac{\partial^4 \Delta w}{\partial y^4}\right) - t\left[\frac{\partial^2 F_{i-1}}{\partial y^2}\frac{\partial^2 \Delta w}{\partial x^2} + \frac{\partial^2 \Delta F}{\partial y^2}\frac{\partial^2 (w_{i-1} + w_0)}{\partial x^2}\right.$$

$$+\frac{\partial^2 F_{i-1}}{\partial x^2}\frac{\partial^2 \Delta w}{\partial y^2} + \frac{\partial^2 \Delta F}{\partial x^2}\frac{\partial^2 (w_{i-1} + w_0)}{\partial y^2} - 2\frac{\partial^2 F_{i-1}}{\partial x \partial y}\frac{\partial^2 \Delta w}{\partial x \partial y} - 2\frac{\partial^2 \Delta F}{\partial x \partial y}\frac{\partial^2 (w_{i-1} + w_0)}{\partial x \partial y}\right]$$

$$+ \sum_{ii=1}^{n_{sx}} \left[EI_{ii} \frac{\partial^4 \Delta w}{\partial x^4} - A_{ii} \left(\frac{\partial^2 F_{i-1}}{\partial y^2} - \nu \frac{\partial^2 F_{i-1}}{\partial x^2} \right) \frac{\partial^2 \Delta w}{\partial x^2} - A_{ii} \left(\frac{\partial^2 \Delta F}{\partial y^2} - \nu \frac{\partial^2 \Delta F}{\partial x^2} \right) \frac{\partial^2 (w_{i-1} + w_0)}{\partial x^2} \right]_{y=y_{ii}}$$

$$+ \sum_{jj=1}^{n_{sy}} \left[EI_{jj} \frac{\partial^4 \Delta w}{\partial y^4} - A_{jj} \left(\frac{\partial^2 F_{i-1}}{\partial x^2} - \nu \frac{\partial^2 F_{i-1}}{\partial y^2} \right) \frac{\partial^2 \Delta w}{\partial y^2} - A_{jj} \left(\frac{\partial^2 \Delta F}{\partial x^2} - \nu \frac{\partial^2 \Delta F}{\partial y^2} \right) \frac{\partial^2 (w_{i-1} + w_0)}{\partial y^2} \right]_{x=x_{jj}}$$

$$- \Delta p = 0$$

$$(11.45a)$$

$$\frac{\partial^4 \Delta F}{\partial x^4} + 2 \frac{\partial^4 \Delta F}{\partial x^2 \partial y^2} + \frac{\partial^4 \Delta F}{\partial y^4} - E \left[2 \frac{\partial^2 (w_{i-1} + w_0)}{\partial x \partial y} \frac{\partial^2 \Delta w}{\partial x \partial y} \right.$$

$$(11.45b)$$

$$\left. - \frac{\partial^2 (w_{i-1} + w_0)}{\partial x^2} \frac{\partial^2 \Delta w}{\partial y^2} - \frac{\partial^2 \Delta w}{\partial x^2} \frac{\partial^2 (w_{i-1} + w_0)}{\partial y^2} \right] = 0$$

where the terms of very small quantities with an order higher than the second order of increments Δw and ΔF are neglected.

At the end of the $(i-1)$th load increment step, the deflection w_{i-1} and the stress function F_{i-1} are obtained as follows:

$$w_{i-1} = \sum_{m=1}^{} \sum_{n=1}^{} A_{mn}^{i-1} f_m(x) g_n(y) \tag{11.46a}$$

$$F_{i-1} = F_H^{i-1} + \sum_{i=1}^{} \sum_{j=1}^{} K_{ij}^{i-1} p_i(x) q_j(y) \tag{11.46b}$$

where A_{mn}^{i-1} and K_{ij}^{i-1} are the known coefficients and F_H^{i-1} is a homogeneous solution for the stress function that satisfies the applied loading condition. The welding induced residual stresses can be included in the stress function F_H^{i-1} as initial stress terms. (The welding residual stresses are also set to zero when welding is not used in fabrication.)

The added deflection increment Δw associated with the load increment at the ith step can be assumed as follows:

$$\Delta w = \sum_{k=1}^{} \sum_{l=1}^{} \Delta A_{kl} f_k(x) g_l(y) \tag{11.47}$$

where ΔA_{kl} is the unknown added deflection increment.

Substituting Equations (11.38b), (11.46a), and (11.47) into Equation (11.45b), the stress function increment ΔF can be obtained by

$$\Delta F = \Delta F_H + \sum_{i=1}^{} \sum_{j=1}^{} \Delta K_{ij} p_i(x) q_j(y) \tag{11.48}$$

where ΔK_{ij} are linear (i.e., first-order) functions in the unknown coefficients ΔA_{kl} and ΔF_H is a homogeneous solution for the stress function increment that satisfies the applied loading condition.

To compute the unknown coefficients ΔA_{kl}, the Galerkin method can then be applied to Equation (11.43a):

$$\iiint \Delta \Phi f_r(x) g_s(y) d\text{Vol} = 0, \quad r = 1,2,3,\ldots, \quad s = 1,2,3,\ldots \tag{11.49}$$

Substituting Equation (11.38b) and Equations (11.46)–(11.48) into Equation (11.49) and performing the integration over the whole volume of the panel, a set of linear simultaneous equations for the unknown coefficients ΔA_{kl} is obtained. Solving these linear simultaneous equations is normally easy. Having obtained ΔA_{kl}, one can then calculate Δw, that is, from Equation (11.47); ΔF, that is, from Equation (11.48); $w_i(= w_{i-1} + \Delta w)$, that is, from Equation (11.44a); and $F_i(= F_{i-1} + \Delta F)$, that is, from Equation (11.44b), at the end of the and ith load increment step.

The elastic large-deflection response for the panel can be obtained by repeating the preceding procedure with increases in the applied loads. In this process, it is apparent that the load increments must be small to obtain more accurate solutions by avoiding non-equilibrating forces. As the computational effort required for this procedure is normally very small, the use of smaller load increments does not lead to any severe penalties, unlike the usual numerical methods.

11.4.3 Application to the Stiffened Panels Simply Supported at Four Edges

In this section, the incremental Galerkin method is applied to analysis of the elastic large-deflection behavior of stiffened panels simply supported at four edges. The simply supported edge conditions for the panel should satisfy the following conditions:

$$w = 0, \quad \frac{\partial^2 w}{\partial y^2} + v\frac{\partial^2 w}{\partial x^2} = 0, \quad \text{at } y = 0,b \tag{11.50a}$$

$$\frac{\partial^2 w}{\partial y^2} = 0 \quad \text{at } x = x_{jj} \text{ and } y = 0,b \tag{11.50b}$$

$$w = 0, \quad \frac{\partial^2 w}{\partial x^2} + v\frac{\partial^2 w}{\partial y^2} = 0 \quad \text{at } x = 0,a \tag{11.50c}$$

$$\frac{\partial^2 w}{\partial x^2} = 0 \quad \text{at } y = y_{ii} \text{ and } x = 0,a \tag{11.50d}$$

The Fourier series deflection functions that need to satisfy the boundary conditions can be assumed as follows:

$$w_0 = \sum_{m=1}^{} \sum_{n=1}^{} A_{omn} \sin\frac{m\pi x}{a} \sin\frac{n\pi y}{b} \tag{11.51a}$$

$$w_{i-1} = \sum_{m=1}^{} \sum_{n=1}^{} A_{mn}^{i-1} \sin\frac{m\pi x}{a} \sin\frac{n\pi y}{b} \tag{11.51b}$$

$$\Delta w = \sum_{k=1}^{} \sum_{l=1}^{} \Delta A_{kl} \sin\frac{k\pi x}{a} \sin\frac{l\pi y}{b} \tag{11.51c}$$

where $A_{omn}(= A_{mn}^o)$ and A_{mn}^{i-1} are the known coefficients and ΔA_{kl} are the unknown coefficients to be calculated for the external load increments.

The condition of combined load application—biaxial loads, biaxial in-plane bending, edge shear, and lateral pressure loads—gives

$$\int_0^b \frac{\partial^2 F}{\partial y^2} t\, dy + \sum_{ii=1} A_{ii} \left(\frac{\partial^2 F}{\partial y^2} - \nu \frac{\partial^2 F}{\partial x^2} \right)_{y=y_{ii}} = P_x \quad \text{at } x = 0, a \tag{11.52a}$$

$$\int_0^b \frac{\partial^2 F}{\partial y^2} t \left(y - \frac{b}{2} \right) dy + \sum_{ii=1} A_{ii} \left(\frac{\partial^2 F}{\partial y^2} - \nu \frac{\partial^2 F}{\partial x^2} \right)_{y=y_{ii}} \left(y_{ii} - \frac{b}{2} \right) = M_x \quad \text{at } x = 0, a \tag{11.52b}$$

$$\int_0^a \frac{\partial^2 F}{\partial x^2} t\, dx + \sum_{jj=1} A_{jj} \left(\frac{\partial^2 F}{\partial x^2} - \nu \frac{\partial^2 F}{\partial y^2} \right)_{x=x_{jj}} = P_y \quad \text{at } x = 0, b \tag{11.52c}$$

$$\int_0^a \frac{\partial^2 F}{\partial x^2} t \left(x - \frac{a}{2} \right) dx + \sum_{jj=1} A_{jj} \left(\frac{\partial^2 F}{\partial x^2} - \nu \frac{\partial^2 F}{\partial y^2} \right)_{x=x_{jj}} \left(x_{jj} - \frac{a}{2} \right) = M_y \quad \text{at } x = 0, b \tag{11.52d}$$

$$\frac{\partial^2 F}{\partial x \partial y} = -\tau \quad \text{at all boundaries} \tag{11.52e}$$

where the axial loads and in-plane bending moments are sustained by the stiffener and by the plate part. For the sake of simplicity in expressing the various functions, the following abbreviations are used from this point:

$$sx(m) = \sin\frac{m\pi x}{a}, \quad sy(n) = \sin\frac{n\pi y}{b}, \quad cx(m) = \cos\frac{m\pi x}{a}, \quad cy(n) = \cos\frac{n\pi y}{b}$$

To obtain the stress function increment ΔF, substituting Equation (11.51) into Equation (11.45b) yields

$$\frac{\partial^4 \Delta F}{\partial x^4} + 2\frac{\partial^4 \Delta F}{\partial x^2 \partial y^2} + \frac{\partial^4 \Delta F}{\partial y^4} = \frac{E\pi^4}{4a^2 b^2} \sum_m \sum_n \sum_k \sum_l \Delta A_{kl} A_{mn}^{i-1}$$

$$\times \left[-(kn-ml)^2 cx(m-k)cy(n-l) + (kn+ml)^2 cx(m-k)cy(n+l) \right.$$

$$\left. + (kn+ml)^2 cx(m+k)cy(n-l) - (kn-ml)^2 cx(m+k)cy(n+l) \right] \tag{11.53}$$

A particular solution for the stress function increment, denoted by ΔF_P, is assumed as follows:

$$\Delta F_P = \sum_m \sum_n \sum_k \sum_l [B_1(m,n,k,l)cx(m-k)cy(n-l) + B_2(m,n,k,l)cx(m-k)cy(n+l)$$

$$+ B_3(m,n,k,l)cx(m+k)cy(n-l) + B_4(m,n,k,l)cx(m+k)cy(n+l)] \tag{11.54}$$

Substituting Equation (11.54) into the left side of Equation (11.53) by using $\Delta F = \Delta F_P$ yields the coefficients $B_1 \cdots B_4$ as follows:

$$B_1(m, n, k, l) = \frac{E\alpha^2 \pi^4}{4} \Delta A_{kl} A_{mn}^{i-1} \frac{-(kn - ml)^2}{\left[(m - k)^2 + \alpha^2 (n - l)^2\right]^2}$$

$$B_2(m, n, k, l) = \frac{E\alpha^2 \pi^4}{4} \Delta A_{kl} A_{mn}^{i-1} \frac{(kn + ml)^2}{\left[(m - k)^2 + \alpha^2 (n + l)^2\right]^2}$$

$$B_3(m, n, k, l) = \frac{E\alpha^2}{4} \Delta A_{kl} A_{mn}^{i-1} \frac{(kn - ml)^2}{(m + k)^2 + \alpha^2 (n - l)^2}$$

$$B_4(m, n, k, l) = \frac{E\alpha^2}{4} \Delta A_{kl} A_{mn}^{i-1} \frac{-(kn - ml)^2}{(m + k)^2 + \alpha^2 (n + l)^2}$$

(11.55)

Substituting Equation (11.55) into Equation (11.54), the particular solution for the stress function increment can be written in a more simplified form as follows:

$$\Delta F_P = \frac{E\alpha^2 \pi^4}{4} \sum_m \sum_n \sum_k \sum_l \Delta A_{kl} A_{mn}^{i-1} \times$$

(11.56)

$$\sum_{r=1}^{2} \sum_{s=1}^{2} (-1)^{r+s+1} \cdot h_1 \left[(-1)^r k, (-1)^r l\right] cx \left(m + (-1)^r k\right) \left(n + (-1)^s l\right)$$

where $h_1[\omega_1, \omega_2] = (-n\omega_1 + m\omega_2)^2 / \left[(m + \omega_1)^2 + \alpha^2 (n + \omega_2)^2\right]^2$, $h_1 = 0$ if $m + \omega_1 = 0$ and $n + \omega_2 = 0$.

By considering the condition of load application, the homogeneous solution for the stress function increment denoted by ΔF_H is given by

$$\Delta F_H = \frac{1}{c}\left[\Delta P_x \left(1 + \frac{A_{sy}}{at}\right) + \Delta P_y \left(\frac{\nu A_{sx}}{at}\right)\right]\frac{y^2}{2bt} + \frac{1}{c}\left[\Delta P_y \left(1 + \frac{A_{sx}}{bt}\right) + \Delta P_x \left(\frac{\nu A_{sy}}{by}\right)\right]\frac{x^2}{2at}$$

$$- \Delta M_x \frac{Z_{px}}{Z_{px} + Z_{sx}} \frac{y^2(2y - 3b)}{b^3 t} - \Delta M_y \frac{Z_{py}}{Z_{py} + Z_{sy}} \frac{x^2(2x - 3a)}{a^3 t} - \Delta \tau_{xy} xy$$

(11.57)

where $c = \left(1 + \frac{A_{sx}}{bt}\right)\left(1 + \frac{A_{sy}}{at}\right) - \nu^2 \frac{A_{sx} A_{sy}}{abt^2}$, $Z_{px} = \frac{b^2 t}{6}$, $Z_{py} = \frac{a^2 t}{6}$, $Z_{sx} = \sum_{ii=1}^{n_{sx}} A_{ii}$

$\left(y_{ii} - \frac{b}{2}\right)^2 \frac{2}{b}$, $Z_{sy} = \sum_{jj=1}^{n_{sy}} A_{jj} \left(x_{jj} - \frac{a}{2}\right)^2 \frac{2}{a}$.

The overall stress function increment may then be expressed by the sum of the particular and homogeneous solutions as follows:

$$\Delta F = \Delta F_P + \Delta F_H$$

(11.58)

In the same manner, to obtain the stress function at the end of the $(i-1)$th load increment step, denoted by F_P^{i-1}, Equation (11.51) is substituted into the compatibility equation at the end of the $(i-1)$th load increment step, resulting in the following equation:

$$\frac{\partial^4 F_{i-1}}{\partial x^4} + 2\frac{\partial^4 F_{i-1}}{\partial x^2 \partial y^2} + \frac{\partial^4 F_{i-1}}{\partial y^4} = \frac{E\pi^4}{4a^2 b^2}\sum_m\sum_n\sum_k\sum_l \left(A_{mn}^{i-1}A_{kl}^{i-1} - A_{mn}^o A_{kl}^o\right) \times$$

$$[ml(kn-ml)cx(m-k)cy(n-l)$$
$$+ ml(kn+ml)cx(m-k)cy(n+l)$$
$$+ ml(kn+ml)cx(m+k)cy(n-l)$$
$$+ ml(kn-ml)cx(m+k)cy(n+l)]$$

(11.59)

A particular solution for the stress function $F_{i-1}(=F^{i-1})$, denoted by F_P^{i-1}, can then be given by

$$F_P^{i-1} = \sum_m\sum_n\sum_k\sum_l [C_1(m,n,k,l)cx(m-k)cy(n-l) + C_2(m,n,k,l)cx(m-k)cy(n+l)$$

$$+ C_3(m,n,k,l)cx(m+k)cy(n-l) + C_4(m,n,k,l)cx(m+k)cy(n+l)]$$

(11.60)

The coefficients $C_1 \cdots C_4$ of Equation (11.60) can be determined by substituting Equation (11.60) into Equation (11.59) as follows:

$$C_1(m,n,k,l) = \frac{E\pi^4}{4\alpha^2}\left(A_{mn}^{i-1}A_{kl}^{i-1} - A_{mn}^o A_{kl}^o\right)\frac{ml(kn-ml)}{\left[(m-k)^2 + (n-l)^2/\alpha^2\right]^2}$$

$$C_2(m,n,k,l) = \frac{E\pi^4}{4\alpha^2}\left(A_{mn}^{i-1}A_{kl}^{i-1} - A_{mn}^o A_{kl}^o\right)\frac{ml(kn+ml)}{\left[(m-k)^2 + (n+l)^2/\alpha^2\right]^2}$$

$$C_3(m,n,k,l) = \frac{E\pi^4}{4\alpha^2}\left(A_{mn}^{i-1}A_{kl}^{i-1} - A_{mn}^o A_{kl}^o\right)\frac{ml(kn+ml)}{\left[(m+k)^2 + (n-l)^2/\alpha^2\right]^2}$$

$$C_4(m,n,k,l) = \frac{E\pi^4}{4\alpha^2}\left(A_{mn}^{i-1}A_{kl}^{i-1} - A_{mn}^o A_{kl}^o\right) \cdot \frac{ml(kn-ml)}{\left[(m+k)^2 + (n+l)^2/\alpha^2\right]^2}$$

(11.61)

The particular solution for the stress function F^{i-1} is then written by substituting the coefficients of Equation (11.61) into Equation (11.60) as follows:

$$F_P^{i-1} = \frac{E\pi^4}{4\alpha^2}\sum_m\sum_n\sum_k\sum_l \left(A_{mn}^{i-1}A_{kl}^{i-1} - A_{mn}^o A_{kl}^o\right)$$

$$\times \sum_{r=1}^{2}\sum_{s=1}^{2}(-1)^{r+s}h_2[(-1)^r k, (-1)^r l]cx(m+(-1)^r k)cy(n+(-1)^s l)$$

(11.62)

where $h_2[\omega_1, \omega_2] = m\omega_2 \cdot (n\omega_1 - m\omega_2)/\left[(m+\omega_1)^2 + (n+\omega_2)^2/\alpha^2\right]^2$, $h_2 = 0$ if $m+\omega_1 = 0$ and $n+\omega_2 = 0$.

The homogeneous solution for the stress function F^{i-1}, denoted by F_H^{i-1}, can be written by considering the condition of load application as follows:

$$F_H^{i-1} = \frac{1}{c}\left[P_x^{i-1}\left(1 + \frac{A_{sy}}{at}\right) + P_y^{i-1}\frac{\nu A_{sx}}{at}\right]\frac{y^2}{2bt} + \sigma_{rx}\frac{y^2}{2} + \sum_{ii=1}^{n_{sx}}\sigma_{rsx}\frac{y^2}{2}\bigg|_{y=y_{ii}}$$

$$+ \frac{1}{c}\left[P_y^{i-1}\left(1 + \frac{A_{sx}}{bt}\right) + P_x^{i-1}\frac{\nu A_{sy}}{bt}\right]\frac{x^2}{2at} + \sigma_{ry}\frac{x^2}{2} + \sum_{jj=1}^{n_{sy}}\sigma_{rsy}\frac{x^2}{2}\bigg|_{x=x_{jj}} \qquad (11.63)$$

$$- M_x^{i-1}\frac{Z_{px}}{Z_{px}+Z_{sx}}\frac{y^2(2y-3b)}{b^3 t} - M_y^{i-1}\frac{Z_{py}}{Z_{py}+Z_{sy}}\frac{x^2(2x-3a)}{a^3 t} - \tau_{xy}^{i-1}xy$$

where the welding induced residual stresses are included as initial stress terms. Therefore, the stress function F^{i-1} at the end of the $(i-1)$th load increment step can be obtained by the sum of Equations (11.62) and (11.63) as follows:

$$F^{i-1} = F_P^{i-1} + F_H^{i-1} \qquad (11.64)$$

To efficiently calculate Equation (11.49), a numerical technique can be used. With the panel subdivided (meshed) into a number of regions in the x, y, and z directions, Equation (11.49) may be expressed by

$$\sum_u \sum_v \sum_w \int_{a_u}^{a_{u+1}} \int_{b_v}^{b_{v+1}} \int_{t_w}^{t_{w+1}} \Delta\Phi(x,y,z)sx(r)sy(s)\,dxdydz = 0, \quad r = 1,2,3,\ldots, \quad s = 1,2,3,\ldots$$

$$(11.65)$$

where $\sum_u$, $\sum_v$, and $\sum_w$ indicate the summation for mesh regions in the x, y, and z directions, respectively.

Substituting Equation (11.45a) into Equation (11.65) yields the following expression, given in detail:

$$\sum_u \sum_v \sum_w \int_{a_u}^{a_{u+1}} \int_{b_v}^{b_{v+1}} \int_{t_w}^{t_{w+1}} \left[\frac{E}{1-\nu^2}\left(\frac{\partial^4\Delta w}{\partial x^4} + 2\frac{\partial^4\Delta w}{\partial x^2\partial y^2} + \frac{\partial^4\Delta w}{\partial y^4}\right)z^2\right.$$

$$- \left(\frac{\partial^2 F_{i-1}}{\partial y^2}\frac{\partial^2\Delta w}{\partial x^2} + \frac{\partial^2\Delta F}{\partial y^2}\frac{\partial^2 w_{i-1}}{\partial x^2} - 2\frac{\partial^2 F_{i-1}}{\partial x\partial y}\frac{\partial^2\Delta w}{\partial x\partial y}\right.$$

$$\left.\left. - 2\frac{\partial^2\Delta F}{\partial x\partial y}\frac{\partial^2 w_{i-1}}{\partial x^2} + \frac{\partial^2 F_{i-1}}{\partial x^2}\frac{\partial^2\Delta w}{\partial y^2} + \frac{\partial^2\Delta F}{\partial x^2}\frac{\partial^2 w_{i-1}}{\partial y^2}\right)\right]sx(r)sy(s)\,dxdydz$$

$$+ \sum_{ii=1}^{n_{sx}}\sum_u \sum_w \int_{a_u}^{a_{u+1}} \int_{t_w}^{t_{w+1}} \left[E\frac{\partial^2\Delta w}{\partial x^4}z^2\right.$$

$$\left. - t_{ii}\left\{\left(\frac{\partial^2 F_{i-1}}{\partial y^2} - \nu\frac{\partial^2 F_{i-1}}{\partial x^2}\right)\frac{\partial^2\Delta w}{\partial x^2} - \left(\frac{\partial^2\Delta F}{\partial y^2} - \nu\frac{\partial^2\Delta F}{\partial x^2}\right)\frac{\partial^2 w_{i-1}}{\partial x^2}\right\}_{y=y_{ii}} sx(r)sy(s)\right]dxdz$$

$$+ \sum_{jj=1}^{n_{sy}}\sum_v \sum_w \int_{b_v}^{b_{v+1}} \int_{t_w}^{t_{w+1}} \left[E\frac{\partial^4\Delta w}{\partial y^4}z^2\right.$$

$$-t_{ij}\left\{\left(\frac{\partial^2 F_{i-1}}{\partial x^2}-\nu\frac{\partial^2 F_{i-1}}{\partial y^2}\right)\frac{\partial^2 \Delta w}{\partial y^2}-\left(\frac{\partial^2 \Delta F}{\partial x^2}-\nu\frac{\partial^2 \Delta F}{\partial y^2}\right)\frac{\partial^2 w_{i-1}}{\partial y^2}\right\}_{x=x_{ij}} sx(r)sy(s)\Bigg]dydz$$

$$-\sum_u\sum_v\int_{a_u}^{a_{u+1}}\int_{b_v}^{b_{v+1}}\Delta psx(r)sy(s)dxdy=0$$

$$(11.66)$$

where $D=\sum_w\int_{-t/2}^{t/2}\frac{E}{1-\nu^2}zdz$ and $t=\sum_w\int_{-t/2}^{t/2}dz$ are used. The lateral pressure loads are considered as distributed on the surface of the panel, and the related integration is performed in the two directions in the xy plane, that is, no integration associated with the lateral pressure loads is undertaken for the z direction.

The integration of Equation (11.66) eventually results in a set of linear (i.e., first-order) simultaneous equations for the unknown coefficients ΔA_{kl}. The equations can be written in matrix form as follows:

$$\{\Delta P\}+\{\Delta P_S\}=([P_O]+[P_{OS}]+[K_B]+[K_{BS}]+[K_M]+[K_{MS}])\{\Delta A\}\qquad(11.67)$$

where

$\{\Delta P\}$ = external load increments applied to the plate part
$\{\Delta P_S\}$ = external load increments applied to the stiffeners
$[P_O]$ = stiffness matrix associated with initial stress for the plate part (including welding induced residual stresses)
$[P_{OS}]$ = stiffness matrix associated with initial stress for the stiffeners (including welding induced residual stresses)
$[K_B]$ = bending stiffness matrix for the plate part
$[K_{BS}]$ = bending stiffness matrix for the stiffeners
$[K_M]$ = stiffness matrix due to membrane action for the plate part
$[K_{MS}]$ = stiffness matrix due to membrane action for the stiffeners
$\{\Delta A\}$ = unknown coefficients

The quantities of various matrices or vectors in Equation (11.67), that is, in the elastic regime, are given in detail as follows.

External Load Increment Vector for the Plate Part

$$\{\Delta P\}=\{\Delta P_1,\Delta P_2,...,\Delta P_{N_i},...,\Delta P_{N_x\times N_y}\}\qquad(11.68a)$$

where each component of the vector $\{\Delta P\}$ in Equation (11.68a) can be calculated by

$$\Delta P_{N_i}=\Delta pH_O(i,j)+\frac{\pi^2}{a^2b^2}\sum_{m=1}^{N_x}\sum_{n=1}^{N_y}A_{mn}^{i-1}\big[m^2b^2H_1(i,j,m,n)(\Delta\sigma_s-\nu\Delta\sigma_{by})$$

$$+n^2a^2H_1(i,j,m,n)(\Delta\sigma_y-\nu\Delta\sigma_{bx})+2m^2bH_2(i,j,m,n)\Delta\sigma_{bx}$$

$$+2n^2aH_3(i,j,m,n)\Delta\sigma_{by}-mnabH_4(i,j,m,n)\Delta\tau\big]$$

for $i=1,2,...,N_x$, $j=1,2,...,N_y$, $N_i=(i-1)\cdot N_x+j$.

Stiffness Matrix Associated with Initial Stress for the Plate Part

$$[P_O] = \begin{bmatrix} P_O(1,1) & P_O(1,2) & \cdots & P_O(1,N_j) & \cdots & P_O(1,N_x \times N_y) \\ P_O(2,1) & P_O(2,2) & \cdots & P_O(2,N_j) & \cdots & P_O(2,N_x \times N_y) \\ P_O(N_i,1) & P_O(N_i,2) & \cdots & P_O(N_i,N_j) & \cdots & P_O(N_i,N_x \times N_y) \\ \vdots & \vdots & \cdots \vdots & \cdots \vdots \\ P_O(N_x \times N_y,1) & P_O(N_x \times N_y,2) & \cdots & P_O(N_x \times N_y,N_j) & \cdots & P_O(N_x \times N_y,N_x \times N_y) \end{bmatrix}$$

$$(11.68b)$$

where each component of the matrix $[P_O]$ in Equation (11.68b) can be calculated by

$$P_O(N_i, N_j) = \frac{m^2 \pi^2}{a^2} \left[-H_1(i,j,m,n)\sigma_x^{i-1} + H_1(i,j,m,n)\sigma_{bx}^{i-1} \right]$$

$$+ \frac{m^2 \pi^2}{a^2} \left[-H_1(i,j,m,n)\sigma_y^{i-1} + H_1(i,j,m,n)\sigma_{by}^{i-1} \right]$$

$$- \frac{2\pi^2}{a^2 \cdot b^2} \left[m^2 b H_2(i,j,m,n)\sigma_{bx}^{i-1} + n^2 a H_3(i,j,m,n)\sigma_{by}^{i-1} \right]$$

$$+ \frac{mn\pi^2}{ab} H_4(i,j,m,n)\tau^{i-1}$$

for $i, m = 1, 2, \ldots, N_x$, $j, n = 1,2, \ldots, N_y$, $N_i = (i-1) \cdot N_x + j$, $N_j = (m-1) \cdot N_x + n$.

Bending Stiffness Matrix for the Plate Part

$$[K_B] = \begin{bmatrix} K_B(1,1) & K_B(1,2) & \cdots & K_B(1,N_j) & \cdots & K_B(1,N_x \times N_y) \\ K_B(2,1) & K_B(2,2) & \cdots & K_B(2,N_j) & \cdots & K_B(2,N_x \times N_y) \\ K_B(N_i,1) & K_B(N_i,2) & \cdots & K_B(N_i,N_j) & \cdots & K_B(N_i,N_x \times N_y) \\ \vdots & \vdots & \cdots \vdots & \cdots \vdots \\ K_B(N_x \times N_y,1) & K_B(N_x \times N_y,2) & \cdots & K_B(N_x \times N_y,N_j) & \cdots & K_B(N_x \times N_y,N_x \times N_y) \end{bmatrix}$$

$$(11.68c)$$

where each component of the matrix $[K_B]$ in Equation (11.68c) can be calculated by

$$K_B(N_i, N_j) = \frac{E\pi^4}{a^4(i - \nu^2)} (m^2 + \alpha^2 n^2)^2 G_O(i,j,m,n)$$

for $i, m = 1,2,\ldots,N_x$, $j, n = 1,2,\ldots,N_y$, $N_i = (i-1) \cdot N_x + j$, $N_j = (m-1) \cdot N_x + n$

Stiffness Matrix Due to Membrane Action for the Plate Part

$$
[K_{\mathrm{M}}] = \begin{bmatrix}
K_{\mathrm{M}}(1,1) & K_{\mathrm{M}}(1,2) & \cdots & K_{\mathrm{M}}(1,N_j) & \cdots & K_{\mathrm{M}}(1,N_x \times N_y) \\
K_{\mathrm{M}}(2,1) & K_{\mathrm{M}}(2,2) & \cdots & K_{\mathrm{M}}(2,N_j) & \cdots & K_{\mathrm{M}}(2,N_x \times N_y) \\
K_{\mathrm{M}}(N_i,1) & K_{\mathrm{M}}(N_i,2) & \cdots & K_{\mathrm{M}}(N_i,N_j) & \cdots & K_{\mathrm{M}}(N_i,N_x \times N_y) \\
\vdots & \vdots & \cdots\ \vdots & & \cdots\ \vdots \\
K_{\mathrm{M}}(N_x \times N_y,1) & K_{\mathrm{M}}(N_x \times N_y,2) & \cdots & K_{\mathrm{M}}(N_x \times N_y,N_j) & \cdots & K_{\mathrm{M}}(N_x \times N_y,N_x \times N_y)
\end{bmatrix}
$$

(11.68d)

where each component of the matrix $[K_{\mathrm{M}}]$ in Equation (11.68d) can be calculated by

$$
K_{\mathrm{M}}(N_i,N_j) = \frac{E\alpha^2\pi^4}{4a^2b^2}\sum_{m=1}^{N_x}\sum_{n=1}^{N_y}\sum_{k=1}^{N_x}\sum_{l=1}^{N_y}[A_{mn}^{i-1}A_{kl}^{i-1}
$$

$$
\times\,[(rn-sm)\{k^2(n+s)^2+l^2(m+r)^2\}G_1(i,j,m,n,k,l,r,s)
$$

$$
+(rn-sm)\{k^2(n-s)^2+l^2(m-r)^2\}G_2(i,j,m,n,k,l,r,s)
$$

$$
-(rn+sm)\{k^2(n-s)^2+l^2(m+r)^2\}G_3(i,j,m,n,k,l,r,s)
$$

$$
-(rn+sm)\{k^2(n+s)^2+l^2(m-r)^2\}G_4(i,j,m,n,k,l,r,s)
$$

$$
-2kl(rn-sm)(m+r)(n+s)G_5(i,j,m,n,k,l,r,s)
$$

$$
-2kl(rn-sm)(m-r)(n-s)G_6(i,j,m,n,k,l,r,s)
$$

$$
+2kl(rn+sm)(m+r)(n-s)G_7(i,j,m,n,k,l,r,s)
$$

$$
-2kl(rn-sm)(m-r)(n+s)G_8(i,j,m,n,k,l,r,s)]
$$

$$
-\left(A_{mn}^{i-1}A_{kl}^{i-1}-A_{mn}^{0}A_{kl}^{0}\right)
$$

$$
\times\,[ml\{r^2(n+l)^2+s^2(m+k)^2\}G_9(i,j,m,n,k,l,r,s)
$$

$$
+ml\{r^2(n-l)^2+s^2(m-k)^2\}G_{10}(i,j,m,n,k,l,r,s)
$$

$$
+ml\{r^2(n-l)^2+s^2(m+k)^2\}G_{11}(i,j,m,n,k,l,r,s)
$$

$$
+ml\{r^2(n+l)^2+s^2(m-k)^2\}G_{12}(i,j,m,n,k,l,r,s)
$$

$$
-2mlrs(m+k)(n+l)G_{13}(i,j,m,n,k,l,r,s)
$$

$$
-2mlrs(m-k)(n-l)G_{14}(i,j,m,n,k,l,r,s)
$$

$$
-2mlrs(m+k)(n-l)G_{15}(i,j,m,n,k,l,r,s)
$$

$$
-2mlrs(m-k)(n+l)G_{16}(i,j,m,n,k,l,r,s)]]
$$

for $i,r = 1,2,\dots,N_x$, $j,s = 1,2,\dots,N_y$, $N_i = (i-1)\cdot N_x + j$, $N_j = (r-1)\cdot N_x + s$

Unknown Coefficient Vector

$$\{\Delta A\} = \left\{\Delta A_1, \Delta A_2, ..., \Delta A_{N_i}, ..., \Delta A_{N_x \times N_y}\right\} \tag{11.68e}$$

External Load Increment Vector for Stiffeners

$$\{\Delta P_S\} = \left\{\Delta P_{S1}, \Delta P_{S2}, ..., \Delta P_{SN_i}, ..., \Delta P_{SN_x \times N_y}\right\} \tag{11.68f}$$

where each component of the vector $\{\Delta P_S\}$ in Equation (11.68f) can be calculated by

$$\Delta P_{SN_i} = \frac{\pi^2}{a^2} \sum_{ii=1}^{mm} \sum_{m=1}^{N_x} \sum_{n=1}^{N_y} A_{mn}^{i-1} m^2 Q_1(i,j,m,n)\left(\Delta\sigma_x - \nu\Delta\sigma_y\right)$$

$$+ \frac{\pi^2}{b^2} \sum_{jj=1}^{nn} \sum_{m=1}^{N_x} \sum_{n=1}^{N_y} A_{mn}^{i-1} n^2 Q_2(i,j,m,n)\left(\Delta\sigma_y - \nu\Delta\sigma_x\right)$$

for $i = 1, 2, ..., N_x$, $j = 1, 2, ..., N_y$, $N_i = (i-1) \cdot N_x + j$

Stiffness Matrix Associated with Initial Stress for Stiffeners

$$[P_{OS}] = \begin{bmatrix} P_{OS}(1,1) & P_{OS}(1,2) & \cdots & P_{OS}(1,N_j) & \cdots & P_{OS}(1,N_x \times N_y) \\ P_{OS}(2,1) & P_{OS}(2,2) & \cdots & P_{OS}(2,N_j) & \cdots & P_{OS}(2,N_x \times N_y) \\ P_{OS}(N_i,1) & P_{OS}(N_i,2) & \cdots & P_{OS}(N_i,N_j) & \cdots & P_{OS}(N_i,N_x \times N_y) \\ \vdots & \vdots & \cdots \vdots & \cdots \vdots \\ P_{OS}(N_x \times N_y,1) & P_{OS}(N_x \times N_y,2) & \cdots & P_{OS}(N_x \times N_y,N_j) & \cdots & P_{OS}(N_x \times N_y,N_x \times N_y) \end{bmatrix} \tag{11.68g}$$

where each component of the matrix $[P_{OS}]$ in Equation (11.68g) can be calculated by

$$P_{OS}(N_i,N_j) = \frac{\pi^2}{a^2} \sum_{ii=1}^{mn} A_{mn}^{i-1} m^2 Q_1(i,j,m,n)\left(\sigma_x^{i-1} - \nu\sigma_y^{i-1}\right)$$

$$+ \frac{\pi^2}{b^2} \sum_{jj=1}^{nn} A_{mn}^{i-1} n^2 Q_2(i,j,m,n)\left(\sigma_y^{i-1} - \nu\sigma_x^{i-1}\right)$$

for $i, m = 1, 2, ..., N_x$, $j, n = 1, 2, ..., N_y$, $N_i = (i-1) \cdot N_x + j$, $N_j = (m-1) \cdot N_x + n$

Bending Stiffness Matrix for Stiffeners

$$[K_{BS}] = \begin{bmatrix} K_{BS}(1,1) & K_{BS}(1,2) & \cdots & K_{BS}(1,N_j) & \cdots & K_{BS}(1,N_x \times N_y) \\ K_{BS}(2,1) & K_{BS}(2,2) & \cdots & K_{BS}(2,N_j) & \cdots & K_{BS}(2,N_x \times N_y) \\ K_{BS}(N_i,1) & K_{BS}(N_i,2) & \cdots & K_{BS}(N_i,N_j) & \cdots & K_{BS}(N_i,N_x \times N_y) \\ \vdots & \vdots & \cdots \vdots & \cdots \vdots \\ K_{BS}(N_x \times N_y,1) & K_{BS}(N_x \times N_y,2) & \cdots & K_{BS}(N_x \times N_y,N_j) & \cdots & K_{BS}(N_x \times N_y,N_x \times N_y) \end{bmatrix}$$

(11.68h)

where each component of the matrix $[K_{BS}]$ in Equation (11.68h) can be calculated by

$$K_{BS}(N_i,N_j) = \sum_{ii=1}^{mm} \frac{E\pi^4}{a^4} m^4 Q_3(i,j,m,n) + \sum_{jj=1}^{nn} \frac{E\pi^4}{b^4} n^4 Q_4(i,j,m,n)$$

for $i,m = 1,2,...,N_x$, $j,n = 1,2,...,N_y$, $N_i = (i-1) \cdot N_x + j$, $N_j = (m-1) \cdot N_x + n$

Stiffness Matrix Due to Membrane Action for Stiffeners

$$[K_{MS}] = \begin{bmatrix} K_{MS}(1,1) & K_{MS}(1,2) & \cdots & K_{MS}(1,N_j) & \cdots & K_{MS}(1,N_x \times N_y) \\ K_{MS}(2,1) & K_{MS}(2,2) & \cdots & K_{MS}(2,N_j) & \cdots & K_{MS}(2,N_x \times N_y) \\ K_{MS}(N_i,1) & K_{MS}(N_i,2) & \cdots & K_{MS}(N_i,N_j) & \cdots & K_{MS}(N_i,N_x \times N_y) \\ \vdots & \vdots & \cdots \vdots & \cdots \vdots \\ K_{MS}(N_x \times N_y,1) & K_{MS}(N_x \times N_y,2) & \cdots & K_{MS}(N_x \times N_y,N_j) & \cdots & K_{MS}(N_x \times N_y,N_x \times N_y) \end{bmatrix}$$

(11.68i)

where each component of the matrix $[K_{MS}]$ in Equation (11.68i) can be calculated by

$$K_{MS}(N_i,N_j) = \frac{E\pi^4}{4a^2b^2} \sum_{ii=1}^{mm} \sum_{m=1}^{N_x} \sum_{n=1}^{N_y} \sum_{k=1}^{N_x} \sum_{l=1}^{N_y} [A_{mn}^{i-1} A_{kl}^{i-1}$$

$$\times \left[k^2(rn-sm)\{\alpha^2(n+s)^2 - \nu(m+r)^2\} R_1(i,j,m,n,k,l,r,s) \right.$$

$$+ k^2(rn-sm)\{\alpha^2(n-s)^2 - \nu(m-r)^2\} R_2(i,j,m,n,k,l,r,s)$$

$$- k^2(rn+sm)\{\alpha^2(n-s)^2 - \nu(m+r)^2\} R_3(i,j,m,n,k,l,r,s)$$

$$\left. - k^2(rn+sm)\{\alpha^2(n+s)^2 - \nu(m-r)^2\} R_4(i,j,m,n,k,l,r,s) \right]$$

$$+ \left(A_{mn}^{i-1} A_{kl}^{i-1} - A_{mn}^0 A_{kl}^0 \right)$$

$$\times \left[-mlr^2\{\alpha^2(n+l)^2 - \nu(m+k)^2\} R_5(i,j,m,n,k,l,r,s) \right]$$

$$-mlr^2\{\alpha^2(n-l)^2 - \nu(m-k)^2\}R_6(i,j,m,n,k,l,r,s)$$

$$-mlr^2\{\alpha^2(n-l)^2 - \nu(m+k)^2\}R_7(i,j,m,n,k,l,r,s)$$

$$-mlr^2\{\alpha^2(n+l)^2 - \nu(m-k)^2\}R_8(i,j,m,n,k,l,r,s)]]$$

$$+\frac{E\pi^4}{4a^2b^2}\sum_{jj=1}^{nn}\sum_{m=1}^{N_x}\sum_{n=1}^{N_y}\sum_{k=1}^{N_x}\sum_{l=1}^{N_y}[A_{mn}^{i-1}A_{kl}^{i-1}$$

$$\times[l^2(rn-sm)\{(m+r)^2 - \nu\alpha^2(n+s)^2\}R_9(i,j,m,n,k,l,r,s)$$

$$+l^2(rn-sm)\{(m-r)^2 - \nu\alpha^2(n-s)^2\}R_{10}(i,j,m,n,k,l,r,s)$$

$$-l^2(rn+sm)\{(m+r)^2 - \nu\alpha^2(n-s)^2\}R_{11}(i,j,m,n,k,l,r,s)$$

$$-l^2(rn+sm)\{(m-r)^2 - \nu\alpha^2(n+s)^2\}R_{12}(i,j,m,n,k,l,r,s)]$$

$$+\left(A_{mn}^{i-1}A_{kl}^{i-1} - A_{mn}^{0}A_{kl}^{0}\right)$$

$$\times[-mls^2\{(m+k)^2 - \nu\alpha^2(n+l)^2\}R_{13}(i,j,m,n,k,l,r,s)$$

$$-mls^2\{(m-k)^2 - \nu\alpha^2(n+l)^2\}R_{14}(i,j,m,n,k,l,r,s)$$

$$-mls^2\{(m+k)^2 - \nu\alpha^2(n-l)^2\}R_{15}(i,j,m,n,k,l,r,s)$$

$$-mls^2\{(m-k)^2 - \nu\alpha^2(n-l)^2\}R_{16}(i,j,m,n,k,l,r,s)]]$$

for $i,r = 1,2,...,N_x$, $j,s = 1,2,...,N_y$, $N_i = (i-1)\cdot N_x + j$, $N_j = (r-1)\cdot N_x + s$

Finally, the coefficients G, H, Q, and R used in the preceding equations are given as follows:

$$H_0(i,j) = \sum_u\sum_v\left[\int_{a_u}^{a_{u+1}}\int_{b_v}^{b_{v+1}}sx(i)sy(j)dxdy\right]$$

$$H_1(i,j,m,n) = \sum_u\sum_v\sum_w\left[\int_{a_u}^{a_{u+1}}\int_{b_v}^{b_{v+1}}\int_{t_w}^{t_{w+1}}sx(m)sy(n)sx(i)sy(j)dxdydz\right]$$

$$H_2(i,j,m,n) = \sum_u\sum_v\sum_w\left[\int_{a_u}^{a_{u+1}}\int_{b_v}^{b_{v+1}}\int_{t_w}^{t_{w+1}}y(sx(m)sy(n)sx(i)sy(j))dxdydz\right]$$

$$H_3(i,j,m,n) = \sum_u\sum_v\sum_w\left[\int_{a_u}^{a_{u+1}}\int_{b_v}^{b_{v+1}}\int_{t_w}^{t_{w+1}}x(sx(m)sy(n)sx(i)sy(j))dxdydz\right]$$

$$H_4(i,j,m,n) = \sum_u\sum_v\sum_w\left[\int_{a_u}^{a_{u+1}}\int_{b_v}^{b_{v+1}}\int_{t_w}^{t_{w+1}}cx(m)cy(n)sx(i)sy(j)dxdydz\right]$$

$$H_1(i,j,m,n) = \sum_u\sum_v\sum_w\left[\int_{a_u}^{a_{u+1}}\int_{b_v}^{b_{v+1}}\int_{t_w}^{t_{w+1}}sx(m)sy(n)sx(i)sy(j)dxdydz\right]$$

$$G_1(i,j,m,n,k,l,r,s) = \frac{(rn-sm)}{\left[(m+r)^2+\alpha^2(n+s)^2\right]^2}$$

$$\sum_u\sum_v\sum_w\left[\int_{a_u}^{a_u+1}\int_{b_v}^{b_v+1}\int_{t_w}^{t_w+1}cx(m+r)cy(n+s)sx(k)sy(l)sx(i)sy(j)dxdydz\right]$$

$$G_2(i,j,m,n,k,l,r,s) = \frac{(rn-sm)}{\left[(m-r)^2+\alpha^2(n-s)^2\right]^2}$$

$$\sum_u\sum_v\sum_w\left[\int_{a_u}^{a_u+1}\int_{b_v}^{b_v+1}\int_{t_w}^{t_w+1}cx(m-r)cy(n-s)sx(k)sy(l)sx(i)sy(j)dxdydz\right]$$

$$G_3(i,j,m,n,k,l,r,s) = \frac{(rn+sm)}{\left[(m+r)^2+\alpha^2(n-s)^2\right]^2}$$

$$\sum_u\sum_v\sum_w\left[\int_{a_u}^{a_u+1}\int_{b_v}^{b_v+1}\int_{t_w}^{t_w+1}cx(m+r)cy(n-s)sx(k)sy(l)sx(i)sy(j)dxdydz\right]$$

$$G_4(i,j,m,n,k,l,r,s) = \frac{(rn+sm)}{\left[(m-r)^2+\alpha^2(n+s)^2\right]^2}$$

$$\sum_u\sum_v\sum_w\left[\int_{a_u}^{a_u+1}\int_{b_v}^{b_v+1}\int_{t_w}^{t_w+1}cx(m-r)cy(n+s)sx(k)sy(l)sx(i)sy(j)dxdydz\right]$$

$$G_5(i,j,m,n,k,l,r,s) = \frac{(rn-sm)}{\left[(m+r)^2+\alpha^2(n+s)^2\right]^2}$$

$$\sum_u\sum_v\sum_w\left[\int_{a_u}^{a_u+1}\int_{b_v}^{b_v+1}\int_{t_w}^{t_w+1}sx(m+r)sy(n+s)cx(k)cy(l)sx(i)sy(j)dxdydz\right]$$

$$G_6(i,j,m,n,k,l,r,s) = \frac{(rn-sm)}{\left[(m-r)^2+\alpha^2(n-s)^2\right]^2}$$

$$\sum_u\sum_v\sum_w\left[\int_{a_u}^{a_u+1}\int_{b_v}^{b_v+1}\int_{t_w}^{t_w+1}sx(m-r)sy(n-s)cx(k)cy(l)sx(i)sy(j)dxdydz\right]$$

$$G_7(i,j,m,n,k,l,r,s) = \frac{(rn+sm)}{\left[(m+r)^2+\alpha^2(n-s)^2\right]^2}$$

$$\sum_u\sum_v\sum_w\left[\int_{a_u}^{a_u+1}\int_{b_v}^{b_v+1}\int_{t_w}^{t_w+1}sx(m+r)sy(n-s)cx(k)cy(l)sx(i)sy(j)dxdydz\right]$$

$$G_8(i,j,m,n,k,l,r,s) = \frac{(rn+sm)}{\left[(m-r)^2+\alpha^2(n+s)^2\right]^2}$$

$$\sum_u\sum_v\sum_w\left[\int_{a_u}^{a_u+1}\int_{b_v}^{b_v+1}\int_{t_w}^{t_w+1}sx(m-r)sy(n+s)cx(k)cy(l)sx(i)sy(j)dxdydz\right]$$

$$G_9(i,j,m,n,k,l,r,s) = \frac{(kn-ml)}{\left[(m+k)^2 + \alpha^2(n+l)^2\right]^2}$$

$$\sum_u \sum_v \sum_w \left[\int_{a_u}^{a_{u+1}} \int_{b_v}^{b_{v+1}} \int_{t_w}^{t_{w+1}} cx(m+k)cy(n+l)sx(r)sy(s)sx(i)sy(j)dxdydz\right]$$

$$G_{10}(i,j,m,n,k,l,r,s) = \frac{(kn-ml)}{\left[(m-k)^2 + \alpha^2(n-l)^2\right]^2}$$

$$\sum_u \sum_v \sum_w \left[\int_{a_u}^{a_{u+1}} \int_{b_v}^{b_{v+1}} \int_{t_w}^{t_{w+1}} cx(m-k)cy(n-l)sx(r)sy(s)sx(i)sy(j)dxdydz\right]$$

$$G_{11}(i,j,m,n,k,l,r,s) = \frac{(kn+ml)}{\left[(m+k)^2 + \alpha^2(n-l)^2\right]^2}$$

$$\sum_u \sum_v \sum_w \left[\int_{a_u}^{a_{u+1}} \int_{b_v}^{b_{v+1}} \int_{t_w}^{t_{w+1}} cx(m+k)cy(n-l)sx(r)sy(s)sx(i)sy(j)dxdydz\right]$$

$$G_{12}(i,j,m,n,k,l,r,s) = \frac{(kn+ml)}{\left[(m-k)^2 + \alpha^2(n+l)^2\right]^2}$$

$$\sum_u \sum_v \sum_w \left[\int_{a_u}^{a_{u+1}} \int_{b_v}^{b_{v+1}} \int_{t_w}^{t_{w+1}} cx(m-k)cy(n+l)sx(r)sy(s)sx(i)sy(j)dxdydz\right]$$

$$G_{13}(i,j,m,n,k,l,r,s) = \frac{(kn-ml)}{\left[(m+k)^2 + \alpha^2(n+l)^2\right]^2}$$

$$\sum_u \sum_v \sum_w \left[\int_{a_u}^{a_{u+1}} \int_{b_v}^{b_{v+1}} \int_{t_w}^{t_{w+1}} sx(m+k)sy(n+l)cx(r)cy(s)sx(i)sy(j)dxdydz\right]$$

$$G_{14}(i,j,m,n,k,l,r,s) = \frac{(kn-ml)}{\left[(m-k)^2 + \alpha^2(n-l)^2\right]^2}$$

$$\sum_u \sum_v \sum_w \left[\int_{a_u}^{a_{u+1}} \int_{b_v}^{b_{v+1}} \int_{t_w}^{t_{w+1}} sx(m-k)sy(n-l)cx(r)cy(s)sx(i)sy(j)dxdydz\right]$$

$$G_{15}(i,j,m,n,k,l,r,s) = \frac{(kn+ml)}{\left[(m+k)^2 + \alpha^2(n-l)^2\right]^2}$$

$$\sum_u \sum_v \sum_w \left[\int_{a_u}^{a_{u+1}} \int_{b_v}^{b_{v+1}} \int_{t_w}^{t_{w+1}} sx(m+k)sy(n-l)cx(r)cy(s)sx(i)sy(j)dxdydz\right]$$

$$G_{16}(i,j,m,n,k,l,r,s) = \frac{(kn+ml)}{\left[(m-k)^2 + \alpha^2(n+l)^2\right]^2}$$

$$\sum_u \sum_v \sum_w \left[\int_{a_u}^{a_{u+1}} \int_{b_v}^{b_{v+1}} \int_{t_w}^{t_{w+1}} sx(m-k)sy(n+l)cx(r)cy(s)sx(i)sy(j)dxdydz\right]$$

$$Q_1(i,j,m,n) = \sum_u \sum_w \left[\int_{a_u}^{a_{u+1}} \int_{h_w}^{h_{w+1}} t_{sii} sx(m) sy(n) sx(i) sy(j) dx dz \right]_{y=y_{ii}}$$

$$Q_2(i,j,m,n) = \sum_v \sum_w \left[\int_{b_v}^{b_{v+1}} \int_{h_w}^{h_{w+1}} t_{sjj} sx(m) sy(n) sx(i) sy(j) dy dz \right]_{x=x_{jj}}$$

$$Q_3(i,j,m,n) = \sum_u \sum_v \left[\int_{a_u}^{a_{u+1}} \int_{h_w}^{h_{w+1}} t_{sii} sx(m) sy(n) sx(i) sy(j) z^2 dx dz \right]_{y=y_{ii}}$$

$$Q_4(i,j,m,n) = \sum_v \sum_w \left[\int_{b_v}^{b_{v+1}} \int_{h_w}^{h_{w+1}} t_{sjj} sx(m) sy(n) sx(i) sy(j) z^2 dy dz \right]_{x=x_{jj}}$$

$$R_1(i,j,m,n,k,l,r,s) = \frac{(rn-sm)}{\left[(m+r)^2 + (n+s)^2/\alpha^2\right]^2}$$

$$\sum_u \sum_w \left[\int_{a_u}^{a_{u+1}} \int_{h_w}^{h_{w+1}} t_{ii} cx(m+r) cy(n+s) sx(k) sy(l) sx(i) sy(j) dx dz \right]$$

$$R_2(i,j,m,n,k,l,r,s) = \frac{(rn-sm)}{\left[(m-r)^2 + (n-s)^2/\alpha^2\right]^2}$$

$$\sum_u \sum_w \left[\int_{a_u}^{a_{u+1}} \int_{h_w}^{h_{w+1}} t_{ii} cx(m-r) cy(n-s) sx(k) sy(l) sx(i) sy(j) dx dz \right]_{y=y_{ii}}$$

$$R_3(i,j,m,n,k,l,r,s) = \frac{(rn+sm)}{\left[(m+r)^2 + (n-s)^2/\alpha^2\right]^2}$$

$$\sum_u \sum_w \left[\int_{a_u}^{a_{u+1}} \int_{h_w}^{h_{w+1}} t_{ii} cx(m+r) cy(n-s) sx(k) sy(l) sx(i) sy(j) dx dz \right]_{y=y_{ii}}$$

$$R_4(i,j,m,n,k,l,r,s) = \frac{(rn+sm)}{\left[(m-r)^2 + (n+s)^2/\alpha^2\right]^2}$$

$$\sum_u \sum_w \left[\int_{a_u}^{a_{u+1}} \int_{h_w}^{h_{w+1}} t_{ii} cx(m-r) cy(n+s) sx(k) sy(l) sx(i) sy(j) dx dz \right]_{y=y_{ii}}$$

$$R_5(i,j,m,n,k,l,r,s) = \frac{(kn-ml)}{\left[(m+k)^2 + (n+l)^2/\alpha^2\right]^2}$$

$$\sum_u \sum_w \left[\int_{a_u}^{a_{u+1}} \int_{h_w}^{h_{w+1}} t_{ii} cx(m+k) cy(n+l) sx(r) sy(s) sx(i) sy(j) dx dz \right]_{y=y_{ii}}$$

$$R_6(i,j,m,n,k,l,r,s) = \frac{(kn-ml)}{\left[(m-k)^2 + (n-l)^2/\alpha^2\right]^2}$$

$$\sum_u \sum_w \left[\int_{a_u}^{a_{u+1}} \int_{h_w}^{h_{w+1}} t_{ii} cx(m-k) cy(n-l) sx(r) sy(s) sx(i) sy(j) dx dz \right]_{y=y_{ii}}$$

$$R_7(i,j,m,n,k,l,r,s) = \frac{(kn+ml)}{\left[(m+k)^2 + (n-l)^2/\alpha^2\right]^2}$$

$$\sum_u \sum_w \left[\int_{a_u}^{a_{u+1}} \int_{h_w}^{h_{w+1}} t_{ii} cx(m+k)cy(n-l)sx(r)sy(s)sx(i)sy(j)dxdz \right]_{y=y_{ii}}$$

$$R_8(i,j,m,n,k,l,r,s) = \frac{(kn+ml)}{\left[(m-k)^2 + (n+l)^2/\alpha^2\right]^2}$$

$$\sum_u \sum_w \left[\int_{a_u}^{a_{u+1}} \int_{h_w}^{h_{w+1}} t_{ii} cx(m-k)cy(n+l)sx(r)sy(s)sx(i)sy(j)dxdz \right]_{y=y_{ii}}$$

$$R_9(i,j,m,n,k,l,r,s) = \frac{(rn-sm)}{\left[(m+r)^2 + (n+s)^2/\alpha^2\right]^2}$$

$$\sum_v \sum_w \left[\int_{b_v}^{b_{v+1}} \int_{h_w}^{h_{w+1}} t_{jj} cx(m+r)cy(n+s)sx(k)sy(l)sx(i)sy(j)dydz \right]_{x=x_{jj}}$$

$$R_{10}(i,j,m,n,k,l,r,s) = \frac{(rn-sm)}{\left[(m+r)^2 + (n+s)^2/\alpha^2\right]^2}$$

$$\sum_v \sum_w \left[\int_{b_v}^{b_{v+1}} \int_{h_w}^{h_{w+1}} t_{jj} cx(m-r)cy(n-s)sx(k)sy(l)sx(i)sy(j)dydz \right]_{x=x_{jj}}$$

$$R_{11}(i,j,m,n,k,l,r,s) = \frac{(rn-sm)}{\left[(m+r)^2 + (n+s)^2/\alpha^2\right]^2}$$

$$\sum_v \sum_w \left[\int_{b_v}^{b_{v+1}} \int_{h_w}^{h_{w+1}} t_{jj} cx(m+r)cy(n-s)sx(k)sy(l)sx(i)sy(j)dydz \right]_{x=x_{jj}}$$

$$R_{12}(i,j,m,n,k,l,r,s) = \frac{(rn+sm)}{\left[(m-r)^2 + (n+s)^2/\alpha^2\right]^2}$$

$$\sum_v \sum_w \left[\int_{b_v}^{b_{vu+1}} \int_{h_w}^{h_{w+1}} t_{jj} cx(m-r)cy(n+s)sx(k)sy(l)sx(i)sy(j)dydz \right]_{x=x_{jj}}$$

$$R_{13}(i,j,m,n,k,l,r,s) = \frac{(kn-ml)}{\left[(m+k)^2 + (n+l)^2/\alpha^2\right]^2}$$

$$\sum_v \sum_w \left[\int_{b_v}^{b_{v+1}} \int_{h_w}^{h_{w+1}} t_{jj} cx(m+k)cy(n+l)sx(r)sy(s)sx(i)sy(j)dydz \right]_{x=x_{jj}}$$

$$R_{14}(i,j,m,n,k,l,r,s) = \frac{(kn-ml)}{\left[(m-k)^2 + (n-l)^2/\alpha^2\right]^2}$$

$$\sum_v \sum_w \left[\int_{b_v}^{b_{v+1}} \int_{h_w}^{h_{w+1}} t_{jj} cx(m-k)cy(n-l)sx(r)sy(s)sx(i)sy(j)dydz \right]_{x=x_{jj}}$$

$$R_{15}(i,j,m,n,k,l,r,s) = \frac{(kn+ml)}{\left[(m+k)^2 + (n-l)^2/\alpha^2\right]^2}$$

$$\times \sum_v \sum_w \left[\int_{b_v}^{b_{v+1}} \int_{h_w}^{h_{w+1}} t_{jj} cx(m+k)cy(n-l)sx(r)sy(s)sx(i)sy(j)dydz\right]_{x=x_{jj}}$$

$$R_{16}(i,j,m,n,k,l,r,s) = \frac{(kn+ml)}{\left[(m-k)^2 + (n+l)^2/\alpha^2\right]^2}$$

$$\times \sum_v \sum_w \left[\int_{b_v}^{b_{v+1}} \int_{h_w}^{h_{w+1}} t_{jj} cx(m-k)cy(n+l)sx(r)sy(s)sx(i)sy(j)dydz\right]_{x=x_{jj}}$$

11.4.4 Treatment of Plasticity

In a manner similar to that of the plates, the progress of plasticity with increases in the applied loads is treated numerically. As shown in Figure 11.3, the stiffened panel is subdivided into a number of mesh regions in the three directions, similar to the conventional finite element method. The average membrane stress components for each mesh region can be calculated at every load increment step. Yielding for each mesh region is checked for both the plate part and the stiffeners with the following yield criteria:

For the plate part:

$$\sigma_x^2 - \sigma_x \sigma_y + \sigma_y^2 + 3\tau^2 \geq \sigma_Y^2 \qquad (11.69a)$$

For the stiffeners:

$$\sigma_{sx} \geq \sigma_{Ys}, \quad \sigma_{sy} \geq \sigma_{Ys} \qquad (11.69b)$$

where σ_{sx} and σ_{sy} are the normal stresses of the stiffeners in the x or y direction, respectively, and σ_Y and σ_{Ys} are the yield stresses of the plating or stiffeners, respectively. For the softened zone, a reduced yield strength of material shall be used.

As the applied loads increase, the stiffness matrices for the panel are redefined by considering the progress of plasticity. In Equation (11.67), the stiffness matrix associated with external loads should be calculated for the whole volume of the panel regardless of the plasticity. However, the bending stiffness will be reduced by the plasticity if any mesh region yields. In the calculation (i.e., integration) of the bending stiffness matrix, therefore, the contribution to the yielded regions is removed. The stiffness for the panel will be progressively reduced by large deflection and local yielding as the applied loads increase. The panel can be considered to have reached the ULS when the panel stiffness eventually becomes zero (or negative).

11.5 Applied Examples

The incremental Galerkin method formulations were implemented in the computer program ALPS/SPINE (2017). The process described in the previous sections to include plasticity effects is carried out numerically within the computer program. In this regard, this method could perhaps be better classified as a semi-analytical (or a semi-numerical) approach. Also, the user of the computer program can adopt an option so that only the elastic large-deflection analysis, that is, without consideration of plasticity, can be undertaken. The program automatically stops if the determinant of the total stiffness matrix for the whole stiffened panel reaches zero (or is negative).

In the following, applied examples for plates and stiffened panels are shown with varying dimensions of plate panels, load applications, and initial imperfections (Paik et al. 2001, Paik & Kang 2005). The boundary conditions of the plates and the stiffened panels are considered simply supported at all (four) edges. The plate panels are made of steel and have a Young's modulus (E) of 205.8 GPa and a Poisson's ratio (v) of 0.3.

11.5.1 A Rectangular Plate Under Longitudinal Axial Compression

Elastic–plastic large-deflection analyses for simply supported plates subject to uniaxial compression are carried out until the ultimate strength is reached, varying the plate aspect ratio. Figure 11.4 shows the average compressive stress versus deflection curves for a rectangular plate with $a/b = 3$, where the initial and added deflection functions of Equation (11.18) are assumed to consist of two terms, with $m = 1$ and 3, while $n = 1$, as follows:

$$w_0 = \left(A_{011} \sin\frac{\pi x}{a} + A_{031} \sin\frac{3\pi x}{a}\right)\sin\frac{\pi y}{b}, \quad w = \left(A_{11} \sin\frac{\pi x}{a} + A_{31} \sin\frac{3\pi x}{a}\right)\sin\frac{\pi y}{b}$$

It is apparent from Figure 11.4 that at the beginning of loading, one half-wave mode is predominant. However, with an increase in the applied loads, the plating collapses with a half-wave number of three, which correctly corresponds to the buckling mode of the plate with an aspect ratio of $a/b = 3$. Figure 11.5 shows the variation in the ultimate strength for the plates plotted versus the aspect ratio. A comparison of the ALPS/SPINE

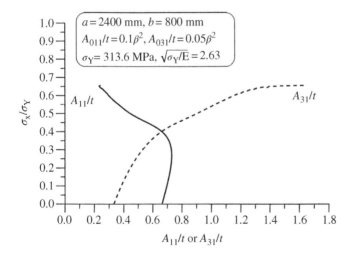

Figure 11.4 Average compressive stress versus deflection curves of a simply supported rectangular plate subject to longitudinal axial compression.

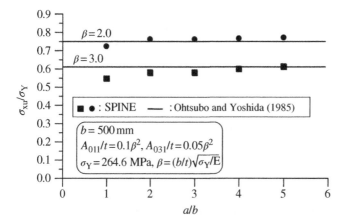

Figure 11.5 Comparison of ultimate strength results for steel plates subject to uniaxial longitudinal compression ($i = 2, 3, 4, 5$ for $a/b = 2, 3, 4, 5$, respectively).

solutions for this case is made against an empirical formula developed by curve fitting based on nonlinear finite element analysis (FEA) (Ohtsubo & Yoshida 1985).

11.5.2 A Rectangular Plate Under Transverse Axial Compression

A long steel plate (of aspect ratio $a/b = 3$) subject to uniaxial transverse compression is considered. Figure 11.6 shows the plate's load versus deflection curves. From the beginning of the load application, the two deflection terms increase together, but one half-wave mode is always clearly predominant.

Figure 11.7 shows the deformed shape of the plate at the ULS. It is apparent from Figure 11.7 that the plate deflection pattern is not sinusoidal but has a "bathtub" (or bulb)

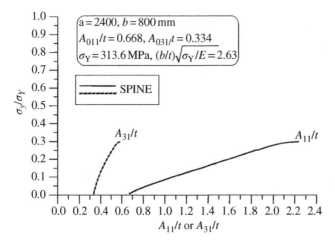

Figure 11.6 Average compressive stress versus deflection curves for a simply supported rectangular plate subject to transverse axial compression.

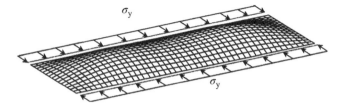

Figure 11.7 A so-called "bathtub"-shaped deflection of a simply supported rectangular plate under transverse axial compression at the ULS, as obtained by ALPS/SPINE, $a/b = 3$.

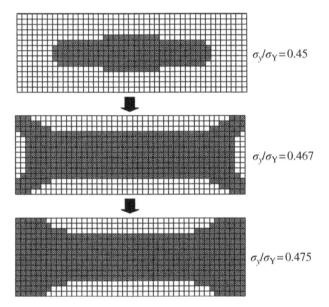

$\sigma_y/\sigma_Y = 0.45$

$\sigma_y/\sigma_Y = 0.467$

$\sigma_y/\sigma_Y = 0.475$

Figure 11.8 Progressive expansion of the plastic region at the top layer of a simply supported rectangular plate ($a/b = 3$) under transverse axial compression, as obtained by ALPS/SPINE.

shape around the edges. Figure 11.8 shows the progressive expansion of the plastic region at the top layer of the plate. Figure 11.9 shows the variation of the ultimate transverse compressive strength for the plates plotted against the plate aspect ratio. The nonlinear FEA results of Ohtsubo and Yoshida (1985) are also compared in the figure. The figure indicates that good agreement is achieved between the SPINE and nonlinear finite element method solutions.

11.5.3 A Rectangular Plate Under Edge Shear

The elastic–plastic large-deflection response of a square plate under edge shear up to the ultimate strength is now analyzed using the ALPS/SPINE method. The initial and added deflection functions in this case are assumed as follows:

$$w_0 = \sum_{m=1}^{3}\sum_{n=1}^{3} A_{0mn}\sin\frac{m\pi x}{a}\sin\frac{n\pi y}{b}, \quad w = \sum_{m=1}^{3}\sum_{n=1}^{3} A_{mn}\sin\frac{m\pi x}{a}\sin\frac{n\pi y}{b}$$

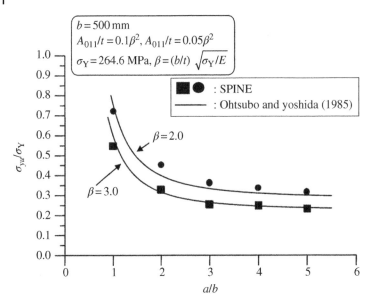

Figure 11.9 Variation of the ultimate strength of a simply supported rectangular plate subject to uniaxial transverse compression ($i = 2, 3, 4, 5$, for $a/b = 2, 3, 4, 5$).

where $A_{0mn} = 0$ is taken except for $A_{011} = 0.1\beta^2 t$ and $A_{033} = 0.05\beta^2 t$.

Figure 11.10 shows the load versus deflection curves for the square plate subject to edge shear. Figure 11.11 shows the deformed shape of the plate under edge shear at the ULS. Figure 11.12 shows the progressive expansion of the plastic region at the top layer of the plate under edge shear.

11.5.4 A Rectangular Plate Under In-Plane Bending

The elastic–plastic large-deflection response of a square plate under in-plane bending moment is now analyzed with the ALPS/SPINE method. The initial and added deflection functions for this case are assumed as follows:

$$w_0 = \sum_{m=1}^{5}\sum_{n=1}^{5} A_{0mn}\sin\frac{m\pi x}{a}\sin\frac{n\pi y}{b}, \quad w = \sum_{m=1}^{5}\sum_{n=1}^{5} A_{mn}\sin\frac{m\pi x}{a}\sin\frac{n\pi y}{b}$$

where $A_{0mn} = 0$ is taken except for $A_{011} = A_{055} = 0.1\beta^2 t$.

Figure 11.13 shows the load versus deflection curve for the square plate under in-plane bending in one direction.

11.5.5 A Rectangular Plate Under Lateral Pressure Loads

The elastic–plastic large-deflection response of a square plate under uniformly distributed lateral pressure loads up to the ultimate strength is now analyzed using the ALPS/SPINE method. The plate deflection function is assumed with only one half wave, that is, $m = n = 1$. Figure 11.14 shows the load versus deflection curves of the simply supported

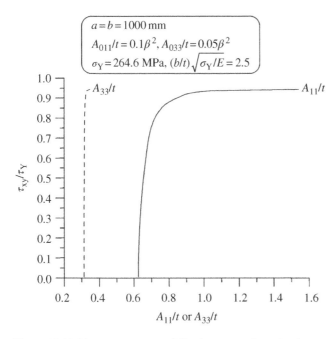

$$a = b = 1000 \text{ mm}$$
$$A_{011}/t = 0.1\beta^2, \ A_{033}/t = 0.05\beta^2$$
$$\sigma_Y = 264.6 \text{ MPa}, \ (b/t)\sqrt{\sigma_Y/E} = 2.5$$

Figure 11.10 Mean stress versus deflection curves for a simply supported square plate subject to edge shear.

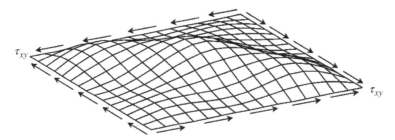

Figure 11.11 Deformed shape of a simply supported square plate under edge shear at the ULS.

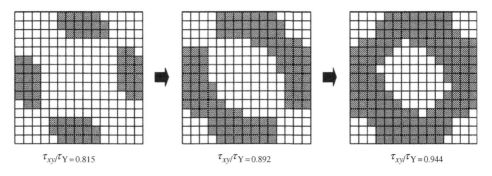

$\tau_{xy}/\tau_Y = 0.815$ $\qquad\qquad$ $\tau_{xy}/\tau_Y = 0.892$ $\qquad\qquad$ $\tau_{xy}/\tau_Y = 0.944$

Figure 11.12 Progressive expansion of the plastic region at the top layer of a simply supported square plate under edge shear.

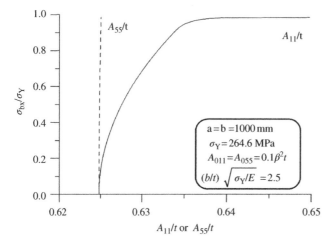

Figure 11.13 Mean stress versus deflection curve for a simply supported square plate subject to in-plane bending.

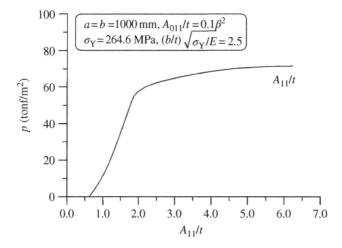

Figure 11.14 Lateral pressure load versus deflection curve for a simply supported square plate.

square plate under lateral pressure load. Figure 11.14 shows that the plate deflection is to some extent resisted by membrane action in the beginning, but progressively increases due to plasticity. Figure 11.15 shows the progressive expansion of the plastic region at the top layer of a square plate under lateral pressure load at the ULS.

11.5.6 A Rectangular Plate Under Combined Transverse Axial Compression and Edge Shear

The elastic–plastic large-deflection response of a rectangular plate (with aspect ratio $a/b = 2$) subject to combined transverse axial compression and edge shear is now analyzed, varying the slenderness ratio. The initial deflection for the plate is given by the following:

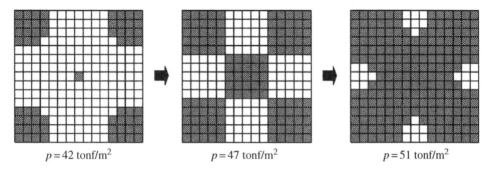

$$p = 42 \text{ tonf/m}^2 \qquad\qquad p = 47 \text{ tonf/m}^2 \qquad\qquad p = 51 \text{ tonf/m}^2$$

Figure 11.15 Progressive expansion of the plastic region at the top layer of a simply supported square plate under lateral pressure load.

$$w_0 = A_{011} \sin \frac{\pi x}{a} \sin \frac{\pi y}{b} + A_{021} \sin \frac{2\pi x}{a} \sin \frac{\pi y}{b}$$

where $A_{011} = 0.1\beta^2 t$, $A_{021} = 0.05\beta^2 t$, and $\beta = (b/t)\sqrt{\sigma_Y/E}$.

The added deflection function for this case requires some special consideration. As the deflection pattern may be complex when the edge shear is a predominant load component, more deflection terms should be used. The added deflection function for the present calculations may thus be assumed to consist of five half-wave terms in the x direction and three half-wave terms in the y direction as follows:

$$w = \sum_{m=1}^{5} \sum_{n=1}^{3} A_{mn} \sin \frac{m\pi x}{a} \sin \frac{n\pi y}{b}$$

In the analysis, the loading ratio that defines the proportion of axial compressive stresses to edge shear stresses is kept constant for each calculation point until the ultimate strength is reached. Figure 11.16 shows the resulting ultimate strength interactive relationship for the plate aspect ratio of $a/b = 2$ in the case of combined transverse compression and edge shear. In the same figure, the corresponding nonlinear FEA solutions obtained by Ohtsubo and Yoshida (1985) are compared with the ALPS/SPINE results. Good agreement is apparent.

11.5.7 A Rectangular Plate Under Other Types of Combined Load Applications

The ALPS/SPINE method can be applied for steel or aluminum plates under any combination of the six load components: longitudinal axial compression or tension, transverse axial compression/tension, longitudinal in-plane bending, transverse in-plane bending, edge shear, and lateral pressure. Figures 11.17 and 11.18 show the ultimate strength interactive relationship of steel plates under combined in-plane bending and edge shear and under all six load components, respectively.

11.5.8 A Stiffened Panel with Flat-Bar Stiffeners Under Uniaxial Compression

A steel square panel stiffened by one flat-bar stiffener in the longitudinal direction subjected to uniaxial compression is considered. The number of mesh regions for the plate

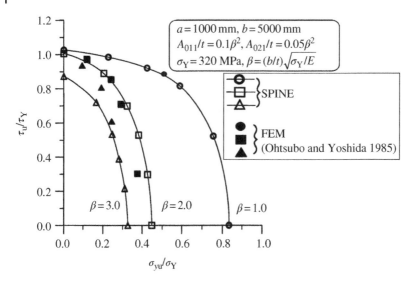

Figure 11.16 Ultimate strength interactive relationship for a simply supported rectangular plate subject to combined transverse compression and edge shear.

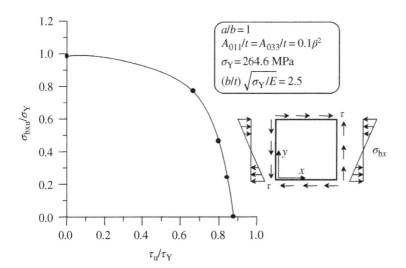

Figure 11.17 Ultimate strength interactive relationship for a simply supported square plate subject to combined in-plane longitudinal bending and edge shear.

part alone is taken as $11 \times 11 \times 9$ in the x, y, and z directions, respectively, and for the stiffener it is taken as 9 in the z (i.e., the stiffener web height) direction. The initial and added deflection functions are assumed by

$$w_0 = A_{011} \sin\frac{\pi x}{a}\sin\frac{\pi y}{b}, \quad w = \sum_{m=1}^{2}\sum_{n=1}^{2} A_{mn} \sin\frac{m\pi x}{a}\sin\frac{n\pi y}{b}$$

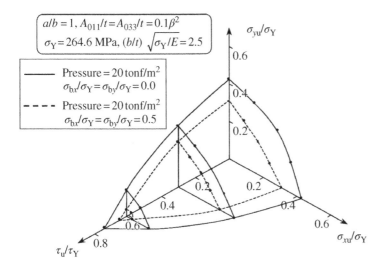

Figure 11.18 Ultimate strength interactive relationship for a simply supported square plate subject to combined biaxial compression, edge shear, biaxial in-plane bending, and lateral pressure loading.

Figure 11.19 shows the variation in the ultimate compressive strength of the stiffened steel panel with increases in the stiffener web height. The results obtained by others are compared in the figure. In the finite strip method calculations of Ohtsubo et al. (1978), unloaded edges were assumed to remain straight, the same condition as in the SPINE analyses. However, in the FEA of Ueda et al. (1976), the unloaded edges were assumed

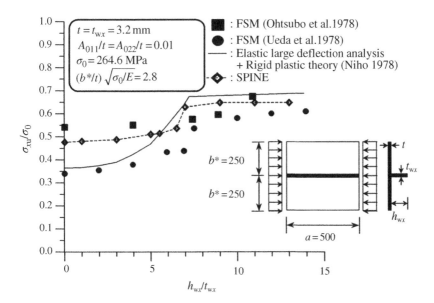

Figure 11.19 Comparison of ultimate strengths for a stiffened panel with one flat-bar stiffener subject to uniaxial compression.

to move freely in the plane direction (even though they would be simply supported). Niho (1978) predicted the ultimate strength for the stiffened panel by applying rigid-plastic theory to account for the large-deflection effects. The plates are stiffened with one-sided stiffeners. The neutral plane of the stiffened plate no longer coincides with the neutral plane of the plate. The effects of eccentricity are not accounted for in the present analysis. It is apparent from Figure 11.19 that the ultimate compressive strength increases as the stiffener web height increases; however, it appears that a critical threshold value of the stiffener web height separates into two distinct collapse modes: overall and local collapse.

11.5.9 A Stiffened Panel with Three Stiffeners Under Combined Axial Compression and Lateral Pressure Loads

A rectangular steel panel stiffened by three T-sections in the longitudinal direction is analyzed. The panel is subjected to combined uniaxial longitudinal (or transverse) compression and lateral pressure loads. The initial and added deflection functions with the odd numbers of half waves are assumed as follows:

$$w_0 = A_{011} \sin\frac{\pi x}{a}\sin\frac{\pi y}{b} + A_{039} \sin\frac{3\pi x}{a}\sin\frac{9\pi y}{b}$$

$$w = \sum_{m=1}^{2}\sum_{n=1}^{5} A_{(2m-1)(2n-1)} \sin\frac{(2m-1)\pi x}{a}\sin\frac{(2n-1)\pi y}{b}$$

where the even numbers of half waves are not included because the panel is subjected to predominantly lateral pressure loads and in this case only the odd numbers of half waves play a dominant role in the large-deflection response.

Figure 11.20 shows ALPS/SPINE results for the variation of the ultimate compressive strength for the panel with increases in the magnitude of lateral pressure loads.

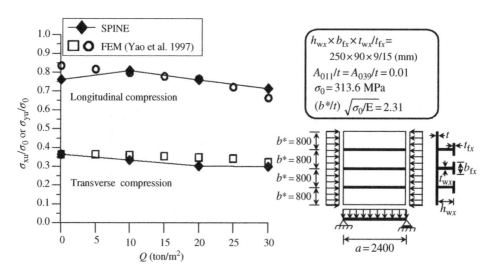

Figure 11.20 Comparison of ultimate strength results for a stiffened panel with three T-stiffeners subject to combined uniaxial longitudinal compression and lateral pressure loads.

Conventional nonlinear finite element results obtained by Yao et al. (1997) are compared in the figure. Yao et al. (1997) calculated the ultimate strength of the panels in this case using the plate-stiffener combination model, that is, using one stiffener with attached plating as representative of the stiffened panel as described in Chapter 2. In contrast, the SPINE program analyzes the nonlinear behavior of the stiffened panel as a whole.

It is apparent from Figure 11.20 that the ultimate compressive strength of the panel decreases almost linearly in a general sense as the magnitude of lateral pressure increases. However, it can also be observed from the ALPS/SPINE results shown in Figure 11.20 that the ultimate longitudinal compressive strength can increase slightly if the magnitude of lateral pressure is comparatively small. With further increases in the lateral pressure loads, the ultimate strength in compression decreases, which is thought to be due to the lateral pressure loads disturbing the inception of buckling of a long plate for the theoretical longitudinal buckling mode in longitudinal compression because of an imposed transverse shell type of curvature when the magnitude of lateral pressure is relatively small. As the magnitude of the lateral pressure increases, the theoretical longitudinal buckling mode appears to remain suppressed.

As long as the magnitudes of the increases in lateral pressure loads are not very large, the buckling strength of a long plating (between the stiffeners) under axial compression may increase because the buckling mode is different from the deflection pattern due to lateral pressure loads alone and because more energy is normally needed to buckle when lateral pressure loading disturbs occurrence of buckling. In contrast, for a wide (and square) plate, lateral pressure loads always decrease the ultimate strength because the buckling mode of plating corresponds to the deflection pattern due to lateral pressure loads alone, and in this case bifurcation (buckling) does not occur because the plate deflects from the beginning of loading. The panel reaches its ultimate strength with one half-wave mode in the transverse direction, which is imposed by the action of lateral pressure loads themselves.

11.5.10 A Very Large Crude Oil Carrier's Deck Structure Under Combined Axial Compression and Lateral Pressure

The incremental Galerkin method is used to investigate the ultimate strength characteristics of a cross-stiffened VLCC (very large crude oil carrier) deck structure under longitudinal axial compression and lateral pressure, the latter caused by an imposed vacuum to hypothetically reduce oil outflow in ship collision and/or grounding accidents. Of course, in this case, no corresponding test data or FEA results are available for comparison, whereas the solutions of the ALPS/SPINE (2017) method are described in Chapters 4 and 6.

Figure 11.21 indicates the relevant information for the VLCC deck stiffened panel used for the example. Figure 11.22 shows the elastic–plastic large-deflection behavior of the structure under lateral pressure loading alone until the ultimate strength is reached, as obtained by the ALPS/SPINE incremental Galerkin method. Figure 11.23 presents the ALPS/ULSAP (2017) and ALPS/SPINE predictions for the ultimate strength interactive relationship.

Two kinds of structural modeling with regard to the extent of calculation are considered for the ALPS/ULSAP ultimate strength predictions: one for an entire structure and one for a longitudinally stiffened panel between two adjacent transverse frames. The real

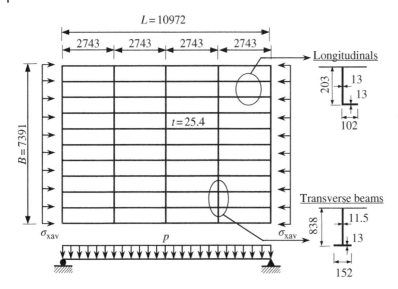

Figure 11.21 A cross-stiffened VLCC deck structure under combined axial compression and lateral vacuum pressure (in mm).

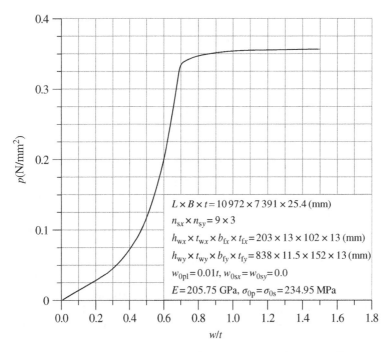

Figure 11.22 The elastic–plastic large deflection behavior of a cross-stiffened VLCC deck structure subject to lateral pressure load, as obtained by the ALPS/SPINE incremental Galerkin method (w, lateral deflection at the center of the structure).

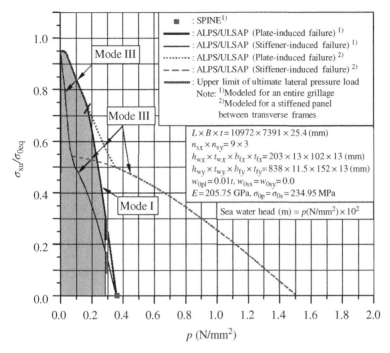

Figure 11.23 Variation of the ultimate axial compressive strength for a cross-stiffened VLCC deck structure plotted against the net lateral pressure, with predictions based on ALPS/SPINE and ALPS/ULSAP methods.

ultimate strength is taken as the smaller of the two results. The latter calculation model significantly overestimates the ultimate strength compared with the former, particularly when the magnitude of lateral pressure loads is large because the ultimate strength characteristics of cross-stiffened structures depend on the dimensions of the transverse frames and those of the longitudinal stiffeners.

When a larger magnitude of lateral pressure loads is applied, the relatively weak transverse frames in the grillage can fail to support the related panel, and therefore the latter type of modeling, that is, only for stiffened panels between two adjacent transverse frames that are assumed to remain straight or not to fail, may provide quite optimistic ultimate strength predictions of the grillage.

The mode III predictions based on stiffener-induced failure, using the Perry–Robertson formulation method as described in Chapter 6, are also shown for comparison, although they are not included in the ALPS/ULSAP ultimate strength computations. The upper limit of the "critical" lateral pressure is found to be at a waterhead of about 30 m. The real deck ultimate strength is then represented by the thicker solid line of Figure 11.23 inside the upper limit of the critical lateral pressure.

It is apparent from Figure 11.23 that the lateral pressure may not in this particular case affect the compressive collapse up to about 5 psi (0.035 MPa or a sea waterhead of 3.5 m), but it then significantly reduces the panel ultimate compressive strength for larger magnitudes of lateral pressure. It is also evident that, under these large vacuum pressures, mode I failure for the entire cross-stiffened structure is indicated. This suggests extreme caution in the potential use of an imposed vacuum as a means of reducing oil outflow, unless the structure is explicitly designed for the performance needed under such conditions with little or no prior experience. It is again important to realize that the ultimate strength calculation model for a cross-stiffened structure must account for the entire extent, that is, including transverse frames and longitudinal stiffeners. This case illustrates a possible use of the ALPS/SPINE and ALPS/ULSAP methods, although a purely hypothetical one.

References

ALPS/SPINE (2017). *A computer program for the elastic-plastic large deflection analysis of plates and stiffened panels using the incremental Galerkin method*. MAESTRO Marine LLC, Stevensville, MD.

ALPS/ULSAP (2017). *A computer program for the ultimate strength analysis of plates and stiffened panels*. MAESTRO Marine LLC, Stevensville, MD.

Fletcher, C.A.J. (1984). *Computational Galerkin method*. Springer-Verlag, New York.

Lancaster, J. (2003). *Handbook of structural welding: processes, materials and methods used in the welding of major structures, pipelines and process plant*. Abington Publishing, Cambridge.

Marguerre, K. (1938). Zur Theorie der gekreummter Platte grosser Formaenderung. *Proceedings of the 5th International Congress for Applied Mechanics*, Cambridge.

Niho, O. (1978). *Ultimate strength of plated structures*. Dr. Eng. Dissertation, Tokyo University, Tokyo (in Japanese).

Ohtsubo, H., Yamamoto, Y. & Lee, Y.J. (1978). Ultimate compressive strength of stiffened plates (part 1). *Journal of the Society of Naval Architects of Japan*, **143**: 316–325 (in Japanese).

Ohtsubo, H. & Yoshida, J. (1985). Ultimate strength of rectangular plates under combination of loads (part 2): interaction of compressive and shear stresses. *Journal of the Society of Naval Architects of Japan*, **158**: 368–375 (in Japanese).

Paik, J.K. & Kang, S.J. (2005). A semi-analytical method for the elastic-plastic large deflection analysis of stiffened panels under combined biaxial compression/tension, biaxial in-plane bending, edge shear and lateral pressure loads. *Thin-Walled Structures*, **43**(2): 375–410.

Paik, J.K., Thayamballi, A.K., Lee, S.K. & Kang, S.J. (2001). A semi-analytical method for the elastic-plastic large deflection analysis of welded steel or aluminum plating under combined in-plane and lateral pressure loads. *Thin-Walled Structures*, **39**: 125–152.

Ueda, Y., Rashed, S.M.H. & Paik, J.K. (1987). An incremental Galerkin method for plates and stiffened plates. *Computers & Structures*, **27**(1): 147–156.

Ueda, Y., Yao, T. & Kikumoto, H. (1976). Minimum stiffness ratio of a stiffener against ultimate strength of a plate. *Journal of the Society of Naval Architects of Japan*, **140**: 199–204 (in Japanese).

Yao, T., Fujikubo, M., Yanagihara, D. & Irisawa, W. (1997). Considerations on FEM modeling for buckling/plastic collapse analysis of stiffened plates. *Journal of the Kansai Society of Naval Architects of Japan*, **95**: 121–128 (in Japanese).

12

The Nonlinear Finite Element Method

12.1 Introduction

The finite element method (Zienkiewicz 1977) is one of the most powerful approaches available for analyzing the nonlinear behavior of structures. In a general case, the method requires considerable computational effort, mostly because of the large number of unknowns that must be addressed in the solution procedure and also because of the fairly complicated numerical integration procedures used, especially to obtain the nonlinear stiffness matrices for the finite elements as they deform.

A comprehensive discussion of nonlinear finite element method analysis would require a full book or more (Wriggers 2008, Belytschko et al. 2014, Borst et al. 2014, Kim 2014, Reddy 2015). It is important to realize that nonlinear finite element method solutions can be completely incorrect if the structural modeling techniques are inadequate in their idealization of a real problem. Several textbooks have dealt with nonlinear finite element method theories, but few have included tips and techniques for the modeling of this method (Paik & Hughes 2007, Hughes & Paik 2013).

This chapter focuses on techniques to develop successful models of nonlinear finite element methods for analyzing nonlinear structural consequences. Some illustrative examples of nonlinear finite element method modeling are presented for analysis of ultimate strength and structural crashworthiness; the former is associated with extreme loads, whereas the latter is associated with accidental actions such as collisions, grounding, fire, and explosions. It is noted that the nonlinear finite element method can of course be commonly applied to steel- and aluminum-plated structures.

12.2 Extent of the Analysis

Figure 12.1 shows a typical plated structure with strong, small support members—a ship hull structure under construction at shipyard. It is desirable to consider the entire structure in the analysis, but if the time or resources for structural modeling and computation are limited, finite element method modeling may consider only a part of the target structure. In this case, it must be realized that an artificial boundary is formed for the target structure, and the solution will only be satisfactory if the boundary conditions (loads, supports, etc.) are idealized appropriately.

Ultimate Limit State Analysis and Design of Plated Structures, Second Edition. Jeom Kee Paik.
© 2018 John Wiley & Sons Ltd. Published 2018 by John Wiley & Sons Ltd.

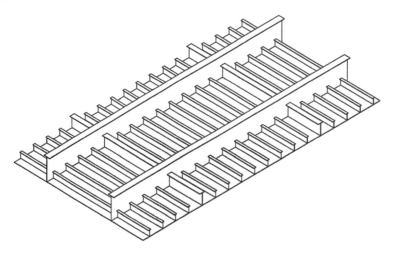

Figure 12.1 A typical example of a plated structure.

The extent of the analysis is typically removed from the target structure with respect to the symmetric envelope in terms of structural deformations and failure modes. Some illustrative examples are shown in Figure 12.2 for plates and stiffened plate structures and in Figure 12.3 for ship hull structures.

12.3 Types of Finite Elements

Many finite element types are available, but it can be difficult to establish specific guidelines for which types are best for a given application. For nonlinear analysis of plated structures, rectangular plate-shell elements are more appropriate than triangular elements because the former make it easier to define the membrane stress components inside each element when the Cartesian coordinate system is applied. This practice is also true for linear structural mechanics and analysis (Paik & Hughes 2007).

Therefore, four-noded plate-shell elements are generally used for nonlinear analysis of plated structures associated with their ultimate limit states and structural crashworthiness. The nodal points in the plate thickness direction are located in the mid-thickness of each element, which indicates that no element mesh is assigned to the thickness layers. To reflect the nonlinear behavior more accurately, plate-shell elements should be used for the webs, flanges, and plating. However, beam elements can be more efficient when modeling these supporting members, or at least the flanges, although they are not recommended for more precise analysis.

12.4 Mesh Size of Finite Elements

Although finer mesh modeling results in more accurate solutions, it may not be the best practice. A similar degree of accuracy can be attained with coarser mesh modeling, which requires considerably less computational cost. A convergence study

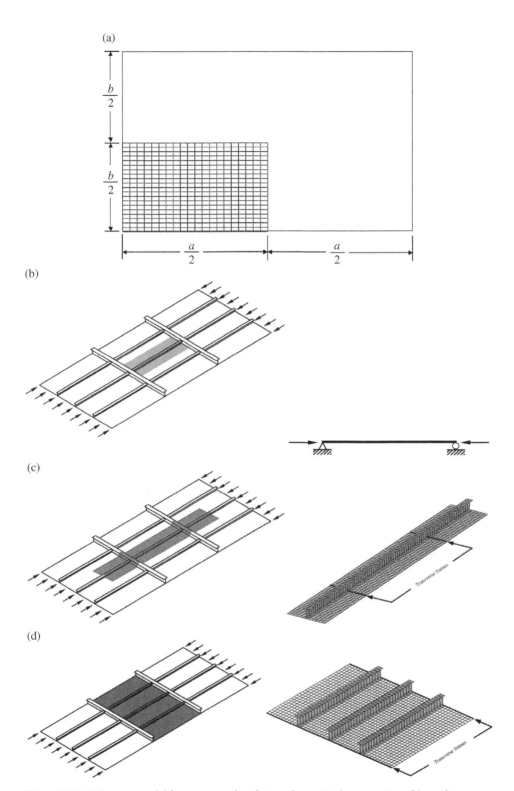

Figure 12.2 (a) Quarter model for a rectangular plate under uniaxial compression; (b) one-bay plate–stiffener combination model for a stiffened plate structure under uniaxial compression; (c) two-bay plate–stiffener combination model for a stiffened plate structure under uniaxial compression; (d) one-bay stiffened panel model for a stiffened plate structure under uniaxial compression; (e) two-bay/one-span stiffened panel model for a stiffened plate structure under uniaxial compression; (f) two-bay/two-span stiffened panel model for a stiffened plate structure under uniaxial compression.

(e)

(f)

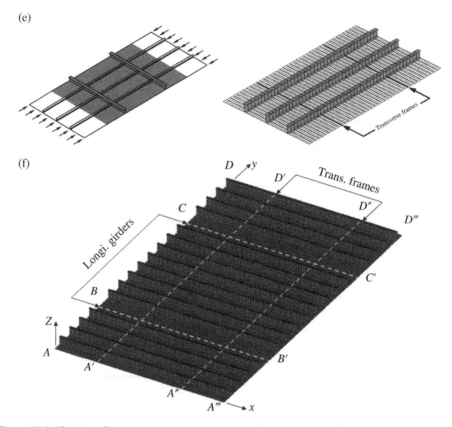

Figure 12.2 (Continued)

(or patch test) is usually performed to determine the best size of finite element mesh based on a compromise between computational cost and accuracy. Sample applications of the corresponding nonlinear analysis are carried out with a variety of element mesh sizes to search for the largest size that provides a sufficient level of accuracy.

Figure 12.4 shows a schematic of the convergence study where the relation between the load-carrying capacity of a structure and the number of finite elements is investigated. Greater number of finite elements indicates the application of finer mesh sizes. It is emphasized that the load-carrying capacity tends to converge to a value with a decrease in the mesh size or with an increase in the number of elements when "nonconforming" finite elements (Shi 2002) are used. The best size of the finite element mesh can then be selected in association with the converged value of the load-carrying capacity as shown in Figure 12.4. Such a convergence study can provide best practice nonlinear finite element method modeling for determining the relevant mesh size. In some cases where nonconforming element is not convergent, some mechanical consideration and computational experiences will be helpful to resolve the issue (Taylor et al. 1986, Irons & Razzaque 1972).

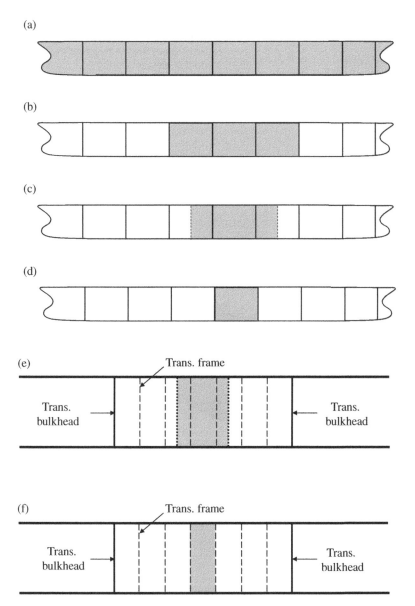

Figure 12.3 Extent of the analysis for the progressive ship's hull girder collapse analysis: (a) entire hull model; (b) three-cargo-hold model; (c) two-cargo-hold model; (d) one-cargo-hold model; (e) two-bay sliced hull cross-section model; (f) one-bay sliced hull cross-section model.

It is however emphasized that a convergence study itself requires considerable computational effort. Therefore, guidance is required to define the finite element mesh size without the need for such a study. For the ultimate strength analysis of stiffened plate structures that involve an elastic–plastic large deflection response, at least eight four-noded plate-shell elements are required to model the plating between the small support members (e.g., the longitudinal stiffeners), as shown in Figure 12.5. The size of these

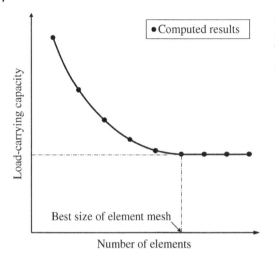

Figure 12.4 The convergence study (or patch test) associated with the load-carrying capacity versus the mesh size (or the number of elements) using "nonconforming" finite elements.

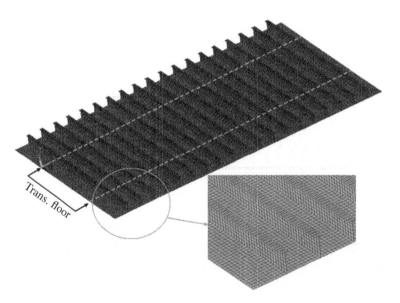

Figure 12.5 Mesh size for plating, stiffener webs, and flanges in a stiffened plate structure used for the ultimate limit state analysis.

plate-shell elements is assigned in the plate length direction to ensure that each finite element's aspect ratio is almost unity, which is desirable. There are likely to be at least six elements in the web height direction and at least four elements across the (full) flange breadth when four-noded plate-shell elements are used.

In analysis of structural crashworthiness that involves the crushing or folding of thin walls, at least eight four-noded plate-shell elements are required to reflect the folding behavior of the single crushing length of a plate, as shown in Figure 12.6 and described in Chapter 10. Theoretical formulations of the plate crushing length for thin-walled

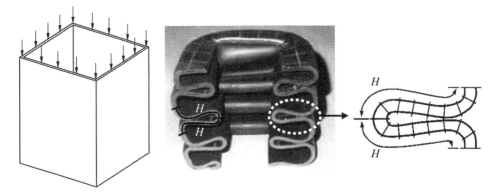

Figure 12.6 Crushing behavior of a plated structure and the necessary size of the finite elements.

structures under crushing loads are available. For example, Wierzbicki and Abramowicz (1983) derived the following plate crushing length formula, as described in Chapter 10:

$$H = 0.983b^{2/3}t^{1/3} \tag{12.1}$$

where b is the plate breadth, t is the plate thickness, and H is the half-fold length.

Therefore, the mesh size of a single finite element for plate crashworthiness analysis can be determined as the crushing length, as predicted by Equation (12.1), divided by eight, that is, the mesh size should be smaller than $H/8$. The element size should again be determined to ensure that the element aspect ratio is almost unity.

Figure 12.7 shows an example of the structural crashworthiness analysis model associated with collisions between a striking ship's bow and a struck ship's side, where the collided areas involving buckling, yielding, crushing, and fracture are modeled by finer meshes, while other areas (far away from the collided areas) are modeled by coarse meshes. As the striking ship's bow is also deformable rather than behaving as a rigid body, it also needs to be modeled by finer meshes in that case.

12.5 Material Modeling

Nonlinear structural consequences usually involve material nonlinearity in association with plasticity or yielding, among other factors. For nonlinear finite element analysis, therefore, the characteristics of material behavior should be defined precisely in terms of the true stress–true strain relationship as described in Section 1.3. It is of course desirable to determine the realistic relationship between these stresses and strains by means of tensile coupon testing, which covers pre-yielding behavior, yielding, post-yielding behavior including the strain-hardening effect, ultimate strength, and post-ultimate strength behavior, including the necking effect described. It is emphasized that the material properties are also significantly affected by temperatures.

The current industry practice for ultimate limit state assessment uses a simpler material model, although the realistic characteristics of the material have been applied to assess the accidental limit state. For example, the effects of strain hardening and necking

(a)

(b)

(c)

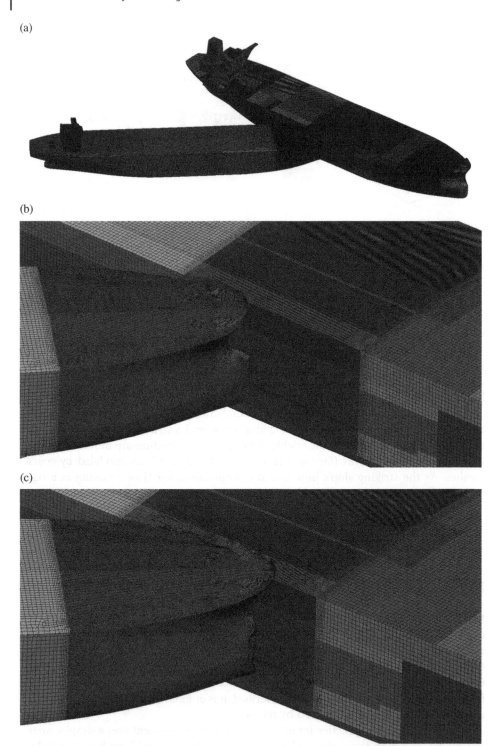

Figure 12.7 An example of the structural crashworthiness analysis model for a ship–ship collision: (a) finite element model with fine meshes in both struck ship side and striking ship bow (areas in black color); (b) zoomed model before collision starts; (c) deformed view during collision; (d) deformed view in struck ship side.

(d)

Figure 12.7 (Continued)

(strain softening) are often not considered in ultimate strength analysis. This simplified type of material model is the elastic–perfectly plastic material model and represents the material's elastic behavior until the yield strength has been reached, as described in Section 1.3.2. Neither strain hardening nor necking is considered in the post-yielding regime. This approximation may be useful for steel when the primary concern is buckling and when the plastic strain is only moderate, in contrast to structural crashworthiness, which involves crushing and rupture with large plastic strains. However, it is cautioned that the elastic–perfectly plastic model does not always give sufficiently accurate solutions for aluminum alloy materials.

In accidental situations, structures may be exposed to fracture caused by progressive cracking damage. In this case, rupture or fracture behavior must be considered as described in Chapter 10. One factor that affects the critical fracture strain of plate-shell-type finite elements is the element size. This is important for analyzing ruptures or ductile fractures. The following formula can be used to predict the critical fracture strain of the material as a function of the finite element size (Paik 2007a, 2007b, Hughes & Paik 2013):

$$\varepsilon_{fc} = \gamma d_1 \left(\frac{t}{s}\right)^{d_2} \varepsilon_f \tag{12.2}$$

where ε_{fc} is the critical fracture strain used in the finite element model for the ultimate strength analysis, ε_f is the fracture strain obtained from the tensile coupon test data, t is the element thickness, s is the mesh size, γ is the correction (knock-down) factor associated with localized bending effects, and d_1 and d_2 are the coefficients which may be taken as $d_1 = 4.1$ and $d_2 = 0.58$ for mild steel at room temperature when $t = 2$ mm (Hughes & Paik 2013). The correction factor, γ, will take much smaller value than unity, such as 0.3–0.4, with increase in the element thickness because the localized bending effect becomes more significant.

The strain-rate sensitivity plays an important role in the analysis of structural crashworthiness and/or impact response. Therefore, material modeling in terms of the dynamic yield strength and dynamic fracture strain must be considered. The Cowper–Symonds equation, as described in Sections 10.3.2 and 10.3.3, is generally used for this purpose. With the Cowper–Symonds equation, the strain rate $\dot{\varepsilon}$ can be calculated approximately by assuming that the initial speed V_0 of the dynamic loads is linearly reduced to zero until the loading is finished, with average displacement δ as indicated in Equation (1.29):

$$\dot{\varepsilon} = \frac{V_0}{2\delta} \tag{12.3}$$

For ship–ship collision accidents, the strain rate, $\dot{\varepsilon}$ (1/s), may be approximately determined as a function of the collision velocity as follows (Paik et al. 2017; Ko et al. 2017):

$$\dot{\varepsilon} = 2.970V_0 - 0.686 \text{ for } V_0 \geq 0.231 \text{ m/s} \tag{12.4}$$

where V_0 is the velocity of the striking ship in m/s which may vary with time, but for the sake of simplicity, V_0 may be taken by the initial collision velocity.

In structural crashworthiness and/or impact response analysis, strain-rate sensitivity plays a significant role and thus it should be taken into account. As described in Chapter 10, the Cowper–Symonds equation is usually applied for this purpose.

$$\sigma_{Yd} = \left\{ 1 + \left(\frac{\dot{\varepsilon}}{C} \right)^{1/q} \right\} \sigma_Y \tag{12.5}$$

$$\varepsilon_{fd} = \left\{ 1 + \left(\frac{\dot{\varepsilon}}{C} \right)^{1/q} \right\}^{-1} \varepsilon_{fc} \tag{12.6}$$

where σ_{Yd} is the dynamic yield stress, σ_Y is the static yield stress, ε_{fd} is the dynamic fracture strain used in the finite element model, ε_{fc} is the static fracture strain used in the finite element model that is obtained from Equation (12.2). C and q are test constants that are described in Chapter 10.

12.6 Boundary Condition Modeling

When the target structure's boundaries are linked to adjacent structures, the condition of these boundaries must be idealized realistically. This problem most often occurs with a partial analysis conducted by cutting a section out of the target structure, thus producing artificial boundaries. A similar situation may occur inside the target structure when certain structural modeling simplifications are attempted. For example, rigid restraints can be replaced with a strong support member that is regarded as undeforming and is thought to prevent displacement and/or rotation, and a weak support member may be ignored (zero restraint). However, when the degree of restraint at the boundaries is neither zero nor infinite, a more detailed set of boundary conditions is required.

A clear understanding of the reality of these boundaries is very important before idealizations are made. If the correct boundary conditions for replacing a portion of structure create uncertainty, it is probably better to include that portion in the structural model, even though it would require more computation.

Table 12.1 Boundary conditions of the stiffened plate structure's finite element model using the two-bay/double-span stiffened panel.

Boundary	Description
A–D and A'''–D'''	Symmetric condition with $R_y = R_z = 0$ and uniform displacement in the x direction (U_x = uniform), coupled with the longitudinal stiffener
A–A''' and D–D'''	Symmetric condition with $R_x = R_z = 0$ and uniform displacement in the y direction (U_y = uniform), coupled with the transverse frame
A'–D', A''–D'', B–B', and C–C'	$U_z = 0$

Note: U_x, U_y and U_z indicate the translational degrees of freedom in the x, y, and z direction, and R_x, R_y and R_z indicate the rotational degrees of freedom in the x, y, and z direction.

As an illustrative example of a stiffened plate structure as modeled in Figure 12.2f, the one-bay model is relevant only when the restraints at the transverse frame location are either zero or infinite (i.e., simply support or fixed). However, the rigidity of these frames is neither zero nor infinite, and the decision depends entirely on the required level of accuracy. When the model of Figure 12.2f is adopted, such as with a two-bay/two-span model, Table 12.1 indicates the boundary conditions to be applied as similar to Table 4.3.

12.7 Initial Imperfection Modeling

As described in Section 1.7, welded metal structures always have initial imperfections in the form of initial distortions and residual stresses. In contrast to steel structures, the yield stress in the heat-affected zone of welded aluminum structures is less than that of the base metal. As shown in Figure 12.8, three types of initial distortions are relevant to welded metal stiffened plate structures:

- Initial deflection of the plating between the support members $w_{0pl} = A_0 \sin\dfrac{m\pi x}{a}\sin\dfrac{\pi y}{b}$
- Column-type initial deflection of the support members $w_{0c} = B_0 \sin\dfrac{\pi x}{a}\sin\dfrac{\pi y}{b}$
- Sideways initial deflection of the support members $w_{0s} = C_0\dfrac{z}{h_w}\sin\dfrac{\pi x}{a}$

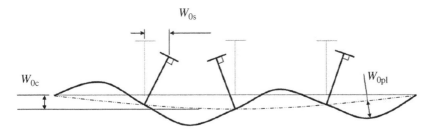

Figure 12.8 Three types of initial distortion in a stiffened plate structure.

The magnitude and shape of each type of initial distortion play important roles in buckling collapse behavior, and thus a better understanding of the actual imperfection configurations in the target structures is necessary. In fact, precise information about the initial distortions of the target structure is desirable before structural modeling even begins. Considering the significant amount of uncertainty involved in fabrication related initial imperfections, current measurements of the initial distortions in welded metal structures are often useful in the development of representative models.

The welding induced residual stress distribution inside the metal plates or stiffeners can be idealized as described in Section 1.7.3, where the welding residual stress comprises the tensile and compressive residual stress blocks. Welding residual stresses may develop in both the longitudinal and transverse directions because the support members are usually attached by welding in these two directions. The softening phenomenon can be modeled to allow the reduced material yield stress in the softened region, that is, in the heat-affected zone, to be applied as described in Section 1.7.4.

Figure 12.9 shows examples of the nonlinear finite element method models for the ultimate hull girder strength analysis of a double hull oil tanker or a container ship under vertical bending moment, accounting for the effects of welding induced initial imperfections, where the one-bay sliced hull cross-section model is adopted as the extent of the analysis, as shown in Figure 12.3f.

12.8 Order of Load Component Application

When combined load components are applied simultaneously, an issue associated with the order of load component application may arise. For example, the bottom panels of a ship's structures are likely subjected to combined lateral pressure and compressive loads; the former are caused by cargo and water, whereas the latter are caused by the hull girder bending moment under hogging conditions, as shown in Figure 12.10. In this situation, lateral pressure is usually applied first, and axial compressive loads are then applied while keeping the lateral pressure constant.

The shape and magnitude of the initial distortions in the plate panels can vary widely with the lateral pressure. Figures 12.11 and 12.12 show examples of plate panels under longitudinal and transverse compression before and after lateral pressure. The pressure causes effective "clamping" of the plating and changes the deflected shape away from the buckling mode shape, which may cause the ultimate strength value of the in-plane compression to be greater than if the pressure was small or absent. Therefore, in panels that may undergo in-plane compression with either high or low lateral pressure (such as underwater panels in a tanker or bulker), the ultimate strength should be calculated both for a full load and for the ballasted condition, and the lower value should be taken as the true ultimate strength.

In linear structural mechanics under a combination of multiple load components, the principle of linear superposition of structural responses by individual load components is satisfied, and the final status of structural response is identical regardless of the load paths. This principle is often adopted even for nonlinear structural mechanics

(a)

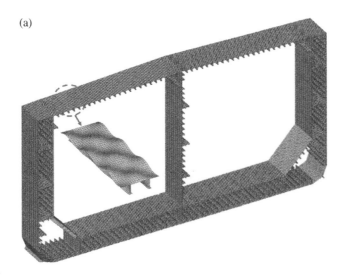

(b)

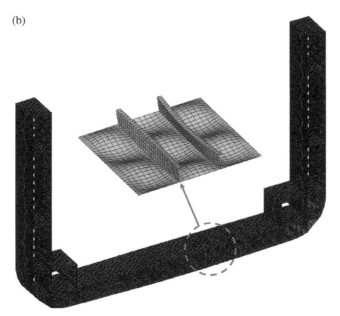

Figure 12.9 Examples of the nonlinear finite element method models for the ultimate hull girder strength analysis of (a) a double hull oil tanker and (b) a container ship, under vertical bending moment accounting for the initial imperfections.

problems that focus on buckling or ultimate strength, in which the load effects or resulting deformations are not large with small strains until the buckling or ultimate strength is reached. In contrast, the problems of structural crashworthiness in accidental situations such as collisions and grounding exhibit large strains associated with crushing and rupture, and therefore the principle of linear superposition is no longer applicable.

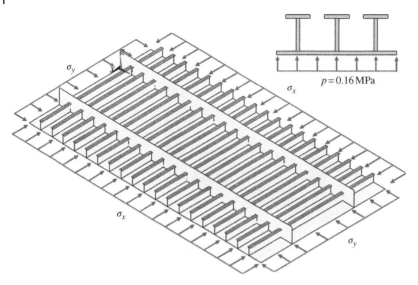

Figure 12.10 Illustrative example of a stiffened panel under combined axial compression and lateral pressure.

(a)

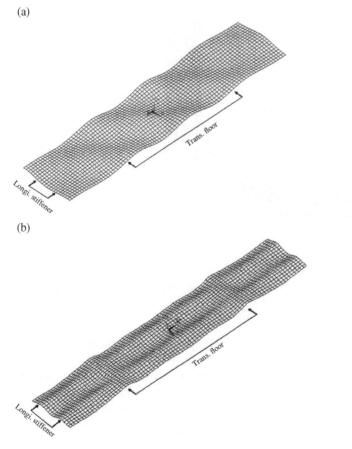

(b)

Figure 12.11 The panel initial deflection shapes under predominantly longitudinal compression for the two-bay panel model: (a) before lateral pressure; (b) after lateral pressure (amplification factor of 30).

Figure 12.12 The panel initial deflection shapes under predominantly transverse compression for the two-bay panel model: (a) before lateral pressure; (b) after lateral pressure (amplification factor of 30).

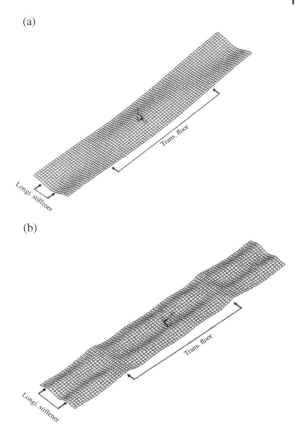

References

Belytschko, T., Liu, W.K., Moran, B. & Elkhodary, K. (2014). *Nonlinear finite elements for continua and structures.* Second Edition, John Wiley & Sons, Ltd, Chichester.

Borst, R., Crisfield, M.A., Remmers, J.C. & Verhoosel, C.V. (2014). *Nonlinear finite element analysis of solids and structures.* Second Edition, John Wiley & Sons, Ltd, Chichester.

Hughes, O.F. & Paik, J.K. (2013). *Ship structural analysis and design.* The Society of Naval Architects and Marine Engineers, Alexandria, VA.

Irons, B.M. & Razzaque, A. (1972). Experience with the patch test. In *Proceedings of the symposium on mathematical foundations of the finite element method*, Edited by Aziz, A.R., Academic Press, New York, 557–587.

Kim, N.H. (2014). *Introduction to nonlinear finite element analysis.* Springer, Berlin.

Ko, Y.G., Kim, S.J., Sohn, J.M. & Paik, J.K. (2017). A practical method to determine the dynamic fracture strain for the nonlinear finite element analysis of structural crashworthiness in ship-ship collisions. *Ships and Offshore Structures*, doi.org/10.1080/17445302.2017.1405584.

Paik, J.K. (2007a). Practical techniques for finite element modeling to simulate structural crashworthiness in ship collisions and grounding (Part I: Theory). *Ships and Offshore Structures*, **2**(1): 69–80.

Paik, J.K. (2007b). Practical techniques for finite element modeling to simulate structural crashworthiness in ship collisions and grounding (Part II: Verification). *Ships and Offshore Structures*, **2**(1): 81–85.

Paik, J.K. & Hughes, O.F. (2007). Ship structures. In *Modeling complex engineering structures*, Edited by Melchers, R.E. & Hough, R., ASCE Press, The American Society of Civil Engineers, Reston, VA.

Paik, J.K., Kim, S.J., Ko, Y.G. & Youssef, S.A.M. (2017). Collision risk assessment of a VLCC tanker. Proceedings of the 2017 SNAME Maritime Convention, Houston.

Reddy, J.N. (2015). *An introduction to nonlinear finite element analysis.* Oxford University Press, Oxford.

Shi, Z.C. (2002). Nonconforming finite element methods. *Journal of Computational and Applied Mathematics*, **149**(1): 221–225.

Taylor, R.L., Simo, T.C., Zienkiewicz, O.C. & Chan, A.H.C. (1986). The patch test: a condition for assessing FEM convergence. *International Journal of Numerical Methods*, **22**: 39–62.

Wierzbicki, T. & Abramowicz, W. (1983). On the crushing mechanics of thin-walled structures. *Journal of Applied Mechanics*, **50**: 727–734.

Wriggers, P. (2008). *Nonlinear finite element methods.* Springer, Berlin.

Zienkiewicz, O.C. (1977). *The finite element method.* Third Edition, McGraw-Hill, London.

13

The Intelligent Supersize Finite Element Method

13.1   Features of the Intelligent Supersize Finite Element Method

The nonlinear finite element method (NLFEM) described in Chapter 12 is a powerful technique used to simulate nonlinear structural response. However, a weak feature of the conventional NLFEM is that it requires enormous modeling effort and computing time for nonlinear analysis of large structures. In this regard, much effort has been devoted to reducing modeling and computing time for the nonlinear analyses of structures. Under extreme or accidental loading, structures can be involved in a highly nonlinear response associated with the yielding, buckling, crushing, and sometimes fracture of individual structural components.

The most obvious way to reduce modeling effort and computing time is to reduce the number of degrees of freedom so that the number of unknowns in the finite element stiffness matrix decreases. Modeling the object structure with very large structural units may offer the best approach. To avoid loss of accuracy, special-purpose finite elements must be used. Properly formulated structural units in such an approach can then be used to efficiently model the actual nonlinear behavior of the corresponding large parts of the structures.

In one such contribution, the Ueda group (Ueda & Rashed 1974, 1984, 1991, Ueda et al. 1983, 1984, 1986a, 1986b) suggested the idealized structural unit method (ISUM). In the ISUM, the nonlinear behavior for the so-called idealized structural units is formulated in closed-form expressions based on analytical solutions that are provided in the incremental form, and the structural stiffness of individual structural units is assembled for the entire target structure with a matrix calculation. With increases in the applied loading, the progressive collapse behavior of the structure can be computed.

In an almost parallel development to the ISUM, Smith (1977) suggested a similar approach to predict the ultimate bending moment of a ship hull. He modeled the ship hull as an assembly of plate-stiffener combination models, that is, stiffeners with attached plating (or beam–column units) as described in Figure 2.2a. The load versus end-shortening relationships for the plate-stiffener combination models were obtained with the NLFEM accounting for initial imperfections. The behavior of the larger structure was then constructed.

In contrast, ISUM is based on analytical engineering models for structural units, and thus the solutions are quite accurate but its use is sometimes limited to structures with a

Ultimate Limit State Analysis and Design of Plated Structures, Second Edition. Jeom Kee Paik.
© 2018 John Wiley & Sons Ltd. Published 2018 by John Wiley & Sons Ltd.

Figure 13.1 A typical example of a plated structure—a ship hull structure under construction at shipyard.

simple shape, and difficulties may be encountered in the analysis of structures with complex three-dimensional shapes.

The intelligent supersize finite element method (ISFEM) proposed by Paik (Hughes & Paik 2013) uses large elements, and the approach for formulating the method is similar to the conventional finite element method. In contrast to the conventional finite element method, the ISFEM is said to be intelligent because the highly nonlinear behavior of the large elements is "educated" or formulated in advance, generating a high level of intelligence in terms of judging the failure status and modes of such a large element. This approach is beneficial for simulating the nonlinear behavior of a structure with a complex shape because the conventional finite element method modeling technique is basically applied. Figure 13.1 shows a typical stiffened plate structure for which the ISFEM can model the plating, webs, or flanges using rectangular plate elements, as described in Figure 2.2d, while the NLFEM model is described in Figure 2.2e.

This chapter describes the formulation of the ISFEM in association with the ultimate limit state analysis using rectangular plate elements. The method can also be developed for the analysis of structural crashworthiness involving crushing and fracture failures. Applied examples are presented to demonstrate the method in association with the progressive collapse analysis of plated structures until and after the ultimate strength is reached.

13.2 Nodal Forces and Nodal Displacements of the Rectangular Plate Element

The combined in-plane and out-of-plane deformation behavior for a rectangular plate element can be expressed by the nodal force vector, $\{R\}$, and the displacement vector,

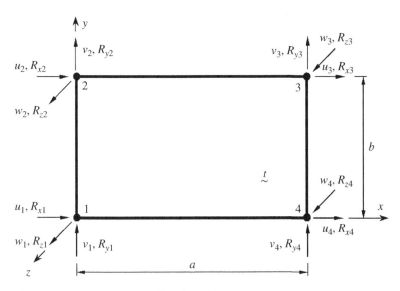

Figure 13.2 An ISFEM rectangular plate element.

$\{U\}$, with six degrees of freedom at each corner nodal point, which is considered to be located in the mid-thickness of the element:

$$\{R\} = \left\{ R_{x1}\ R_{y1}\ R_{z1}\ M_{x1}\ M_{y1}\ M_{z1} \cdots R_{x4}\ R_{y4}\ R_{z4}\ M_{x4}\ M_{y4}\ M_{z4} \right\}^{T} \tag{13.1a}$$

$$\{U\} = \left\{ u_1\ v_1\ w_1\ \theta_{x1}\ \theta_{y1}\ \theta_{z1} \cdots u_4\ v_4\ w_4\ \theta_{x4}\ \theta_{y4}\ \theta_{z4} \right\}^{T} \tag{13.1b}$$

where R_x, R_y, and R_z are the translational nodal forces in the x, y, and z directions, respectively. M_x and M_y are the out-of-plane bending moments for the x and y directions, respectively. M_z is the torsional moment for the z direction. u, v, and w are the translational displacements in the x, y, and z directions, respectively. θ_x ($= -\partial w/\partial y$), θ_y ($= \partial w/\partial x$), and θ_z are the rotations with regard to the x, y, and z directions, respectively. $\{\}^{T}$ represents the transpose of the vector. A digit in subscript indicates the node number of the rectangular plate element as defined in Figure 13.2.

13.3 Strain versus Displacement Relationship

The strain versus displacement relationship accounting for out-of-plane and in-plane large deformation effects for the ISFEM rectangular plate element is given in the Cartesian coordinate system by the following:

$$\varepsilon_x = \frac{\partial u}{\partial x} - z\frac{\partial^2 w}{\partial x^2} + \frac{1}{2}\left\{ \left(\frac{\partial u}{\partial x}\right)^2 + \left(\frac{\partial v}{\partial x}\right)^2 \right\} + \frac{1}{2}\left(\frac{\partial w}{\partial x}\right)^2 \tag{13.2a}$$

$$\varepsilon_y = \frac{\partial v}{\partial y} - z\frac{\partial^2 w}{\partial y^2} + \frac{1}{2}\left\{ \left(\frac{\partial u}{\partial y}\right)^2 + \left(\frac{\partial v}{\partial y}\right)^2 \right\} + \frac{1}{2}\left(\frac{\partial w}{\partial y}\right)^2 \tag{13.2b}$$

$$\gamma_{xy} = \left(\frac{\partial u}{\partial y} + \frac{\partial v}{\partial x}\right) - 2z\frac{\partial^2 w}{\partial x \partial y} + \left\{\left(\frac{\partial u}{\partial x}\right)\left(\frac{\partial u}{\partial y}\right) + \left(\frac{\partial v}{\partial x}\right)\left(\frac{\partial v}{\partial y}\right)\right\} + \left(\frac{\partial w}{\partial x}\right)\left(\frac{\partial w}{\partial y}\right)$$

(13.2c)

where ε_x, ε_y, and γ_{xy} are the generalized strain components for a plane stress state.

The first term on the right side of the preceding equation represents the small deformation in-plane strain. The second term denotes the small deformation out of plane. The third and fourth terms are nonlinear strain components due to large deflections in plane and out of plane, respectively. It is apparent from Equation (13.2) that the component for rotation with respect to the z axis, which is normal to the plane of the element, does not affect the strains of the element.

The incremental expressions corresponding to Equation (13.2) are written as follows:

$$\Delta\varepsilon_x = \frac{\partial \Delta u}{\partial x} - z\frac{\partial^2 \Delta w}{\partial x^2} + \left(\frac{\partial u}{\partial x}\right)\left(\frac{\partial \Delta u}{\partial x}\right) + \left(\frac{\partial v}{\partial x}\right)\left(\frac{\partial \Delta v}{\partial x}\right) + \left(\frac{\partial w}{\partial x}\right)\left(\frac{\partial \Delta w}{\partial x}\right)$$

$$+ \frac{1}{2}\left\{\left(\frac{\partial \Delta u}{\partial x}\right)^2 + \left(\frac{\partial \Delta v}{\partial x}\right)^2\right\} + \frac{1}{2}\left(\frac{\partial \Delta w}{\partial x}\right)^2$$

(13.3a)

$$\Delta\varepsilon_y = \frac{\partial \Delta v}{\partial y} - z\frac{\partial^2 \Delta w}{\partial y^2} + \left(\frac{\partial u}{\partial y}\right)\left(\frac{\partial \Delta u}{\partial y}\right) + \left(\frac{\partial v}{\partial y}\right)\left(\frac{\partial \Delta v}{\partial y}\right) + \left(\frac{\partial w}{\partial y}\right)\left(\frac{\partial \Delta w}{\partial y}\right)$$

$$+ \frac{1}{2}\left\{\left(\frac{\partial \Delta u}{\partial y}\right)^2 + \left(\frac{\partial \Delta v}{\partial y}\right)^2\right\} + \frac{1}{2}\left(\frac{\partial \Delta w}{\partial y}\right)^2$$

(13.3b)

$$\Delta\gamma_{xy} = \left(\frac{\partial \Delta u}{\partial y} + \frac{\partial \Delta v}{\partial x}\right) - 2z\frac{\partial^2 \Delta w}{\partial x \partial y} + \left(\frac{\partial u}{\partial x}\right)\left(\frac{\partial \Delta u}{\partial y}\right) + \left(\frac{\partial u}{\partial y}\right)\left(\frac{\partial \Delta u}{\partial x}\right) + \left(\frac{\partial v}{\partial x}\right)\left(\frac{\partial \Delta v}{\partial y}\right) + \left(\frac{\partial v}{\partial y}\right)\left(\frac{\partial \Delta v}{\partial x}\right)$$

$$+ \left(\frac{\partial w}{\partial x}\right)\left(\frac{\partial \Delta w}{\partial y}\right) + \left(\frac{\partial w}{\partial y}\right)\left(\frac{\partial \Delta w}{\partial x}\right) + \left(\frac{\partial \Delta u}{\partial x}\right)\left(\frac{\partial \Delta u}{\partial y}\right) + \left(\frac{\partial \Delta v}{\partial x}\right)\left(\frac{\partial \Delta v}{\partial y}\right) + \left(\frac{\partial \Delta w}{\partial x}\right)\left(\frac{\partial \Delta w}{\partial y}\right)$$

(13.3c)

where the prefix, Δ, denotes an infinitesimal increase in the variable.

For the sake of convenience in the formulations of the ISFEM rectangular plate element, the nodal displacement vector, $\{U\}$, is divided into three components: the in-plane component, $\{S\}$, the out-of-plane component, $\{W\}$, and the component for the rotations about the axis z. Thus, equation (13.3) can be rewritten in matrix form using the vectors, $\{S\}$ and $\{W\}$, as follows:

$$\{\Delta\varepsilon\} = [B_p]\{\Delta S\} - z[B_b]\{\Delta W\} + [C_p][G_p]\{\Delta S\} + [C_b][G_b]\{\Delta W\}$$

$$+ \frac{1}{2}[\Delta C_p][G_p]\{\Delta S\} + \frac{1}{2}[\Delta C_b][G_b]\{\Delta W\}$$

(13.4)

$$= [B]\{\Delta U\}$$

where $\{\Delta\varepsilon\} = \left\{\Delta\varepsilon_x \Delta\varepsilon_y \Delta\gamma_{xy}\right\}^T$ is the increase in the strain vector, $\{U\} = \{S\ W\}^T$ is the nodal displacement vector, $\{S\} = \{u_1\ v_1\ u_2\ v_2\ u_3\ v_3\ u_4\ v_4\}^T$ is the in-plane displacement vector, $\{W\} = \{w_1\ \theta_{x1}\ \theta_{y1}\ w_2\ \theta_{x2}\ \theta_{y2}\ w_3\ \theta_{x3}\ \theta_{y3}\ w_4\ \theta_{x4}\ \theta_{y4}\}^T$ is the out-of-plane displacement vector, and $[B]$ is the strain-displacement matrix

$$\left\{\frac{\partial u}{\partial x}\frac{\partial v}{\partial y}\frac{\partial u}{\partial y}+\frac{\partial v}{\partial x}\right\}^T = [B_p]\{S\}$$

$$\left\{\frac{\partial^2 w}{\partial x^2}\frac{\partial^2 w}{\partial y^2}2\frac{\partial^2 w}{\partial x\partial y}\right\}^T = [B_b]\{W\}$$

$$\left\{\frac{\partial u}{\partial x}\frac{\partial v}{\partial x}\frac{\partial u}{\partial y}\frac{\partial v}{\partial y}\right\}^T = [G_p]\{S\}$$

$$\left\{\frac{\partial w}{\partial x}\frac{\partial w}{\partial y}\right\}^T = [G_b]\{W\}$$

$$[C_p] = \begin{bmatrix} \dfrac{\partial u}{\partial x} & \dfrac{\partial v}{\partial x} & 0 & 0 \\[2ex] 0 & 0 & \dfrac{\partial u}{\partial y} & \dfrac{\partial v}{\partial y} \\[2ex] \dfrac{\partial u}{\partial y} & \dfrac{\partial v}{\partial y} & \dfrac{\partial u}{\partial x} & \dfrac{\partial v}{\partial x} \end{bmatrix}$$

$$[C_b] = \begin{bmatrix} \dfrac{\partial w}{\partial x} & 0 \\[2ex] 0 & \dfrac{\partial w}{\partial y} \\[2ex] \dfrac{\partial w}{\partial y} & \dfrac{\partial w}{\partial x} \end{bmatrix}$$

13.4 Stress versus Strain Relationship

The membrane stress increments, $\{\Delta\sigma\}$, due to strain increments, $\{\Delta\varepsilon\}$, can be calculated for a plane stress state as follows:

$$\{\Delta\sigma\} = [D]\{\Delta\varepsilon\} \tag{13.5}$$

where $\{\Delta\sigma\} = \{\Delta\sigma_x\ \Delta\sigma_y\ \Delta\tau_{xy}\}^T$ is the increase in the average membrane stress components for a plane stress state and $\{\Delta\varepsilon\} = \{\Delta\varepsilon_x\ \Delta\varepsilon_y\ \Delta\tau_{xy}\}^T$ is the increase in the average membrane strain components for a plane stress state.

In Equation (13.5), $[D]$ is the average stress-average strain matrix, which can be determined differently depending on the failure state, as described in Section 4.13. It is specified as follows:

- In the pre-buckling or undeflected regime: $[D] = [D_p]^E$, as defined in Equation (4.94b)
- In the post-buckling or deflected regime: $[D] = [D_p]^B$, as defined in Equation (4.102)
- In the post-ultimate strength regime: $[D] = [D_p]^U$, as defined in Equation (4.110)

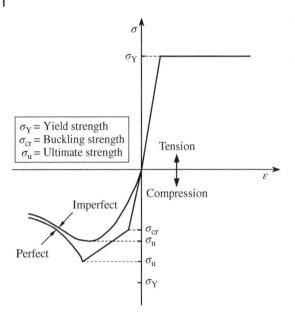

Figure 13.3 The behavior of the ISFEM rectangular plate element with the focus on predominantly axial compressive or tensile loads.

The failure status, such as buckling or ultimate strength, is checked by the theory described in Chapter 4, in which the average membrane stresses and lateral pressure loads are used for the failure checks as necessary. Figure 13.3 illustrates the behavior of the ISFEM rectangular plate element with the focus on predominantly axial compressive or tensile loads. It is noted that the elastic-perfectly plastic model of material as described in Section 1.3.2 is applied by neglecting the strain-hardening effects. The effects of welding induced initial imperfections and structural damages are accounted for in association with the average stress-average strain relationship indicated in Equation (13.5).

13.5 Tangent Stiffness Equation

Two approaches are widely used to calculate the nonlinear finite element stiffness matrix: the total and updated Lagrangian formulations. The latter can be used for the ISFEM rectangular plate element formulations. In the following, it is convenient to deal separately with the matrix components related to the rotations with regard to the axis z, which is normal to the plane of the element.

13.5.1 The Total Lagrangian Approach

Consider that an elastic structure under the nodal forces, $\{R\}$, resulting in the internal stresses, $\{\sigma\}$, is in equilibrium. Assume that the structure remains in equilibrium even after the increase of the virtual displacement increments, $\delta\{\Delta U\}$, corresponding to the virtual strain increments, $\delta\{\Delta\varepsilon\}$, which develop the nodal forces, $\{\Delta R\}$, and the resultant stresses, $\{\Delta\sigma\}$.

By applying the principle of the virtual work, the following equation should be satisfied:

$$\delta\{\Delta U\}^T\{R+\Delta R\} = \int_V \delta\{\Delta\varepsilon\}^T\{\sigma+\Delta\sigma\}d\text{Vol} \tag{13.6}$$

where the term on the left side represents the external work undertaken by the virtual displacement increments and the term on the right side denotes the strain energy dissipated by deformation during the applied loading. $\int_V ()d\text{Vol}$ indicates the integration over the entire volume of the element, and δ indicates a virtual value.

The virtual value of strain components, $\delta\{\Delta\varepsilon\}$, can be obtained by the differentiation of Equation (13.4) with respect to the increase in displacement as follows:

$$\begin{aligned}\delta\{\Delta\varepsilon\} = [B_p]\delta\{\Delta S\} - z[B_b]\delta\{\Delta W\} + [C_p + \Delta C_p][G_p]\delta\{\Delta S\} \\ + [C_b + \Delta C_b][G_b]\delta\{\Delta W\}\end{aligned} \tag{13.7}$$

Substituting Equations (13.5) and (13.7) into Equation (13.6) and neglecting the infinitesimal terms that have higher than second-order increments, the elastic stiffness equation for the element is given by

$$\{L\} + \{\Delta R\} = [K]\{\Delta U\} \tag{13.8}$$

where $[K]$ is the tangent stiffness matrix for the element. $\{L\} = \{R\} - \{r\}$ is the unbalanced forces caused by the differences between the total external forces, $\{R\}$, and the total internal forces, $\{r\}$, which in turn is calculated by

$$\{r\} = \int_V [B_p]^T\{\sigma\}d\text{Vol} + \int_V [G_p]^T[C_p]^T\{\sigma\}d\text{Vol} + \int_V [G_b]^T[C_b]^T\{\sigma\}d\text{Vol} \tag{13.9}$$

where $\{\sigma\} = \{\sigma_x\ \sigma_y\ \tau_{xy}\}^T$ are the total average membrane stress components.

The unbalanced forces should be eliminated at every step of the load increments. The tangent stiffness matrix, $[K]$, in Equation (13.8) can generally be subdivided into four terms:

$$[K] = [K_p] + [K_b] + [K_g] + [K_\sigma] \tag{13.10}$$

On the right side of the preceding equation, the first and second terms represent the stiffness matrices related to the in-plane and out-of-plane small deformations, respectively. The third term is the so-called initial deformation stiffness matrix, which in turn consists of three terms that represent the geometric nonlinear effects associated with the in-plane and out-of-plane deformations and their interactions. The fourth term is the so-called initial stress stiffness matrix, which is produced by the initial stresses for the element, in which no term related to their interactions appears.

Each term can be developed in more detail as follows:

$$[K_p] = \begin{bmatrix} [K_1] & 0 \\ 0 & 0 \end{bmatrix},$$

$$[K_b] = \begin{bmatrix} 0 & 0 \\ 0 & [K_2] \end{bmatrix},$$

$$[K_g] = \begin{bmatrix} [K_3] & [K_4] \\ [K_4]^T & [K_5] \end{bmatrix},$$

$$[K_\sigma] = \begin{bmatrix} [K_6] & 0 \\ 0 & [K_7] \end{bmatrix}$$

(13.11)

where $[K_1] = \int_V [B_p]^T [D]^E [B_p] d\text{Vol}, \ [K_2] = \int_V [B_b]^T [D]^e [B_b] z^2 d\text{Vol}, \ [K_3] = \int_V [G_p]^T [C_p]^T$

$[D]^E [B_p] d\text{Vol} + \int_V [B_p]^T [D]^E [C_p] [G_p] d\text{Vol} + \int_V [G_p]^T [C_p]^T [D]^e [C_p] [G_p] d\text{Vol},$

$[K_4] = \int_V [B_p]^T [D]^E [C_b] [G_b] d\text{Vol} + \int_V [G_p]^T [C_p]^T [D]^E [C_b] [G_b] d\text{Vol},$

$[K_5] = \int_V [G_b]^T [C_b]^T [D]^E [C_b] [G_b] d\text{Vol}, \ [K_6] = \int_V [G_p]^T [\sigma_p] [G_p] d\text{Vol},$

$[K_7] = \int_V [G_b]^T [\sigma_b] [G_b] d\text{Vol}, \ [\sigma_p] = \begin{bmatrix} \sigma_x & 0 & \tau_{xy} & 0 \\ 0 & \sigma_x & 0 & \tau_{xy} \\ \tau_{xy} & 0 & \sigma_y & 0 \\ 0 & \tau_{xy} & 0 & \sigma_y \end{bmatrix}, \ [\sigma_b] = \begin{bmatrix} \sigma_x & \tau_{xy} \\ \tau_{xy} & \sigma_y \end{bmatrix}.$

In calculating Equation (13.11), the terms involving the first order of the variable, z, become zero after completing the integration for the entire volume of the element in the elastic regime. Even in the elastic–plastic regime, the plasticity is condensed into the plastic nodes, and the inside of the element, with the exception of the plastic nodes, is in the present method, which is assumed to be elastic. These terms can thus be eliminated from the expressions.

13.5.2 The Updated Lagrangian Approach

The tangent stiffness matrix, $[K]$, in Equation (13.10) is derived by the total Lagrangian approach considering that the local coordinate system for the element is fixed with regard to the global coordinate system, which makes possible the use of an identical transformation matrix throughout the entire incremental loading process.

In contrast, the so-called updated Lagrangian approach requires one to update the local coordinate system at every incremental loading process, such that the transformation matrix from the local coordinate to the global system is newly established each time. The benefit of the updated Lagrangian approach is that the initial deformation matrix,

$[K_g]$, can be eliminated from Equation (13.10) because the initial deformation at the beginning of every incremental loading process can be set to zero. Therefore, the tangent elastic stiffness matrix, $[K]$, can be simplified to

$$[K] = [K_p] + [K_b] + [K_\sigma] \tag{13.12}$$

13.6 Stiffness Matrix for the Displacement Component, θ_z

The stiffness matrix components for the rotations with regard to the z axis may normally be set to zero, but this can in some cases produce numerical instability in the computation of the structural stiffness equation. To obtain a stabilizing effect in the numerical computation, the stiffness matrix components for the displacement component, θ_z, can be added to the stiffness matrix in Equation (13.12). The stiffness equation for the displacement component, θ_z, may be given by

$$
\begin{Bmatrix} M_{z1} \\ M_{z2} \\ M_{z3} \\ M_{z4} \end{Bmatrix} = \alpha E A t
\begin{bmatrix}
1 & -\dfrac{1}{2} & -\dfrac{1}{2} & -\dfrac{1}{2} \\
-\dfrac{1}{2} & 1 & -\dfrac{1}{2} & -\dfrac{1}{2} \\
-\dfrac{1}{2} & -\dfrac{1}{2} & 1 & -\dfrac{1}{2} \\
-\dfrac{1}{2} & -\dfrac{1}{2} & -\dfrac{1}{2} & 1
\end{bmatrix}
\begin{Bmatrix} \theta_{z1} \\ \theta_{z2} \\ \theta_{z3} \\ \theta_{z4} \end{Bmatrix}
\tag{13.13}
$$

where t is the plate thickness, A is the surface area of the element, and α is a constant that may normally be taken to be a very small value, for example, 5.0×10^{-5} (Zienkiewicz 1977).

13.7 Displacement (Shape) Functions

To attain a uniform state of shear stresses inside the element, a nonlinear function is in the present finite element method assumed for the in-plane displacements, u and v, whereas a polynomial function is assumed for the out-of-plane displacement, w, which is expressed in terms of 12 parameters. We thus have

$$u = a_1 + a_2 x + a_3 y + a_4 xy + \frac{b_4}{2}\left(b^2 - y^2\right) \tag{13.14a}$$

$$v = b_1 + b_2 x + b_3 y + b_4 xy + \frac{a_4}{2}\left(a^2 - x^2\right) \tag{13.14b}$$

$$w = c_1 + c_2 x + c_3 y + c_4 x^2 + c_5 xy + c_6 y^2 + c_7 x^3 + c_8 x^2 y$$
$$+ c_9 xy^2 + c_{10} y^3 + c_{11} x^3 y + c_{12} xy^3 \tag{13.14c}$$

where $a_1, a_2, ..., c_{12}$ are unknown coefficients that are expressed in terms of nodal displacements, $\{U\}$.

For a rectangular plate element with a length of a and a breadth of b, the coefficients of the displacement functions can be obtained by substituting the local coordinates and displacements at the nodes into Equation (13.14).

13.8 Local to Global Transformation Matrix

An exact formulation of the transformation matrix for a rectangular plate element is difficult to define. In the approximate formulation, it is normally considered that the element is in a plane that contains at least three nodal points of the element. The transformation matrix, $[T]$, from the local coordinate system to the global coordinate system can then be obtained in Cartesian terms (i.e., as functions of the global coordinates at the nodal points). Therefore, the element stiffness matrix in the local coordinate system can be transformed to the global system coordinate as follows:

$$[K]_G = [T]^T [K]_L [T] \qquad (13.15)$$

where $[K]_L$ and $[K]_G$ are the element stiffness matrices in the local and global coordinates, respectively, and $[T]$ is the transformation matrix from the local coordinate system to the global coordinate system.

All of the element stiffness matrices in the global coordinate system are then assembled in the usual manner for the finite element procedure to obtain the stiffness matrix for the entire structure. By solving the resulting stiffness equations for the prescribed load increments and boundary conditions, the structural response is obtained.

The stiffness equation of the target structure in the global coordinate system can then be given by assembling all of the elemental global stiffness matrices in the incremental form as follows:

$$\{\Delta R\} = \sum [K]_G \{\Delta U\} \qquad (13.16)$$

13.9 Modeling of Flat Bar Stiffener Web and One-Sided Stiffener Flange

The web of flat bar stiffener or one-sided (asymmetrical) flange of angle or Tee stiffener is supported at three edges and free at one edge, as shown in Figure 5.8. The web of flat bar stiffener can fail by lateral–torsional buckling (or tripping). The critical strength of lateral–torsional buckling that may occur in the web of flat bar stiffener can be evaluated as described in Sections 5.6 and 5.9, the latter section being with the Johnson–Ostenfeld formulation method as described in Section 2.9.5.1. When the one-sided flange of angle or Tee stiffener buckles locally, the critical buckling strength of the stiffener flange can be evaluated as described in Sections 5.7 and 5.9.

Until such local buckling occurs in the flat bar stiffener web or one-sided stiffener flange, the stress–strain matrix $[D]$ of the plate element associated with Equation (13.5) is taken as $[D] = [D_p]^E$, which is available in the pre-buckling strength or undeflected regime, as defined in Equation (4.94b). In reality, the stiffener flange should be designed to be stocky enough with a sufficiently low plate slenderness ratio. In this case, such local buckling may rarely take place, whereas the ultimate strength is reached by gross yielding. In such a case, the stress–strain matrix $[D]$ of the ISFEM rectangular plate element associated with Equation (13.5) may follow the elastic-perfectly plastic material model without local buckling effect.

13.10   Applied Examples

ISFEM theory has been implemented into the computer program ALPS/GENERAL (2017) to analyze the progressive collapse of large plated structures. A special version of the ALPS/GENERAL is ALPS/HULL (2017), which can deal with analysis of the progressive collapse of a ship's hull under combined vertical bending, horizontal bending, shearing force, and torsional moment. In the following, applied examples of the ISFEM theory are demonstrated for the progressive collapse analysis of a plate, a box column, a ship's hull girder, and a bridge structure until and after the ultimate strength is reached.

For more applied examples of the ISFEM, interested readers may refer to Magoga and Flockhart (2014), Zhang et al. (2016), Faisal et al. (2017), and Youssef et al. (2016, 2017), among others.

13.10.1   A Rectangular Plate

The ultimate strength behavior of a rectangular plate in longitudinal or transverse compression is analyzed. The dimension of the plate is length, $a = 3000$ mm, and breadth, $b = 1000$ mm, with a Young's modulus of $E = 205.8$ GPa, a yield stress of $\sigma_Y = 355$ MPa, and a Poisson's ratio of $\nu = 0.3$, while the plate thickness t is varied at $t = 15$, 20, and 25 mm, which is equivalent to the plate slenderness ratio of $\beta = (b/t)\sqrt{\sigma_Y/E} = 2.768$, 2.076, and 1.661, respectively.

Initial deflection distribution in the plate is assumed to be $w_{0pl} = 0.05t\ \sin(m\pi x/a)\sin(\pi y/b)$, where m is the buckling mode half-wave number in the x direction, that is, $m = 3$ under uniaxial compression in the x direction and $m = 1$ under uniaxial compression in the y direction. It is assumed that no residual stress exists. The plate is considered to be simply supported along all (four) edges keeping them straight, although unloaded edges can move freely in plane. This boundary condition is more practical because plate edges are more likely to keep straight in a continuous plate structure.

Figure 13.4 shows the analysis models by ISFEM ALPS/GENERAL and NLFEM. For the NLFEM analysis, a quarter of the plate is taken as the extent of the analysis considering the symmetric condition in terms of geometry and resulting behavior. Figure 13.5a and b compares the ultimate strength behavior of the plate under uniaxial compressive loads in the longitudinal or transverse direction, with varying plate thickness. It is apparent from Figure 13.5 that the ISFEM solutions are in good agreement with the NLFEM results.

13.10.2   A Box Column

The ISFEM ALPS/GENERAL is now applied to the progressive collapse analysis of a thin-walled box column under axial compressive loads, and the computational results are compared with NLFEM solutions.

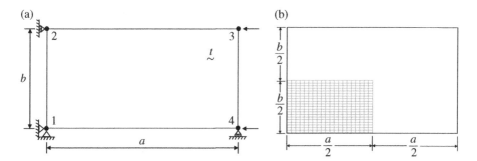

Figure 13.4 Structural models for a simply supported rectangular plate by (a) ISFEM with one rectangular plate element; (b) NLFEM with 400 four-noded plate-shell elements.

Figure 13.6 shows a half of a box column with respect to the mid-span in the length direction, which is composed of unstiffened rectangular plate elements together with several diaphragms. The edges of individual plate elements at both ends of the box column are simply supported, and subsequently both joint translation and moment restraint can occur after buckling at the ends of the box column, implying that the end condition of the box column as a system is more likely to be clamped.

In the analysis, the entire length L of the box column is varied at $L = 500$, 8 000, and 21 000 mm, and the dimensions, material properties, and initial imperfections of the structure are as follows:

- Geometry of plate elements: $a \times b \times t = 500 \times 500 \times 7.5$ (mm)
- Thickness of transverse diaphragms = 3.0 mm
- Young's modulus: $E = 205.8$ GPa
- Yield stress: $\sigma_Y = 352.8$ MPa
- Poisson's ratio: $\nu = 0.3$
- Maximum plate initial deflection: $w_{0pl} = 0.05t$
- Column-type initial deflection function of the entire box column: $w_{0c} = \delta_0 \sin(\pi x / L)$ where $\delta_0 = 0.0015L$
- Welding residual stresses that do not exist: $\sigma_{rcx} = \sigma_{rcy} = 0$

In the box column, the unloaded edge conditions along the corners of individual plate elements may play a role in the progressive collapse behavior. Therefore, three different conditions of unloaded plate edges are considered for the NLFEM analysis, while the ISFEM analysis presumes that the unloaded plate edges are always kept straight although they may move in plane of the corresponding plate elements. The unloaded edge conditions along the corners of two adjacent plate elements positioned in the right angle, considered for the NLFEM analysis, are as follows:

- Case 1: The unloaded plate edges are left free.
- Case 2: The unloaded plate edges are kept straight in the y (horizontal) direction, while they are left free in the z (vertical) direction, although they can move in plane of the corresponding plate elements in the y direction.
- Case 3: The unloaded plate edges are kept straight in the y and z directions, although they can move in plane of the corresponding plate elements in the y or z direction.

(a)

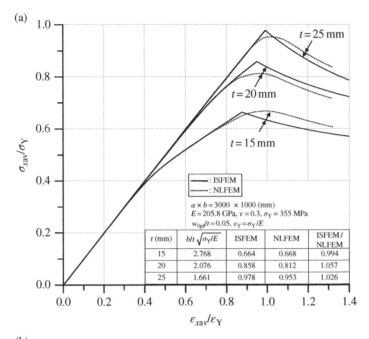

(b)

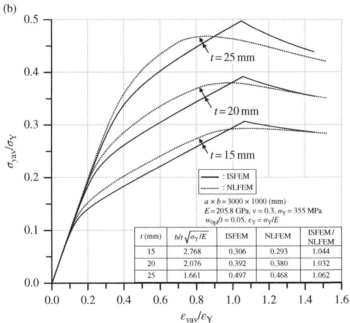

Figure 13.5 The ultimate strength behavior of a simply supported plate under (a) longitudinal compression and (b) transverse compression (σ_{xav} and σ_{yav} = average axial compressive stresses in the x or y direction, ε_{xav} and ε_{yav} = average axial compressive strain in the x or y direction, ε_Y = yield strain).

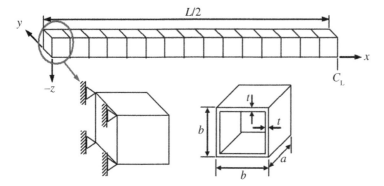

Figure 13.6 A box column simply supported with both joint translation and moment restraint at both ends.

It is noted that axial displacements (i.e., by displacement control) are applied uniformly over the cross section of the box columns. Also, the global deflection of the box column will occur in the z direction, and the edge condition of Case 3 should not be applied when global buckling (as well as local buckling) of the box column is aimed at.

13.10.2.1 A Short Box Column with $L = 500$ mm

The progressive collapse behavior of a short box column with $L = 500$ mm is analyzed. This structure corresponds to one bay box column that is composed of four rectangular plate elements. Figure 13.7 shows the structural models used for the NLFEM and ISFEM analyses. A quarter of the structure is taken as the extent of the NLFEM analysis considering the symmetric condition in terms of geometry and resulting behavior, while the entire structure is taken as the extent of the ISFEM analysis. Although a number

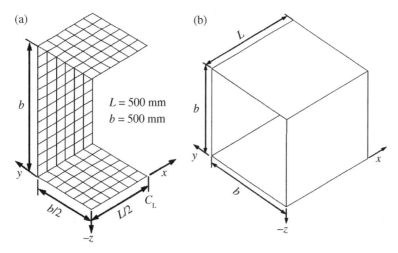

Figure 13.7 Structural models in one bay (short) box column used for (a) the NLFEM analysis and (b) the ISFEM analysis.

of four-noded plate-shell elements with fine mesh need to be used for the NLFEM analysis, each plate element of the structure is modeled as one ISFEM plate element for the ISFEM analysis, resulting in a total of four ISFEM rectangular plate elements only. It is obvious that the computational efforts of the ISFEM analysis are very small compared with those of the NLFEM analysis.

Figure 13.8 shows the deformed shape of the short box column at ultimate limit state. In this case, only local failure of individual plate elements of course takes place. Figure 13.9 shows the progressive collapse behavior of the short box column obtained by the NLFEM and ISFEM analyses. It is apparent from Figure 13.9 that the progressive collapse behavior depends on the unloaded plate edge conditions, among others, but the ISFEM solutions are in between Cases 2 and 3 of the NLFEM results in terms of ultimate strength predictions.

13.10.2.2   A Medium Box Column with $L = 8000$ mm

The progressive collapse behavior is now computed by NLFEM and ISFEM for a medium box column with $L = 8000$ mm or $\lambda =$ column slenderness ratio $= L/(\pi r) \times \sqrt{\sigma_Y/E} = 0.504$ where $r = \sqrt{I/A}$ is the radius of gyration, A is the cross-sectional area, and I is the moment of inertia of the box column. Figure 13.10 shows the structure models used for the NLFEM and ISFEM analyses. Again, a quarter of the structure is taken as the extent of the NLFEM analysis considering the symmetric condition, but the entire structure is included in the ISFEM structural modeling.

Figure 13.11 shows the deformed shape of the structure at ultimate limit state, as obtained by the NLFEM analysis. Figure 13.12 shows the ultimate strength behavior

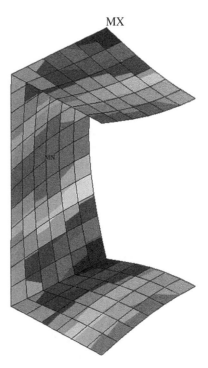

Figure 13.8 Deformed shape of a short box column with $L = 500$ mm at ultimate limit state (shown for a quarter of the structure), as obtained by the NLFEM analysis.

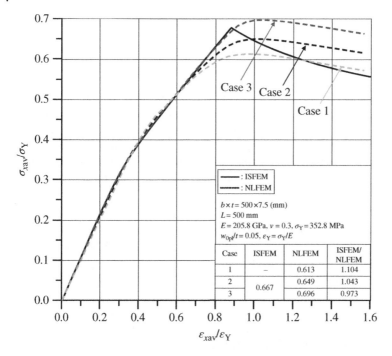

Figure 13.9 The progressive collapse behavior of a short box column with $L = 500$ mm (ε_{xav}, average axial compressive strain in the x direction; ε_Y, yield strain; σ_{xav}, average axial compressive stresses in the x direction).

of local plate elements on compressed flanges of the medium box column at the center or at the end. Figure 13.13 shows the progressive collapse behavior of the medium box column until and after the ultimate strength is reached. The behavior of the structure in the post-ultimate strength regime obtained by the NLFEM analysis is very unstable for Case 1 where the unloaded plate edges are left entirely free in terms of the finite element modeling. The ISFEM solutions of ultimate strength are in between Cases 2 and 3 of the NLFEM results. Again, the unloaded plate edge conditions affect the progressive collapse behavior of the structure. It is apparent from Figures 13.11 and 13.12 that the progressive collapse behavior of the box column is still governed primarily by local failure of individual plate elements, but the interacting effects between local and global failure modes of the structure seem to be small.

13.10.2.3 A Long Box Column with $L = 21\,000$ mm

The progressive collapse behavior is now computed by NLFEM and ISFEM for a long box column with $L = 21\,000$ mm or λ = column slenderness ratio = $L/(\pi r) \times \sqrt{\sigma_Y/E} = 1.323$ where $r = \sqrt{I/A}$ is the radius of gyration, A is the cross-sectional area, and I is the moment of inertia of the box column.

Figure 13.14 shows the analysis models used for NLFEA and ISFEM. Again, a quarter of the structure is taken as the extent of the NLFEM analysis considering the symmetric condition, but the entire structure is included in the ISFEM structural modeling.

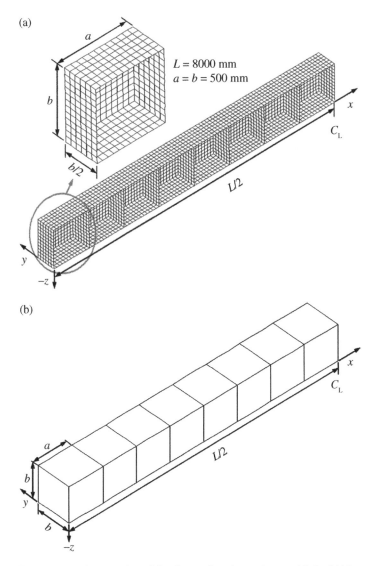

(a)

$L = 8000$ mm
$a = b = 500$ mm

(b)

Figure 13.10 Structural models of a medium box column with $L = 8000$ mm used for (a) the NLFEM analysis and (b) the ISFEM analysis.

Figure 13.15 shows the deformed shape of the structure at ultimate limit state, obtained by the NLFEM analysis. Figure 13.16 shows the progressive collapse behavior of the long box column until and after the ultimate strength is reached. It is apparent from the figures that the progressive collapse behavior of the long box column is significantly affected by local failure of plate elements, global system failure, and their interacting effects. In this case, the unloaded edge condition of Case 3 is not applied for the NLFEM analysis, while the inception of global buckling is allowed in the z (vertical) direction. The unloaded plate edge conditions again affect the progressive collapse behavior of the structure.

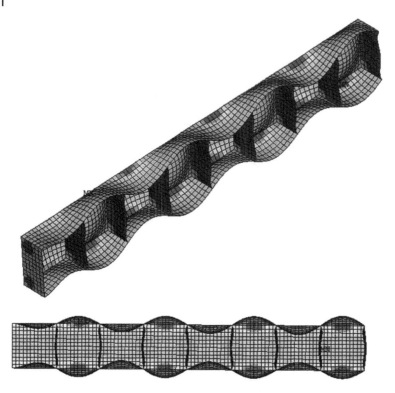

Figure 13.11 Deformed shape of a medium box column with $L = 8000$ mm at ultimate limit state (shown for a quarter of the structure), obtained by the NLFEM analysis.

The behavior of the structure in the post-ultimate strength regime obtained by NLFEM is very unstable and shows "snap-through" behavior for both Cases 1 and 2. The ISFEM solutions of ultimate strength are in good agreement with more refined NLFEM analysis.

13.10.2.4   Global Buckling of a Box Column

The Euler (elastic) global buckling stress of a box column clamped at both ends can be predicted from Section 7.4 or Section 2.9.3, as follows:

$$\sigma_{EG} = \frac{4\pi^2 EI}{AL^2} \tag{13.17}$$

where σ_{EG} is the elastic global buckling stress, I is the moment of inertia, A is the cross-sectional area, and L is the length of the box column. The critical buckling stress accounting for the effect of plasticity can be calculated by the plasticity correction of the elastic buckling stress defined in Equation (13.17) using the Johnson–Ostenfeld formulation method described in Section 2.9.5.1.

   NLFEM can also be used to predict the global buckling strength. In this section, two cases are considered, one case for only global buckling mode and the other case for both

(a)

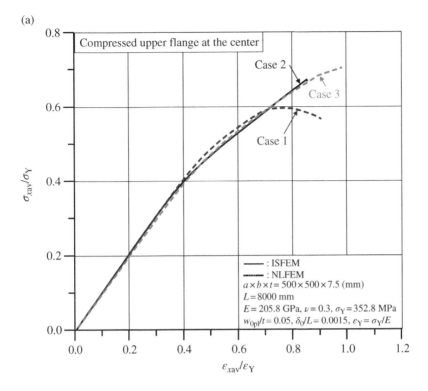

(b)

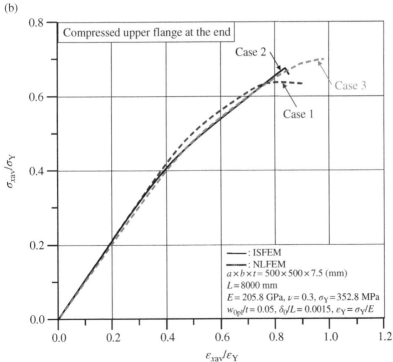

Figure 13.12 Average stress versus average strain relations of plate elements under axial compression in a medium box column with $L = 8000$ mm on (a) compressed upper-flange at center and (b) compressed lower-flange at end (ε_{xav} = average axial compressive strain in the x direction; ε_Y, yield strain; σ_{xav}, average axial compressive stresses in the x direction).

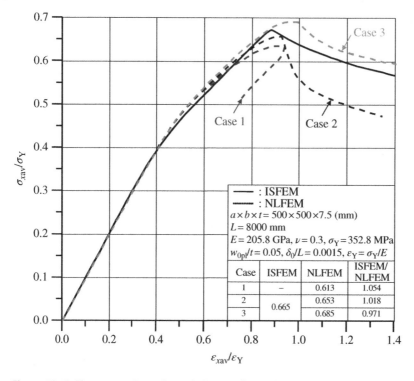

Figure 13.13 The progressive collapse behavior of a medium box column with $L = 8000$ mm (ε_{xav}, average axial compressive strain in the x direction; ε_Y, yield strain; σ_{xav}, average axial compressive stresses in the x direction).

local and global buckling modes, although the effect of plasticity is taken into account in the two cases. For the purpose of only global buckling analysis, relatively coarse mesh size of finite elements is adopted so that local buckling of plate elements is not allowed. It is found from Figure 13.16 that the load-carrying capacity (stress) without local failure of plate elements is $0.858\sigma_Y$ that is much larger than the actual ultimate strength considering local buckling, global buckling, and their interacting effects. The critical buckling stress σ_{crG} of the long box column with $L = 21\,000$ mm is obtained as $\sigma_{crG} = 0.882\sigma_Y$, which is very close to $0.858\sigma_Y$, obtained by NLFEM with 2.8% error.

On the other hand, the actual value of ultimate strength for the box column is smaller than the load-carrying capacity ($0.858\sigma_Y$) obtained by neglecting the effect of local buckling with the error of 31%. This implies that the effects of local failure, global failure, and their interaction must be taken into account upon calculating the ultimate strength of box columns.

13.10.3 A Ship's Hull Girder: The Dow Test Model

The extent of the analysis of the ship hull structures between two transverse frames is taken as shown in Figure 12.3f in association with one-bay sliced hull cross-section model.

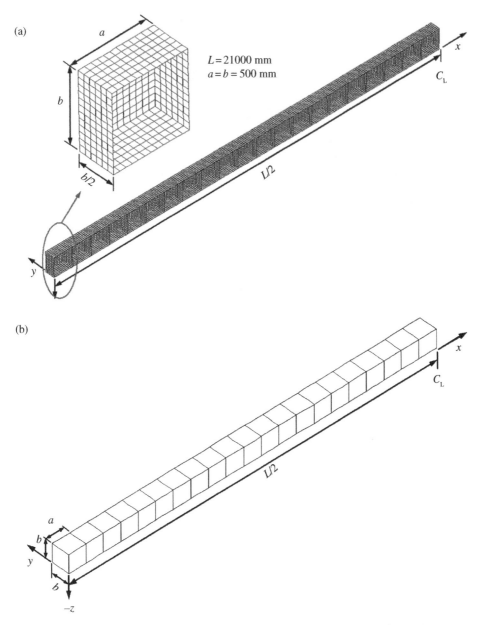

(a)

$L = 21000$ mm
$a = b = 500$ mm

C_L

(b)

C_L

Figure 13.14 Structural models of a long box column with $L = 21\,000$ mm used for (a) the NLFEM analysis and (b) the ISFEM analysis.

This method is applied to the one-third-scale Frigate hull model tested at the sagging bending moment (Dow 1991). Figure 13.17 shows the cross-sectional profiles of the test hull. Table 13.1 lists the details of the structure, including its dimensions and coordinates. The structure certainly has initial imperfections, and measurements of the initial deflection and residual stresses due to fabrication were reported by Dow (1991). However, for the

Figure 13.15 Deformed shape of a long box column with $L = 21\,000$ mm at ultimate limit state (shown for a quarter of the structure), as obtained by NLFEM analysis.

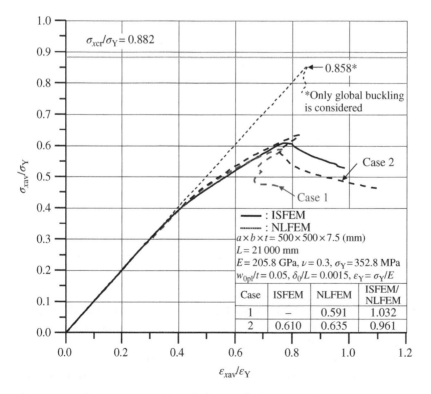

The plot shows σ_{xav}/σ_Y on the vertical axis versus $\varepsilon_{xav}/\varepsilon_Y$ on the horizontal axis.

$\sigma_{xcr}/\sigma_Y = 0.882$

0.858*

*Only global buckling is considered

Case 2

Case 1

——— : ISFEM

·········· : NLFEM

$a \times b \times t = 500 \times 500 \times 7.5$ (mm)

$L = 21\,000$ mm

$E = 205.8$ GPa, $\nu = 0.3$, $\sigma_Y = 352.8$ MPa

$w_{0pl}/t = 0.05$, $\delta_0/L = 0.0015$, $\varepsilon_Y = \sigma_Y/E$

Case	ISFEM	NLFEM	ISFEM/ NLFEM
1	–	0.591	1.032
2	0.610	0.635	0.961

Figure 13.16 The progressive collapse behavior of a long box column with $L = 21\,000$ mm.

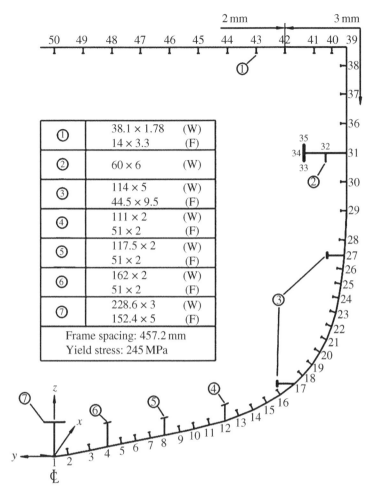

Figure 13.17 The one-third-scale Frigate hull model tested in the sagging condition (Dow 1991).

sake of simplicity, an average level of initial imperfections of the plating and the stiffeners in the test model is assumed in the present computations as follows:

$$w_{0pl} = 0.1t, \quad \sigma_{rcx} = -0.1\sigma_Y, \quad w_{0c} = w_{0s} = 0.0015a$$

where w_{0pl} is the maximum initial deflection of the plating, σ_{rcx} is the compressive residual stress of the plating in the longitudinal (x) direction, w_{0c} is the column-type initial distortion of the stiffeners, w_{0s} is the sideways initial distortion of the stiffeners, a is the length of the stiffener between the transverse frames, and b is the breadth of the plating between the stiffeners.

Figure 13.18 shows the NLFEM model that uses four-noded plate-shell elements with a total of 36 432 elements, where 10 elements in the plate width direction, four elements in the stiffener web height direction, and two elements in the stiffener flange side are applied. Figure 13.19 shows the model of ISFEM ALPS/HULL (2017) in which the

Table 13.1 Coordinates of the plate-stiffener intersections together with the structural dimensions of the plating and stiffeners in the one-third-scale Frigate test hull model (Dow 1991).

No.	x	y (mm)	z (mm)	Portion	Plate (mm)	No.	Web (mm)	Flange (mm)
1	0.0	0.0	0.0	1–2	99.2 × 3	1	228.6 × 3	152.4 × 5
2	0.0	−98.4	13.9	2–3	153.7 × 3	2	38.1 × 1.78	14 × 3.3
3	0.0	−249.3	41.9	3–4	127.2 × 3	3	38.1 × 1.78	14 × 3.3
4	0.0	−373.9	67.7	4–5	100.3 × 3	4	162 × 2	51 × 2
5	0.0	−472.3	87.1	5–6	103.5 × 3	5	38.1 × 1.78	14 × 3.3
6	0.0	−574.0	106.5	6–7	103.5 × 3	6	38.1 × 1.78	14 × 3.3
7	0.0	−675.7	125.8	7–8	100.3 × 3	7	38.1 × 1.78	14 × 3.3
8	0.0	−774.1	145.2	8–9	110.5 × 3	8	117.5 × 2	51 × 2
9	0.0	−882.3	167.7	9–10	104.2 × 3	9	38.1 × 1.78	14 × 3.3
10	0.0	−984.0	190.3	10–11	108.1 × 3	10	38.1 × 1.78	14 × 3.3
11	0.0	−1089.0	216.1	11–12	111.2 × 3	11	38.1 × 1.78	14 × 3.3
12	0.0	−1,197.0	241.9	12–13	101.5 × 3	12	111 × 2	51 × 2
13	0.0	−1292.0	277.4	13–14	108.8 × 3	13	38.1 × 1.78	14 × 3.3
14	0.0	−1394.0	316.1	14–15	109.6 × 3	14	38.1 × 1.78	14 × 3.3
15	0.0	−1492.0	364.5	15–16	109.8 × 3	15	38.1 × 1.78	14 × 3.3
16	0.0	−1588.0	419.4	16–17	123.2 × 3	16	38.1 × 1.78	14 × 3.3
17	0.0	−1686.0	493.5	17–18	78.3 × 3	17	114 × 5	44.5 × 9.5
18	0.0	−1742.0	548.4	18–19	99.0 × 3	18	38.1 × 1.78	14 × 3.3
19	0.0	−1807.0	622.6	19–20	103.4 × 3	19	38.1 × 1.78	14 × 3.3
20	0.0	−1863.0	709.7	20–21	95.6 × 3	20	38.1 × 1.78	14 × 3.3
21	0.0	−1909.0	793.5	21–22	97.3 × 3	21	38.1 × 1.78	14 × 3.3
22	0.0	−1945.0	883.9	22–23	98.1 × 3	22	38.1 × 1.78	14 × 3.3
23	0.0	−1975.0	977.4	23–24	101.9 × 3	23	38.1 × 1.78	14 × 3.3
24	0.0	−1994.0	1077.4	24–25	98.2 × 3	24	38.1 × 1.78	14 × 3.3
25	0.0	−2011.0	1174.2	25–26	100.9 × 3	25	38.1 × 1.78	14 × 3.3
26	0.0	−2024.0	1274.2	26–27	94.0 × 3	26	38.1 × 1.78	14 × 3.3
27	0.0	−2034.0	1367.7	27–28	103.5 × 3	27	114 × 5	44.5 × 9.5
28	0.0	−2040.0	1471.0	28–29	200.2 × 3	28	38.1 × 1.78	14 × 3.3
29	0.0	−2050.0	1671.0	29–30	196.7 × 3	29	38.1 × 1.78	14 × 3.3
30	0.0	−2050.0	1867.7	30–31	196.8 × 3	30	38.1 × 1.78	14 × 3.3
31	0.0	−2050.0	2064.5	31–32	146 × 6	31	–	–
32	0.0	−1904.0	2064.5	32–33	146 × 6	32	60 × 6	–
33	0.0	−1758.0	2004.5	33–34	60 × 10	33	–	–
34	0.0	−1758.0	2064.5	34–35	60 × 10	34	–	–
35	0.0	−1758.0	2124.4	31–36	200 × 3	35	–	–
36	0.0	−2050.0	2264.5	36–37	200 × 3	36	38.1 × 1.78	14 × 3.3

Table 13.1 (Continued)

No.	x	y (mm)	z (mm)	Portion	Plate (mm)	No.	Web (mm)	Flange (mm)
37	0.0	−2050.0	2464.5	37–38	193.6 × 3	37	38.1 × 1.78	14 × 3.3
38	0.0	−2050.0	2658.1	38–39	141.9 × 3	38	38.1 × 1.78	14 × 3.3
39	0.0	−2050.0	2800.0	39–40	101.7 × 3	39	–	–
40	0.0	−1948.3	2800.0	40–41	124 × 3	40	38.1 × 1.78	14 × 3.3
41	0.0	−1824.3	2800.0	41–42	202.7 × 3	41	38.1 × 1.78	14 × 3.3
42	0.0	−1621.6	2800.0	42–43	202.7 × 2	42	38.1 × 1.78	14 × 3.3
43	0.0	−1418.9	2800.0	43–44	202.7 × 2	43	38.1 × 1.78	14 × 3.3
44	0.0	−1216.2	2800.0	44–45	202.7 × 2	44	38.1 × 1.78	14 × 3.3
45	0.0	−1013.5	2800.0	45–46	202.7 × 2	45	38.1 × 1.78	14 × 3.3
46	0.0	−810.8	2800.0	46–47	202.7 × 2	46	38.1 × 1.78	14 × 3.3
47	0.0	−608.1	2800.0	47–48	202.7 × 2	47	38.1 × 1.78	14 × 3.3
48	0.0	−405.4	2800.0	48–49	202.7 × 2	48	38.1 × 1.78	14 × 3.3
49	0.0	−202.7	2800.0	49–50	202.7 × 2	49	38.1 × 1.78	14 × 3.3
50	0.0	0.0	2800.0	–	–	50	38.1 × 1.78	14 × 3.3

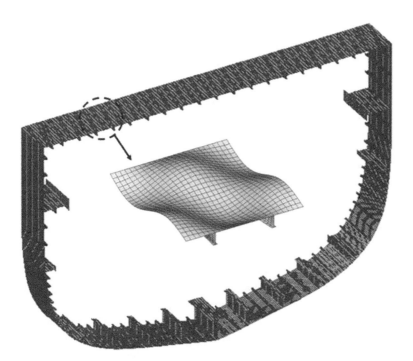

Figure 13.18 The NLFEM one-bay sliced hull cross-section model with a total of 36 432 four-noded plate-shell elements (assigned with 10 elements in the plate width direction, four elements in the web height direction, and two elements in the flange with one element for each side of Tee stiffener flange).

(a)

(b)

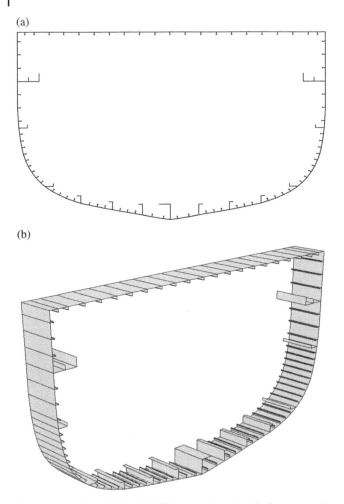

Figure 13.19 The ISFEM ALPS/HULL one-bay sliced hull cross-section model for the Dow test hull with a total of 345 intelligent supersize rectangular plate elements: (a) plan view; (b) three-dimensional view.

plating, webs, and flanges are all modeled using the ISFEM rectangular plate elements with a total of 345 elements. Figure 13.20 shows the vertical sagging bending moment versus the curvature curve by comparing the test results of NLFEM and ISFEM. Table 13.2 summarizes the ultimate sagging moments obtained by the test of NLFEM and ISFEM. The ISFEM ALPS/HULL method computations agree well with the test results. Figure 13.21 shows the changes in the neutral axis position of the hull cross section. It is apparent from Figure 13.21 that the neutral axis moves down as the sagging bending moment increases, resulting from the loss of effectiveness at the deck plate panels due to buckling collapse. The benefit of the ISFEM is obvious in terms of computational cost compared with that of the NLFEM. Another benefit of the ISFEM is that the progressive failure status and collapse modes of the individual plate elements can be explicitly captured as the external forces increase. On the other hand, the one (cargo)

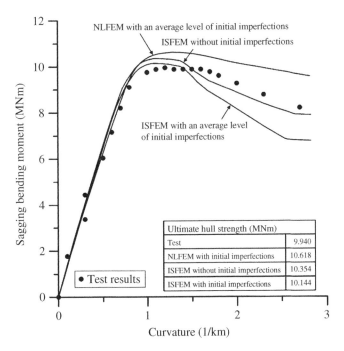

Figure 13.20 The progressive collapse behavior of the Dow's one-third-scale Frigate test hull under vertical sagging bending moment.

Table 13.2 Ultimate sagging bending moments obtained by the test of NLFEM and ISFEM.

Test	NLFEM	NLFEM/test	ISFEM	ISFEM/test
9.940 MNm	10.618 MNm	1.068	10.144 MNm	1.020

hold model between two transverse bulkheads as described in Figure 12.3d can be used if there are possibilities of failures or even deformations at transverse frames before the entire structure reaches the ultimate strength. Figure 13.22 shows the one hold model of ISFEM ALPS/HULL for the Dow test hull with a total of 4990 intelligent supersize plate elements.

13.10.4 A Corroded Steel-Bridge Structure

The ultimate bending capacity of a steel bridge supported by pillars as shown in Figure 13.23 is now analyzed. Figure 13.24 shows the ISFEM model using ALPS/GENERAL for a hypothetical steel-bridge structure, which is composed of 1845 plate elements, where $L = 30a = 75$ m (span of the bridge), $B = 15b = 15$ m (breadth), $H = 1.5$ m (height), a is the spacing of transverse frames, b is the spacing of longitudinal girders, $t_d = 15$ mm (thickness of deck plate), $t_b = 10$ mm (thickness of bottom plate), $t_w = 15$ mm (web thickness

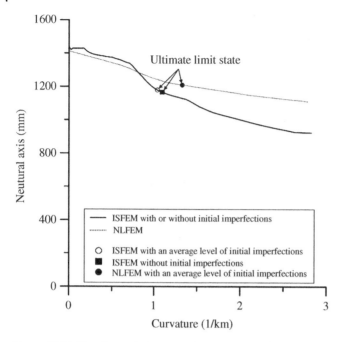

Figure 13.21 The change in the neutral axis position of the hull cross section with an increase in the vertical sagging bending moment.

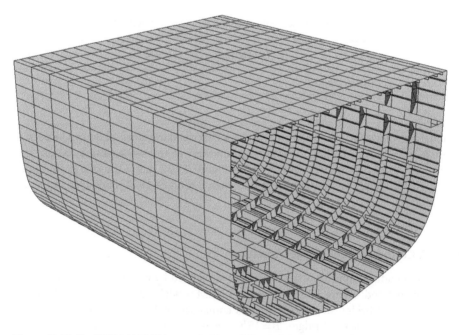

Figure 13.22 The ISFEM ALPS/HULL one hold model including transverse frames for the Dow test hull with a total of 4990 intelligent supersize rectangular plate elements.

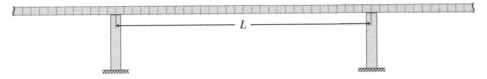

Figure 13.23 A steel bridge simply supported by pillars.

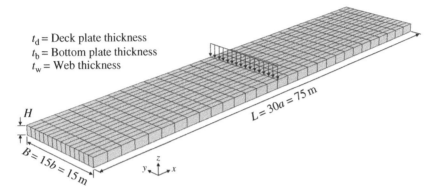

t_d = Deck plate thickness
t_b = Bottom plate thickness
t_w = Web thickness

H

$B = 15b = 15\,m$

$L = 30a = 75\,m$

Figure 13.24 The ISFEM ALPS/GENERAL model of a hypothetical steel-bridge structure.

of longitudinal girder or transverse frame), σ_Y = 320 MPa (yield stress of material), E = 205.8 GPa (elastic modulus), and ν = 0.3 (Poisson's ratio).

For a deck plate simply supported by longitudinal girders and transverse frames with b = 1000 mm and t_d = 15 mm, the elastic compressive buckling strength σ_{xE} and the critical compressive buckling strength σ_{xcr} are calculated from Equations (3.2) and (3.33), respectively, as follows:

$$\sigma_{xE} = \left[\frac{a}{m_o b} + \frac{m_o b}{a} \right]^2 \frac{\pi^2 E}{12(1-\nu^2)} \left(\frac{t_d}{b} \right)^2 = 172.8 \, \text{MPa}$$

$$\sigma_{xcr} = \sigma_Y \left(1 - \frac{\sigma_Y}{4\sigma_{xE}} \right) = 171.9 \, \text{MPa}$$

where m_o is the buckling half-wave number, which is determined from Equation (3.6b) to be 3 for the deck plate with a = 2500 mm and b = 1000 mm.

It is considered that the structure has suffered uniform corrosion wastage in the region indicated in Figure 13.25 where the diminution in plate thickness is 1 mm. It is assumed

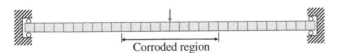

Corroded region

Figure 13.25 Corroded region and loading and boundary conditions applied for the progressive collapse analysis.

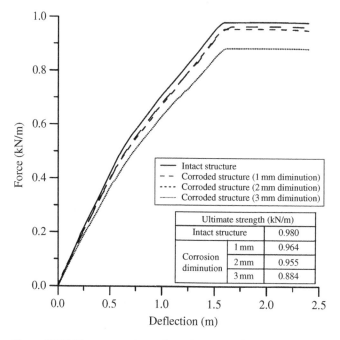

Figure 13.26 The progressive collapse behavior of the bridge structure obtained by the ISFEM ALPS/GENERAL computer program.

that the plate elements have an average level of initial imperfections as those considered in Section 13.10.3. Figure 13.26 shows the progressive collapse behavior of the bridge structure clamped at both ends and under lateral loads at the mid-span (see Figure 13.25), obtained by ISFEM ALPS/GENERAL computer program, which is represented by the relationship between the applied line load versus the deflection at midspan. It is confirmed that the corrosion wastage can significantly reduce the ultimate bending strength of the bridge structure. Also, the ISFEM is useful to compute the progressive collapse behavior of large-sized plated structures.

References

ALPS/GENERAL (2017). *A computer program for the progressive collapse analysis of general plated structures*. MAESTRO Marine LLC, Stevensville, MD.

ALPS/HULL (2017). *A computer program for the progressive collapse analysis of ship's hull structures*. MAESTRO Marine LLC, Stevensville, MD.

Dow, R.S. (1991). Testing and analysis of 1/3-scale welded steel frigate model. Proceedings of the International Conference on Advances in Marine Structures, Dunfermline, Scotland: 749–773.

Faisal, M., Noh, S.H., Kawsar, M.R.U., Youssef, S.A.M., Seo, J.K., Ha, Y.C. & Paik, J.K. (2017). Rapid hull collapse strength calculations of double hull oil tankers after collisions. *Ships and Offshore Structures*, **12**(5): 624–639.

Hughes, O.F. & Paik, J.K. (2013). *Ship structural analysis and design*. The Society of Naval Architects and Marine Engineers, Alexandria, VA.

Magoga, T. & Flockhart, C. (2014). Effect of weld-induced imperfections on the ultimate strength of an aluminium patrol boat determined by the ISFEM rapid assessment method. *Ships and Offshore Structures*, 9(2): 218–235.

Smith, C.S. (1977). Influence of local compressive failure on ultimate longitudinal strength of ship's hull. Proceedings of the International Symposium on Practical Design in Shipbuilding, Tokyo: 73–79.

Ueda, Y. & Rashed, S.M.H. (1974). An ultimate transverse strength analysis of ship structures. *Journal of the Society of Naval Architects of Japan*, **136**: 309–324 (in Japanese).

Ueda, Y. & Rashed, S.M.H. (1984). The idealized structural unit method and its application to deep girder structures. *Computers & Structures*, **18**(2): 277–293.

Ueda, Y. & Rashed, S.M.H. (1991). Advances in the application of ISUM to marine structures. Proceedings of the International Conference on Advances in Marine Structures, Dunfermline, Scotland: 628–649.

Ueda, Y., Rashed, S.M.H., Nakacho, K. & Sasaki, H. (1983). Ultimate strength analysis of offshore structures: application of idealized structural unit method. *Journal of the Kansai Society of Naval Architects of Japan*, **190**: 131–142 (in Japanese).

Ueda, Y., Rashed, S.M.H. & Paik, J.K. (1984). Plate and stiffened plate units of the idealized structural unit method (1st report): under in-plane loading. *Journal of the Society of Naval Architects of Japan*, **156**: 389–400 (in Japanese).

Ueda, Y., Rashed, S.M.H. & Paik, J.K. (1986a). Plate and stiffened plate units of the idealized structural unit method (2nd report): under in-plane and lateral loading considering initial deflection and residual stress. *Journal of the Society of Naval Architects of Japan*, **160**: 321–339 (in Japanese).

Ueda, Y., Rashed, S.M.H., Paik, J.K. & Masaoka, K. (1986b). The idealized structural unit method including global nonlinearities: idealized rectangular plate and stiffened plate elements. *Journal of the Society of Naval Architects of Japan*, **159**: 283–293 (in Japanese).

Youssef, S.A.M., Faisal, M., Seo, J.K., Kim, B.J., Ha, Y.C., Kim, D.K., Paik, J.K., Cheng, F. & Kim, M.S. (2016). Assessing the risk of ship hull collapse due to collision. *Ships and Offshore Structures*, **11**(4): 335–350.

Youssef, S.A.M., Noh, S.H. & Paik, J.K. (2017). A new method for assessing the safety of ships damaged by collisions. *Ships and Offshore Structures*, **12**(6): 862–872.

Zhang, X.M., Paik, J.K. & Jones, N. (2016). A new method for assessing the shakedown limit state associated with the breakage of a ship's hull girder. *Ships and Offshore Structures*, **11**(1): 92–104.

Zienkiewicz, O.C. (1977). *The finite element method*. Third Edition, McGraw-Hill, New York.

Appendices

A.1   Source Listing of the FORTRAN Computer Program CARDANO

Consider a third-order equation with regard to an unknown variable, W, given by

$$C_1 W^3 + C_2 W^2 + C_3 W + C_4 = 0$$

This equation can be solved by the so-called Cardano method. The following is the source listing of FORTRAN subroutine CARDANO to solve the above equation:

```
      SUBROUTINE CARDANO(C1,C2,C3,C4,W)
      IMPLICIT REAL*8(A-H,O-Z)
C
C***  C1*W**3+C2*W**2+C3*W+C4=0
C***  INPUT: C1,C2,C3,C4
C***  OUTPUT: W
C     PROGRAMMED BY PROF. J.K. PAIK
C     (C) J.K. PAIK. ALL RIGHTS RESERVED.
C
      S1=C2/C1
      S2=C3/C1
      S3=C4/C1
      P=S2/3.0-S1**2/9.0
      Q=S3-S1*S2/3.0+2.0*S1**3/27.0
      Z=Q**2+4.0*P**3
      IF(Z.GE.0.0) THEN
      AZ=(-Q+SQRT(Z))*0.5
      BZ=(-Q-SQRT(Z))*0.5
      AM=ABS(AZ)
      BM=ABS(BZ)
      IF(AM.LT.1.0E-10) THEN
      CA=0.0
      ELSE
      CA=AZ/AM
      END IF
      IF(BM.LT.1.0E-10) THEN
```

Ultimate Limit State Analysis and Design of Plated Structures, Second Edition. Jeom Kee Paik.

```
CB=0.0
ELSE
CB=BZ/BM
END IF
W=CA*AM**(1.0/3.0)+CB*BM**(1.0/3.0)-S1/3.0
ELSE
TH=ATAN(SQRT(-Z)/(-Q))
W=2.0*(-P)**0.5*COS(TH/3.0)-S1/3.0
END IF
RETURN
END
```

A.2 SI Units

A.2.1 Conversion Factors

Quantity	SI unit	Other unit	Inverse factor
Length	1 m = 1000 mm	3.28084 feet (ft)	1 ft = 0.3048 m
	1 cm = 10 mm	0.393701 inch (in)	1 in = 2.54 cm
	1 km = 1000 m	0.539957 nautical mile (nm)	1 nm = 1.852 km
		0.621371 mile	1 mile = 1.609344 km
Area	$1\,m^2$	$10.7639\,ft^2$	$1\,ft^2 = 0.09290304\,m^2$
	$1\,mm^2$	$0.00155\,in^2$	$1\,in^2 = 645.16\,mm^2$
Volume	$1\,m^3$	$35.3147\,ft^3$	$1\,ft^3 = 0.0283168\,m^3$
	$1000\,cm^3 = 1\,L$	0.219969 gal (UK)	1 gal (UK) = 4.54609 L
		0.264172 gal (US)	1 gal (US) = 3.78541 L
		1 bushel (UK) = 8 gal (UK)	1 gal (UK) = 0.125 bushel (UK)
		1 barrel (US) = 42 gal (US)	1 gal (US) = 0.02381 barrel (US)
Mass	1 kg	2.20462 pound (lb)	1 lb = 0.45359237 kg
	1 mg	0.0154323 grain (gr)	1 gr = 64.79891 mg
	1 g	0.035274 ounce (oz)	1 oz = 28.3495 g
	1 tonne	0.984204 long tonne (LT) (UK)	1 LT = 1.01605 tonne
		1.10231 short tonne (ST) (US)	1 ST = 0.907185 tonne
Velocity, speed	1 m/s = 3.6 km/h	3.28084 ft/s	1 ft/s = 0.3048 m/s
		2.23694 mile/h	1 mile/h = 0.44704 m/s
		1.94384 knot (kt) (meter system)	1 kt (meter system) = 0.514444 m/s

(Continued)

Quantity	SI unit	Other unit	Inverse factor
		1.94260 knot (kt) (yard–pound system)	1 kt (Yard–Pound system) = 0.514773 m/s
Speed–length ratio	$1 \dfrac{m/s}{\sqrt{m}}$	0.31933 Froude no. $(V/\sqrt{Lg})$	1 Froude no. = $3.13156 \dfrac{m/s}{\sqrt{m}}$
		$1.94384 \dfrac{kt}{\sqrt{m}}$	$1 \dfrac{kt}{\sqrt{m}} = 0.51444 \dfrac{m/s}{\sqrt{m}}$
		$1.07249 \dfrac{kt}{\sqrt{ft}}$	$1 \dfrac{kt}{\sqrt{ft}} = 0.93241 \dfrac{m/s}{\sqrt{m}}$
Acceleration	$1\ m/s^2$	$100\ cm/s^2$ (Gal)	$1\ Gal = 0.01\ m/s^2$
		0.101972 G	$1\ G = 9.80665\ m/s^2$
Density	$1\ kg/m^3$	$3.61273 \times 10^{-5}\ lb/in^3$	$1\ lb/in^3 = 2.76799 \times 10^4\ kg/m^3$
		$1.00224 \times 10^{-2}\ lb/gal$ (UK)	$1\ lb/gal\ (UK) = 99.7764\ kg/m^3$
		$8.3454 \times 10^{-3}\ lb/gal$ (US)	$1\ lb/gal\ (US) = 119.826\ kg/m^3$
Kinematic viscosity	$1\ m^2/s$	$10.7639\ ft^2/s$	$1\ ft^2/s = 9.2903 \times 10^{-2}\ m^2/s$
Force	1 N	0.101972 kgf	1 kgf = 9.80665 N
		0.1 Mdyn	l Mdyn = 10 N
		0.224809 lbf	1 lbf = 4.44822 N
Pressure	$1\ Pa = 1\ N/m^2 = 1.01972 \times 10^{-5}\ kgf/cm^2$	$1.45038 \times 10^{-4}\ lbf/in^2$ (psi)	1 psi = 6894.76 Pa
		1.0×10^{-5} bar	$1\ bar = 1.0 \times 10^5\ Pa$
		9.86923×10^{-6} atm	$1\ atm = 1.01325 \times 10^5\ Pa$
Stress	$1\ N/mm^2 = 1\ MPa = 0.101972\ kgf/mm^2$, $1\ kgf/mm^2 = 9.80665\ MPa$		
	$1\ N/mm^2$	$145.038\ lbf/in^2$	$1\ lbf/in^2 = 6.89476 \times 10^{-3}\ MPa$
Impact value	$1\ J/cm^2 = 0.101972\ kgf{\cdot}m/cm^2$	$4.75845\ lbf{\cdot}ft/in^2$	$1\ lbf{\cdot}ft/in^2 = 0.210152\ J/cm^2$
Energy	$1\ J = 1\ N{\cdot}m$, $1\ kJ = 101.972\ kgf{\cdot}m$, $1\ kgf{\cdot}m = 9.80665\ J$		
	1 kJ	737.563 lbf·ft	$1\ lbf{\cdot}ft = 1.35582 \times 10^{-3}\ kJ$
		0.238889 kcal	1 kcal = 4.18605 kJ
Power	$1\ kW = 101.972\ kgf{\cdot}m/s$, $1\ kgf{\cdot}m/s = 9.80665 \times 10^{-3}\ kW$		
	1 kW	1.35962 PS (meter system)	1 PS (meter system) = 0.7355 kW
		1.34102 HP (yard–pound system)	1 HP (yard–pound system) = $7.457 \times 10^{-3}\ kW$
		737.562 lbf·ft/s	$1\ lbf{\cdot}ft/s = 1.35582 \times 10^{-3}\ kW$
		0.238889 kcal/s	1 kcal/s = 4.18605 kW
Temperature	$°C = (°F - 32) \times \dfrac{5}{9}$, $°F = °C \times \dfrac{9}{5} + 32$, $K = °C + 273.15$		

A.2.2   SI Unit Prefixes

Exa (E) = 10^{18}	Deci (d) = 10^{-1}
Peta (P) = 10^{15}	Centi (c) = 10^{-2}
Tera (T) = 10^{12}	Milli (m) = 10^{-3}
Giga (G) = 10^{9}	Micro (μ) = 10^{-6}
Mega (M) = 10^{6}	Nano (n) = 10^{-9}
Kilo (k) = 10^{3}	Pico (p) = 10^{-12}
Hecto (h) = 10^{2}	Femto (f) = 10^{-15}
Deca (da) = 10	Atto (a) = 10^{-18}

A.3   Density and Viscosity of Water and Air

Temperature (°C)	Density (kg/m³)			Kinematic viscosity (m²/s)		
	Fresh water	Salt water	Dry air	Fresh water	Salt water	Dry air
0	999.8	1028.0	1.293	1.79×10^{-6}	1.83×10^{-6}	1.32×10^{-5}
10	999.7	1026.9	1.247	1.31×10^{-6}	1.35×10^{-6}	1.41×10^{-5}
20	998.2	1024.7	1.205	1.00×10^{-6}	1.05×10^{-6}	1.50×10^{-5}
30	995.6	1021.7	1.165	0.80×10^{-6}	0.85×10^{-6}	1.60×10^{-5}

A.4   Scaling Laws for Physical Model Testing

Physical model testing is usually required for the limit state analysis and design of structures although theoretical and numerical simulations are today becoming increasingly adopted with reasonable confidence. It is highly desirable to perform the physical model tests using a full scale prototype or at least large-scale models. It is of course essential to consider and keep the correct scaling laws for both structural mechanics and hydrodynamics model tests when small-scale models are used.

A.4.1   Structural Mechanics Model Tests

In physical model testing for the purpose of structural mechanics, a small-scale test model and a full scale prototype must have certain similarities including the geometric similarity and the values of Young's modulus (modulus of elasticity) (E), mass density (ρ), and Poisson's ratio (ν). The relationship between the characteristics of a small-scale model and a full scale prototype is given by the geometric scale factor that is defined as follows:

$$\frac{\ell}{L} = \alpha \tag{A.4.1}$$

Table A.4.1 Similarity considerations for a small-scale model and a full scale prototype.

Quantity	Full scale prototype	Small-scale model	Relationship
Length	L	ℓ	$\ell = \alpha L$
Displacement	Δ	δ	–
Strain	$E = \Delta/L$	ε	$\varepsilon = E$
Stress	$\Sigma = EE$	$\sigma = E\varepsilon$	$\sigma = \Sigma$
Pressure[a]	P	p	$p = P$
Dynamic force	$F = M\Delta/T^2$	$f = m\delta/t^2$	$f = \alpha^2 F$
Mass	M	m	$m = \alpha^3 M$
Time	T	t	$t = \alpha T$
Velocity	$V = \Delta/T$	v	$v = V$
Acceleration	$A = \Delta/T^2$	$a = \delta/t^2$	$a = A/\alpha$
Stress wave speed	C	c	$c = C$

Note:
a) Force on the boundary of the body under hydrostatic pressure is related to PL^2 in the full scale prototype, while it is related to $p\ell^2 = p\alpha^2 L^2$.

where α is the geometric scaling factor, ℓ is the length dimension for the small-scale model, L is the length dimension for the full scale prototype. Typically, $\alpha \leq 1$ will be considered although the test model can of course be larger in size than the full scale prototype. Table A.4.1 indicates the relationships of various quantities for a small-scale model and a full scale prototype.

A.4.2 Hydrodynamics Model Tests

A.4.2.1 Froude's Scaling Law
The Froude scaling law is to ensure the correct relationship between inertial and gravitational forces (except for viscous roll damping forces) when the full scale vessel is scaled down to model dimensions. It is recognized that the Froude law is ensured if the following Froude number F_n is the same at both a small-scale model and a full-scale prototype, namely,

$$F_n = \frac{V}{\sqrt{gL}} \tag{A.4.2}$$

where F_n is the Froude number, V is the velocity, L is the length, and g is the acceleration of gravity.

To get the correct Froude number scaling, all lengths in a particular model test must be scaled by the same factor, as indicated in Table A.4.2. For example, if the water depth is considered at a scale of $1 : \kappa$, then the same scale should be considered for the vessel's length, breadth, draught, wave height, and wave length. The model test is usually undertaken in fresh water, while the full scale unit will be used in salt water. The density ratio of salt water to fresh water is considered to be $r = 1.025$.

Table A.4.2 The Froude scaling laws for various physical quantities.

Quantity	Typical units	Scaling parameter
Length	m	κ
Time	s	$\kappa^{0.5}$
Frequency	1/s	$\alpha^{-0.5}$
Velocity	m/s	$\kappa^{0.5}$
Acceleration	m/s^2	1
Volume	m^3	κ^3
Water density	ton/m^3	r
Mass	ton	$r \times \kappa^3$
Force	kN	$r \times \kappa^3$
Moment	kNm	$r \times \kappa^4$
Extension stiffness	kN/m	$r \times \kappa^2$

Note: κ is the scale factor, r is the density ratio of fresh water to sea water that may usually be taken as $r = 1.025$ when the model testing is performed in the freshwater, while the full scale unit will be used in the salt water.

A.4.2.2 Reynolds Scaling Law

The scaling effect associated with viscous forces, for example, viscous roll damping moments on a vessel, risers, and mooring lines, is not consistent with the Froude scaling law, but follows the Reynolds scaling law, which must be the same at both model and full scales, namely,

$$R_e = \frac{VL}{\mu} \tag{A.4.3}$$

where R_e is the Reynolds number, μ is the kinematic viscosity, and L and V are defined in Equation (A.4.2).

In reality, it is not straightforward to achieve both the Froude and Reynolds scaling laws simultaneously in a particular model test. This is because the Froude scaling requires the model velocity to vary with the square root of length, while the Reynolds law requires an inverse relationship. In practice, the model testing may need to be performed with a high Reynolds number so that a larger model with a faster flow speed must be applied specifically when the free surface condition is not relevant due to currents and wind. However, this is again not easy to achieve because of physical limitations on the model flow speed and also the model Reynolds number. It is noted that the differences between model- and full-scale Reynolds numbers may not be significant if the Reynolds numbers at model and full scales are both high enough.

A.4.2.3 Vortex-Shedding Effects

Vortex-shedding effects are important in flow around bluff bodies where instabilities in the wake flow result in the periodic creation and shedding of eddies and vortices. Due to vortex shedding, the body can be subjected to the forces that have the largest components in the direction transverse to the flow and the smallest components in the line of the flow. Flexible structures with low damping such as risers and mooring lines in offshore installations can be subject to the phenomenon of vortex-induced vibration when the excitation frequency corresponds to one of the natural modes of the flexible structure. Vortex-shedding effects may be considered in model testing by keeping the following quantity the same at both model and full scales, namely,

$$S = \frac{\lambda D}{V} \quad \text{or} \quad V_r = \frac{1}{S} \tag{A.4.4}$$

where S is the so-called Strouhal number, V is the flow speed, D is the diameter of the body, λ is the frequency of the eddy shedding, and V_r is the reduced flow speed.

A.4.2.4 Surface-Tension Effects
The effects of surface tension can be important in model testing with a very small scale compared to the full scale prototype. The primary source of the surface-tension effects arises from the properties of small waves. This is because a significant straightening effect on the surface of the water can be caused when the waves become small enough. This can change the relationship between wave length and phase velocity so that the surface tension behaves as if an additional effect of gravity. The surface-tension effects are considered to be important when the wave length of the model is less than 0.1 m, and the waves are referred to as ripples. In offshore engineering, the waves with a period shorter than 4 s equivalent to a wave length less than 25 m may generally not be of interest. Therefore, the surface tension effects may not be important as long as the model scale is larger than 1 : 250.

A.4.2.5 Compressibility Effects
The effects of water and air compressibility are usually not considered for design of offshore structures, although they may be important for propellers of trading ships and thrusters of dynamic positioning systems.

Index

Ultimate Limit State Analysis and Design of Plated Structures, Second Edition. Jeom Kee Paik.
© 2018 John Wiley & Sons Ltd. Published 2018 by John Wiley & Sons Ltd.

Printed and bound by CPI Group (UK) Ltd, Croydon, CR0 4YY

16/04/2025

14658385-0005